Lecture Notes in Computer Scier

Edited by G. Goos, J. Hartmanis, and J. van L

Springer
Berlin
Heidelberg
New York
Barcelona
Hong Kong
London
Milan
Paris
Tokyo

Sophie Tison (Ed.)

Rewriting Techniques and Applications

13th International Conference, RTA 2002
Copenhagen, Denmark, July 22-24, 2002
Proceedings

Springer

Volume Editors

Sophie Tison
LIFL Bâtiment M3
Université de Lille 1, Cité Scientifique
59655 Villeneuve d'Ascq cedex, France
E-mail: tison@lifl.fr

Cataloging-in-Publication Data applied for

Die Deutsche Bibliothek - CIP-Einheitsaufnahme

Rewriting techniques and applications : 13th international conference ; proceedings / RTA 2002, Copenhagen, Denmark, July 22 - 24, 2002. Sophie Tison (ed.). - Berlin ; Heidelberg ; New York ; Barcelona ; Hong Kong ; London ; Milan ; Paris ; Tokyo : Springer, 2002
(Lecture notes in computer science ; Vol. 2378)
ISBN 3-540-43916-1

CR Subject Classification (1998): F.4, F.3.2, D.3, I.2.2-3, I.1

ISSN 0302-9743
ISBN 3-540-43916-1 Springer-Verlag Berlin Heidelberg New York

Springer-Verlag Berlin Heidelberg New York
a member of BertelsmannSpringer Science+Business Media GmbH

http://www.springer.de

Typesetting: Camera-ready by author, data conversion by PTP-Berlin, Stefan Sossna e.K.
Printed on acid-free paper SPIN: 10870449 06/3142 5 4 3 2 1 0

Preface

This volume contains the proceedings of the 13th International Conference on Rewriting Techniques and Applications (RTA 2002), which was held July 22-24, 2002 in Copenhagen as part of the 3rd Federated Logic Conference (FLoC 2002). RTA is the major international forum for the presentation of research on all aspects of rewriting. Previous RTA conferences took place in Dijon (1985), Bordeaux (1987), Chapel Hill (1989), Como (1991), Montreal (1993), Kaiserslautern (1995), Rutgers (1996), Sitges (1997), Tsukuba (1998), Trento (1999), Norwich (2000), and Utrecht (2001).

A total of 20 regular papers, 2 application papers and 4 system descriptions were selected for presentation from 49 submissions from Argentina (1), Brazil ($\frac{2}{3}$), Czech Republic (1), France (13), Germany (8), Israel ($\frac{1}{3}$), Italy ($1\frac{5}{6}$), Japan (6), The Netherlands (2), Poland (1), Portugal ($1\frac{1}{3}$), Rumania (1), Spain (4), UK ($\frac{5}{6}$), Uruguay ($\frac{1}{2}$), USA ($5\ \frac{1}{2}$), Venezuela (1). The program committee awarded the *best paper* prize to Paul-André Melliès for his paper *Residual Theory Revisited*. This paper presents an elegant and subtle generalization of Jean-Jacques Lévy's residual theory.

I am especially grateful to the invited speakers Franz Baader, John Mitchell, and Natarajan Shankar for accepting our invitation to present us their insights into their research areas.

Many people helped to make RTA 2002 a success. I thank the members of the program committee for their demanding work in the evaluation process. I thank also the numerous external referees for reviewing the submissions and maintaining the high standards of the RTA conference. It is a great pleasure to thank Thomas Arts who was in charge of the local organization. I would also express my gratitude to Anne-Cécile Caron for her assistance in many of my tasks as the program chair. Finally, I thank the FLoC 2002 Organizing Committee and the Local Organization Committee for all their hard work.

May 2002 Sophie Tison

Preface

Conference Organization

Program Chair

Sophie Tison	University of Lille

Conference Chair

Thomas Arts	Ericsson, Computer Science Laboratory

Program Committee

Andrea Corradini	University of Pisa
Daniel J. Dougherty	Wesleyan University
Jürgen Giesl	RWTH Aachen
Bernhard Gramlich	Vienna University of Technology
Thérèse Hardin	University of Paris VI
Christopher Lynch	Clarkson University
Jerzy Marcinkowski	University of Wroclaw
Aart Middeldorp	University of Tsukuba
Joachim Niehren	University of Saarland
Femke van Raamsdonk	Free University Amsterdam
Albert Rubio	Technical University of Catalonia
Sophie Tison	University of Lille
Ralf Treinen	University of Paris XI

Steering Committee

Franz Baader	Dresden
Leo Bachmair	Stony Brook
Hélène Kirchner	Nancy
Pierre Lescanne	Lyon (publicity chair)
Jose Meseguer	Menlo Park
Aart Middeldorp	Tsukuba (chair)

List of External Referees

Yves André
Jürgen Avenhaus
Steffen van Bakel
Sebastian Bala
Paolo Baldan
Inge Bethke
Stefano Bistarelli
Roel Bloo
Manuel Bodirsky
Benedikt Bollig
Eduardo Bonelli
Sander Bruggink
Roberto Bruni
Venanzio Capretta
Anne-Cécile Caron
Witold Charatonik
Horatiu Cirstea
René David
Roberto Di Cosmo
Gilles Dowek
Irène Durand
Joost Engelfriet
Katrin Erk
Maribel Fernández
Jean-Christophe Filliâtre
Andrea Formisano
Fabio Gadducci
Silvia Ghilezan
Isabelle Gnaedig
Guillem Godoy
Philippe de Groote
Stefano Guerrini
Rolf Hennicker
Miki Hermann
Thomas Hillenbrand
Dieter Hofbauer
Patricia Johann
Gueorgui Jojgov
Wolfram Kahl
Łukasz Kaiser
Zurab Khasidashvili
Alexander Koller
Pierre Lescanne
Martin Leucker
Jean-Jacques Lévy
Jordi Levy
James B. Lipton
Sébastien Limet
Daniel Loeb
Salvador Lucas
Denis Lugiez
Luc Maranget
Massimo Marchiori
Ralph Matthes
Paul-André Melliès
Oskar Miś
Markus Mohnen
Benjamin Monate
César A.Muñoz
Robert Nieuwenhuis
Thomas Noll
Enno Ohlebusch
Paweł Olszta
Vincent van Oostrom
Dmitrii V. Pasechnik
Jorge Sousa Pinto
Detlef Plump
Jaco van de Pol
Tim Priesnitz
Pierre Réty
Alexandre Riazanov
Christophe Ringeissen
Andreas Rossberg
Michael Rusinowitch
Cesar Sanchez
Christelle Scharff
Manfred Schmidt-Schauß
Aleksy Schubert
Klaus U. Schulz
Jan Schwinghammer
Helmut Seidl
Géraud Sénizergues
Jürgen Stuber
Zhendong Su
Taro Suzuki
Gabriele Taentzer
Jean-Marc Talbot
Ashish Tiwari
Hélène Touzet
Yoshihito Toyama
Tomasz Truderung
Xavier Urbain
Paweł Urzyczyn
Sándor Vágvölgyi
Laurent Vigneron
Mateu Villaret
Christian Vogt
Johannes Waldmann
Freek Wiedijk
Herbert Wiklicky
Claus-Peter Wirth
Hans Zantema
Wieslaw Zielonka

Table of Contents

Invited Talks

Regular Papers, Application Papers

System Descriptions

Combining Shostak Theories*

Natarajan Shankar and Harald Rueß

SRI International Computer Science Laboratory
Menlo Park CA 94025 USA
{shankar,ruess}@csl.sri.com
http://www.csl.sri.com/{~shankar,~ruess}
Phone: +1 (650) 859-5272 Fax: +1 (650) 859-2844

Abstract. Ground decision procedures for combinations of theories are used in many systems for automated deduction. There are two basic paradigms for combining decision procedures. The Nelson–Oppen method combines decision procedures for disjoint theories by exchanging equality information on the shared variables. In Shostak's method, the combination of the theory of pure equality with canonizable and solvable theories is decided through an extension of congruence closure that yields a canonizer for the combined theory. Shostak's original presentation, and others that followed it, contained serious errors which were corrected for the basic procedure by the present authors. Shostak also claimed that it was possible to combine canonizers and solvers for disjoint theories. This claim is easily verifiable for canonizers, but is unsubstantiated for the case of solvers. We show how our earlier procedure can be extended to combine multiple disjoint canonizable, solvable theories within the Shostak framework.

1 Introduction

Consider the sequent

$$\begin{array}{l} 2 * car(x) - 3 * cdr(x) = f(cdr(x)) \\ \vdash f(cons(4 * car(x) - 2 * f(cdr(x)), y)) = f(cons(6 * cdr(x), y)). \end{array}$$

* This work was funded by NSF Grant CCR-0082560, DARPA/AFRL Contract F33615-00-C-3043, and NASA Contract NAS1-00079. During a phone conversation with the first author on 2nd April 2001, Rob Shostak suggested that the problem of combining Shostak solvers could be solved through variable abstraction. His suggestion is the key inspiration for the combination of Shostak theories presented here. We thank Clark Barrett, Sam Owre, and Ashish Tiwari for their meticulous reading of earlier drafts. We also thank Harald Ganzinger for pointing out certain limitations of our original definition of solvability with respect to σ-models. The first author is grateful to the program committees and program chairs of the FME, LICS, and RTA conferences at FLoC 2002 for their kind invitation.

S. Tison (Ed.): RTA 2002, LNCS 2378, pp. 1–18, 2002.

It involves symbols from three different theories. The symbol f is uninterpreted, the operations $*$ and $-$ are from the theory of linear arithmetic, and the pairing and projection operations *cons*, *car*, and *cdr*, are from the theory of lists. There are two basic methods for building combined decision procedures for disjoint theories, i.e., theories that share no function symbols. Nelson and Oppen [NO79] gave a method for combining decision procedures through the use of variable abstraction for replacing subterms with variables, and the exchange of equality information on the shared variables. Thus, with respect to the example above, decision procedures for pure equality, linear arithmetic, and the theory of lists can be composed into a decision procedure for the combined theory. The other combination method, due to Shostak, yields a decision procedure for the combination of canonizable and solvable theories, based on the congruence closure procedure. Shostak's original algorithm and proof were seriously flawed. His algorithm is neither terminating nor complete (even when terminating). These flaws went unnoticed for a long time even though the method was widely used, implemented, and studied [CLS96,BDL96,Bjø99]. In earlier work [RS01], we described a correct algorithm for the *basic* combination of a single canonizable, solvable theory with the theory of equality over uninterpreted terms. That correctness proof has been mechanically verified using PVS [FS02]. The generality of the basic combination rests on Shostak's claim that it is possible to combine solvers and canonizers from disjoint theories into a single canonizer and solver. This claim is easily verifiable for canonizers, but fails for the case of solvers. In this paper, we extend our earlier decision procedure to the combination of uninterpreted equality with multiple canonizable, solvable theories. The decision procedure does not require the combination of solvers. We present proofs for the termination, soundness, and completeness of our procedure.

2 Preliminaries

We introduce some of the basic terminology needed to understand Shostak-style decision procedures. Fixing a countable set of variables X and a set of function symbols F, a term is either a variable x from X or an n-ary function symbol f from F applied to n terms as in $f(a_1, \ldots, a_n)$. Equations between terms are represented as $a = b$. Let $vars(a)$, $vars(a = b)$, and $vars(T)$ represent the sets of variables in a, $a = b$, and the set of equalities T, respectively. We are interested in deciding the validity of sequents of the form $T \vdash c = d$ where c and d are terms, and T is a set of equalities such that $vars(c = d) \subseteq vars(T)$. The condition $vars(c = d) \subseteq vars(T)$ is there for technical reasons. It can always be satisfied by padding T with reflexivity assertions $x = x$ for any variables x in $vars(c = d) - vars(T)$. We write $[\![a]\!]$ for the set of subterms of a, which includes a.

The semantics for a term a, written as $M[\![a]\!]\rho$, is given relative to an interpretation M over a domain D and an assignment ρ. For an n-ary function f, the interpretation $M(f)$ of f in M is a map from D^n to D. For an *uninterpreted*

n-ary function symbol f, the interpretation $M(f)$ may be any map from D^n to D, whereas only restricted interpretations might be suitable for an interpreted function symbol like the arithmetic $+$ operation. An assignment ρ is a map from variables in X to values in D. We define $M[\![a]\!]\rho$ to return a value in D by means of the following equations.

$$M[\![x]\!]\rho = \rho(x)$$
$$M[\![f(a_1, \ldots, a_n)]\!]\rho = M(f)(M[\![a_1]\!]\rho, \ldots, M[\![a_n]\!]\rho)$$

We say that $M, \rho \models a = b$ iff $M[\![a]\!]\rho = M[\![b]\!]\rho$, and $M \models a = b$ iff $M, \rho \models a = b$ for all assignments ρ. We write $M, \rho \models S$ when $\forall a, b : a = b \in S \Rightarrow M, \rho \models a = b$, and $M, \rho \models (T \vdash a = b)$ when $(M, \rho \models T) \Rightarrow (M, \rho \models a = b)$. A sequent $T \vdash c = d$ is valid, written as $\models (T \vdash c = d)$, when $M, \rho \models (T \vdash c = d)$, for all M and ρ.

There is a simple pattern underlying the class of decision procedures studied here. Let ψ be the state of the decision procedure as given by a set of formulas.[1] Let τ be a family of state transformations so that we write $\psi \xrightarrow{\tau} \psi'$ if ψ' is the result of applying a transformation in τ to ψ, where $vars(\psi) \subseteq vars(\psi')$ (variable preservation). An assignment ρ' is said to extend ρ over $vars(\psi') - vars(\psi)$ when it agrees with ρ on all variables except those in $vars(\psi') - vars(\psi)$ for $vars(\psi) \subseteq vars(\psi')$. We say that ψ' *preserves* ψ if $vars(\psi) \subseteq vars(\psi')$ and for all interpretations M and assignments ρ, $M, \rho \models \psi$ holds iff there exists an assignment ρ' extending ρ such that $M, \rho' \models \psi'$.[2] When preservation is restricted to a limited class of interpretations ι, we say that ψ' ι-preserves ψ. Note that the *preserves* relation is transitive. When the operation τ is deterministic, $\tau(\psi)$ represents the result of the transformation, and we call τ a *conservative* operation to indicate that $\tau(\psi)$ preserves ψ for all ψ. Correspondingly, τ is said to be ι-conservative when $\tau(\psi)$ ι-preserves ψ. Let τ^n represent the n-fold iteration of τ, then τ^n is a conservative operation. The composition $\tau_2 \circ \tau_1$ of conservative operations τ_1 and τ_2, is also a conservative operation. The operation $\tau^*(\psi)$ is defined as $\tau^i(\psi)$ for the least i such that $\tau^{i+1}(\psi) = \tau^i(\psi)$. The existence of such a bound i must be demonstrated for the termination of τ^*. If τ is conservative, so is τ^*.

If τ is a conservative operation, it is sound and complete in the sense that for a formula ϕ with $vars(\phi) \subseteq vars(\psi)$, $\models (\psi \vdash \phi)$ iff $\models (\tau(\psi) \vdash \phi)$. This is clear since τ is a conservative operation and $vars(\phi) \subseteq vars(\psi)$.

[1] In our case, the state is actually represented by a list whose elements are sets of equalities. We abuse notation by viewing such a state as the set of equalities corresponding to the union of the sets of equalities contained in it.

[2] In general, one could allow the interpretation M to be extended to M' in the transformation from ψ to ψ' to allow for the introduction of new function symbols, e.g., skolem functions. This abstract design pattern then also covers skolemization in addition to methods like prenexing, clausification, resolution, variable abstraction, and Knuth-Bendix completion.

If $\tau^*(\psi)$ returns a state ψ' such that $\models (\psi' \vdash \bot)$, where $\bot$ is an unsatisfiable formula, then ψ' and ψ are both clearly unsatisfiable. Otherwise, if ψ' is *canonical*, as explained below, $\models (\psi' \vdash \phi)$ can be decided by computing a canonical form $\psi'[\![\phi]\!]$ for ϕ with respect to ψ'.

3 Congruence Closure

In this section, we present a warm-up exercise for deciding equality over terms where all function symbols are uninterpreted, i.e., the interpretation of these operations is unconstrained. This means that a sequent $T \vdash c = d$ is valid, i.e., $\models (T \vdash c = d)$ iff for all interpretations M and assignments ρ, the satisfaction relation $M, \rho \models (T \vdash c = d)$ holds. Whenever we write $f(a_1, \ldots, a_n)$, the function symbol f is uninterpreted, and $f(a_1, \ldots, a_n)$ is then said to be uninterpreted. Later on, we will extend the procedure to allow interpreted function symbols from disjoint Shostak theories such as linear arithmetic and lists. The congruence closure procedure sets up the template for the extended procedure in Section 5.

The congruence closure decision procedure for *pure equality* has been studied by Kozen [Koz77], Shostak [Sho78], Nelson and Oppen [NO80], Downey, Sethi, and Tarjan [DST80], and, more recently, by Kapur [Kap97]. We present the congruence closure algorithm in a Shostak-style, i.e., as an online algorithm for computing and using canonical forms by successively processing the input equations from the set T. For ease of presentation, we make use of variable abstraction in the style of the abstract congruence closure technique due to Bachmair, Tiwari, and Vigneron [BTV02]. Terms of the form $f(a_1, \ldots, a_n)$ are variable-abstracted into the form $f(x_1, \ldots, x_n)$ where the variables $x_1, \ldots, x_n$ abstract the terms $a_1, \ldots, a_n$, respectively. The procedure shown here can be seen as a specific strategy for applying the abstract congruence closure rules. In Section 5, we make essential use of variable abstraction in the Nelson–Oppen style where it is not merely a presentation device.

Let $T = \{a_1 = b_1, \ldots, a_n = b_n\}$ for $n \geq 0$ so that T is empty when $n = 0$. Let x and y be metavariables that range over variables. The state of the algorithm consists of a *solution state* S and the input equalities T. The solution state S will be maintained as the pair $(S_V; S_U)$, where $(l_1; l_2; \ldots; l_n)$ represents a list with n elements and semi-colon is an associative separator for list elements. The set S_U then contains equalities of the form $x = f(x_1, \ldots, x_n)$ for an n-ary uninterpreted function f, and the set S_V contains equalities of the form $x = y$ between variables. We blur the distinction between the equality $a = b$ and the singleton set $\{a = b\}$. Syntactic identity is written as $a \equiv b$ as opposed to semantic equality $a = b$.

A set of equalities R is *functional* if $b \equiv c$ whenever $a = b \in R$ and $a = c \in R$, for any a, b, and c. If R is functional, it can be used as a lookup table for obtaining the right-hand side entry corresponding to a left-hand side expression. Thus $R(a) = b$ if $a = b \in R$, and otherwise, $R(a) = a$. The domain of R, $dom(R)$

is defined as $\{a \mid a = b \in R \text{ for some } b\}$. When R is not necessarily functional, we use $R(\{a\})$ to represent the set $\{b \mid a = b \in R \vee b \equiv a\}$ which is the image of $\{a\}$ with respect to the reflexive closure of R. The inverse of R, written as R^{-1}, is the set $\{b = a \mid a = b \in R\}$. A functional set R of equalities can be applied as in $R[a]$.

$$\begin{aligned} R[x] &= R(x) \\ R[f(a_1, \ldots, a_n)] &= R(f(R[a_1], \ldots, R[a_n])) \\ R[\{a_1 = b_1, \ldots, a_n = b_n\}] &= \{R[a_1] = R[b_1], \ldots, R[a_n] = R[b_n]\} \end{aligned}$$

In typical usage, R will be a *solution set* where the left-hand sides are all variables, so that $R[a]$ is just the result of applying R as a substitution to a.

When S_V is functional, then S given by $(S_V; S_U)$ can also be used to compute the canonical form $S[\![a]\!]$ of a term a with respect to S. Hilbert's epsilon operator is used in the form of the *when* operator: $F(\overline{x})$ when $\overline{x} : P(\overline{x})$ is an abbreviation for $F(\epsilon \overline{x} : \ P(\overline{x}))$, if $\exists \overline{x} : P(\overline{x})$.

$$\begin{aligned} S[\![x]\!] &= S_V(x) \\ S[\![f(a_1, \ldots, a_n)]\!] &= S_V(x), \text{ when } x : x = f(S[\![a_1]\!], \ldots, S[\![a_n]\!]) \in S_U \\ S[\![f(a_1, \ldots, a_n)]\!] &= f(S[\![a_1]\!], \ldots, S[\![a_n]\!]), \text{ otherwise.} \end{aligned}$$

The set S_V of variable equalities will be maintained so that $vars(S_V) \cup vars(S_U) = dom(S_V)$. The set S_V partitions the variables in $dom(S_V)$ into equivalence classes. Two variables x and y are said to be in the same equivalence class with respect to S_V if $S_V(x) \equiv S_V(y)$. If R and R' are solution sets and R' is functional, then $R \triangleright R' = \{a = R'[b] \mid a = b \in R\}$, and $R \circ R' = R' \cup (R \triangleright R')$. The set S_V is maintained in idempotent form so that $S_V \circ S_V = S_V$. Note that S_U need not be functional since it can, for example, simultaneously contain the equations $x = f(y)$, $x = f(z)$, and $x = g(y)$.

We assume a strict total ordering $x \prec y$ on variables. The operation $orient(x = y)$ returns $\{x = y\}$ if $x \prec y$, and returns $\{y = x\}$, otherwise. The solution state S is said to be *congruence-closed* if $S_U(\{x\}) \cap S_U(\{y\}) = \emptyset$ whenever $S_V(x) \not\equiv S_V(y)$. A solution set S is *canonical* if S is congruence-closed, S_V is functional and idempotent, and S_U is normalized, i.e., $S_U \triangleright S_V = S_U$.

In order to determine if $\models (T \vdash c = d)$, we check if $S'[\![c]\!] \equiv S'[\![d]\!]$ for $S' = process(S; T)$, where $S = (S_V; S_U)$, $S_V = id_T$, $id_T = \{x = x \mid x \in vars(T)\}$, and $S_U = \emptyset$. The congruence closure procedure *process* is defined in Figure 1.

Explanation. We explain the congruence closure procedure using the validity of the sequent $f(f(f(x))) = x$, $x = f(f(x)) \vdash f(x) = x$ as an example. Its validity will be verified by constructing a solution state S' equal to $process(S_V; S_U; T)$ for $T = \{f(f(f(x))) = x, x = f(f(x))\}$, $S_V = id_T$, $S_U = \emptyset$, and checking $S'[\![f(x)]\!] \equiv S'[\![x]\!]$. Note that id_T is $\{x = x\}$. In processing $f(f(f(x))) = x$ with respect to S, the canonization step, $S[\![f(f(f(x))) = x]\!]$

$$\begin{aligned}
process(S;\emptyset) &= S\\
process(S;\{a=b\}\cup T) &= process(S';T),\ \text{where},\\
&\quad S' = close^*(merge(abstract^*(S;S[\![a=b]\!]))).
\end{aligned}$$

$$\begin{aligned}
close(S) &= merge(S; S_V(x) = S_V(y)),\\
&\quad \text{when } x,y : S_V(x) \not\equiv S_V(y), (S_U(\{x\}) \cap S_U(\{y\}) \neq \emptyset)\\
close(S) &= S,\ \text{otherwise.}
\end{aligned}$$

$$\begin{aligned}
merge(S; x=x) &= S\\
merge(S; x=y) &= (S'_V;\ S'_U),\ \text{where } x \not\equiv y, R = orient(x=y),\\
&\qquad S'_V = S_V \circ R, S'_U = S_U \rhd R.
\end{aligned}$$

$$\begin{aligned}
abstract(S; x=y) &= (S;\ x=y)\\
abstract(S; a=b) &= (S';\ a'=b'), \text{when } S', a', b', x_1, \ldots, x_n :\\
&\qquad f(x_1,\ldots,x_n) \in \lceil\!\lceil a=b \rceil\!\rceil\\
&\qquad x \notin vars(S; a=b)\\
&\qquad R = \{x = f(x_1,\ldots,x_n)\},\\
&\qquad S' = (S_V \cup \{x=x\};\ S_U \cup R),\\
&\qquad a' = R^{-1}[a], b' = R^{-1}[b].
\end{aligned}$$

Fig. 1. Congruence closure

yields $f(f(f(x))) = x$, unchanged. Next, the variable abstraction step computes $abstract^*(f(f(f(x))) = x)$. First $f(x)$ is abstracted to v_1 yielding the state $\{x = x, v_1 = v_1\}; \{v_1 = f(x)\}; \{f(f(v_1)) = x\}$. Variable abstraction eventually terminates renaming $f(v_1)$ to v_2 and $f(v_2)$ to v_3 so that S is $\{x = x, v_1 = v_1, v_2 = v_2, v_3 = v_3\}; \{v_1 = f(x), v_2 = f(v_1), v_3 = f(v_2)\}$. The variable abstracted input equality is then $v_3 = x$. Let $orient(v_3 = x)$ return $v_3 = x$. Next, $merge(S; v_3 = x)$ yields the solution state $\{x = x, v_1 = v_1, v_2 = v_2, v_3 = x\}; \{v_1 = f(x), v_2 = f(v_1), v_3 = f(v_2)\}$. The congruence closure step $close^*(S)$ leaves S unchanged since there are no variables that are merged in S_U and not in S_V. The next input equality $x = f(f(x))$ is canonized as $x = v_2$ which can be oriented as $v_2 = x$ and merged with S to yield the new value $\{x = x, v_1 = v_1, v_2 = x, v_3 = x\}; \{v_1 = f(x), v_2 = f(v_1), v_3 = f(x)\}$ for S. The congruence closure step $close^*(S)$ now detects that v_1 and v_3 are merged in S_U but not in S_V and generates the equality $v_1 = v_3$. This equality is merged to yield the new value of S as $\{x = x, v_1 = x, v_2 = x, v_3 = x\}; \{v_1 = f(x), v_2 = f(x), v_3 = f(x)\}$, which is congruence-closed.

With respect to this final value of the solution state S, it can be checked that $S[\![f(x)]\!] \equiv x \equiv S[\![x]\!]$.

Invariants. The Shostak-style congruence closure algorithm makes heavy use of canonical forms and this requires some key invariants to be preserved on the solution state S. If $vars(S_V) \cup vars(S_U) \subseteq dom(S_V)$, then $vars(S'_V) \cup vars(S'_U) \subseteq$

$dom(S'_V)$, when S' is either $abstract(S; a = b)$ or $close(S)$. If S is canonical and $a' = S[\![a]\!]$, then $S_V[a'] = a'$. If $S_U \triangleright S_V = S_U$, $S_V[a] = a$, and $S_V[b] = b$, then $S'_U \triangleright S'_V = S'_U$ where $S'; a' = b'$ is $abstract(S; a = b)$. Similarly, if $S_U \triangleright S_V = S_U$, $S_V(x) \equiv x$, $S_V(y) \equiv y$, then $S'_U \circ S'_V = S'_U$ for $S' = merge(S; x = y)$. If S_V is functional and idempotent, then so is S'_V, where S' is either of $abstract(S; a = b)$ or $close(S)$. If $S' = close^*(S)$, then S' is congruence-closed, and if S_V is functional and idempotent, S_U is normalized, then S' is canonical.

Variations. In the *merge* operation, if S'_U is computed as $R[S_U]$ instead of $S_U \triangleright R$, this would preserve the invariant that S_U^{-1} is always functional and $S_V[S_U] = S_U$. If this is the case, the canonizer can be simplified to just return $S_U^{-1}(f(S[\![a_1]\!], \ldots, S[\![a_n]\!]))$.

Termination. The procedure $process(S;\ T)$ terminates after each equality in T has been asserted into S. The operation $abstract^*$ terminates because each recursive call decreases the number of occurrences of function applications in the given equality $a = b$ by at least one. The operation $close^*$ terminates because each invocation of the *merge* operation merges two distinct equivalence classes of variables in S_V. The *process* operation terminates because the number of input equations in T decreases with each recursive call. Therefore the computation of $process(S;\ T)$ terminates returning a canonical solution set S'.

Soundness and Completeness. We need to show that $\models (T \vdash c = d) \iff S'[\![c]\!] \equiv S'[\![d]\!]$ for $S' = process(id_T;\ \emptyset;\ T)$ and $vars(c = d) \subseteq vars(T)$. We do this by showing that S' preserves $(id_T;\ \emptyset;\ T)$, and hence $\models (T \vdash c = d) \iff \models (S' \vdash c = d)$, and $\models (S' \vdash c = d) \iff S'[\![c]\!] \equiv S'[\![d]\!]$. We can easily establish that if $process(S;\ T) = S'$, then S' preserves $(S;\ T)$. If $a' = b'$ is obtained from $a = b$ by applying equality replacements from S, then $(S;\ a' = b')$ preserves $(S;\ a = b)$. In particular, $\models (S \vdash S[\![c]\!] = c)$ holds. The following claims can then be easily verified.

1. $(S;\ S[\![a = b]\!])$ preserves $(S;\ a = b)$.
2. $abstract(S;\ a = b)$ preserves $(S;\ a = b)$.
3. $merge(S;\ a = b)$ preserves $(S;\ a = b)$.
4. $close(S)$ preserves S.

The only remaining step is to show that if S' is canonical, then $\models (S' \vdash c = d) \iff S'[\![c]\!] \equiv S'[\![d]\!]$ for $vars(c = d) \subseteq vars(S)$. Since we know that $\models S' \vdash S'[\![c]\!] = c$ and $\models S' \vdash S'[\![d]\!] = d$, hence $\models (S' \vdash c = d)$ follows from $S'[\![c]\!] \equiv S'[\![d]\!]$. For the *only if* direction, we show that if $S'[\![c]\!] \not\equiv S'[\![d]\!]$, then there is an interpretation $M_{S'}$ and assignment $\rho_{S'}$ such that $M_{S'}, \rho_{S'} \models S$ but $M_{S'}, \rho_{S'} \not\models c = d$. A *canonical* term (in S') is a term a such that $S'[\![a]\!] \equiv a$. The domain $D_{S'}$ is taken to be the set of canonical terms built from the function symbols F and variables from $vars(S')$. We constrain $M_{S'}$ so that $M_{S'}(f)(a_1, \ldots, a_n) = S'_V(x)$ when there is an x such that $x = f(a_1, \ldots, a_n) \in S'_U$, and $f(a_1, \ldots, a_n)$, otherwise. Let $\rho_{S'}$ map x in $vars(S')$ to $S'_V(x)$; the mappings for the variables outside $vars(S')$ are irrelevant. It is easy to see that $M_{S'}[\![c]\!]\rho_{S'} = S'[\![c]\!]$ by induction on the structure of c. In particular, when S' is canonical, $M_{S'}(f)(x_1 \dots, x_n) = x$ for $x = f(x_1, \dots, x_n) \in S'_U$, so that one can easily verify that $M_{S'}, \rho_{S'} \models S'$. Hence, if $S'[\![c]\!] \not\equiv S'[\![d]\!]$, then $\not\models (S' \vdash c = d)$.

4 Shostak Theories

A Shostak theory [Sho84] is a theory that is canonizable and solvable. We assume a collection of Shostak theories $\theta_1, \dots, \theta_N$. In this section, we give a decision procedure for a single Shostak theory θ_i, but with i as a parameter. This background material is adapted from Shankar [Sha01]. Satisfiability $M, \rho \models a = b$ is with respect to i-models M. The equality $a = b$ is i-valid, i.e., $\models_i a = b$, if for all i-models M and assignments ρ, $M[\![a]\!]\rho = M[\![b]\!]\rho$. Similarly, $a = b$ is i-unsatisfiable, i.e., $\models_i a \neq b$, when for all i-models M and assignments ρ, $M[\![a]\!]\rho \neq M[\![b]\!]\rho$. An i-term a is a term whose function symbols all belong to θ_i and $vars(a) \subseteq X \cup X_i$.

A canonizable theory θ_i admits a computable operation σ_i on terms such that $\models_i a = b$ iff $\sigma_i(a) \equiv \sigma_i(b)$, for i-terms a and b. An i-term a is canonical if $\sigma_i(a) \equiv a$. Additionally, $vars(\sigma_i(a)) \subseteq vars(a)$ and every subterm of $\sigma_i(a)$ must be canonical. For example, a canonizer for the theory θ_A of linear arithmetic can be defined to convert expressions into an ordered sum-of-monomials form. Then, $\sigma_A(y + x + x) \equiv 2 * x + y \equiv \sigma_A(x + y + x)$.

A solvable theory admits a procedure $solve_i$ on equalities such that $solve_i(Y)(a = b)$ for a set of variables Y with $vars(a = b) \subseteq Y$, returns a solved form for $a = b$ as explained below. $solve_i(Y)(a = b)$ might contain fresh variables that do not appear in Y. A functional solution set R is in *i-solved form* if it is of the form $\{x_1 = t_1, \dots, x_n = t_n\}$, where for j, $1 \leq j \leq n$, t_j is a canonical i-term, $\sigma_i(t_j) \equiv t_j$, and $vars(t_j) \cap dom(R) = \emptyset$ unless $t_j \equiv x_j$. The i-solved form $solve_i(Y)(a = b)$ is either $\bot_i$, when $\models_i a \neq b$, or is a solution set of equalities which is the union of sets R_1 and R_2. The set R_1 is the solved form $\{x_1 = t_1, \dots, x_n = t_n\}$ with $x_j \in vars(a = b)$ for $1 \leq j \leq n$, and for any i-model M and assignment ρ, we have that $M, \rho \models a = b$ iff there is a ρ' extending ρ over $vars(solve_i(Y)(a = b)) - Y$ such that $M, \rho' \models x_j = t_j$, for $1 \leq j \leq n$. The set R_2 is just $\{x = x \mid x \in vars(R_1) - Y\}$ and is included in order to preserve variables. In other words, $solve_i(Y)(a = b)$ i-preserves $a = b$. For example, a solver for linear arithmetic can be constructed to isolate a variable on one side of the equality through scaling and cancellation. We assume that the fresh variables generated by $solve_i$ are from the set X_i. We take $vars(\bot_i)$ to be $X \cup X_i$ so as to maintain variable preservation, and indeed $\bot_i$ could be represented as just $\bot$ were it not for this condition.

We now describe a decision procedure for sequents of the form $T \vdash c = d$ in a single Shostak theory with canonizer σ_i and solver $solve_i$. Here the solution state S is just a functional solution set of equalities in i-solved form. Given a solution set S, we define $S\langle\!\langle a\rangle\!\rangle_i$ as $\sigma_i(S[a])$. The composition of solutions sets is defined so that $S \circ_i \bot_i = \bot_i \circ_i S = \bot_i$ and $S \circ_i R = R \cup \{a = R\langle\!\langle b\rangle\!\rangle_i \mid a = b \in S\}$. Note that solved forms are idempotent with respect to composition so that $S \circ_i S = S$. The solved form $solveclose_i(id_T;\ T)$ is obtained by processing the equations in T to build up a solution set S. An equation $a = b$ is first canonized with respect to S as $S\langle\langle a\rangle\rangle_i = S\langle\langle b\rangle\rangle_i$ and then solved to yield the solution R. If R is $\bot_i$, then T is i-unsatisfiable and we return the solution state with $S_i = \bot_i$ as the result. Otherwise, the composition $S \circ_i R$ is computed and used to similarly process the remaining formulas in T.

$$\begin{aligned}
solveclose_i(S;\ \emptyset) &= S \\
solveclose_i(\bot_i;\ T) &= \bot_i \\
solveclose_i(S;\ \{a = b\} \cup T) &= solveclose_i(S', T), \\
&\qquad \text{where } S' = S \circ_i solve_i(vars(S))(S\langle\langle a\rangle\rangle_i = S\langle\langle b\rangle\rangle_i)
\end{aligned}$$

To check i-validity, $\models_i (T \vdash c = d)$, it is sufficient to check that either $solveclose_i(id_T;\ T) = \bot$ or $S'\langle\langle c\rangle\rangle_i \equiv S'\langle\langle d\rangle\rangle_i$, where $S' = solveclose_i(id_T;\ T)$.

Soundness and Completeness. As with the congruence closure procedure, each step in $solveclose_i$ is i-conservative. Hence $solveclose_i$ is sound and complete: if $S' = solveclose_i(S;\ T)$, then for every i-model M and assignment ρ, $M, \rho \models S \cup T$ iff there is a ρ' extending ρ over the variables in $vars(S') - vars(S)$ such that $M, \rho' \models S'$. If $\sigma_i(S'[a]) \equiv \sigma_i(S'[b])$, then $M, \rho' \models a = S'[a] = \sigma_i(S'[a]) = \sigma_i(S'[b]) = S'[b] = b$, and hence $M, \rho \models a = b$. Otherwise, when $\sigma_i(S'[a]) \not\equiv \sigma_i(S'[b])$, we know by the condition on σ_i that there is an i-model M and an assignment ρ' such that $M[\![S'[a]]\!]\rho' \neq M[\![S'[b]]\!]\rho'$. The solved form S' divides the variables into independent variables x such that $S'(x) = x$, and dependent variables y where $y \neq S'(y)$ and the variables in $vars(S'(y))$ are all independent. We can therefore extend ρ' to an assignment ρ where the dependent variables y are mapped to $M[\![S'(y)]\!]\rho'$. Clearly, $M, \rho \models S'$, $M, \rho \models a = S'[a]$, and $M, \rho \models b = S'[b]$. Since S' i-preserves $(id_T;\ T)$, $M, \rho \models T$ but $M, \rho \not\models a = b$ and hence $T \vdash a = b$ is not i-valid, so the procedure is complete. The correctness argument is thus similar to that of Section 3 but for the case of a single Shostak theory considered here, there is no need to construct a canonical term model since $\models_i a = \sigma_i(a)$, and $\sigma_i(a) \equiv \sigma_i(b)$ iff $\models_i a = b$.

Canonical term model. The situation is different when we wish to combine Shostak theories. It is important to resolve potential semantic incompatibilities between two Shostak theories. With respect to some fixed notion of i-validity for θ_i and j-validity for θ_j with $i \neq j$, a formula A in the union of θ_i and θ_j may be satisfiable in an i-interpretation of only a specific finite cardinality for which there might be no corresponding satisfying j-interpretation for the formula. Such an incompatibility can arise even when a theory θ_i is extended with uninterpreted function symbols. For example, if ϕ is a formula with variables x and y that is satisfiable only in a two-element model M where $\rho(x) \neq \rho(y)$, then the set of formulas Γ where $\Gamma = \{\phi, f(x) = x, f(u) = y, f(y) = x\}$ additionally requires $\rho(x) \neq \rho(u)$ and $\rho(y) \neq \rho(u)$. Hence, a model for Γ must have at least three elements, so that Γ is unsatisfiable. However, there is no way to detect this kind of unsatisfiability purely through the use of solving and canonization.

We introduce a canonical term model as a way around such semantic incompatibilities. The set of canonical i-terms a such that $\sigma_i(a) \equiv a$ yields a domain for a *term model* M_i where $M_i(f)(a_1, \ldots, a_n) = \sigma_i(f(a_1, \ldots, a_n))$. If M_i is (isomorphic to) an i-model, then we say that the theory θ_i is *composable*. Note that the *solve* operation is conservative with respect to the model M_i as well, since M_i is taken as an i-model.

Given the usual interpretation of disjunction, a notion of validity is said to be *convex* when $\models (T \vdash c_1 = d_1 \vee \ldots \vee c_n = d_n)$ implies $\models (T \vdash c_k = d_k)$ for some k, $1 \leq k \leq n$. If a theory θ_i is composable, then i-validity is convex. Recall that $\models_i (T \vdash c_1 = d_1 \vee \ldots \vee c_n = d_n)$ iff $\models_i (S \vdash c_1 = d_1 \vee \ldots \vee c_n = d_n)$ for $S = solveclose_i(id_T; T)$. If $S = \bot_i$, then $\models_i (T \vdash c_k = d_k)$, for $1 \leq k \leq n$. If $S \neq \bot_i$, then since S i-preserves T, $\models_i (S \vdash c_1 = d_1 \vee \ldots \vee c_n = d_n)$, but (by assumption) $\not\models_i (S \vdash c_k = d_k)$. An assignment ρ_S can be constructed so that for independent (i.e., where $S(x) = x$) variables $x \in vars(S)$, $\rho_S(x) = x$, and for dependent variables $y \in vars(S)$, $\rho_S(y) = M_i[\![S(y)]\!]\rho_S$. If for $S \neq \bot_i$, $\not\models_{\sigma_i} (S \vdash c_k = d_k)$, then $M_i, \rho_S \models S$ and $M_i, \rho_S \not\models c_k = d_k$. Hence $M_i, \rho_S \not\models (S \vdash c_k = d_k)$, for $1 \leq k \leq n$. This yields $M_i, \rho_S \not\models (T \vdash c_1 = d_1 \vee \ldots \vee c_n = d_n)$, contradicting the assumption.

5 Combining Shostak Theories

We now examine the combination of the theory of equality over uninterpreted function symbols with several disjoint Shostak theories. Examples of interpreted operations from Shostak theories include + and − from the theory of linear arithmetic, *select* and *update* from the theory of arrays, and *cons*, *car*, and *cdr* from the theory of lists. The basic Shostak combination algorithm covers the union of equality over uninterpreted function symbols and a single canonizable and solvable equational theory [Sho84,CLS96,RS01]. Shostak [Sho84] had claimed that the basic combination algorithm was sufficient because canonizers and solvers for disjoint theories could be combined into a single canonizer and solver for their union. This claim is incorrect.[3] We present a combined decision procedure for multiple Shostak theories that overcomes the difficulty of combining solvers.

Two theories θ_1 and θ_2 are said to be disjoint if they have no function symbols in common. A typical subgoal in a proof can involve interpreted symbols from several theories. Let σ_i be the canonizer for θ_i. A term $f(a_1, \ldots, a_n)$ is said to be in θ_i if f is in θ_i even though some a_i might contain function symbols outside θ_i. In processing terms from the union of pairwise disjoint theories $\theta_1, \ldots, \theta_N$, it is quite easy to combine the canonizers so that each theory treats terms in the other theory as variables. Since σ_i is only applicable to i-terms, we first

[3] The difficulty with combining Shostak solvers was observed by Jeremy Levitt [Lev99].

have to extend the canonizer σ_i to treat terms in θ_j for $j \neq i$, as variables. We treat uninterpreted function symbols as belonging to a special theory θ_0 where $\sigma_0(a) = a$ for $a \in \theta_0$. The extended operation σ'_i is defined below.

$$\begin{aligned}\sigma'_i(a) = R[\sigma_i(a')], \text{ when } a', b, R :\ & a' \text{ is an } i\text{-term},\\ & R \text{ is functional},\\ & dom(R) \subseteq vars(a'),\\ & R(x) \in \theta_j, \text{ for } x \in dom(R), \text{ some } j \neq i,\\ & R[a'] \equiv a\end{aligned}$$

Note that the *when* condition in the above definition can always be satisfied. The combined canonizer σ can then be defined as

$$\begin{aligned}\sigma(x) &= x\\ \sigma(f(a_1, \ldots, a_n)) &= \sigma'_i(f(\sigma(a_1), \ldots, \sigma(a_n))), \text{ when } i : f \text{ is in } \theta_i.\end{aligned}$$

This canonizer is, however, not used in the remainder of the paper.

We now discuss the difficulty of combining the solvers $solve_1$ and $solve_2$ for θ_1 and θ_2, respectively, into a single solver. The example uses the theory θ_A of linear arithmetic and the theory θ_L of the pairing and projection operations *cons*, *car*, *cdr*, where, somewhat nonsensically, the projection operations also apply to numerical expressions. Shostak illustrated the combination using the example

$$5 + car(x + 2) = cdr(x + 1) + 3.$$

Since the top-level operation on the left-hand side is $+$, we can treat $car(x + 2)$ and $cdr(x + 1)$ as variables and use $solve_A$. This might yield a partially solved equation of the form $car(x + 2) = cdr(x + 1) - 2$. Now since the top-level operation on the left-hand side is from the theory of lists, we use $solve_L$ to obtain $x + 2 = cons(cdr(x + 1) - 2, u)$ with a fresh variable u. We once again apply $solve_A$ to obtain $x = cons(cdr(x + 1) - 2, u) - 2$. This is, however, not in solved form: the left-hand side variable occurs in an interpreted context in its solution. There is no way to prevent this from happening as long as each solver treats terms from another theory as variables. Therefore the union of Shostak theories is not necessarily a Shostak theory.

The problem of combining disjoint Shostak theories actually has a very simple solution. There is no need to combine solvers. Since the theories are disjoint, the canonizer can tolerate multiple solutions for the same variable as long as there is at most one solution from any individual theory. This can be illustrated on the same example: $5 + car(x + 2) = cdr(x + 1) + 3$. By variable abstraction, we obtain the equation $v_3 = v_6$, where $v_1 = x + 2, v_2 = car(v_1), v_3 = v_2 + 5, v_4 = x + 1, v_5 = cdr(v_4), v_6 = v_5 + 3$. We can separate these equations out into the respective theories so that S is $(S_V;\ S_U;\ S_A;\ S_L)$, where S_V contains the variable equalities in canonical form, S_U is as in congruence closure but is always $\emptyset$ since there are no uninterpreted operations in this example, and S_A and S_L are the solution sets for θ_A and θ_L, respectively. We then get $S_V = \{x = x, v_1 = v_1, v_2 =$

$v_2, v_3 = v_6, v_4 = v_4, v_5 = v_5, v_6 = v_6\}$, $S_A = \{v_1 = x+2, v_3 = v_2+5, v_4 = x+1, v_6 = v_5+3\}$, and $S_L = \{v_2 = car(v_1), v_5 = cdr(v_4)\}$. Since v_3 and v_6 are merged in S_V, but not in S_A, we solve the equality between $S_A(v_3)$ and $S_A(v_6)$, i.e., $solve_A(v_2 + 5 = v_5 + 3)$ to get $v_2 = v_5 - 2$. This result is composed with S_A to get $\{v_1 = x+2, v_3 = v_5+3, v_4 = x+1, v_6 = v_5+3, v_2 = v_5-2\}$ for S_A. There are no new variable equalities to be propagated out of either S_A, S_L, or S_V. Notice that v_2 and v_5 both have different solved forms in S_A and S_L. This is tolerated since the solutions are from disjoint theories and the canonizer can pick a solution that is appropriate to the context. For example, when canonizing a term of the form $f(x)$ for $f \in \theta_i$, it is clear that the only relevant solution for x is the one from S_i.

We can now check whether the resulting solution state verifies the original equation $5 + car(x+2) = cdr(x+1) + 3$. In canonizing $f(a_1, \ldots, a_n)$ we return $S_V(y)$ whenever the term $f(S_i(S[\![a_1]\!]), \ldots, S_i(S[\![a_n]\!]))$ being canonized is such that $y = f(S_i(S[\![a_1]\!]), \ldots, S_i(S[\![a_n]\!])) \in S_i$ for $f \in \theta_i$. Thus $x+2$ canonizes to v_1 using S_A, and $car(v_1)$ canonizes to v_2 using S_L. The resulting term $5+v_2$, using the solution for v_2 from S_A, simplifies to $v_5 + 3$, which returns the canonical form v_6 by using S_A. On the right-hand side, $x + 1$ is equivalent to v_4 in S_A, and $car(v_4)$ simplifies to v_5 using S_L The right-hand side therefore simplifies to v_5+3 which is canonized to v_6 using S_A. The canonized left-hand and right-hand sides are identical.

We present a formal description of the procedure used informally in the above example. We show how *process* from Section 3 can be extended to combine the union of disjoint solvable, canonizable, composable theories. We assume that there are N disjoint theories $\theta_1, \ldots, \theta_N$. Each theory θ_i is equipped with a canonizer σ_i and solver $solve_i$ for i-terms. If we let I represent the interval $[1, N]$, then an I-model is a model M that is an i-model for each $i \in I$. We will ensure that each inference step is conservative with respect to I-models, i.e., I-conservative. We represent the uninterpreted part of S as S_0 instead of S_U. The solution state S of the algorithm now consists of a list of sets of equations $(S_V;\ S_0;\ S_1;\ \ldots;\ S_N)$. Here S_V is a set of variable equations of the form $x = y$, and S_0 is the set of equations of the form $x = f(x_1, \ldots, x_n)$ where f is uninterpreted. Each S_i is in i-solved form and is the solution set for θ_i.

Terms now contain a mixture of function symbols that are uninterpreted or are interpreted in one of the theories θ_i. A solution state S is *confluent* if for all $x, y \in dom(S_V)$ and i, $0 \le i \le N$: $S_V(x) \equiv S_V(y) \iff S_i(\{x\}) \cap S_i(\{y\}) \neq \emptyset$. A solution state S is canonical if it is confluent; S_V is functional and idempotent, i.e., $S_V \circ S_V = S_V$; the uninterpreted solution set S_0 is normalized, i.e., $S_0 \rhd S_V = S_0$; each S_i, for $i > 0$, is functional, idempotent, i.e., $S_i \circ_i S_i = S_i$, normalized i.e., $S_i \rhd S_V = S_i$, and in i-solved form. The canonization of expressions with respect to a canonical solution set S is defined as follows.

$$S[\![x]\!] = S_V(x)$$
$$S[\![f(a_1, \ldots, a_n)]\!] = S_V(x), \text{ when } i, x :$$

$$
\begin{aligned}
&abstract(S;\ x = y) = (S;\ x = y),\\
&abstract(S;\ a = b) = (S';\ a' = b'),\\
&\qquad\qquad\text{when } S', c, i:\ c \in max(\llbracket a = b \rrbracket_i),\\
&\qquad\qquad\qquad\qquad x \notin vars(S \cup a = b),\\
&\qquad\qquad\qquad\qquad S'_V = S_V \cup \{x = x\},\\
&\qquad\qquad\qquad\qquad S'_i = S_i \cup \{x = c\},\\
&\qquad\qquad\qquad\qquad S'_j = S_j,\ \text{for }, i \neq j\\
&\qquad\qquad\qquad\qquad a' = \{c = x\}[a],\\
&\qquad\qquad\qquad\qquad b' = \{c = x\}[b].
\end{aligned}
$$

Fig. 2. Variable abstraction step for multiple Shostak theories

$$
\frac{i \geq 0, f \in \theta_i, x = \sigma'_i(f(S_i(S\llbracket a_1 \rrbracket), \ldots, S_i(S\llbracket a_n \rrbracket))) \in S_i}{S\llbracket f(a_1, \ldots, a_n) \rrbracket = \sigma'_i(f(S_i(S\llbracket a_1 \rrbracket), \ldots, S_i(S\llbracket a_n \rrbracket))),\ \text{when } i : f \in \theta_i, i \geq 0.}
$$

Since variables are used to communicate between the different theories, the canonical variable x in S_V is returned when the term being canonized is known to be equivalent to an expression a such that $y = a$ in S_i, where $x \equiv S_V(y)$. The definition of the above global canonizer is one of the key contributions of this paper. This definition can be applied to the example above of computing $S\llbracket 5 + car(x + 2) \rrbracket$.

Variable Abstraction. The variable abstraction procedure $abstract(S; a = b)$ is shown in Figure 2. If a is an i-term such that $a \notin X$, then a is said to be a pure i-term. Let $\llbracket a = b \rrbracket_i$ represent the set of subterms of $a = b$ that are pure i-terms. The set $max(M)$ of maximal terms in M is defined to be $\{a \in M | a \equiv b \vee a \notin \llbracket b \rrbracket,$ for any $b \in M\}$. In a single variable abstraction step, $abstract(S;\ a = b)$ picks a maximal pure i-subterm c from the canonized input equality $a = b$, and replaces it with a fresh variable x from X while adding $x = c$ to S_i. By abstracting a maximal pure i-term, we ensure that S_i remains in i-solved form.

Explanation. The procedure in Figure 3 is similar to that of Figure 1. Equations from the input set T are processed into the solution state S of the form $S_V; S_0; \ldots, S_N$. Initially, S must be canonical. In processing the input equation $a = b$ into S, we take steps to systematically restore the canonicity of S. The first step is to compute the canonical form $S\llbracket a = b \rrbracket$ of $a = b$ with respect to S. It is easy to see that $(S; S\llbracket a = b \rrbracket)$ I-preserves $(S; a = b)$.

The result of the canonization step $a' = b'$ is then variable abstracted as $abstract^*(a' = b')$ (shown in Figure 2) so that in each step, a maximal, pure i-subterm c of $a' = b'$ is replaced by a fresh variable x, and the equality $x = c$ is added to S_i. This is also easily seen to be an I-conservative step. The equality $x = y$ resulting from the variable abstraction of $a' = b'$ is then merged into S_V and S_0. This can destroy confluence since there may be variables w and z such

$$
\begin{aligned}
process(S;\ \emptyset) &= S\\
process(S;\ T) &= S, \text{ when } i : S_i = \bot_i\\
process(S;\ \{a = b\} \cup T) &= process(S';\ T), \text{ where}\\
&\quad S' = close^*(merge_V(abstract^*(S;\ S[\![a = b]\!]))).
\end{aligned}
$$

$$
\begin{aligned}
close(S) &= S, \text{ when } i : S_i = \bot_i\\
close(S) &= S', \text{ when } S', i, x, y :\\
&\quad x, y \in dom(S_V),\\
&\quad (i > 0, S_V(x) \equiv S_V(y), S_i(x) \not\equiv S_i(y), \text{ and}\\
&\quad\ S' = merge_i(S;\ x = y))\\
&\quad \text{or}\\
&\quad (i \geq 0, S_V(x) \not\equiv S_V(y), S_i(\{x\}) \cap S_i(\{y\}) \neq \emptyset, \text{ and}\\
&\quad\ S' = merge_V(S;\ S_V(x) = S_V(y)))\\
close(S) &= normalize(S), \text{ otherwise.}
\end{aligned}
$$

$$
normalize(S) = (S_V;\ S_0;\ S_1 \rhd S_V;\ \ldots;\ S_N \rhd S_V).
$$

$$
\begin{aligned}
merge_i(S; x = y) &= S', \text{ where } i > 0,\\
&\quad S'_i = S_i \circ_i solve_i(vars(S_i))(S_i(x) = S_i(y)),\\
&\quad S'_j = S_j, \text{ for } i \neq j,\\
&\quad S'_V = S_V.
\end{aligned}
$$

$$
\begin{aligned}
merge_V(S; x = x) &= S\\
merge_V(S; x = y) &= (S_V \circ R;\ S_0 \rhd R;\ S_1;\ \ldots;\ S_N), \text{ where } R = orient(x = y).
\end{aligned}
$$

Fig. 3. Combining Multiple Shostak Theories

that w and z are merged in S_V (i.e., $S_V(w) \equiv S_V(z)$) that are unmerged in some S_i (i.e., $S_i(\{w\}) \cap S_i(\{z\}) = \emptyset$), or vice-versa.[4] The number of variables in $dom(S_V)$ remains fixed during the computation of $close^*(S)$. Confluence is restored by $close^*(S)$ which finds a pair of variables that are merged in some S_i but not in S_V, and merging them in S_V, or that are merged in S_V and not in some S_i and merging them in S_i. Each such merge step is also I-conservative. When this process terminates, S is once again canonical. The solution sets S_i are normalized with respect to S_V in order to ensure that the entries are in the normalized form for lookup during canonization.

Invariants. As with congruence closure, several key invariants are needed to ensure that the solution state S is maintained in canonical form whenever it is given as the argument to *process*. If S is canonical and a and b are canonical with respect to S, then for $(S'; a' = b') = abstract(S;\ a = b)$, S' is canonical, and a' and b' are canonical with respect to S'. The state $abstract(S;\ a = b)$ I-preserves $(S;\ a = b)$. A solution state is said to be well-formed if S_V is functional and idempotent, S_0 is normalized, and each S_i is functional, idempotent, and in

[4] For $i > 0$, S_i is maintained in i-solved form and hence, $S_i(\{x\}) = \{x, S_i(x)\}$.

solved form. Note that if S is well-formed, confluent, and each S_i is normalized, then it is canonical. When S is well-formed, and $S' = merge_V(S; x = y)$ or $S' = merge_i(S; x = y)$, then S' is well-formed and I-preserves $(S; x = y)$. If S is well-formed and congruence-closed, and $S' = normalize(S)$, then S' is well-formed and each S'_i is normalized. If $S' = normalize(S)$, then each S'_i is in solved form because if x replaces y on the right-hand side of a solution set S_i, then $S_i(y) \equiv y$ since S_i is in i-solved form. By congruence closure, we already have that $S_i(x) \equiv S_i(y) \equiv y$. Therefore, the uniform replacement of y by x ensures that $S'_i(x) \equiv x$, thus leaving S in solved form. If $S' = close^*(S)$, where S is well-formed, then S' is canonical.

Variations. As with congruence closure, once S is confluent, it is safe to strengthen the normalization step to replace each S_i by $S_V[S_i]$. This renders S_0^{-1} functional, but S_i^{-1} may still be non-functional for $i > 0$, since it might contain left-hand side variables that are local. However, if $\hat{S}_i$ is taken to be S_i restricted to $dom(S_V)$, then $\hat{S}_i^{-1}$ with the strengthened normalization is functional and can be used in canonization. The solutions for local variables can be safely discarded in an actual implementation. The canonization and variable abstraction steps can be combined within a single recursion.

Termination. The operations $S[\![a = b]\!]$ and $abstract^*(S; a = b)$ are easily seen to be terminating. The operation $close^*(S)$ also terminates because the sum of the number of equivalence classes of variables in $dom(S_V)$ with respect to each of the solution sets $S_V, S_0, S_1, \ldots, S_N$, decreases with each *merge* operation.

Soundness and Completeness. We have already seen that each of the steps: canonization, variable abstraction, composition, merging, and normalization, is I-conservative. It therefore follows that if $S' = process(S; T)$, then S' I-preserves S. Hence, if $S'[\![c]\!] \equiv S'[\![d]\!]$, then clearly $\models_I (S' \vdash c = d)$, and hence $\models_I (S; T \vdash c = d)$.

The completeness argument requires the demonstration that if $S'[\![c]\!] \not\equiv S'[\![d]\!]$, then $\not\models_I (S' \vdash c = d)$ when S' is canonical. This is done by means of a construction of $M_{S'}$ and $\rho_{S'}$ such that $M_{S'}, \rho_{S'} \models S'$ but $M_{S'}, \rho_{S'} \not\models c = d$. The domain D consists of canonical terms e such that $S'[\![e]\!] = e$. As with congruence closure, $M_{S'}$ is defined so that $M_{S'}(f)(e_1, \ldots, e_n) = S'[\![f(e_1, \ldots, e_n)]\!]$. The assignment $\rho_{S'}$ is defined so that $\rho_{S'}(x) = S_V(x)$. By induction on c, we have that $M_{S'}[\![c]\!]\rho_{S'} = S'[\![c]\!]$. We can also easily check that $M_{S'}, \rho_{S'} \models S'$.

It is also the case that $M_{S'}$ is an I-model since $M_{S'}$ is isomorphic to M_i for each i, $1 \leq i \leq N$. This can be demonstrated by constructing a bijective map μ_i between D and the domain D_i corresponding to M_i. Let P_i be the set of pure i-terms in D, and let γ be a bijection between $D - P_i$ and X such that $\gamma(x) = x$ if $S'_i(x) = x$ for $x \in dom(S'_V)$. Define μ_i so that $\mu_i(x) = S'_i(x)$ for $x \in dom(S'_V)$ and $S'_V(x) = x$, $\mu_i(y) = y$ for $y \in X_i$, $\mu_i(f(a_1, \ldots, a_n)) = f(\mu_i(a_1), \ldots, \mu_i(a_n))$ for $f \in \theta_i$, and $\mu_i(a) = \gamma(a)$, otherwise. It can then be verified that for an i-term a, $\mu_i(M_{S'}[\![a]\!]\rho) = M_i[\![a]\!]\rho_i$, where $\rho_i(x) = \mu_i(\rho(x))$. This concludes the proof of completeness.

Convexity revisited. As in Section 4, the term model construction of $M_{S'}$ once again establishes that I-validity is convex. In other words, a sequent $\models_I (T \vdash c_1 = d_1 \vee \ldots \vee c_n = d_n)$ iff $\models_I (T \vdash c_k = d_k)$ for some k, $1 \leq k \leq n$.

6 Conclusions

Ground decision procedures for equality are crucial for discharging the myriad proof obligations that arise in numerous applications of automated reasoning. These goals typically contain operations from a combination of theories, including uninterpreted symbols. Shostak's basic method deals only with the combination of a single canonizable, solvable theory with equality over uninterpreted function symbols. Indeed, in all previous work based on Shostak's method, only the basic combination is considered. Though Shostak asserted that the basic combination was adequate to cover the more general case of multiple Shostak theories, this claim has turned out to be unsubstantiated. We have given here the first Shostak-style combination method for the general case of multiple Shostak theories. The algorithm is quite simple and is supported by straightforward arguments for termination, soundness, and completeness.

Shostak's combination method, as we have described it, is clearly an instance of a Nelson–Oppen combination [NO79] since it involves the exchange of equalities between variables through the solution set S_V. The added advantage of a Shostak combination is that it combines the canonizers of the individual theories into a global canonizer. The definition of such a canonizer for multiple Shostak theories is the key contribution of this paper. The technique of achieving confluence across the different solution sets is unique to our method. Confluence is needed for obtaining useful canonical forms, and is therefore not essential in a general Nelson–Oppen combination. The global canonizer $S[\![a]\!]$ can be applied to input formulas to discharge queries and simplify input formulas. The reduction to canonical form with respect to the given equalities helps keep the size of the term universe small, and makes the algorithm more efficient than a black box Nelson–Oppen combination. The decision algorithm for a Shostak theory given in Section 4 fits the requirements for a black box procedure that can be used within a Nelson–Oppen combination. The Nelson–Oppen combination of Shostak theories with other decision procedures has been studied by Tiwari [Tiw00], Barrett, Dill, and Stump [BDS02], and Ganzinger [Gan02], but none of these methods includes a general canonization procedure as is required for a Shostak combination.

Variable abstraction is also used in the combination unification procedure of Baader and Schulz [BS96], which addresses a similar problem to that of combining Shostak solvers. In our case, there is no need to ensure that solutions are compatible across distinct theories. Furthermore, variable dependencies can be cyclic across theories so that it is possible to have $y \in vars(S_i(x))$ and $x \in vars(S_j(y))$ for $i \neq j$. Our algorithm can be easily and usefully adapted for combining unification and matching algorithms with constraint solving in Shostak theories.

Insights derived from the Nelson–Oppen combination method have been crucial in the design of our algorithm and its proof. Our presentation here is different from that of our previous algorithm for the basic Shostak combination [RS01] in the use of variable abstraction and the theory-wise separation of solution sets. Our proof of the basic algorithm additionally demonstrated the existence of proof objects in a sound and complete proof system. This can easily be replicated for the general algorithm studied here. The soundness and completeness proofs given here are for composable theories and avoid the use of σ-models.

Our Shostak-style algorithm fits modularly within the Nelson–Oppen framework. It can be employed within a Nelson–Oppen combination (as suggested by Rushby [CLS96]) in which there are other decision procedures that generate equalities between variables. It is also possible to combine it with decision procedures that are not disjoint, as for example with linear arithmetic inequalities. Here, the existence of a canonizer with respect to equality is useful for representing inequality information in a canonical form. A variant of the procedure described here is implemented in ICS [FORS01] in exactly such a combination.

References

[BDL96] Clark Barrett, David Dill, and Jeremy Levitt. Validity checking for combinations of theories with equality. In Mandayam Srivas and Albert Camilleri, editors, *Formal Methods in Computer-Aided Design (FMCAD '96)*, volume 1166 of *Lecture Notes in Computer Science*, pages 187–201, Palo Alto, CA, November 1996. Springer-Verlag.

[BDS02] Clark W. Barrett, David L. Dill, and Aaron Stump. A generalization of Shostak's method for combining decision procedures. In A. Armando, editor, *Frontiers of Combining Systems, 4th International Workshop, FroCos 2002*, number 2309 in Lecture Notes in Artificial Intelligence, pages 132–146, Berlin, Germany, April 2002. Springer-Verlag.

[Bjø99] Nikolaj Bjørner. *Integrating Decision Procedures for Temporal Verification.* PhD thesis, Stanford University, 1999.

[BS96] F. Baader and K. Schulz. Unification in the union of disjoint equational theories: Combining decision procedures. *J. Symbolic Computation*, 21:211–243, 1996.

[BTV02] Leo Bachmair, Ashish Tiwari, and Laurent Vigneron. Abstract congruence closure. *Journal of Automated Reasoning*, 2002. To appear.

[CLS96] David Cyrluk, Patrick Lincoln, and N. Shankar. On Shostak's decision procedure for combinations of theories. In M. A. McRobbie and J. K. Slaney, editors, *Automated Deduction—CADE-13*, volume 1104 of *Lecture Notes in Artificial Intelligence*, pages 463–477, New Brunswick, NJ, July/August 1996. Springer-Verlag.

[DST80] P.J. Downey, R. Sethi, and R.E. Tarjan. Variations on the common subexpressions problem. *Journal of the ACM*, 27(4):758–771, 1980.

[FORS01] J.-C. Filliâtre, S. Owre, H. Rueß, and N. Shankar. ICS: Integrated Canonization and Solving. In G. Berry, H. Comon, and A. Finkel, editors, *Computer-Aided Verification, CAV '2001*, volume 2102 of *Lecture Notes in Computer Science*, pages 246–249, Paris, France, July 2001. Springer-Verlag.

[FS02] Jonathan Ford and Natarajan Shankar. Formal verification of a combination decision procedure. In A. Voronkov, editor, *Proceedings of CADE-19*, Berlin, Germany, 2002. Springer-Verlag.

[Gan02] Harald Ganzinger. Shostak light. In A. Voronkov, editor, *Proceedings of CADE-19*, Berlin, Germany, 2002. Springer-Verlag.

[Kap97] Deepak Kapur. Shostak's congruence closure as completion. In H. Comon, editor, *International Conference on Rewriting Techniques and Applications, RTA '97*, number 1232 in Lecture Notes in Computer Science, pages 23–37, Berlin, 1997. Springer-Verlag.

[Koz77] Dexter Kozen. Complexity of finitely presented algebras. In *Conference Record of the Ninth Annual ACM Symposium on Theory of Computing*, pages 164–177, Boulder, Colorado, 2–4 May 1977.

[Lev99] Jeremy R. Levitt. *Formal Verification Techniques for Digital Systems.* PhD thesis, Stanford University, 1999.

[NO79] G. Nelson and D. C. Oppen. Simplification by cooperating decision procedures. *ACM Transactions on Programming Languages and Systems*, 1(2):245–257, 1979.

[NO80] G. Nelson and D. C. Oppen. Fast decision procedures based on congruence closure. *Journal of the ACM*, 27(2):356–364, 1980.

[RS01] Harald Rueß and Natarajan Shankar. Deconstructing Shostak. In *16th Annual IEEE Symposium on Logic in Computer Science*, pages 19–28, Boston, MA, July 2001. IEEE Computer Society.

[Sha01] Natarajan Shankar. Using decision procedures with a higher-order logic. In *Theorem Proving in Higher Order Logics: 14th International Conference, TPHOLs 2001*, volume 2152 of *Lecture Notes in Computer Science*, pages 5–26, Edinburgh, Scotland, September 2001. Springer-Verlag. Available at `ftp://ftp.csl.sri.com/pub/users/shankar/tphols2001.ps.gz`.

[Sho78] R. Shostak. An algorithm for reasoning about equality. *Comm. ACM*, 21:583–585, July 1978.

[Sho84] Robert E. Shostak. Deciding combinations of theories. *Journal of the ACM*, 31(1):1–12, January 1984.

[Tiw00] Ashish Tiwari. *Decision Procedures in Automated Deduction.* PhD thesis, State University of New York at Stony Brook, 2000.

Multiset Rewriting and Security Protocol Analysis

John C. Mitchell*

Stanford University, Stanford, CA 94305
http://www.stanford.edu/~jcm

Abstract. The Dolev-Yao model of security protocol analysis may be-formalized using a notation based on multi-set rewriting with existential quantification. This presentation describes the multiset rewriting approach to security protocol analysis, algorithmic upper and lower bounds on specific forms of protocol analysis, and some of the ways this model is useful for formalizing sublte properties of specific protocols.

Background

Many methods for security protocol analysis are based on essentially the same underlying model. This model, which includes both protocol steps and steps by a malicious attacker, appears to have developed from positions taken by Needham and Schroeder [NS78] and a model presented by Dolev and Yao [DY83]; it is often called the *Dolev-Yao model.* The Dolev-Yao model is used in theorem proving [Pau97], model-checking methods [Low96,Mea96a,MMS97,Ros95, Sch96], constraint solving [CCM01,MS01], and symbolic search tools [KMM94, Mea96b,DMR00]. Strand spaces [FHG98] and Spi-calculus [AG99] also use versions of the Dolev-Yao model. Historically, however, there has not been a standard presentation of the Dolev-Yao model independent of the many tools in which it is used.

The work described in this invited talk, carried out over several years in collaboration with Iliano Cervesato, Nancy Durgin, Patrick Lincoln, and Andre Scedrov, began with an attempt to formulate the Dolev-Yao model in a general way [Mit98,DM99]. There were several goals in doing so:

- Identify the basic assumptions of the model, independent of limitations associated with specific tools, such as finite-state limitations,
- Make it easier for additional researchers to study security protocols using additional tools,
- Study the power and limitations of the model itself.

The Dolev-Yao model seems most easily characterized by a symbolic computation system. Messages are composed of indivisible values, representable by

* Partially supported by DoD MURI "Semantic Consistency in Information Exchange," ONR Grant N00014-97-1-0505.

S. Tison (Ed.): RTA 2002, LNCS 2378, pp. 19–22, 2002.

constants of some vocabulary, and encryption is modeled as a non-invertible function that hides all information about the plaintext. The initial state of a system is represented by a set of private keys and other information available only to honest participants in the protocol, and public and compromised information known to the protocol adversary. The protocol adversary has the following capabilities:

- overhear every message,
- prevent a message from reaching the intended recipient,
- decompose a message comprised of several fields into parts and decrypt any field for which the intruder has already obtained the encryption key,
- remember all parts of all messages that have been intercepted or overheard,
- encrypt using encryption keys known to the attacker,
- generate messages and send them to any other protocol participant.

The messages generated by the intruder may be composed of any information supplied to the intruder initially (such as public keys of protocol participants) and data obtained by overhearing or intercepting messages. In this model, the intruder is not allowed to guess encryption keys or other values that would normally be produced by randomized procedures.

Many aspects of this model are easily represented by standard rewrite rules. The initial state of a system can be characterized by a set of formulas containing the "facts" known to the honest participants and the adversary. Representing each possible step in the exectution of a protocol or attack by a rewrite rule, we obtain the set of possible traces and reachable states. Two subtleties are linearity and the generation of new values. Linearity, in the sense of linear logic [Gir87], not the way the term is used in rewrite systems, arises because protocol participants change state. If we represent the fact that Alice is in an initial state with key k by the atomic formula $A_0(k)$, then this formula should be removed from the state of the system after Alice leaves her initial state. Furthermore, if the system is in a state with two participants in identical states, this is different from a state in which there is only one such participant. Therefore, we represent the state of a system by a multiset of formulas rather than a set.

Many security protocols involve steps that generate new values. In actual implementations, the new values are generally produced by some form of random number generator or pseudo-random generator. For example, a challenge-response protocol begins with one participant, say Alice, generating a random challenge. Alice sends the challenge to Bob, possibly encrypted, and Bob computes and sends a response. If Alice's challenge is *fresh,* meaning that it is a new value not known initially and not used previously in the protocol, then Alice may be sure that the response she receives is the result of computation by Bob. If Alice were to issue an old challenge, then some attacker could impersonate Bob by sending a saved copy of Bob's old response. In our multiset rewriting framework, new values are generated as a result of instantiating existentially quantified variables with fresh constants. This is a form of a standard existential instantiation rule, applied to symbolic computation.

The multiset rewriting approach to security protocol analysis consists of several modeling conventions using multisets of formulas and rewrite rules that may contain existential quantification on the right-hand-side of the rule. The modeling conventions include ways of separating the knowledge of one participant from the knowledge of another, separating private knowledge from public knowledge, and separating the state of a participant from a network message, all based on conventions for naming predicates used in atomic formulas. There are also conventions for separating an initialization phase, in which initial public and privante information is generated, from a key distribution and role-assignment phase and a protocol execution phase.

This talk will present three aspects of the framework:

- Basic concepts in protocol analysis and their formulation using multiset rewriting,
- Algorithmic upper and lower bounds on specific forms of protocol analysis,
- Some of the ways the model is useful for formalizing sublte properties of specific protocols.

Our algorithmic results include undecidability for a class of protocols that only allow each participant a finite number of steps and impose a bound on the length of each message, exponential-time completeness when the number of new values (generated by existential instantiation) is bounded, and NP-completeness when the number of protocol roles used in any run is bounded. These results have been announced previously [DLMS99,CDL+99] and are further elaborated in a submitted journal article available from the author's web site.

Two directions for current and future work are the development of logics for reasoning about protocols and formalization of specific protocol properties. A logic described in [DMP01] uses a trace-based semantics of protocol runs, inspired by work on the multiset-rewriting framework. In ongoing work on contract-signing protocols, we have found it essential to specify clearly how each agent may behave and to use adaptions of game-theoretic concepts to characterize desired correctness conditions. Put as simply as possible, the essential property of any formalism for studing security protocol analysis is that we must define precisely the set of actions available to any attacker and the ways that each protocol participant may respond to any possible message. It is not adequate to simply characterize the actions that take place when no attacker is present.

References

[AG99] M. Abadi and A. Gordon. A calculus for cryptographic protocols: the spi calculus. *Information and Computation*, 148(1):1–70, 1999.

[CCM01] H. Comon, V. Cortier, and J. Mitchell. Tree automata with one memory, set constraints and ping-pong protocols. In *Proc. 28th Int. Coll. Automata, Languages, and Programming (ICALP'2001)*, pages 682–693. Lecture Notes in Computer Science, vol 2076, Springer, July 2001.

[CDL+99] I. Cervesato, N. Durgin, P. Lincoln, J. Mitchell, and A. Scedrov. A meta-notation for protocol analysis. In P. Syverson, editor, *12-th IEEE Computer Security Foundations Workshop*. IEEE Computer Society Press, 1999.

[DLMS99] N. Durgin, P. Lincoln, J. Mitchell, and A. Scedrov. Undecidability of bounded security protocols. In *Proc. Workshop on formal methods in security protocols*, Trento, Italy, 1999.

[DM99] N. Durgin and J. Mitchell. Analysis of security protocols. In *Calculational System Design, Series F: Computer and Systems Sciences, Vol. 173.* IOS Press, 1999.

[DMP01] N. Durgin, J. Mitchell, and D. Pavlovic. A compositional logic for protocol correctness. In *CSFW: Proceedings of The 14th Computer Security Foundations Workshop*. IEEE Computer Society Press, 2001.

[DMR00] G. Denker, J. Millen, and H. Ruess. The CAPSL integrated protocol environment. Technical Report SRI-CSL-2000-02, Computer Science Laboratory, SRI International, October 2000.

[DY83] D. Dolev and A. Yao. On the security of public-key protocols. *IEEE Transactions on Information Theory*, 2(29), 1983.

[FHG98] F. Javier Thayer Fábrega, Jonathan C. Herzog, and Joshua D. Guttman. Strand spaces: Why is a security protocol correct? In *Proceedings of the 1998 IEEE Symposium on Security and Privacy*, pages 160–171, Oakland, CA, May 1998. IEEE Computer Society Press.

[Gir87] J.-Y. Girard. Linear logic. *Theoretical Computer Science*, 50:1–102, 1987.

[KMM94] R. Kemmerer, C. Meadows, and J. Millen. Three systems for cryptographic protocol analysis. *J. Cryptology*, 7(2):79–130, 1994.

[Low96] G. Lowe. Breaking and fixing the Needham-Schroeder public-key protocol using CSP and FDR. In *2nd International Workshop on Tools and Algorithms for the Construction and Analysis of Systems*. Springer-Verlag, 1996.

[Mea96a] C. Meadows. Analyzing the Needham-Schroeder public-key protocol: a comparison of two approaches. In *Proc. European Symposium On Research In Computer Security*. Springer Verlag, 1996.

[Mea96b] C. Meadows. The NRL protocol analyzer: an overview. *J. Logic Programming*, 26(2):113–131, 1996.

[Mit98] J.C. Mitchell. Analysis of security protocols. Slides for invited talk at CAV '98, available at `http://www.stanford.edu/~jcm`, July 1998.

[MMS97] J.C. Mitchell, M. Mitchell, and U. Stern. Automated analysis of cryptographic protocols using Murφ. In *Proc. IEEE Symp. Security and Privacy*, pages 141–151, 1997.

[MS01] J. Millen and V. Shmatikov. Constraint solving for bounded-process cryptographic protocol analysis. In *8th ACM Conference on Computer and Communication Security*, pages 166–175, November 2001.

[NS78] R.M. Needham and M.D. Schroeder. Using encryption for authentication in large networks of computers. *Communications of the ACM*, 21(12):993–999, 1978.

[Pau97] L.C. Paulson. Proving properties of security protocols by induction. In *10th IEEE Computer Security Foundations Workshop*, pages 70–83, 1997.

[Ros95] A. W. Roscoe. Modelling and verifying key-exchange protocols using CSP and FDR. In *8th IEEE Computer Security Foundations Workshop*, pages 98–107. IEEE Computer Soc Press, 1995.

[Sch96] S. Schneider. Security properties and CSP. In *IEEE Symp. Security and Privacy*, 1996.

Engineering of Logics for the Content-Based Representation of Information

Franz Baader

TU Dresden, Germany, baader@inf.tu-dresden.de

Abstract. Storage and transfer of information as well as interfaces for accessing this information have undergone a remarkable evolution. Nevertheless, information systems are still not "intelligent" in the sense that they "understand" the information they store, manipulate, and present to their users. A case in point is the world wide web and search engines allowing to access the vast amount of information available there. Web-pages are mostly written for human consumption and the mark-up provides only rendering information for textual and graphical information. Search engines are usually based on keyword search and often provide a huge number of answers, many of which are completely irrelevant, whereas some of the more interesting answers are not found. In contrast, the vision of a "semantic web" aims for machine-understandable web resources, whose content can then be comprehended and processed both by automated tools, such as search engines, and by human users.

The content-based representation of information requires representation formalisms with a well-defined formal semantics since otherwise there cannot be a common understanding of the represented information. This semantics can elegantly be provided by a translation into an appropriate logic or the use of a logic-based formalism in the first place. This logical approach has the additional advantage that logical inferences can then be used to reason about the represented information, thus detecting inconsistencies and computing implicit information. However, in this setting there is a fundamental tradeoff between the expressivity of the representation formalism on the one hand, and the efficiency of reasoning with this formalism on the other hand.

This motivates the "engineering of logics", i.e., the design of logical formalisms that are tailored to specific representation tasks. This also encompasses the formal investigation of the relevant inference problems, the development of appropriate inferences procedures, and their implementation, optimization, and empirical evaluation. Another important topic in this context is the combination of logics and their inference procedures since a given application my require the use of more than one specialized logic. The talk will illustrate this approach with the example of so-called Description Logics and their application as ontology languages for the semantic web.

S. Tison (Ed.): RTA 2002, LNCS 2378, p. 23, 2002.

Axiomatic Rewriting Theory VI: Residual Theory Revisited

Paul-André Melliès

Equipe PPS, CNRS, Université Paris 7

Abstract. Residual theory is the algebraic theory of confluence for the λ-calculus, and more generally *conflict-free* rewriting systems (=without critical pairs). The theory took its modern shape in Lévy's PhD thesis, after Church, Rosser and Curry's seminal steps. There, Lévy introduces a *permutation equivalence* between rewriting paths, and establishes that among all confluence diagrams $P \longrightarrow N \longleftarrow Q$ completing a span $P \longleftarrow M \longrightarrow Q$, there exists a *minimum* such one, modulo permutation equivalence. Categorically, the diagram is called a *pushout.*
In this article, we extend Lévy's residual theory, in order to enscope "border-line" rewriting systems, which admit critical pairs but enjoy a strong Church-Rosser property (=existence of pushouts.) Typical examples are the *associativity rule* and the *positive braid* rewriting systems. Finally, we show that the resulting theory reformulates and clarifies Lévy's optimality theory for the λ-calculus, and its so-called "extraction procedure".

1 Introduction

Knuth and Bendix [KB] define a *critical pair* as a pair of rewriting rules which induces overlapping redexes. The definition applies with little variation to word, term, or higher-order term rewriting systems [KB,K_2,Ni,MN].

The syntactical definition captures the idea that a critical pair is a "singularity" of the system, and a *risk of non-confluence.* Indeed, the definition is mainly motivated by Knuth and Bendix's theorem, which states that a (word, term, higher order) rewriting system is *locally* confluent, see [K_2], iff all its critical pairs are *convergent* (=confluent).

For instance, the two rules:

$$A \longrightarrow B \qquad\qquad A \longrightarrow C \tag{1}$$

form a critical pair, because they induce overlapping redexes $A \longrightarrow B$ and $A \longrightarrow C$. The critical pair is not convergent, and thus, the term rewriting system is not locally confluent.

S. Tison (Ed.): RTA 2002, LNCS 2378, pp. 24–50, 2002.

On the other hand, the two rules L and R expressing the logical connective "parallel or"

$$\mathrm{por}(\mathbf{true}, \mathrm{x}) \xrightarrow{\mathrm{L}} \mathbf{true} \qquad \mathrm{por}(\mathrm{x}, \mathbf{true}) \xrightarrow{\mathrm{R}} \mathbf{true} \tag{2}$$

form also a critical pair, because they induce overlapping redexes l and r from the term $\mathrm{por}(\mathbf{true}, \mathbf{true})$:

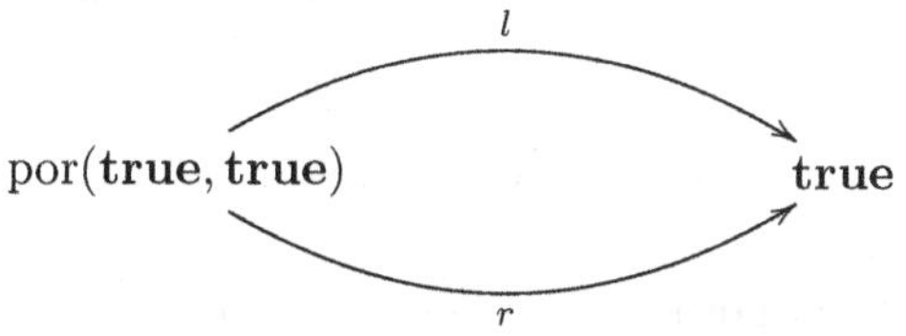

(3)

But this critical pair (2) is convergent, and thus, the rewriting system is locally confluent. It is even *confluent* — this is proved by applying Newman's lemma, or better, by observing that the system is an instance of van Oostrom and van Raamsdonk's *weakly orthogonal* rewriting systems [OR].

Does this mean that the pair (1) is more "critical" than the pair (2) ? Well, from the strict point of view of confluence, yes: the pair (1) is "more" critical than the pair (2). But on the other hand, the standardization theory developed in [HL,Bo,M$_2$] provides excellent reasons to consider the pair (2) as "critical" as (1). We explain this briefly. Suppose that one extends the term rewriting system (2) with the rule

$$\Omega \longrightarrow \mathbf{true}$$

The extended rewriting system is still confluent. Let us count the number of *head* rewriting paths from $\mathrm{por}(\Omega, \Omega)$ to the "value" term $V = \mathbf{true}$. These head rewriting paths are characterized in [M$_3$] for a class of rewriting systems enscoping all (possibly conflicting, possibly higher-order) term rewriting systems. The characterization shows that there exists at most *one* head rewriting path in a *conflict-free* (that is: critical-pair free, and left-linear) term rewriting system; think of the λ-calculus. In contrast, one counts *two* head rewriting paths in the system (2):

$$\begin{array}{l} \mathrm{por}(\Omega, \Omega) \xrightarrow{\Omega} \mathrm{por}(\mathbf{true}, \Omega) \xrightarrow{\mathrm{L}} \mathbf{true} \\ \mathrm{por}(\Omega, \Omega) \xrightarrow{\Omega} \mathrm{por}(\Omega, \mathbf{true}) \xrightarrow{\mathrm{R}} \mathbf{true} \end{array} \tag{4}$$

Each path mirrors the causal cascade leading to the rule L or R which computes the "parallel or" operation to **true**.

By this point, we hope to convince the reader that the pair (2) is not "critical" for syntactical reasons only (as in Knuth and Bendix's definition) but also for *structural* reasons related to causality of computations.

In [M$_1$,M$_2$,M$_3$], we analyze causality of computations by purely *diagrammatic* methods. Rewriting systems are replaced by their reduction graphs, which we equip with 2-dimensional *tiles*, in order to reflect redex permutations. Typically, the reduction graph of $\mathrm{por}(\mathbf{true}, \mathbf{true})$ in (2) is equipped with three tiles:

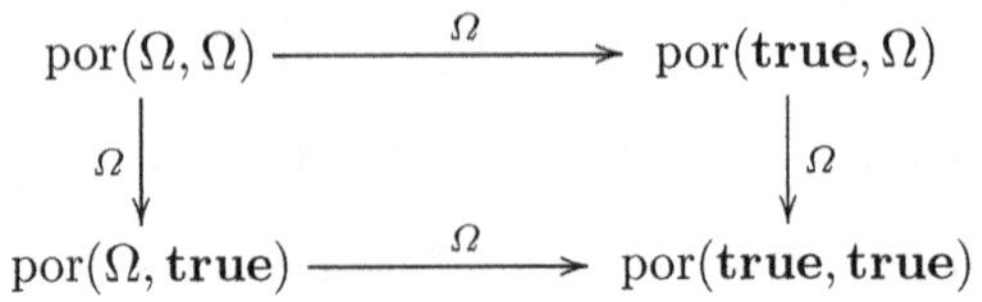

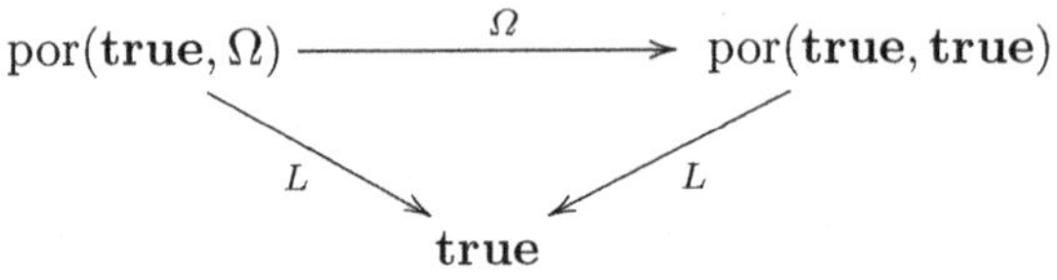

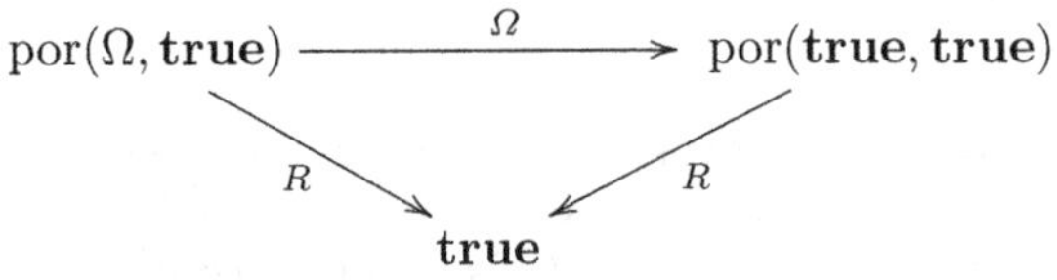

That way, *generic* properties of rewriting systems (seen as 2-dimensional graphs) may be established. For instance, the *standardization* theorem [CF,Lé$_1$,HL,Bo] states that there exists a unique *standard* (= outside-in) path in each "homotopy" equivalence class. It is proved diagrammatically in [GLM,M$_1$,M$_2$] from fairly general axioms on redex permutations in rewriting systems.

It is worth noting here that tiling diagram (3) in the 2-dimensional graph of por(Ω, Ω) would break the standardization theorem, since the two standard paths (4) would appear then in the same "homotopy" class. This dictates to keep diagram (3) untiled in the 2-dimensional graph of por(Ω, Ω), and to understand it as a *2-dimensional hole.*

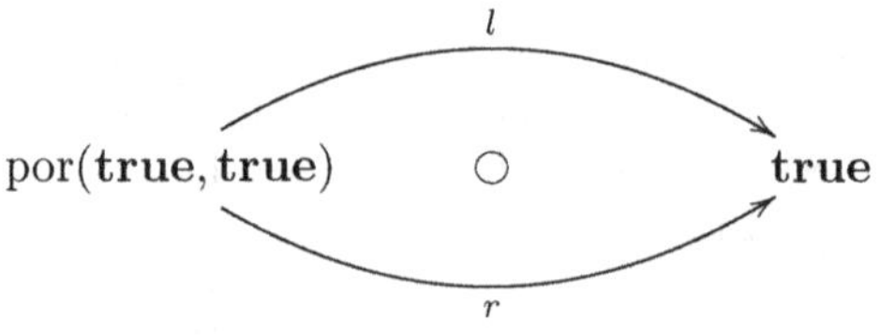

At this point, each path of (4) appears to be the canonical (that is: standard) representative of one homotopy class from por(Ω, Ω) to **true**. Reformulated this way, rewriting theory looks simpler, and closer to algebraic topology. This is certainly promising [1].

But let us come back to our main point. On one hand, Knuth and Bendix's definition of a critical pair is purely syntactic; on the other hand, it captures subtle diagrammatic and causal singularities, like the critical pair (2).

[1] A few readers will find example (2) funny. Indeed, tiling the critical pair (3) defines a 2-dimensional graph which happens to enjoy *universal confluence.* But this is pure chance. In particular, this universality property does not extend to weakly orthogonal rewriting systems.

However, and this is the starting point of the article, we claim that in one case at least, Knuth and Bendix's syntactic definition of critical pair is too crude.

Consider the term rewriting system with a unique binary symbol: the tensor product noted $\otimes$, and a unique rewriting rule: the *associativity rule* below.

$$x \otimes (y \otimes z) \xrightarrow{\alpha} (x \otimes y) \otimes z \tag{5}$$

This rewriting system and various extensions have been studied thoroughly by category theorists, in order to establish so-called *coherence theorems* [Mc]. We recall that the rewriting system (5) is confluent and strongly normalizing; and that its normal forms are the variables, or the terms $x \otimes V$ obtained by "appending" a variable x to a normal form V.

Here, we consider the local confluence diagrams induced by (5). They are of two kinds, depending whether the co-initial redexes u and v overlap, or do not overlap. First, the rewriting rule being right-linear (= it does not duplicate) two non-overlapping redexes u and v induce a square diagram, completed by the obvious *residual redexes* u' and v':

$$\begin{array}{ccc} M & \xrightarrow{u} & P \\ {\scriptstyle v}\downarrow & & \downarrow{\scriptstyle v'} \\ Q & \xrightarrow{u'} & N \end{array} \tag{6}$$

Then, two redexes u and v overlap in one case only:

$$\begin{array}{c} w \otimes (x \otimes (y \otimes z)) \xrightarrow{w \otimes \alpha} w \otimes ((x \otimes y) \otimes z) \\ w \otimes (x \otimes (y \otimes z)) \xrightarrow{\alpha} (w \otimes x) \otimes (y \otimes z) \end{array} \tag{7}$$

The local confluence diagram below is famous in the categorical litterature: it is called Mac Lane's pentagon.

$$\begin{array}{ccc} w \otimes (x \otimes (y \otimes z)) & \xrightarrow{w \otimes \alpha} & w \otimes ((x \otimes y) \otimes z) \\ & & \downarrow{\scriptstyle \alpha} \\ {\scriptstyle \alpha}\Big\downarrow & & (w \otimes (x \otimes y)) \otimes z \\ & & \downarrow{\scriptstyle \alpha \otimes z} \\ (w \otimes x) \otimes (y \otimes z) & \xrightarrow{\alpha} & ((w \otimes x) \otimes y) \otimes z \end{array} \tag{8}$$

Mac Lane introduced it in order to axiomatize the concept of *monoidal category*. We write $\equiv$ for the "homotopy" equivalence relation on paths obtained by tiling every confluence diagrams (6) and (8). By strong normalization, and Newman's lemma [Ne,K_2], every two sequences f, g of associativity rules from a term M to its normal form V are equal modulo homotopy equivalence $\equiv$. This is roughly the content of Mac Lane's coherence theorem for monoidal category, see [M_5] for further details.

At this point, we stress that the rewriting system (5) and homotopy equivalence $\equiv$ verify a remarkable property, which goes far beyond confluence. We call this property *universal confluence*. It states that, for every two coinitial rewriting paths $f : M \longrightarrow P$ and $g : M \longrightarrow Q$, there exists two paths $f' : Q \longrightarrow N$ and $g' : P \longrightarrow N$ such that:

– the rewriting paths $f; g' : M \longrightarrow N$ and $g; f' : M \longrightarrow N$ are equal modulo $\equiv$,

$$\begin{array}{ccc}
M & \xrightarrow{\ f\ } & P \\
{\scriptstyle g}\Big\downarrow & \equiv & \Big\downarrow{\scriptstyle g'} \\
Q & \dashrightarrow[f'] & N
\end{array}$$

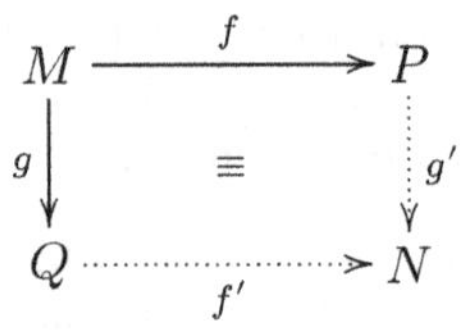

– for every diagram path $f'' : Q \longrightarrow O$ and $g'' : P \longrightarrow O$ such that $f; g'' \equiv g; f''$, there exists a unique rewriting path h (unique modulo homotopy) such that $f'' \equiv f'; h$ and $g'' \equiv g'; h$

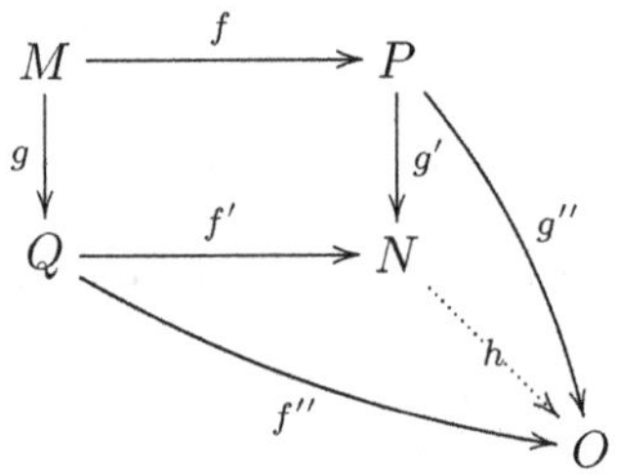

The property is established in Dehornoy's theory of Garside groups, using "reversing word" techniques [D_1,DL]. Categorically speaking, the property states that the category of terms and rewriting paths modulo homotopy $\equiv$ enjoys existence of *pushout diagrams*.

Quite interestingly, the same universal confluence property holds in the λ-calculus, and more generally in any conflict-free rewriting system. There, the homotopy equivalence relation $\equiv$ is obtained by tiling all confluence diagrams generated by Lévy's residual theory. In the λ-calculus, the equivalence relation $\equiv$ is best illustrated by three paradigmatic confluence diagrams:

$$\begin{array}{ccc}
((\lambda x.x)a)((\lambda y.y)b) & \xrightarrow{\hspace{3cm}} & a((\lambda y.y)b) \\
\Big\downarrow & & \Big\downarrow \\
((\lambda x.x)a)b & \xrightarrow{\hspace{3cm}} & ab
\end{array} \qquad (9)$$

and

$$\begin{array}{ccc}
(\lambda x.a)((\lambda y.y)b) & & \\
\downarrow & \searrow & \\
 & & a \\
 & \nearrow & \\
(\lambda x.a)b & &
\end{array}$$

and

$$\begin{array}{ccc}
(\lambda x.xx)((\lambda y.y)a) & \xrightarrow{u} & ((\lambda y.y)a)((\lambda y.y)a) \\
 & & \downarrow {\scriptstyle v_1} \\
{\scriptstyle v} \downarrow & & a((\lambda y.y)a) \\
 & & \downarrow {\scriptstyle v_2} \\
(\lambda x.xx)a & \longrightarrow & aa
\end{array} \qquad (10)$$

So, the associativity rule (5) on one hand, and all conflict-free rewriting systems on the other hand, enjoy the same remarkable universal confluence property.

However, Knuth and Bendix's definition considers that the associativity rule (5) forms a critical pair (with itself) because the two redexes (7) *overlap*. Obviously, this syntactic reason does not take into account Mac Lane's pentagon (8) and its striking similarity to, say, diagram (10).

So, one feels that Lévy's residual theory should be extended to incorporate example (5). The main task of the article is precisely to formulate this extended theory. One technical difficulty on the way is that, despite what we have just suggested, Mac Lane's pentagon is not *exactly* similar to diagram (10). Let us explain why.

In diagram (10), the path $v_1; v_2$ *develops* the set of *residuals* of the β-redex v after the β-redex u. In particular, the β-redex v_2 is residual after v_1 of a β-redex

$$w : ((\lambda y.y)a)((\lambda y.y)a) \longrightarrow ((\lambda y.y)a)a$$

itself residual of the β-redex v after u. This enables Lévy to deduce all local confluence diagrams from his theory of *redexes*, *residuals* and *developments*.

On the contrary, in diagram (8) the path

$$w \otimes ((x \otimes y) \otimes z) \xrightarrow{\alpha} (w \otimes (x \otimes y)) \otimes z \xrightarrow{\alpha \otimes z} ((w \otimes x) \otimes y) \otimes z \qquad (11)$$

is not *generated* by the set of residuals of the redex

$$w \otimes (x \otimes (y \otimes z)) \xrightarrow{\alpha} (w \otimes x) \otimes (y \otimes z) \qquad (12)$$

after the redex

$$w \otimes (x \otimes (y \otimes z)) \xrightarrow{w \otimes \alpha} w \otimes ((x \otimes y) \otimes z) \qquad (13)$$

In particular, the redex

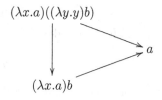

$$(w \otimes (x \otimes y)) \otimes z \xrightarrow{\alpha \otimes z} ((w \otimes x) \otimes y) \otimes z$$

is *not* the residual of any redex after the redex:

$$w \otimes ((x \otimes y) \otimes z) \xrightarrow{\alpha} (w \otimes (x \otimes y)) \otimes z$$

In fact, after a short thought, one must admit that the path (11) *as a whole* is residual of the redex (12) after the redex (13).

Admitting that *paths* may be residuals of *redexes* is not a small change in Lévy's residual theory. If one sees redexes as the computational *atoms* of rewriting theory, then these rewriting paths deduced as residuals of redexes, appear as its *molecules.* We shall call them *treks* to carry the intuition that in most concrete case studies, these rewriting paths implement a unique elementary task, however the number of rewriting steps necessary to complete it.

A few years ago, the author considered such "treks" and proved the (universal) confluence of a positive presentation of Artin braid group [A]. However, the proof [M$_4$,KOV$_1$] was too involved to generalize properly to other rewriting systems. Later discussions with Patrick Dehornoy, Jan Willem Klop and Vincent van Oostrom helped the author to develop the theory exposed here.

So, the theory will keep nice and tractable, in sharp contrast to [M$_4$]. The trick is to avoid defining the residuals of a trek *after a trek* $(*)$ because this is too difficult to formalize in most rewriting systems. Instead, we limit ourselves to defining the residuals of a trek *after a redex,* then deduce $(*)$ from a few coherence properties.

This makes all the difference!

Synopsis: we expose the classical theory of redexes and residuals in section 2, and revisit the theory in section 3. For comparison's sake, we wrote sections 2 and 3 in a very similar fashion. We illustrate the theory on associativity and Mac Lane's pentagon in section 4; on braids and Garside's theorem in section 5; on Lévy's optimality theory for the λ-calculus in section 6. Finally, we conclude and indicate further research directions.

2 Classical Residual Theory (Lévy)

In this section, we recall the residual theory developed by J.-J. Lévy in his PhD thesis [Lé$_1$], see also [HL,M$_1$].

Consider a graph $\mathcal{G}$, that is a set V of *terms* and a set E of *redexes*, together with two functions $\partial_0, \partial_1 : E \longrightarrow V$ indicating the *source* and *target* of each redex. We write

$$u : M \longrightarrow M'$$

when u is a redex and $\partial_0 u = M$ and $\partial_1 u = M'$. A path is sequence

$$f \;=\; (M_1, u_1, M_2, ..., M_m, u_m, M_{m+1}) \tag{14}$$

where M and N are terms, and for every $1 \leq i \leq m$

$$u_i : M_i \longrightarrow M_{i+1}$$

Remark: every term M induce a path (M), also noted $\mathbf{id}_M$, called the *identity* path of M.

We overload the arrow notation, and write $f : M_1 \longrightarrow M_{m+1}$ in a situation like (14). Moreover, we write $f; g : M \longrightarrow N$ for the composite of two paths $f : M \longrightarrow P$ and $g : P \longrightarrow N$.

In any graph $\mathcal{G}$, we write $R(M)$ for the set of redexes outgoing from the term M.

Definition 1 (residual relation). *A* residual relation *over a graph $\mathcal{G}$ is a relation $/_u$ for every redex $u : M \longrightarrow N$, relating redexes in $R(M)$ to redexes in $R(N)$.*

The residual relation $/_\_$ is extended to paths, by declaring that the relation $u \,/_f\, v$ holds when

- either $f = \mathbf{id}_M$ and $u = v$ are redexes outgoing from M,
- or $f = u_1; \cdots; u_n$ and there exists redexes $w_2, ..., w_n$ such that $u \,/_{u_1}\, w_2 \,/_{u_2} \cdots /_{u_{n-1}}\, w_n \,/_{u_n}\, v$.

Notation: given a redex u and a path f, we write $u \,/_f$ for the set $\{\, v \mid u \,/_f\, v \,\}$.

Every residual relation is supposed to verify four properties, formulated in this section. The two first properties are quite elementary:

1. **self-destruction:** for every redex u, the set $u \,/_u$ is empty,
2. **finiteness:** for every two coinitial redexes u and v, the set $u \,/_v$ is finite,

Definition 2 (multi-redex). *A* multi-redex *is a pair (M, U) consisting of:*

- *a term M,*
- *a finite set U of redexes outgoing from M.*

Remark: every redex $u : M \longrightarrow N$ may be seen alternatively as the multi-redex $(M, \{u\})$.

Notation: we say that a redex u is element of a multi-redex (M, U) when $u \in U$.

The residual relation $/_f$ is extended to multi-redexes by declaring that $(M, U) \,/_f\, (N, V)$ when:

- $f : M \longrightarrow N$,
- V is the set of residuals of elements of U after f, that is:

$$V = \{v \mid \exists u \in U,\ u /_f v\}$$

Definition 3 (development). *A rewriting path*

$$M = M_1 \xrightarrow{u_1} M_2 \xrightarrow{u_2} \cdots \xrightarrow{u_{n-1}} M_n \xrightarrow{u_n} M_{n+1}$$

develops *a multi-redex* (M, U) *when*

- *every redex* u_i *is element of* $(M, U) /_{u_1;\cdots;u_{i-1}}$,
- $(M, U) /_f (M_{n+1}, \emptyset)$.

Notation: we write

$$U \Vdash_M f$$

when the rewriting path f develops the multi-redex (M, U). We like this notation because it captures particularly well the idea that, in some sense, the path f "realizes" the multi-redex (M, U).

Definition 4 (partial development). *A path* $f : M \longrightarrow P$ *develops a multi-redex* (M, U) *partially when there exists a path* $g : P \longrightarrow N$ *such that* $U \Vdash_M f; g$.

The two last properties required on a residual relation $/\_$ appear in [GLM,M$_1$]. They axiomatize the *finite development lemma* [Hyl,K$_1$,Ba,M$_1$,O$_4$] and the *cube lemma* [Lé$_1$,HL,Bo,Ba,P,OV].

3. **finite developments:** for every multi-redex (M, U), there does not exist any infinite sequence
$$M_1 \xrightarrow{u_1} M_2 \xrightarrow{u_2} \ldots M_n \xrightarrow{u_n} M_{n+1} \ldots$$
of redexes such that, for every index i, the path
$$M_1 \xrightarrow{u_1} M_2 \xrightarrow{u_2} \ldots M_{i-1} \xrightarrow{u_{i-1}} M_i$$
develops the multi-redex (M, U) partially.
4. **permutation:** for every two coinitial redexes $u : M \longrightarrow P$ and $v : M \longrightarrow Q$, there exists two paths h_u and h_v such that:
   - $u /_v \Vdash_Q h_u$ and $v /_u \Vdash_P h_v$,
   - h_u and h_v are cofinal and induce the same residual relation $/_{u;h_v} = /_{v;h_u}$.

   Diagrammatically

$$\begin{array}{ccc} M & \xrightarrow{u} & P \\ {\scriptstyle v}\downarrow & & \downarrow{\scriptstyle h_v} \\ Q & \xrightarrow{h_u} & N \end{array} \qquad (15)$$

In his PhD thesis, Lévy introduces a *permutation equivalence* between rewriting paths. The idea is to tile all diagrams (15) and consider the induced "homotopy equivalence".

Definition 5 (Lévy). *We write $f \equiv^1 g$ when the paths f and g factor as*

$$f = M' \xrightarrow{f_1} M \xrightarrow{u} P \xrightarrow{h_v} N \xrightarrow{f_2} N'$$
$$g = M' \xrightarrow{f_1} M \xrightarrow{v} Q \xrightarrow{h_u} N \xrightarrow{f_2} N'$$

where

- *u and v are coinitial redexes,*
- *$u \,/_v \, \Vdash_Q h_u$ and $v \,/_u \, \Vdash_P h_v$,*
- *h_u and h_v induce the same residual relation $/_{u;h_v} = /_{v;h_u}$.*

Definition 6 (Lévy). *The* permutation *or* homotopy *equivalence relation $\equiv$ is the least equivalence relation (reflexive, transitive, symmetric) on rewriting paths, containing the relation $\equiv^1$.*

Lévy proves his two main theorems about residual systems in [Lé$_1$,HL], see also [Ba,K$_1$,P,St,M$_1$]:

Theorem 1 (pushouts). *For every two coinitial rewriting paths $f : M \longrightarrow P$ and $g : M \longrightarrow Q$, there exists two paths $f' : Q \longrightarrow N$ and $g' : P \longrightarrow N$ such that:*

- *the rewriting paths $f; g' : M \longrightarrow N$ and $g; f' : M \longrightarrow N$ are equal modulo $\equiv$,*

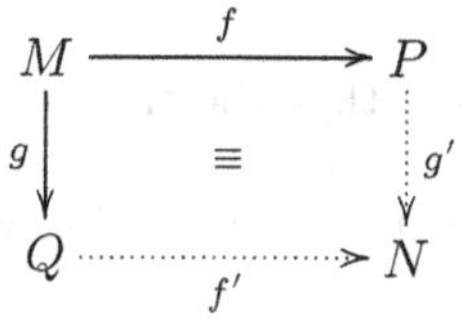

- *for every diagram path $f'' : Q \longrightarrow O$ and $g'' : P \longrightarrow O$ such that $f; g'' \equiv g; f''$, there exists a unique rewriting path h (unique modulo homotopy) such that $f'' \equiv f'; h$ and $g'' \equiv g'; h$*

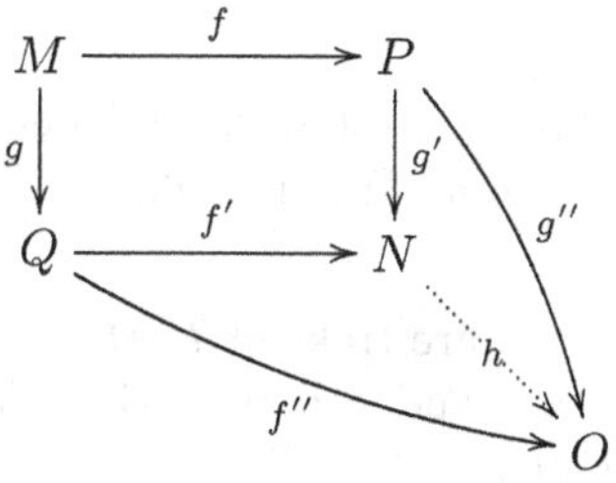

Theorem 2 (epis). *Every path $f : M \longrightarrow P$ enjoys the left-simplification rule below:*

$$f;g \equiv f;h \ \Rightarrow \ g \equiv h$$

for every two paths $g, h : P \longrightarrow N$.

Theorem 1 states that the category $\mathcal{G}/\equiv$ of terms and rewriting paths modulo homotopy has *pushouts*, while Theorem 2 states that every morphism of the category is an epimorphism, see [Mc] for a nice exposition of categorical concepts.

3 Classical Theory Revisited: Treks and Residuals

The residual theory we develop here repeats the pattern of section 2, except that *redexes* are replaced by *treks* in the definition of residuals. We also change the notation $/_u$ into $[\![u]\!]$ to help the reader put the two theories in contrast.

In any graph $\mathcal{G}$, we write $R(M)$ for the set of redexes outgoing from the term M.

Definition 7 (trek graph). *A trek graph $\mathcal{R} = (\mathcal{G}, T, \vdash)$ is a graph equipped with, for every term M*

- *a set $T(M)$ of* treks,
- *a* generation *relation $\vdash$ between treks of $T(M)$ and redexes of $R(M)$,*
- *a function $\underline{\cdot}$ assigning a trek $\underline{u} \in T(M)$ to every redex $u \in R(M)$, such that $\underline{u} \vdash u$.*

Notation: we say alternatively that the trek $\mathfrak{t} \in T(M)$ *generates* the redex $u \in R(M)$, or that the redex $u \in R(M)$ *starts* the trek $\mathfrak{t} \in T(M)$ when $\mathfrak{t} \vdash u$. Remark: every graph defines a trek graph, by setting $T(M) = R(M)$ and $\vdash$ the identity.

Definition 8 (residual relation). *A* residual relation *over a trek graph $\mathcal{R}$ is a relation $[\![u]\!]$ for every redex $u : M \longrightarrow N$, relating treks of $T(M)$ to treks of $T(N)$.*

Just like in section 2, the residual relation $[\![u]\!]$ is extended from redexes to paths, by declaring that the relation $\mathfrak{s}[\![f]\!]\mathfrak{t}$ holds when

- either $f = \mathbf{id}_M$ and $\mathfrak{s} = \mathfrak{t}$ are treks of $T(M)$,
- or $f = u_1; \cdots; u_n$ and there exists treks $\mathfrak{r}_2, ..., \mathfrak{r}_n$ such that $\mathfrak{s}[\![u_1]\!]\mathfrak{r}_2[\![u_2]\!]...[\![u_{n-1}]\!]\mathfrak{r}_n[\![u_n]\!]\mathfrak{t}$.

Notation: given a trek $\mathfrak{s} \in T(M)$ and a path $f : M \longrightarrow N$, we write

$$\mathfrak{s}[\![f]\!] \;=\; \{\, \mathfrak{t} \in T(N) \mid \mathfrak{s}[\![f]\!]\mathfrak{t} \,\}$$

Just like in section 2, four properties are required of every residual relation. The two first properties appear below.

1. **self-destruction:** for every redex $u : M \longrightarrow N$, the set $\underline{u}[\![u]\!]$ is empty,
2. **finiteness:** for every trek $\mathfrak{t} \in T(M)$ and redex $u \in R(M)$, the set $\mathfrak{t}[\![u]\!]$ is finite,

Definition 9 (multi-trek). *A* multi-trek *is a pair* $(M, \mathfrak{S})$ *consisting of:*

- *a term* M*,*
- *a finite subset* $\mathfrak{S}$ *of* $T(M)$*.*

Remark: every redex $u : M \longrightarrow N$ may be seen alternatively as the multi-trek $(M, \{\underline{u}\})$.

The residual relation $[\![f]\!]$ is extended from treks to multi-treks by declaring that $(M, \mathfrak{S})[\![f]\!](N, \mathfrak{T})$ when:

- $f : M \longrightarrow N$,
- $\mathfrak{T} \;=\; \{\mathfrak{t} \in T(N) \mid \exists \mathfrak{s} \in \mathfrak{S},\; \mathfrak{s}[\![f]\!]\mathfrak{t}\} \;=\; \bigcup_{\mathfrak{s} \in \mathfrak{S}} \mathfrak{s}[\![f]\!]$

Remark: the definition above is only possible thanks to property 2. (finiteness.)

Definition 10 (development). *A rewriting path*

$$M = M_1 \xrightarrow{u_1} M_2 \xrightarrow{u_2} \cdots \xrightarrow{u_{n-1}} M_n \xrightarrow{u_n} M_{n+1}$$

develops *a multi-trek* $(M, \mathfrak{S})$ *when*

- *every redex* u_i *starts one of the treks of* $(M, \mathfrak{S})[\![u_1; \cdots; u_{i-1}]\!]$,
- $(M, \mathfrak{S})[\![f]\!](M_{n+1}, \emptyset)$.

Notation: as in section 2, we write

$$\mathfrak{S} \Vdash_M f$$

when the rewriting path f develops the multi-trek $(M, \mathfrak{S})$.

Definition 11 (partial development). *A path* $f : M \longrightarrow P$ *develops a multi-trek* $(M, \mathfrak{S})$ *partially when there exists a path* $g : P \longrightarrow N$ *such that* $\mathfrak{S} \Vdash_M f; g$.

The two properties below repeat the axioms of finite developments and permutation formulated in section 2.

3. **finite developments:** for every multi-trek $(M, \mathfrak{S})$, there does not exist any infinite sequence
$$M_1 \xrightarrow{u_1} M_2 \xrightarrow{u_2} \ldots M_n \xrightarrow{u_n} M_{n+1} \ldots$$
of redexes such that, for every index i, the path
$$M_1 \xrightarrow{u_1} M_2 \xrightarrow{u_2} \ldots M_{i-1} \xrightarrow{u_{i-1}} M_i$$
develops the multi-trek $(M, \mathfrak{S})$ partially.
4. **permutation:** for every two coinitial redexes $u : M \longrightarrow P$ and $v : M \longrightarrow Q$, there exists two paths h_u and h_v such that:
   - $\underline{u}[\![v]\!] \Vdash_Q h_u$ and $\underline{v}[\![u]\!] \Vdash_P h_v$,
   - h_u and h_v are cofinal and induce the same residual relation $[\![u; h_v]\!] = [\![v; h_u]\!]$.

   Diagrammatically
$$\begin{array}{ccc} M & \xrightarrow{\quad u \quad} & P \\ {\scriptstyle v}\big\downarrow & & \big\downarrow{\scriptstyle h_v} \\ Q & \xrightarrow{\quad h_u \quad} & N \end{array} \qquad (16)$$

Here, we adapt Lévy permutation equivalence of section 2 to the new setting. Again, the idea is to tile all diagrams (16) and consider the "homotopy equivalence" on path.

Definition 12 (tile). *We write $f \equiv^1 g$ when the paths f and g factor as*
$$f = M' \xrightarrow{f_1} M \xrightarrow{u} P \xrightarrow{h_v} N \xrightarrow{f_2} N'$$
$$g = M' \xrightarrow{f_1} M \xrightarrow{v} Q \xrightarrow{h_u} N \xrightarrow{f_2} N'$$
where

- *u and v are coinitial redexes,*
- *$\underline{u}[\![v]\!] \Vdash_Q h_u$ and $\underline{v}[\![u]\!] \Vdash_P h_v$,*
- *h_u and h_v induce the same residual relation $[\![u; h_v]\!] = [\![v; h_u]\!]$.*

Definition 13 (homotopy). *The* permutation *or* homotopy *equivalence relation $\equiv$ is the least equivalence relation (reflexive, transitive, symmetric) on rewriting paths, containing the relation $\equiv^1$.*

The axiomatization extends Lévy's theorems 1 and 2 to "border-line" systems like associativity or braids.

Theorem 3 (pushouts). *The category $\mathcal{G}/\equiv$ of terms and rewriting paths modulo homotopy equivalence, has pushouts.*

Theorem 4 (epis). *Every morphism of $\mathcal{G}/\equiv$ is an epimorphism.*

Proof. The proof follows the pattern of [Lé1,HL,M1]. The first step is to prove that two paths f and g are equal modulo homotopy when they develop the same multi-trek. This is done by a simple induction on the depth of the multi-trek, thanks to finite developments property 4. Then, one applies the usual commutation techniques, see also [St], except that multi-redexes are replaced now by multi-treks.

4 Illustration 1: The Associativity Rule and Mac Lane's Pentagon

Here, we apply the theory of section 3 to the rewriting system (5) introduced in section 1, with unique rewriting rule:

$$x \otimes (y \otimes z) \xrightarrow{\alpha} (x \otimes y) \otimes z \tag{17}$$

First, let us define what we mean by *trek* in that case. Given $n \geq 1$ variables $y_1, ..., y_n$, we define the normal form $[y_1, ..., y_n]$ by induction on n:

$$[y_1] = y_1 \qquad [y_1, ..., y_n] = [y_1, ..., y_{n-1}] \otimes y_n$$

A trek $\mathfrak{t}$ of $T(M)$ is defined as a pattern of

$$x \otimes [y_1, ..., y_n] \tag{18}$$

for $n \geq 2$, and its occurrence o in the term M. So, a trek $\mathfrak{t} \in T(M)$ is a pair (o, n) where the occurrence o is a sequence $\epsilon_1 \cdots \epsilon_k$ of L's and R's (for left and right) identifying the position of the "root" tensor product of the trek in the term.

Notation: We say that the trek $\mathfrak{t}$ has a *head*: the tensor product $x \otimes [y_1, ..., y_n]$; and a *spine*: the $n - 1$ tensor products inside $[y_1, ..., y_n]$.

We will construct our residual theory in such a way that $\mathfrak{t}$ is "realized" or "developed" by the path f:

$$x \otimes [y_1, ..., y_n] \longrightarrow [x, y_1, ..., y_n]$$

In our theory, every trek $\mathfrak{t} \in T(M)$ generates a unique redex $u \in R(M)$, defined as the redex with same occurrence as $\mathfrak{t}$. This redex rewrites

$$x \otimes [y_1, ..., y_n] \longrightarrow (x \otimes [y_1, ..., y_{n-1}]) \otimes y_n$$

The left-hand side of the rewrite rule (17) is the particular case $n = 2$ of pattern (18). Thus, every redex $u \in R(M)$ defines a trek $\underline{u} \in T(M)$ of same occurrence in M, satisfying $\underline{u} \vdash u$.

This defines a trek graph $(\mathcal{G}, T, \vdash)$. Now, we define the residual structure $[\![-]\!]$ on it. Consider a redex $u : M \longrightarrow N$ and two treks $\mathsf{t} \in T(M)$ and $\mathsf{t}' \in T(N)$. The redex u induces a one-to-one function φ between the occurrences of M not overlapping the left-hand side of u, and the occurrence of N not overlapping the right-hand side of u.

The two treks t and t' verify $\mathsf{t}[\![u]\!]\mathsf{t}'$ precisely when:

- $\mathsf{t} = (o, n)$ and $\mathsf{t}' = (\varphi(o), n)$ when the redex u and the trek t do not overlap,
- $\mathsf{t} = (o, n)$ and $\mathsf{t}' = (o \cdot L, n - 1)$ when $n \geq 3$ and the redex u overlaps with both the head and the spine of t. Observe that $\mathsf{t} \vdash u$ in that case.
- $\mathsf{t} = (o \cdot R, n)$ and $\mathsf{t}' = (o, n)$ when the redex u overlap with the head (and not the spine) of t,
- $\mathsf{t} = (o, n)$ and $\mathsf{t}' = (o, n + 1)$ when the redex u overlap with the spine (and not the head) of t.

There remains to check that $[\![-]\!]$ verifies the four properties required of residual systems. Property 1. (self-destruction) is obvious. Here, property 2. (finiteness) may be strengthened to the property that a trek has one residual at most. Accordingly, we only need to establish property 3. (finite developments) on singleton multi-treks only. Anyway, property 3. follows from strong normalization of the rewriting system.

There remains to prove property 4. (permutation.) for every pair of coinitial redexes $u : M \longrightarrow P$ and $v : M \longrightarrow Q$. We proceed by case analysis.

Suppose first that u and v do not overlap, and let the redex u' be the residual of u after v, and the redex v' be the residual of v after u, in the usual residual theory (for redexes.) Then, obviously, in our extended residual theory, the trek $\underline{u'}$ is residual of the trek $\underline{u}$ after v, and the trek $\underline{v'}$ is residual of the trek $\underline{v}$ after u. Moreover, $\{\underline{u'}\} \Vdash_Q u'$ and $\{\underline{v'}\} \Vdash_P v'$. Thus, diagram (6) is an instance of (16). For lack of space, we will only sketch how one proves that $[\![u; v']\!] = [\![v; u']\!]$. The proof is not very difficult, quite pleasant to proceed in fact, but requires to consider a few cases, based on whether u or v overlaps with the head and spine of a given trek $\mathsf{t} \in T(M)$.

Now, suppose that u and v do overlap. We may suppose wlog that v has instance o, that u has instance $o \cdot R$, and that the pattern of the critical pair is (7):

$$u : \quad w \otimes (x \otimes (y \otimes z)) \xrightarrow{w \otimes \alpha} w \otimes ((x \otimes y) \otimes z)$$

$$v : \quad w \otimes (x \otimes (y \otimes z)) \xrightarrow{\alpha} (w \otimes x) \otimes (y \otimes z)$$

On one hand, the redex v overlaps with the head of the trek $\underline{u}$. Thus $\underline{u} = (o \cdot R, 2)$ has residual the trek $\mathfrak{t}_u = (o, 2)$:

$$\mathfrak{t}_u : \quad (w \otimes x) \otimes [y, z]$$

developed by the redex u':

$$(w \otimes x) \otimes [y, z] \xrightarrow{\alpha} [w, x, y, z]$$

On the other hand, the redex u overlaps with the spine of the trek $\underline{v}$. Thus, $\underline{v} = (o, 2)$ has residual the trek $\mathfrak{t}_v = (o, 3)$.

$$\mathfrak{t}_v : \quad w \otimes [x, y, z]$$

developed by the path h_v:

$$w \otimes [x, y, z] \xrightarrow{\alpha} (w \otimes [x, y]) \otimes z \xrightarrow{\alpha \otimes z} [w, x, y, z]$$

So, Mac Lane's pentagon (8) is an instance of diagram (16). There remains to prove $[\![u; h_v]\!] = [\![v; u']\!]$. Just as in case of non-overlapping redexes u and v, the proof proceeds by case analysis, which we omit here for lack of space.

Since the square and pentagon diagrams (6) and (8) induce the same homotopy equivalence $\equiv$ as our residual theory, we conclude from theorems 3 and 4 that:

Theorem 5 (universal confluence). *The rewriting paths generated by (17) and quotiented by the homotopy diagrams (6) and (8) define a category of epimorphisms with pushouts.*

Remark: all λ-calculi with explicit substitutions like [ACCL] go back to Pierre-Louis Curien's Categorical Combinatory Logic [C] — whose main rewriting engine is the associativity rule. We are currently experimenting a theory of treks for a λ-calculus with explicit substitutions, extending what we have done here, and inducing the first universal confluence theorem for a calculus of that kind.

5 Illustration 2: Braids and Garside's Theorem

We will be *very* concise here, for lack of space. For information, the reader is advised to read the crisp course notes [KOV$_1$] where a nice exposition of [M$_4$] appears, along with other materials.

The braids group $\mathbb{B}$ was introduced by Artin [A]. The group is presented by the alphabet

$$\Sigma = \{\sigma_i \mid i \in \mathbb{Z}\}$$

and the relations

$$\sigma_i \sigma_{i+1} \sigma_i = \sigma_{i+1} \sigma_i \sigma_{i+1} \quad \text{for every } i \in \mathbb{Z} \tag{19}$$

$$\sigma_i \sigma_j = \sigma_j \sigma_i \qquad \text{when } |i-j| \geq 2 \tag{20}$$

In order to solve the word problem in $\mathbb{B}$, Garside [Ga] considered the monoid $\mathbb{B}^+$ of *positive* braids, with same presentation. He established that every two elements of $\mathbb{B}^+$ have a lower common multiple. The theorem may be restated as the universal confluence property of the rewriting system with elements b of $\mathbb{B}^+$ as terms, arrows $M \xrightarrow{\sigma_i} M\sigma_i$ as redexes, and commutation diagrams (19) and (20) as tiles.

Each σ_i is interpreted diagrammatically as the operation of permuting the strand i over the strand $i+1$ in a braid diagram:

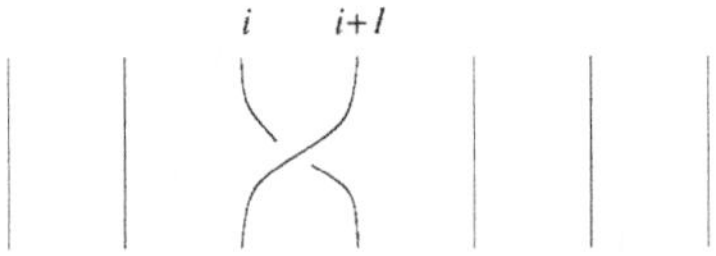

In particular, the equation (19) mirrors the topological equality:

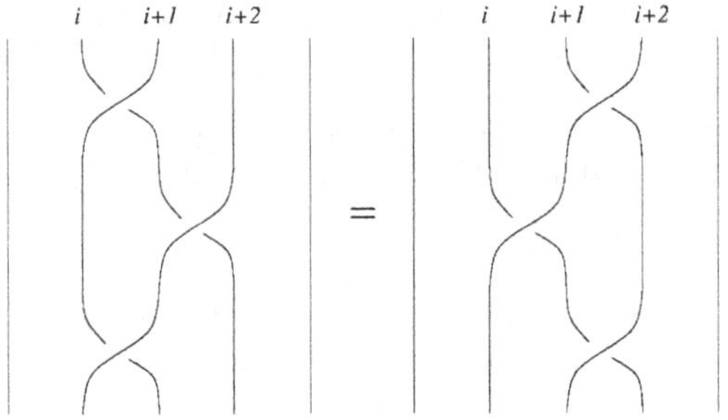

In our residual system, a trek t of M is a pair (I, J) of subsets of $\mathbb{Z}$, such that[2]

- $I \cap J$ is empty,
- $I \cup J$ is a finite interval $[i, j]$ with smaller bound $i \in I$ and larger bound $j \in J$.

Intuitively, a trek (I, J) is intended to "permute" all strands of J over all strands of I, without altering the order of strands internal to I and J. So, we declare that a trek $(I, J) \in T(M)$ generates a redex $\sigma_i \in R(M)$ when $i \in I$ and $j \in J$. Observe that every redex $\sigma_i \in R(M)$ defines the trek $\underline{\sigma_i} = (\{i\}, \{i+1\}) \in T(M)$, which verifies $\underline{\sigma_i} \vdash \sigma_i$.

This defines a trek graph. The residual relation $[\![-]\!]$ is defined on it as follows. Let $(I, J) \in T(M)$ and $\sigma_k \in R(M)$. We write $\dot{\sigma}_k$ for the permutation $(k, k+1)$ in $\mathbb{Z}$; and i, j for the two integers $i \in I$ and $j \in J$ such that $I \cup J = [i, j]$. Then,

[2] The treks (I, J) define a well-known class of generators for braids, with remarkable permutation properties. Another interesting class of generators appears in [BKL].

1. $(I,J)[\![\sigma_k]\!](I,J)$ when k is not element of $[i-1,j]$,
2. $(I,J)[\![\sigma_k]\!](\dot{\sigma}_k(I),\dot{\sigma}_k(J))$ when
   - either $k \in [i+1, j-2]$,
   - or $k = i$ and $i+1 \in I$,
   - or $k = j-1$ and $j-1 \in J$,
3. $(I,J)[\![\sigma_{i-1}]\!](I + \{i-1\}, J)$,
4. $(I,J)[\![\sigma_i]\!](\dot{\sigma}_i(I), J - \{i+1\})$ when $i+1 \in J$ and J is not singleton,
5. $(I,J)[\![\sigma_{j-1}]\!](I - \{j-1\}, \dot{\sigma}_{j-1}(J))$ when $j-1 \in I$ and I is not singleton,
6. $(I,J)[\![\sigma_j]\!](I, J + \{j+1\})$.

The definition verifies the four properties of residual systems: *self-destruction* is immediate while *finiteness* may be strengthened to the property that a trek has a residual at most, just as for the associativity rule in section 4. *Finite development* is easily established too, by weighting treks.

The property of *permutation* is more interesting. Suppose that σ_i and σ_j are redexes from M. Two cases occur. Either $|j-i| \geq 2$, and then

$$\underline{\sigma_i}[\![\sigma_j]\!]\underline{\sigma_i} \quad \text{and} \quad \underline{\sigma_j}[\![\sigma_i]\!]\underline{\sigma_j}$$

induce the commutative diagram:

$$\begin{array}{ccc} M & \xrightarrow{\sigma_i} & M\sigma_i \\ {\scriptstyle\sigma_j}\downarrow & & \downarrow{\scriptstyle\sigma_j} \\ M\sigma_j & \xrightarrow{\sigma_i} & M\sigma_i\sigma_j \end{array} \tag{21}$$

Or $j = i+1$, and then

$$\underline{\sigma_i}[\![\sigma_{i+1}]\!](\{i\}, \{i+1, i+2\})$$

$$\underline{\sigma_{i+1}}[\![\sigma_i]\!](\{i, i+1\}, \{i+2\})$$

induce

$$\begin{array}{ccccc} M & & \xrightarrow{\sigma_i} & & M\sigma_i \\ & & & & \downarrow{\scriptstyle\sigma_{i+1}} \\ {\scriptstyle\sigma_{i+1}}\downarrow & & & & M\sigma_i\sigma_{i+1} \\ & & & & \downarrow{\scriptstyle\sigma_i} \\ M\sigma_{i+1} & \xrightarrow{\sigma_i} & M\sigma_i\sigma_{i+1} & \xrightarrow{\sigma_{i+1}} & M\sigma_i\sigma_{i+1}\sigma_i \end{array} \tag{22}$$

which mirror exactly equations (19) and (20).

There remains to prove that the two paths f and g along the confluence diagrams (21) and (22) define the same residual relation $[\![f]\!] = [\![g]\!]$. This is easy to establish for diagram (21) and an inch more tedious for diagram (22). We skip the proof here for lack of space.

6 Illustration 3: Lévy's Optimality and Extraction Theory

In his PhD thesis, Lévy [Lé$_1$,Lé$_2$] develops a theory of *optimality* for the λ-calculus. After defining *families* of β-redexes "of a common nature" which may be shared by an implementation, Lévy calls *optimal* any implementation which computes one such family at each step.

The moving part of the story is that no optimal implementation existed at the time. Lévy formulated its principle, and the implementation was designed twelve years later [Lam].

In [Lé$_1$,Lé$_2$], families are defined in three different ways, which turn out to be equivalent: zigzag, extraction, labelling. Here, we focus on zigzag and extraction, and forget labelling for lack of space.

Definition 14 (Lévy). *A* β-redex with history *is a pair* (f, u) *where* $f : M \longrightarrow P$ *is a rewriting path and* $u : P \longrightarrow P'$ *is a* β*-redex.*

For notation's sake, we feel free to identify the β-redex with history (f, u) and the path $M \xrightarrow{f} P \xrightarrow{u} P'$.

Definition 15 (residual). *Let* $h : P \longrightarrow Q$ *be a rewriting path, and let*

$$M \xrightarrow{f} P \xrightarrow{u} P' \text{ and } M \xrightarrow{g} Q \xrightarrow{v} Q'$$

be two β*-redexes with history. We write*

$$(f, u) \;/_h\; (g, v)$$

when

$$u \;/_h\; v \quad \text{and} \quad g \equiv f; h$$

In that case, we say that (f, u) *is an ancestor of* (g, v)*.*

The zig-zag relation $\leftrightarrows$ is the least equivalence relation between redexes with history, containing this ancestor relation.

Definition 16 (family). *A family is an equivalence class of the equivalence relation* $\leftrightarrows$ *between* β*-redexes with history.*

One unexpected point about families is the lack of *common ancestor* property. We illustrate this with an example, adapted from [Lé$_2$]. Consider the three λ-terms

$$P = (\lambda x.xa)(\lambda y.y) \qquad Q = (\lambda y.y)a \qquad R = a$$

and the redex with history $\overline{v}$:

$$\overline{v} : \quad P \xrightarrow{u} Q \xrightarrow{v} R \tag{23}$$

The two redexes with history $\overline{v_1}$ and $\overline{v_2}$:

$$\overline{v_1}: \quad \Delta P \xrightarrow{\Delta} PP \xrightarrow{u_1} QP \xrightarrow{v_1} RP$$

$$\overline{v_2}: \quad \Delta P \xrightarrow{\Delta} PP \xrightarrow{u_2} PQ \xrightarrow{v_2} PR$$

are in the same family, but do not have an ancestor $\overline{v_0}$ in common. Intuitively, this common ancestor should be $\overline{v}$. But $\overline{v}$ computes the β-redex u, which is not a subcomputation of $\Delta; u_1$ and not a subcomputation of $\Delta; u_2$.

This means that the theory has to be slightly refined. Lévy [Lé$_2$] introduces a *non-deterministic* extraction procedure $\triangleright$ acting on redexes with history.

At this point, it is useful to recall that the set $R(M)$ of β-redexes from a λ-term M may be ordered by prefix ordering $\preceq_M$ on occurrences. We also introduce a refinement of Lévy's homotopy relation on paths [GLM,M$_1$,M$_2$].

Definition 17 (reversible homotopy). *We write $f \simeq^1 g$ when f and g factor as*

$$f = M' \xrightarrow{f_1} M \xrightarrow{u} P \xrightarrow{v'} N \xrightarrow{f_2} N'$$
$$g = M' \xrightarrow{f_1} M \xrightarrow{v} Q \xrightarrow{u'} N \xrightarrow{f_2} N'$$

where

- *u and v are* disjoint *β-redexes, that is incomparable wrt. $\preceq_M$.*
- *u' is the unique residual of u after v,*
- *v' is the unique residual of v after u.*

The reversible homotopy *equivalence relation $\simeq$ is the reflexive transitive closure of $\simeq^1$.*

So, considering $\simeq$ amounts to tiling all diagrams of the form (9) in the reduction graph of the λ-calculus.

Definition 18 (Lévy). *The* extraction relation *is the union of the four relations below.*

1. $M \xrightarrow{f} P \xrightarrow{u} P' \xrightarrow{v'} N \;\triangleright\; M \xrightarrow{f} P \xrightarrow{v} Q$ when $v \preceq_N u$ and $v /_u v'$,

2. $M \xrightarrow{f} P \xrightarrow{u} P' \xrightarrow{h} Q' \xrightarrow{v'} R' \;\triangleright\; M \xrightarrow{f} P \xrightarrow{g} Q \xrightarrow{v} R$ when u and $g; v$ compute in "disjoint" occurrences, and $h; v'$ is the path residual of $g; v$ after u. In that case, $u /_g u' /_v u''$ and $u; h; v' \simeq g; v; u''$ induce the diagram below.

$$\begin{array}{ccccccc}
M & \xrightarrow{f} & C[A,B] & \xrightarrow{g} & C[A,B'] & \xrightarrow{v} & C[A,B''] \\
 & & {\scriptstyle u}\downarrow & \simeq & {\scriptstyle u'}\downarrow & \simeq & \downarrow{\scriptstyle u''} \\
 & & C[A',B] & \xrightarrow{h} & C[A',B'] & \xrightarrow{v'} & C[A',B'']
\end{array}$$

3. $M \xrightarrow{f} P \xrightarrow{u} P' \xrightarrow{h} Q' \xrightarrow{v'} R' \rhd M \xrightarrow{f} P \xrightarrow{g} Q \xrightarrow{v} R$ *when the path* $g; v$ *computes inside the body* A *of the* β*-redex* $u : C[(\lambda x.A)B] \longrightarrow C[A[B/x]]$ *and the path* $h; v'$ *is the path residual of* $g; v$ *after* u*. In that case,* $u \,/_g\, u' \,/_v\, u''$ *induces the diagram below:*

$$\begin{array}{ccccc} C[(\lambda x.A)B] & \xrightarrow{g} & C[(\lambda x.A')B] & \xrightarrow{v} & C[(\lambda x.A'')B] \\ {\scriptstyle u}\downarrow & \equiv & {\scriptstyle u'}\downarrow & \equiv & \downarrow{\scriptstyle u''} \\ C[A[B/x]] & \xrightarrow{h} & C[A'[B/x]] & \xrightarrow{v'} & C[A''[B/x]] \end{array}$$

4. $M \xrightarrow{f} P \xrightarrow{u} P' \xrightarrow{h} Q' \xrightarrow{v'} R' \rhd M \xrightarrow{f} P \xrightarrow{g} Q \xrightarrow{v} R$ *when* $u : C[(\lambda x.A)B] \longrightarrow C[A[B/x]]$ *and the path* $g; v$ *computes inside the argument* B *of* u*, and the path* $h; v'$ *computes just as* g *inside* one *copy of the argument* B *of* $A[B/x]$*. In the diagram, we write* A_l *(the subscript* l *stands for* linear*) for the* λ*-term* A *where all instances of* x *are substituted by the* λ*-term* B*, except one instance which is substituted by a fresh variable* l*.*

$$\begin{array}{ccccc} C[(\lambda x.A)B] & \xrightarrow{g} & C[(\lambda x.A)B'] & \xrightarrow{v} & C[(\lambda x.A)B''] \\ {\scriptstyle u}\downarrow & & & & \\ C[A_l[B/l]] & \xrightarrow{h} & C[A_l[B'/l]] & \xrightarrow{v'} & C[A_l[B''/l]] \end{array}$$

Lévy shows that the procedure $\rhd$ is strongly normalizing, and confluent. Suppose moreover that (f, u) and (g, v) are two redexes with history whose paths f and g are *left-standard*, that is standard wrt. the *leftmost-outermost* order on β-redexes, see [CF,Lé$_1$,K$_1$,Ba,GLM,M$_2$]. Then, Lévy shows that (f, u) and (g, v) are in the same family iff they have the same $\rhd$-normal-form.

This result provides a pertinent *canonical form* theorem for Lévy families. Typically, stepping back to example (23) the redex with history $\overline{v}$ is not the common ancestor, but the common $\rhd$-normal-form of the two redexes with history $\overline{v_1}$ and $\overline{v_2}$.

We reformulate this part of Lévy optimality theory using the language of treks.

The set $P(M)$ of paths outgoing from a λ-term M, may be ordered as follows:

$$M \xrightarrow{f} P \precsim M \xrightarrow{g} Q \iff \exists h : P \to Q, \quad f; h \simeq g$$

The order $(P(M), \precsim)$ is a *local lattice*[3]: every two bounded elements f and g have a least upper bound $f \vee g$ and a greatest least bound $f \wedge g$. We recall that two paths $f, g \in P(M)$ are *bounded* in $(P(M), \precsim)$ when there exists a path $h \in P(M)$ such that $f \precsim h$ and $g \precsim h$.

[3] The order $(P(M), \precsim)$ is a bit more than a local lattice, because the lub and glb of f and g do not depend on the choice of slice h.

Definition 19 (coprime). *A coprime $f \in P(M)$ is a path such that, for every two bounded paths $g, h \in P(M)$,*

$$f \precsim g \vee h \quad \Rightarrow \quad f \precsim g \text{ or } f \precsim h$$

So, a path f is coprime when it cannot be decomposed into more elementary computations $f \simeq f_1 \vee \ldots \vee f_k$.

Standard coprimes enjoy an important "enclave" or "localization" property. Here, by *standard*, we mean standard wrt. the *prefix-ordering* $\preceq$ on β-redexes, see [GLM,M$_2$]. Let $v \in R(M)$ be the first β-redex computed by a coprime $f = v; g$. The localization property tells that if the β-redex $v \in R(M)$ occurs inside the body (resp. the argument) of a β-redex $u \in R(M)$, then the whole standard coprime f computes inside the body (resp. the argument) of the β-redex u.

This property leads to a new theory of residuals in the λ-calculus.

Definition 20 (β-trek). *A β-trek $\mathfrak{t} \in T(M)$ is a non-empty standard coprime of $(P(M), \precsim)$, considered modulo $\simeq$.*

This defines

Definition 21 (trek graph). *A β-trek $\mathfrak{t} \in T(M)$* generates *a β-redex $u \in R(M)$ when $u \precsim \mathfrak{t}$. Also, every β-redex $u \in R(M)$ defines the singleton β-trek $\underline{u} = \{u\} \in T(M)$ which verifies $\underline{u} \vdash u$.*

The residual relation $\mathfrak{t}[\![u]\!]\mathfrak{t}'$ induced by a β-redex $u : M \longrightarrow N$ on treks $\mathfrak{t} \in T(M)$ and $\mathfrak{t}' \in T(N)$, is defined by induction on the length of $\mathfrak{t}$.

Definition 22 (residual).

1. *when the trek $\mathfrak{t}$ generates the β-redex u, then the β-trek $\mathfrak{t}'$ is defined as the unique β-trek such that $\mathfrak{t} \simeq u; \mathfrak{t}'$,*
2. *when $\mathfrak{t}$ generates a β-redex v occurring inside the body of u; then, $\mathfrak{t}'$ is the residual of $\mathfrak{t}$ after u, modulo $\simeq$,*
3. *when $\mathfrak{t}$ generates a β-redex v occurring inside the argument of u : $C[(\lambda x.A)B] \longrightarrow C[A[B/x]]$; then, $\mathfrak{t}'$ computes just as $\mathfrak{t}$ inside one copy of B in $A[B/x]$,*
4. *when $\mathfrak{t}$ generates a β-redex v not occurring inside the β-redex u, this β-redex v has a unique residual v' after u, and $\mathfrak{t}$ has the residual $\mathfrak{s}$ after v; the β-trek $\mathfrak{t}'$ is any path $v'; \mathfrak{s}'$ where $\mathfrak{s}'$ is a residual of $\mathfrak{s}$ after the development of $u\ /_v$.*

The definition is fine because a β-trek has always residuals of smaller length.

The four properties of residual systems are established as follows. The properties of *self-destruction* and *finiteness* are immediate, while *finite development* follows from Lévy's finite family developments theorem [Lé$_1$,O$_3$]. The last property of *permutation* follows from elementary combinatorial arguments, that we do not expose here for lack of space.

At this point, two β-treks $\mathfrak{t}_1, \mathfrak{t}_2 \in T(M)$ may have the same β-trek $\mathfrak{t}_3 \in T(N)$ as residual after a β-redex $u : M \longrightarrow N$. Typically, in our example, the β-treks $\underline{v}$ and $\underline{v_1}$ have the β-trek

$$PP \xrightarrow{u_1} QP \xrightarrow{v_1} RP$$

as residual after the β-redex $\Delta P \longrightarrow PP$.

This motivates to limit oneself to *normal* β-treks.

Definition 23. *A trek is* normal *when it is a* $\triangleright$*-normal-form.*

It appears that normal β-treks define a sub-residual system of the β-trek system defined above, in the sense that normal β-treks have only normal β-treks as residuals, and that every β-trek $\underline{u}$ is normal.

The main property of normal β-treks is that, for every normal β-trek $\mathfrak{t} \in T(N)$, and morphism $u : M \longrightarrow N$, there exists a *unique* normal β-trek $\mathfrak{s} \in T(M)$ such that $\mathfrak{t}$ is residual of $\mathfrak{s}$ after u. This normal β-trek is simply the $\triangleright$-normal-form of $u; \mathfrak{t}$.

So in a sense, there is no "creation" in the residual theory of normal treks. A *canonical ancestor* theorem for normal treks follows, "replacing" the lack of such a theorem for β-redexes.

Theorem 6 (canonical ancestor). *Every family is the set of residuals of a normal trek* $\mathfrak{t}$ *— modulo homotopy equivalence of the redexes with history.*

Remark: in contrast to section 4 and 5, where a redex may have a non-trivial trek as residual, β-redexes have only β-redexes as residuals. This hides the concept of β-trek very much, explains why the theory of β-treks was not formulated earlier.

Since Lévy's early steps, the optimality theory has been extended to term rewriting systems [Ma] to interaction systems [AL] and to higher-order systems [O$_3$]. For instance, Asperti and Laneve gave several examples of higher-order rewriting systems, where the "zig-zag" notion of Lévy family does not match with the (more correct) "labelling" and "extraction" definitions. We conjecture that our language of treks and residuals repairs this, and provides the relevant "residual theory" to discuss optimality in higher-order rewriting systems.

7 Conclusion

In this article, we revisit Lévy's residual theory, in order to enscope "border-line" rewriting systems admitting "innocuous" critical pairs, like the associativity rule, or the positive braids rewriting systems. We show in addition that the resulting residual theory clarifies Lévy's theory of optimality.

In future work, we would like to express an effective syntactic criterion $(*)$ indicating when critical pairs are "innocuous" in a given rewriting system, or equivalently, when the language of treks applies. The criterion would refine Knuth and Bendix's definition of critical pair, and formulate it in a more "combinatorial" style.

A nice line of research initiated by Gérard Huet [Hu] will certainly guide us. Its main task is to devise conditions on critical pairs, ensuring confluence. As far as we know, three such *syntactical* conditions were already expressed on term rewriting systems:

1. Gérard Huet [Hu] devises a liberal criterion on critical pairs, inducing confluence of left and right-linear term rewriting systems. His class of *strongly closed* rewriting systems is limited by the right-linearity (= non-duplication) condition on rewriting rules.
2. in the same article, Gérard Huet introduces a *critical pair condition* which avoids the right-linearity condition. Since then, the condition was generalized by Yoshihito Toyama [To] and Vincent van Oostrom [O_4].
3. Bernhard Gramlich [Gr] introduces a third criterion on critical pairs, which he calls a *parallel critical pair condition.* As in Huet-Toyama-van Oostrom's work, the condition implies confluence of left-linear term rewriting systems.

In his article, Bernhard Gramlich shows that the three conditions 1. 2. and 3. are not comparable with one another. For instance, the associativity rule (5) is an instance of 1. but not of 2. and 3.

This line of research has not only produced interesting syntactic conditions. Two *abstract* conditions were also exhibited:

1. Vincent van Oostrom [O_1,O_4] and Femke van Raamsdonk [R,OR] introduce *weakly orthogonal* rewriting systems and establish a confluence theorem for them.
2. Vincent van Oostrom and collaborators [O_2,BOK,KOV_2] establish a diagrammatic confluence theorem for a wide class of rewriting systems with *decreasing diagrams.*

Validity of these conditions, either syntactical or abstract, may be decided on any finite rewriting system. We are looking for a criterion $(*)$ just as effective.

There also remains to unify the approaches developed here and in Patrick Dehornoy's work on Garside groups [D_1]. At this stage, one main difference is the choice of "local" strong normalization property. Instead of our finite development hypothesis, Dehornoy requires strong normalization of each homotopy equivalence class of paths. This choice enables a simpler axiomatics, closer to

Newman's spirit, which enjoys an effective criterion [D_2]. But it limits the scope of the theory. Typically, the pure λ-calculus is rejected because every path

$$(\lambda x.a)(\Delta\Delta) \longrightarrow \cdots \longrightarrow (\lambda x.a)(\Delta\Delta) \longrightarrow a$$

lies in the homotopy equivalence class of

$$(\lambda x.a)(\Delta\Delta) \longrightarrow a$$

where Δ is the duplicating combinator $\Delta = \lambda y.yy$.

Acknowledgements. To Jan Willem Klop and Vincent van Oostrom for introducing me to braids rewriting and Garside's theorem; to Patrick Dehornoy for explaining me his universal confluence theorems on monoids, and defying me to express them inside rewriting theory.

References

[A] E. Artin. Theory of braids. *Annals of Mathematics.* 48(1):101–126, 1947.

[ACCL] M. Abadi, L. Cardelli, P.-L. Curien, J.-J. Lévy. "Explicit substitutions". *Journal of Functional Programming*, 1(4):375–416, 1991.

[AL] A. Asperti, C. Laneve. Interaction Systems I: the theory of optimal reductions. In *Mathematical Structures in Computer Science,* Volume 4(4), pp. 457 - 504, 1995.

[Ba] H. Barendregt. *The Lambda Calculus: Its Syntax and Semantics.* North Holland, 1985.

[BKL] J. Birman, K.H. Ko and S.J. Lee. A new approach to the word problem in the braid groups. *Advances in Math.* vol 139-2, pp. 322-353. 1998.

[Bo] G. Boudol. Computational semantics of term rewriting systems. In Maurice Nivat and John C. Reynolds (eds), *Algebraic methods in Semantics.* Cambridge University Press, 1985.

[BOK] M. Bezem, V. van Oostrom and J. W. Klop. Diagram Techniques for Confluence. *Information and Computation, Volume 141, No. 2,* pp. 172 - 204, March 15, 1998.

[C] P.-L. Curien. An abstract framework for environment machines. *Theoretical Computer Science,* 82(2):389-402, 31 May 1991.

[CF] H.B. Curry, R. Feys. *Combinatory Logic I.* North Holland, 1958.

[D_1] P. Dehornoy. On completeness of word reversing. *Discrete Math. 225* pp 93–119, 2000.

[D_2] P. Dehornoy. *Complete positive group presentations.* Preprint. Université de Caen, 2001.

[DL] P. Dehornoy, Y. Lafont. *Homology of gaussian groups.* Preprint. Université de Caen, 2001.

[Ga] F. A. Garside. The braid grup and other groups. *Quarterly Journal of Mathematic* 20:235–254, 1969.

[Gr] B. Gramlich. Confluence without Termination via Parallel Critical Pairs. *Proc. 21st Int. Coll. on Trees in Algebra and Programming (CAAP'96),* ed. H. Kirchner. LNCS 1059, pp. 211-225, 1996.

[GLM] G. Gonthier, J.-J. Lévy, P.-A. Melliès. An abstract standardization theorem. *Proceedings of the 7th Annual IEEE Symposium on Logic In Computer Science*, Santa Cruz, 1992.

[Hin] J.R. Hindley. An abstract form of the Church-Rosser theorem. *Journal of Symbolic Logic,* vol 34, 1969.

[Hu] G. Huet. Confluent reductions: abstract properties and applications to term rewriting systems. *Journal of the ACM,* 27(4):797–821, 1980.

[HL] G. Huet, J.-J. Lévy. Computations in orthogonal rewriting systems. In J.-L. Lassez and G. D. Plotkin, editors, *Computational Logic; Essays in Honor of Alan Robinson*, pages 394–443. MIT Press, 1991.

[Hyl] J.M.E. Hyland. A simple proof of the Church-Rosser theorem. Manuscript, 1973.

[K_1] J.W. Klop. *Combinatory Reduction Systems.* Volume 127 of *Mathematical Centre Tracts*, CWI, Amsterdam, 1980. PhD thesis.

[K_2] J.W. Klop. Term Rewriting Systems. In S. Abramsky, D. Gabbay, T. Maibaum, eds. *Handbook of Logic in Computer Science,* vol. 2, pp. 2–117. Clarendon Press, Oxford, 1992.

[KOV_1] J.W. Klop, V. van Oostrom and R. de Vrijer. *Course notes on braids.* University of Utrecht and CWI. Draft, 45 pages, July 1998.

[KOV_2] J.W. Klop, V. van Oostrom and R. de Vrijer. A geometric proof of confluence by decreasing diagrams *Journal of Logic and Computation, Volume 10, No. 3,* pp. 437 - 460, June 2000

[KB] D. E. Knuth, P.B. Bendix. Simple word problems in universal algebras. In J. Leech (ed) *Computational Problems in Abstract Algebra.* Pergamon Press, Oxford, 1970. Reprinted in *Automation of Reasoning 2.* Springer Verlag, 1983.

[Lam] J. Lamping. An algorithm for optimal lambda-calculus reductions. In *proceedings of Principles of Programming Languages*, 1990.

[$Lé_1$] J.-J. Lévy. *Réductions correctes et optimales dans le λ-calcul.* Thèse de Doctorat d'Etat, Université Paris VII, 1978.

[$Lé_2$] J.-J. Lévy. *Optimal reductions in the lambda-calculus.* In J.P. Seldin and J.R. Hindley, editors, *To H.B. Curry, essays on Combinatory Logic, Lambda Calculus, and Formalisms*, pp. 159–191. Academic Press, 1980.

[Ma] L. Maranget. Optimal Derivations in Weak Lambda-calculi and in Orthogonal Term Rewriting Systems. *Proceedings POPL'91.*

[Mc] S. Mac Lane. *Categories for the working mathematician.* Springer Verlag, 1971.

[M_1] P.-A. Melliès. *Description abstraite des systèmes de réécriture.* Thèse de l'Université Paris VII, December 1996.

[M_2] P.-A. Melliès. Axiomatic Rewriting Theory I: A diagrammatic standardization theorem. 2001. Submitted.

[M_3] P.-A. Melliès. Axiomatic Rewriting Theory IV: A stability theorem in Rewriting Theory. *Proceedings of the 14th Annual Symposium on Logic in Computer Science.* Indianapolis, 1998.

[M_4] P.-A. Melliès. Braids as a conflict-free rewriting system. Manuscript. Amsterdam, 1995.

[M_5] P.-A. Melliès. Mac Lane's coherence theorem expressed as a word problem. Manuscript. Paris, 2001.

[MN] R. Mayr, T. Nipkow. Higher-Order Rewrite Systems and their Confluence. *Theoretical Computer Science,* 192:3-29, 1998.

[Ne] M. H. A. Newman. On theories with a combinatorial definition of "equivalence". *Annals of Math.* 43(2):223-243, 1942.

[Ni] T. Nipkow. Higher-order critical pairs. In *Proceedings of Logic in Computer Science,* 1991.

[O1] V. van Oostrom. *Confluence for Abstract and Higher-Order Rewriting.* PhD thesis, Vrije Universiteit, Amsterdam, 1994.

[O2] V. van Oostrom. Confluence by decreasing diagrams. *Theoretical Computer Science,* 121:259–280, 1994.

[O3] V. van Oostrom. Higher-order family. In *Rewriting Techniques and Applications 96,* Lecture Notes in Computer Science 1103, Springer Verlag, 1996.

[O4] V. van Oostrom. Finite family developments. In *Rewriting Techniques and Applications 97,* Lecture Notes in Computer Science 1232, Springer Verlag, 1997.

[O5] V. van Oostrom. Developing developments. *Theoretical Computer Science,* Volume 175, No. 1, pp. 159 - 181, March 30, 1997

[OR] V. van Oostrom and F. van Raamsdonk. Weak orthogonality implies confluence: the higher-order case. In *Logical Foundations of Computer Science 94,* Lecture Notes in Computer Science 813, Springer Verlag, 1994.

[OV] V. van Oostrom and R. de Vrijer. Equivalence of reductions, and Strategies. Chapter 8 of *Term Rewriting Systems,* a book by TeReSe group. To appear this summer 2002.

[P] G.D. Plotkin, *Church-Rosser categories.* Unpublished manuscript. 1978.

[R] F. van Raamsdonk. *Confluence and Normalisation for Higher-Order Rewriting.* PhD thesis, Vrije Universiteit van Amsterdam, 1996.

[St] E.W. Stark. Concurrent transition systems. *Theoretical Computer Science 64.* May 1989.

[To] Y. Toyama. Commutativity of term rewriting systems. In K. Fuchi and L. Kott (eds) *Programming of Future Generation Computer,* vol II, pp. 393–407. North-Holland, 1988.

Static Analysis of Modularity of β-Reduction in the Hyperbalanced λ-Calculus

Richard Kennaway[1], Zurab Khasidashvili[2], and Adolfo Piperno[3]

[1] School of Information Systems, University of East Anglia,
Norwich NR4 7TJ, England
jrk@sys.uea.ac.uk

[2] Department of Mathematics and Computer Science, Bar-Ilan University,
Ramat-Gan 52900, Israel
khasidz@macs.biu.ac.il

[3] Dipartimento di Informatica, Università di Roma "La Sapienza",
Via Salaria 113, 00198 Roma, Italy
piperno@dsi.uniroma1.it

Abstract. We investigate the degree of *parallelism (or modularity)* in the *hyperbalanced λ-calculus*, λ_H, a subcalculus of λ-calculus containing all simply typable terms (up to a restricted η-expansion). In technical terms, we study the family relation on redexes in λ_H, and the contribution relation on redex-families, and show that the latter is a forest (as a partial order). This means that hyperbalanced λ-terms allow for maximal possible parallelism in computation. To prove our results, we use and further refine, for the case of hyperbalanced terms, some well known results concerning *paths*, which allow for static analysis of many fundamental properties of β-reduction.

1 Introduction

Barendregt et al [7] have shown that there does not exist an *optimal* recursive one-step β-reduction strategy that, given a λ-term, would construct its shortest normalizing reduction path. To achieve optimality, Lévy [20, 21] proposed to consider multi-step normalizing strategies, where each multi-step contracts simultaneously a set of redexes of the same *family*. Lamping [18] and Kathail [9] proved that such multi-steps can indeed be implemented by single (sharing-) graph reduction steps. Since then, there have been many interesting discoveries both in the theory and practice of optimal implementation, and we refer the reader to the recent book on the subject by Asperti and Guerrini [4].

Lévy [20, 21] introduced the family concept in the λ-calculus in three different but equivalent ways: via a suitable notion of *labelling* that records the history of a reduction, via *extraction* procedure that eliminates from a reduction all steps that do not 'contribute' to the creation of the last contracted redex, and by *zig-zag*, which is an equivalence closure of the residual relation. Asperti and Laneve [1] and Asperti et al [2] described a fourth way to characterize families via a concept of *path*, allowing in particular to detect future redexes statically.

Here we continue our investigation of the Hyperbalanced λ-calculus, λ_H [15, 16]; in particular, we study the degree of parallelism that λ_H can offer, by investigating

S. Tison (Ed.): RTA 2002, LNCS 2378, pp. 51–65, 2002.

the family relation in this restricted setting. The hyperbalanced λ-terms have a nice structure that makes it easy to analyze their normalization behavior statically. All simply typable terms are hyperbalanced up to a restricted η-expansion, and moreover all strongly normalizable terms can be made hyperbalanced by means of β-reduction and η-expansion [16]. We will see that the family relation (and the derived *contribution* relation) in the λ_H-calculus has several attractive properties that the family relation in the λ-calculus in general does not posses.

In particular, in hyperbalanced terms, well balanced paths [1] (wbp) are determined uniquely by their initial and ending nodes, and all wbp are legal. Recall that the paths that define families are legal. From this we immediately obtain that the number of families in any hyperbalanced term is finite. We claim (we are not reporting the proof here) that this is at most quadratic in the size of the term, and that such a bound is tight. These results imply strong normalization of λ_H, and of the simply typed λ-calculus. Further, we prove a virtual Church-Rosser property for λ_H, allowing us to prove that redexes of the same family are residuals of a unique virtual redex. Finally, we prove that extraction normal forms are chains and that the contribution relation is a forest. This shows that λ_H allows for a maximal possible parallelism in computation.

In our proofs, we employ substitution paths [15]. They allow for simple proofs of the fact that, in the λ_H-calculus, wbp=legal paths=virtual redexes. This equivalence of different definitions of future redexes shows that it is largely a matter of convenience (and taste) whether one decides to choose wbp or substitution paths in the proofs of the above properties of the contribution relation. We find substitution paths simpler because the residual concept for them is based on a concept of descendant, while for wbp the ancestor concept is more complicated (the ancestor of a single edge may be a complex path, and this makes definition of the residual concept for wbp harder). Moreover, substitution paths allow for a simple characterization of the total number of families, as well as the number of needed families, in hyperbalanced terms (see Corollary 14).

Notation and Terminology We mainly follow [5,6] for the basic concepts and [21, 4] for advanced concepts related to the family relation, but with some deviations. All missing definitions can be found there. We write $M \xrightarrow{u} N$ if N is obtained from N by contracting a β-redex u in M. For any terms N and M, $N \subseteq M$ denotes that N is a subterm occurrence in M. We use w, v to denote terms which ar formed by application, as well as redexes. For any reductions P and Q, $P+Q$ denotes the concatenation of P and Q, $P \sqcup Q$ denotes $P+Q/P$, where Q/P is the residual of Q under P, $\approx_L$ denotes Lévy- or strong-equivalence, and $\simeq$ denotes the family relation.

Due to the lack of space, some proofs have been omitted in the text: they appear in [10]. The proofs are by combining induction with case analysis, and are rather routine anyway.

2 The λ_H-Calculus

The set Λ of λ-terms is generated using the grammar

$$\Lambda ::= x \mid (\lambda x.\Lambda) \mid (\Lambda_1 \Lambda_2),$$

where x belongs to a countable set *Var* of term variables. Terms are considered modulo renaming of bound variables, and the *variable convention* [5] is assumed; namely, bound variables are always chosen to be different from free ones.

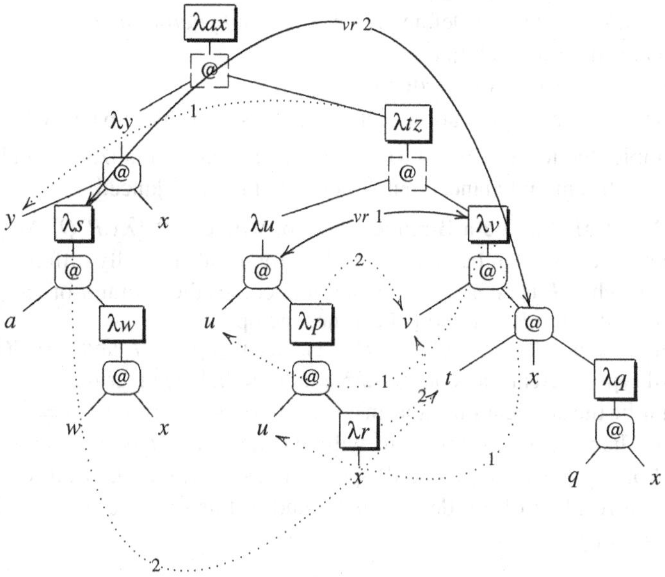

Fig. 1. Structure of λ-terms: abstraction terms correspond to subtrees rooted in a shadowed box, application terms correspond to subtrees rooted in a dashed box, while head-variable terms correspond to subtrees rooted in a rounded box. Arrow decoration is shown up to rank 2. vr i=virtual redex of rank i (see below).

We can see from the example in Figure 1 that lambda terms can be defined using the following grammar, where terms of the shape $Q_{i,j}$, respectively $\lambda x_1 \ldots x_n.S$, respectively x or R_i, will be called *application terms, abstraction terms,* and *head variable terms.*

$$
\begin{array}{lcl}
\Lambda & ::= & S \mid \lambda x_1 \ldots x_n.S \;\; (n > 0); \\
S & ::= & x \mid R_i \;\; (i > 0) \mid Q_{i,j} \;\; (i,j > 0); \\
R_i & ::= & x\Lambda_1 \ldots \Lambda_i; \\
Q_{i,j} & ::= & (\lambda x_1 \ldots x_j.S)\Lambda_1 \ldots \Lambda_i.
\end{array}
\tag{1}
$$

Notation 1 We will often need to write the application operators explicitly in application and head variable terms. For example, the application and head variable terms above are written as $@(\ldots @(\lambda x_1 \ldots \lambda x_j.L, L_1), \ldots, L_i)$ and $@(\ldots @(z, N_1), \ldots, N_j)$, respectively. The *patterns* of these terms are the contexts obtained from them by removing the body and the arguments, and the head and the arguments, respectively.

Consider the set of λ-terms whose generic element is such that every application subterm occurring in it has as many binding variables as arguments.

Definition 2 (i) *(Balanced terms)* A lambda term is called *balanced* iff, in every application subterm $Q_{i,j}$ in it, one has $i = j$.

(ii) An application subterm $N \equiv (\lambda x_1 \ldots x_n.P)N_1 \ldots N_n$ of a balanced term $M \in \widehat{\Lambda}$ is *semi-hyperbalanced* iff whenever N_1 has k (≥ 0) initial abstractions, every free occurrence of x_1 in P has exactly k arguments.

(iii) *(Hyperbalanced terms)* Define the set $\widehat{\widehat{\Lambda}} \subset \widehat{\Lambda}$ of *hyperbalanced* terms as the greatest set of balanced terms such that

- $\widehat{\widehat{\Lambda}}$ is closed with respect to β-reduction;
- for any $M \in \widehat{\widehat{\Lambda}}$, every application subterm in M is semi-hyperbalanced.

For example, the term $\Omega = (\lambda x.xx)(\lambda x.xx)$ is a (non-normalizable!) balanced term whose reducts all remain balanced but which is not hyperbalanced.

Definition 3 Let $M \xrightarrow{u} M'$ be a β-reduction step, where $u \equiv (\lambda x.P)Q$. As pointed out in [19], a new β-redex v can be created in M' in three substantially different ways:

(1) *upwards*: when P is an abstraction and it becomes the operator of v;
(2) *downwards*: when $P \not\equiv x$ and Q becomes the operator of v;
(3) *upwards*: when $u \equiv (\lambda x.x)Q$ (i.e., $P \equiv x$), $(\lambda x.x)QQ' \subseteq M$, and $v = QQ'$.

The third type of creation is not possible in hyperbalanced terms.

Now let w be the application subterm corresponding to u, written $w = aps(u)$, and let $w' = aps(v)$. Then, correspondingly, w' will be called a *created application subterm of type* 1 or 2. For any other (non-created) application subterm w^* in M', its corresponding redex u^* is a u-residual of a redex u'' in M, and in this case we will say that w^* is a *residual* of $w'' = aps(u'')$.

3 Virtual Redexes in the λ_H-Calculus

In this section, we shall show that redex families in the λ_H-calculus are determined uniquely by 'matching pairs' of @- and λ-nodes. To do this, we give an inductive definition of matching pairs, which we call virtual redexes, via *substitution paths* [15]. Then we show (using substitution paths) that every virtual redex uniquely determines a well balanced path (wbp) connecting the corresponding @- and λ-occurrences, and that all wbp are legal. We conclude by using the $1-1$ correspondence between legal paths and families obtained in [1].

Definition 4 (Well balanced paths [1]) Let M be a λ-term.

(1) Any edge connecting a @-occurrence with its left son, ?, whose 'type' ? may be λ, @ or v (variable), is a *well balanced path* (wbp for short) of type @ − ?.

(2) Let ψ be a wbp of type $@ - v$ whose ending variable is bound by a λ-node λ' and ϕ be a wbp of type $@ - \lambda$ coming into λ'. Then $\psi \cdot (\phi)^r \rho$ is a wbp, where ρ is an edge outgoing the second argument of the initial node of ϕ. The type of $\psi \cdot (\phi)^r \rho$ depends on the node connected to ρ.

(3) Let ψ be a wbp of type $@ - @$ ending in a node $@'$ and ϕ be of type $@ - \lambda$ leading from $@'$ to some λ-node λ'. Then $\psi\phi\rho$ is a wbp, where ρ is the edge that outgoes λ' towards its body. The type of $\psi\phi\rho$ depends on the connection of the edge ρ.

Lemma 5 If $@'\phi\lambda'$ is a wbp of type $@ - \lambda$, in a λ-term, connecting a @-occurrence $@'$ with a λ-occurrence λ', then there is no other wbp of type $@ - \lambda$ that starts at $@'$ and properly extends ϕ.

Proof. Immediate from Definition 4.

Definition 6 (Arrows, ranks, degrees) Let $M \in \hat{\hat{\Lambda}}$.

Step 1: For any application subterm $w \equiv (\lambda y_1 \ldots y_l.N)M_1 \ldots M_l$ in M, draw arrows connecting every M_i with all free occurrences of y_i in N (if any), $i = 1,\ldots,l$. The *rank* of all such arrows is 1, and the arrows are *relative* to w – any arrow coming out of M_i is called an i-th arrow relative to w. If $w = aps(u)$, arrows coming out of M_1 are also called *relative* to u. The *degree* of any arrow ι coming out of M_i is $deg(\iota) = i$.

Further, for any $k \geq 2$, do (until applicable):

Step k: For any head-variable subterm $zN_1 \ldots N_m$ such that z has received an arrow κ of rank $k-1$ from an abstraction $L \equiv \lambda z_1 \ldots z_m.P$ in the $(k-1)$-th step [1], draw arrows connecting every N_j with all free occurrences of z_j in P, $j = 1,\ldots,m$. The rank of any such arrow is k, and the *degree* of any arrow ι coming out of N_j is $deg(\iota) = deg(\kappa) + j$. Call ι a *successor*, or a j-th *successor*, of κ, and call κ the *predecessor* of ι. If κ is relative to an application subterm $w \subseteq M$ (respectively, to a β-redex u), then so is ι.

An arrow ι from a subterm $N \subseteq M \in \hat{\hat{\Lambda}}$ to an occurrence $x \in M$ is written $\iota : N \mapsto x$, and $rank(\iota)$ denotes its rank. We have shown in [15] that $rank(\iota)$ determines the minimal number of *superdevelopments* [22] needed to perform the substitution corresponding to ι. Similarly, we will see that $deg(\iota)$ is the number of β-steps needed to perform the substitution corresponding to ι. See Figure 1 for a graphical example.

Definition 7 (Virtual redexes) Let λ' be a λ-occurrence in $M \in \hat{\hat{\Lambda}}$. Assume that M is marked with arrows (according to Definition 6). We define the *matching application* $@'$ for λ', the *degree* of the *virtual redex* $(@',\lambda')$, and the arrows that $(@',\lambda')$ *contracts*, as follows:

(1) Let λ' occur as the λ-occurrence of λx_i in an application subterm

$$w = @(\ldots @(\lambda x_1 \ldots \lambda x_m.S, T_1),\ldots,T_m).$$

Then the matching application $@'$ for λ' is the i-th from the right occurrence of @ in the pattern of w. Further, $(@',\lambda')$ is called the *i-th virtual redex relative to w*, and the degree of $(@',\lambda')$ is $deg(@',\lambda') = i-1$. Finally, we say that $(@',\lambda')$ contracts all arrows coming out of T_i.

(2) Let λ' occur as the λ-occurrence of λx_i in an abstraction subterm

$$L = \lambda x_1 \ldots \lambda x_m.S.$$

For any arrow $\iota : L \mapsto z$ in M, the matching application $@'$ for λ', *relative* to ι, is the i-th from the right occurrence of @ in the pattern of the head-variable subterm

$$R = @(\ldots @(z, T_1),\ldots,T_m)$$

corresponding to z. (Thus λ' has as many matching applications as the number of arrows coming out of L.) Further, $(@',\lambda')$ is called the *i-th virtual redex relative to* ι, $(@',\lambda')$ is *relative* to the same application subterm as ι, and the degree of $(@',\lambda')$ is $deg(@',\lambda') = deg(\iota) + i - 1$. Finally, the say that $(@',\lambda')$ contracts all arrows coming out of T_i (such arrows are i-th successors of ι).

[1] We observe that such step cannot be in general successfully carried out in the unrestricted λ-calculus, where the initial abstractions of L are not required to be exactly m.

The degree of a virtual redex is the number of steps needed to created the 'corresponding' redex. This will be made precise in Section 5. A virtual redex is illustrated in Figure 1.

Convention: Clearly, a virtual redex v with degree 0 constitutes the pattern of a unique β-redex, and in the sequel we will identify v with the latter redex. The set of virtual redexes in a term $M \in \hat{\hat{\Lambda}}$ will be denoted by $vr(M)$. We will use again v, u and w to denote virtual redexes.

Lemma 8 Let ϕ be a wbp of type $@ - \lambda$ in $M \in \hat{\hat{\Lambda}}$, connecting an occurrence $@'$ with a λ-occurrence λ'. Then $(@', \lambda')$ is a virtual redex.

Proof. (See [10] for a detailed proof) By induction on the length of ϕ.

Lemma 9 Let $M \in \hat{\hat{\Lambda}}$ and $u = (@', \lambda') \in vr(M)$, and let u contract an arrow $L \mapsto x$. Then L is the second argument of $@'$ and λ' binds x.

Proof. The lemma is an immediate consequence of Definitions 6 and 7.

Lemma 10 Let $v' = (@', \lambda')$ be a virtual redex in $M \in \hat{\hat{\Lambda}}$. Then there is a wbp of type $@ - \lambda$ that connects $@'$ with λ'.

Proof. (See [10] for a detailed proof) By induction on $n = deg(v')$.

Along the proof of Lemma 8, we have proved the following characterization of wbp in the λ_H-calculus:

Lemma 11 Let ϕ be a wbp of type $@ - \lambda$ connecting the occurrences $@'$ and λ' in $M \in \hat{\hat{\Lambda}}$.

(1) If the left son of $@'$ is a λ-occurrence, then it coincides with λ' and $\phi = @'\lambda'$.

(2) If the left son of $@'$ is a $@$-occurrence, $@''$, then $\phi = @'@''\psi\lambda''\lambda'$, where $@''\psi\lambda''$ is a wbp and λ' is a son of λ''.

(3) If the left son of $@'$ is a variable-occurrence, x, then $\phi = @'x\lambda''(\psi)^r@''\lambda'$, where $@''\psi\lambda''$ is a wbp and λ' is the right son of $@''$.

From this characterization of wbp we conclude that wbp in hyperbalanced terms do not contain *cycles*, and that all wbp are *legal paths*. (So there is no need to recall the definition of cycles or legal paths from [1].) Furthermore, we can prove that virtual redexes determine their corresponding wbp uniquely:

Theorem 12 Let ϕ and ψ be wbp connecting the same occurrences in $M \in \hat{\hat{\Lambda}}$. Then $\phi = \psi$.

Proof. Suppose on the contrary that ϕ and ψ are two different wbp connecting the same occurrences, $@'$ and λ', in M, and assume that ϕ is shortest such path. If the left son of $@'$ in ϕ is a λ, then $\phi = \psi = @', \lambda'$ by Lemma 5. If the left son of $@'$ is a $@$-occurrence $@''$, then by Lemma 11 $\phi = @'@''\phi''\lambda''\lambda$ and $\psi = @'@''\psi''\lambda''\lambda'$, where $@''\phi\lambda''$ and $@''\psi\lambda''$ are wbp and λ' is a son of λ''. But $\phi'' = \psi''$ by the minimality of ϕ, implying $\phi = \psi$. The case when the left son of $@'$ is a variable is similar to the previous case.

Now, from the theorem of Asperti and Laneve [1] stating that there is a $1-1$ correspondence between the families in a λ-term and wbp of type $@ - \lambda$, we get the following theorem:

Theorem 13 Let $M \in \widehat{\widehat{\Lambda}}$. Then there is $1-1$ correspondence between families relative to M and virtual redexes in M.

Let the *arity* of an arrow $\iota : \lambda x_1 \ldots x_n.L \mapsto x$ be n. Then, by Theorem 13 and the definition of virtual redexes (Definition 7), any arrow of arity n in $M \in \widehat{\widehat{\Lambda}}$ determines n families relative to M. Further, recall from [15] that the garbage-collected form $M_\circ$ of a term $M \in \widehat{\widehat{\Lambda}}$ can be constructed from M statically. Now we can state and prove the following corollaries:

Corollary 14 (1) The number of families in any hyperbalanced λ-term M coincides with the sum of the arities of application subterms and arrows in M. Thus the number is finite, and it can be determined statically.
(2) The number of multi-steps in the normalizing leftmost-outermost complete family-reduction starting from M coincides with the sum of the arities of application subterms and arrows in the garbage-collected form $M_\circ$ of M, thus it can be determined statically.

Proof. (1) By Definition 7, all virtual redexes in M can be found statically, and it remains to apply Theorem 13.
(2) By [15, Theorem 4.7], the garbage-collected form $M_\circ$ of M can be found statically. Since the set of garbage-collected terms is closed under reduction [15], any reduction starting from $M_\circ$ is needed.[2] Hence the length of complete family-reduction starting from $M_\circ$ coincides with the number of families relative to $M_\circ$, i.e., the number of needed families in M, and it remains to observe that the leftmost-outermost normalizing complete family-reductions starting from M and $M_\circ$ have the same length.

Corollary 15 (1) The λ_H-calculus is strongly normalizing wrt β-reduction.
(2) The simply typed λ-calculus is strongly normalizing wrt β-reduction.

Proof. (1) Exactly as in [3]: Immediate from Corollary 14.1 and the Generalized Finite Developments theorem [21, Theorem 5.1], stating that any reduction that contracts redexes in finitely many families is terminating.
(2) Exactly as in [15], by combining (1) with the following two facts: (a) any simply typable λ-term has a hyperbalanced η-expansion; and (b) contraction of η-redexes can be postponed after contraction of β-redexes in any βη-reduction (since η-steps do not create β-redexes) [5, Corollaries 15.1.5-6].

4 Virtual Church-Rosser Theorem

Let us recall the concept of *descendant* from [11], allowing the tracing of subterms along β-reductions. In a β-step $M \xrightarrow{u} M'$, where $u \equiv (\lambda x.L)N$, the contracted redex u does not have a residual, but it has a descendant: the descendant of u, as well as of its operator $\lambda x.L$ and its body L, is the contractum of u. The descendants of free occurrences of x in L are the substituted occurrences of N. The descendants of other subterms of M

[2] Closure of the set of garbage-free terms (called *good* terms in [15]) under reduction implies that $M_\circ \in \widehat{\widehat{\Lambda}}$: Indeed, $M_\circ$ is balanced since it is obtained from M by replacing a number of *components* with fresh variables; and every application subterm in $M_\circ$ is semi-hyperbalanced since there are no arrows in $M_\circ$ coming out of the replaced subterms.

are intuitively clear (see [14] for a more precise definition). This definition extends to arbitrary β-reductions by transitivity.

We recall from [21,5] that any (finite) strongly equivalent reductions $P : M \twoheadrightarrow N$ and $Q : M \twoheadrightarrow N$ are *H-equivalent*: for any redex $w \subseteq M$, $w/P = w/Q$. Further, it follows from the results in [11] that $P \approx_L Q$ implies also that P and Q are *strictly-equivalent*: for any subterm L of M, the descendants of L under P and under Q are the same subterms of N. We will refer to these two properties as HCR and SCR, respectively.

Our aim in this section is to generalize HCR to virtual redexes. The result will be used in Section 6 to prove a new characterization of families via virtual redexes. We first prove an analog of (a weaker form of) HCR for residuals of arrows.

The following definition of *residuals* of arrows is equivalent to the one in [15], but is different in that the current definition is inductive, and the residuals of arrows are *defined* as arrows, so unlike [15] there is no need to *prove* this by a tedious case analysis. Pictures illustrating the residual concept for arrows can be found in [10] (see also [15]).

Definition 16 (Residuals of arrows)[3] Let $M \xrightarrow{u} M'$, let $u \equiv (\lambda y.P_0)P$, and let $\iota : N \mapsto x$ be an arrow in $M \in \hat{\hat{\Lambda}}$. We define the *u-residuals* of ι in M' by induction on $n = rank(\iota)$.

(1) $(n = 1)$ Then ι is the j-th arrow relative to an application subterm w (for some $1 \leq j \leq arity(w)$).

(a) If $w = aps(u)$ and $j = 1$ (i.e., if u contracts ι), then ι does not have u-residuals.

(b) If $w = aps(u)$ and $j > 1$, then u creates a unique application subterm w' of type 1, and the $(j-1)$-th arrow $\iota' : N' \mapsto x'$ relative to w' such that x' is a u-descendant of x is the unique u-residual of ι

(c) If $w \neq aps(u)$, then for any u-residual $w' \subseteq M'$ of w, any j-th arrow $\iota' : N' \mapsto x'$ relative to w' such that x' is a u-descendant of x is a u-residual of ι.

(2) $(n > 1)$ Let ι be say m-th successor of $\kappa : L \mapsto z$.

(a) If $n = 2$ and u contracts κ (and thus $L = P$ and z is a free occurrence of y in P_0), then the unique u-residual of ι is the arrow $\iota' : N' \mapsto x'$ in M' such that: ι' is relative to the application subterm w' of type 2 created by contraction of κ; and x' is the unique u-descendant of x that is in the copy of L that substitutes z.

(b) If u does not contract κ and κ does not have a u-residual, then ι does not have a u-residual either.

(c) Otherwise, for any u-residual κ of κ, any m-th successor of κ' in M' such that x' is a u-descendant of x is a u-residual of ι.

Definition 17 (Residuals of virtual redexes) Let $M \in \hat{\hat{\Lambda}}$, let $M \xrightarrow{u} M'$, and let $v \in vr(M)$ be relative to an application subterm w. We define *residuals* of v as virtual redexes in M' by induction on $n = deg(v)$.

- $(n = 0)$ Then v is a redex, and the residuals of the redex v are declared as the residuals of the virtual redex v.

- $(n > 0)$ We distinguish two subcases:

(a) Let v be the i-th virtual redex relative to w (for some $i > 1$). If $w = aps(u)$, then u creates a unique application subterm w' of type 1, and $arity(w') = arity(w) - 1$. Then the $(i-1)$-th virtual redex relative to w' is the only u-residual of v. Otherwise, any u-residual w^* of w is an application subterm such that $arity(w^*) = arity(w)$, and the i-th virtual redex relative to w^* is a u-residual of v (that is, v has as many u-residuals as w).

[3] [15], as well as [10], contains figures illustrating the definition.

(b) Let v be the i-th virtual redex relative to an arrow $\iota : L \mapsto z$ relative to w. If $rank(\iota) = 1$ and u contracts ι, then ι collapses and the u-descendant L' of L that substitutes z becomes the operator of a created application subterm w' of type 2 in M'. (In this case we say that w' is created *by contracting* ι.) In this case, the i-th virtual redex relative to w' is the only u-residual of v. If ι does not collapse and does not have a u-residual, then v does not have a u-residual either. Otherwise, for any u-residual ι' of ι, the i-th virtual redex relative to ι' is an u-residual of v.

Remark 18 One can easily verify (by induction on $rank(\iota)$) that, in the notation of Definition 16, if ι is relative to $aps(u) \subseteq M$ and u does not contract ι, then ι has exactly one u-residual in M'. From this and Definition 17 it is immediate that any virtual redex in M relative to $aps(u)$ and different from u has exactly one u-residual in M'.

Notation 19 We will use the same notation for residuals of arrows and virtual redexes as for residuals of redexes. For example, ι/P will denote the residuals of ι after the reduction P.

Lemma 20 Let $u_1, u_2 \in M \in \hat{\hat{\Lambda}}$, and let ι be an arrow in M. Then the residuals of ι under $u_1 \sqcup u_2$ and $u_2 \sqcup u_1$ are the same arrows.

Proof. (See [10] for a detailed proof) Induction on $rank(\iota)$, using Defs 3 and 16.

Lemma 21 Let $u_1, u_2 \in M \in \hat{\hat{\Lambda}}$, and let $v \in vr(M)$. Then $v/(u_1 \sqcup u_2) = v/(u_2 \sqcup u_1)$.

Proof. (See [10] for a detailed proof) Case analysis, according to Definition 17.

Theorem 22 (Virtual Church-Rosser, VCR) Let $M \in \hat{\hat{\Lambda}}$, and let $w \in vr(M)$, and let $P : M \twoheadrightarrow N$ and $Q : M \twoheadrightarrow N$ be Lévy-equivalent. Then $w/P = w/Q$.

Proof. Lemma 21 implies the theorem exactly as in the case when w is a redex, see e.g. [17].

5 Extraction Normal Forms and Chains

In this section we show that if Pv is an extraction normal form, in the λ_H-calculus, then P is a *chain*, that is, every step of P except the last creates the next contracted redex. We show that the converse is also true. We will use this characterization of extraction normal forms to study the contribution relation in the λ_H-calculus. Actually, we use this characterization also to show that redexes of the same family are residuals of a unique virtual redex, but we expect that this result can be proved for the λ-calculus in general, where extraction normal forms need not be chains.

Lemma 23 Let $M \in \hat{\hat{\Lambda}}$, let $M \xrightarrow{u} M'$, and let $v = aps(u)$.

(1) Let ι' be a u-residual of an arrow ι relative to an application subterm w in $M \in \hat{\hat{\Lambda}}$. If $w = v$, then $deg(\iota') = deg(\iota) - 1$; otherwise, $deg(\iota') = deg(\iota)$.
(2) Any arrow in M' is a residual of an arrow in M.

Proof. (See [10] for a detailed proof)
(1) By induction on $n = rank(\iota)$.

(2) By induction on the rank $n = rank(\iota')$ of an arrow $\iota' : N' \mapsto x'$ in M', it is shown that $\iota : N \mapsto x$ is an arrow in M whose u-residual is ι', where x is the only u-ancestor of x' and N is the only u-ancestor of N' that is an abstraction (but is not the operator of a redex).[4]

Using Lemma 23, we now prove its analog for virtual redexes.

Lemma 24 Let $M \in \widehat{\widehat{\Lambda}}$, let $M \xrightarrow{u} M'$, and let $v' \in vr(M')$ be a u-residual of a virtual redex $v \subseteq M$ relative to an application subterm w. If $w = aps(u)$, then $deg(v') = deg(v) - 1$; otherwise, $deg(v') = deg(v)$.

Proof. If $deg(v) = 0$, then $w \neq aps(u)$, both v and v' are redexes, and $deg(v') = 0 = deg(v)$ by Definition 7. So let $deg(v) > 0$. We distinguish two subcases:
(a) Let v be the i-th virtual redex relative to w ($i > 1$ since $n > 0$).
If $w = aps(u)$, then v has exactly one residual, v', which is the $(i-1)$-th virtual redex relative to the created application subterm w' of type 1, hence $deg(v') = i - 1 = deg(v) - 1$ by Definition 7.
And if $w \neq aps(u)$, then any residual v' of v is the i-th virtual redex relative to some u-residual w^* of w (if any), hence $deg(v') = i = deg(v)$ by Definition 7.
(b) Now let v be the i-th virtual redex relative to an arrow ι relative to w. Then $deg(v) = deg(\iota) + i - 1$ by Definition 7.
If $deg(\iota) = 1$ and u contracts ι (in this case, $w = aps(u)$), then v' is the i-th virtual redex relative to the application subterm w' of type 2 created by the contraction of ι, thus $deg(\iota') = i - 1 = deg(v) - 1$. Otherwise, v' is i-th virtual redex relative to a u-residual ι' of ι, and $deg(v') = deg(\iota') + i - 1$ by Definition 7. Now it remains to apply Lemma 23.

Lemma 25 Let $M \in \widehat{\widehat{\Lambda}}$, let $M \xrightarrow{u} M'$, and let $v' \in vr(M')$. Then v' is a u-residual of a unique virtual redex in M.

Proof. By Definition 3, we need to consider the following four cases:
(1) Let v' be i-th virtual redex relative to a residual $w' \subseteq M'$ of an application subterm $w \subseteq M$ (hence $w \neq v$). Then, by Definition 17, v' is the u-residual of (only) the i-th virtual redex relative to w.
(2) Let v' be i-th virtual redex relative to a created application subterm $w' \subseteq M'$ of type 1. Then, by Definition 17, v' is the u-residual of (only) the $(i+1)$-th virtual redex relative to $aps(u)$ (in this case, $w = aps(u)$).
(3) Let v' be i-th virtual redex relative to the application subterm $w' \subseteq M'$ of type 2 created by contraction of an arrow ι. Then, by Definition 17, v' is the u-residual of (only) the i-th virtual redex relative to ι (in this case, $w = aps(u)$).
(4) Let v' be i-th virtual redex relative to an arrow ι'. Then, by Lemma 23, ι' is a u-residual of an arrow ι in M, and by Definition 17, v' is the u-residual of (only) the i-th virtual redex relative to ι.

Definition 26 We call a reduction P a *chain* if every step in P except the last creates the next contracted redex.

[4] The only two different abstractions in M that may have the same descendant abstraction in M' are the operator $\lambda y.P_0$ and the body P_0 of the contracted redex u (when P_0 is an abstraction), and it can be shown that there is no arrow $\iota' : N' \mapsto x'$ in M' such that N' is the u-descendant of P_0.

Definition 27 Let $w \in vr(M)$, let $P : M \twoheadrightarrow N$, and let $v \in N$ be a P-residual of w. Then we call v, or rather, Pv, an *w-redex*.

The following two corollaries are now immediate from Lemma 24 and Lemma 25, respectively:

Corollary 28 Let v_0 be a virtual redex in $M_0 \in \hat{\hat{\Lambda}}$ with $deg(v_0) = n$. Then there is a reduction $P : M_0 \xrightarrow{u_0} M_1 \xrightarrow{u_1} \ldots \xrightarrow{u_{n-1}} M_n$ and a redex $u_n \subseteq M_n$ such that v_0 has a unique residual v_i in M_i $(i = 0, \ldots, n)$, $deg(v_i) = n - i$, v_i is relative to $aps(u_i)$ (hence $u_n = v_n$ and Pu_n is a v_0-redex), and $P + u_n$ is a chain.

Corollary 29 Let $P : M \twoheadrightarrow N$, where $M \in \hat{\hat{\Lambda}}$, and let $v \subseteq N$. Then there is a unique virtual redex $w \in vr(M)$ such that Pv is an w-redex.

We will now give a characterization of extraction normal forms in the λ_H-calculus. We refer to [21,4] for the definition of *extraction* relation and extraction normal forms in the λ-calculus. We denote $enf(M)$ the set of extraction normal forms relative to M.

Lemma 30 Let $P : M \twoheadrightarrow N$ be a chain, where M is any (i.e., not necessarily hyperbalanced) term. Then $P \in enf(M)$.

Proof. It is trivial to check that none of the four types of extraction steps in Definition 4.7 of [21] can be applied to a chain P, which means that P is in extraction normal form.

The converse is not true in the λ-calculus in general, but is true in the λ_H-calculus (Lemma 34). We need two simple lemmas first.

Lemma 31 (See [10]) Let v_1 be in the argument of a redex $u = (\lambda x.L)N \subseteq M \in \hat{\hat{\Lambda}}$, let $v'_1 \in v_1/u$, and let v'_1 create a redex v'_2. Then v_1 creates a redex v_2 such that $v'_2 \in v_2/u'$, where $u' = u/v_1$.

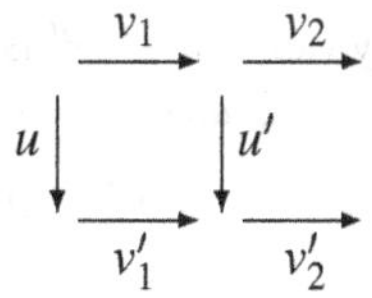

Definition 32 let $N_0 \subseteq M_0$. We call a reduction $P : M_0 \xrightarrow{u_0} M_1 \xrightarrow{u_1} \ldots \xrightarrow{u_{n-1}} M_n$ *internal* to N_0 if every u_i is in the unique descendant $N_i \subseteq M_i$ of N_0 along P.

Lemma 33 Let $M \in \hat{\hat{\Lambda}}$ and let $P : M \twoheadrightarrow N$ be a chain. Then P is internal to the application subterm corresponding to the first redex contracted in P.

Proof. Immediate from the fact that if $M \xrightarrow{u} M'$ creates a redex $v \subseteq M'$ (of type 1 or 2), then the application subterm corresponding to v is inside the descendant of the application subterm corresponding to u.

Now we can prove the converse of Lemma 30.

Lemma 34 Let P be a reduction in the λ_H-calculus in extraction normal form. Then P is a chain.

Proof. Suppose on the contrary that there is an extraction normal form P which is not a chain, and assume P is shortest among such reductions. Then P has the form

$$P : M_0 \xrightarrow{u_0} M_1 \xrightarrow{u_1} M_2 \twoheadrightarrow M_n$$

where u_1 is a u_0-residual of some redex $v_0 \subseteq M_0$. Let $w_0 = aps(v_0)$, and let

$$P' : M_1 \xrightarrow{u_1} M_2 \twoheadrightarrow M_n$$

By Definition 4.7.1 of [21], $n > 2$. Since P is standard, there are three possible relative positions of u_0 and v_0.

(1) The redexes u_0 and v_0 are disjoint, and v_0 is to the right of u_0. Then w_0 is disjoint from u_0, and by Lemma 33, P' is internal to the residual w_1 of w_0. Hence where is a reduction $Q : M_0 \twoheadrightarrow N$ internal to w_0 such that $P' = Q/u_0$, and by Definition 4.7.2 of [21], u_0 can be extracted from P – a contradiction.

(2) The redex v_0 is in the body of u_0. By Lemma 31, there exist redexes $v_1, \ldots, v_{n-2}$ such that the following diagram commutes

$$\begin{array}{ccccccccccc}
M_0 & \xrightarrow{v_0} & N_1 & \xrightarrow{v_1} & N_2 & \twoheadrightarrow & N_{n-3} & \xrightarrow{v_{n-3}} & N_{n-2} & \xrightarrow{v_{n-2}} & N_{n-1} \\
u_0 \downarrow & & u'_1 \downarrow & & u'_2 \downarrow & & u'_{n-3} \downarrow & & u'_{n-2} \downarrow & & \\
M_1 & \xrightarrow[u_1]{} & M_2 & \xrightarrow[u_2]{} & M_3 & \twoheadrightarrow & M_{n-2} & \xrightarrow[u_{n-2}]{} & M_{n-1} & \xrightarrow[u_{n-1}]{} & M_n
\end{array}$$

where $u'_j = u_0/(v_0 + \ldots + v_{j-1})$ $(j = 1, \ldots, n-2)$, $u_{i+1} = v_i/u'_i$ $(i = 0, \ldots, n-2)$ and $Q : v_0 + \ldots + v_{n-1}$ is a chain. By Lemma 33, Q is internal to w_0, hence is internal to the body of u_0, and we have $P' = Q/u_0$. Hence by Definition 4.7.3 of [21], u_0 can be extracted from P – a contradiction.

(3) The redex v_0 is in the argument S of u_0. Further, u_1 is in a substituted copy S' of S. Then the application subterm w_0 has a residual $w_1 \subseteq M_1$ which is again an application subterm (since $M_0, M_1 \in \hat{\hat{\Lambda}}$), and by Lemma 33, P' is internal to w_1 (and thus is internal to S'). Hence by Definition 4.7.4 of [21], u_0 can be extracted from P – a contradiction.

Corollary 35 In the λ_H-calculus, a reduction P is a chain iff it is in extraction normal form.

6 From Zig-Zag Back to Residuals

We show that redexes with co-initial histories are in the same family iff they are residuals of the same virtual redex in the initial term. This is actually a new characterization of families in the λ_H-calculus. (Asperti and Laneve [1] only showed how the path corresponding to a family can be traced back along the history of the canonical representative of the family.) We will make use of the virtual Church-Rosser property, which in particular implies that residuals of virtual redexes under reductions are invariant under Lévy equivalence. We expect that such a characterization can also be given for the λ-calculus in general.

Lemma 36 (See [10]) Let $M \in \hat{\hat{\Lambda}}$, and let $Pv, P'v' \in enf(M)$ be v^*-redexes for some $v^* \in vr(M)$. Then $Pv = P'v'$.

Remark 37 Using Lemma 36 we can actually give another proof of the $1-1$ correspondence of families and virtual redexes in the λ-calculus (Theorem 13). That proof does not use the properties of wbp from [1].

Recall from [21] that every family relative to M has exactly one member in extraction normal form, and the latter are different for different families. Hence it is enough to prove that there is a $1-1$ correspondence between $enf(M)$ and $vr(M)$.

By Corollary 29, there is a function $f : enf(M) \to vr(M)$ assigning to any $Pv \in enf(M)$ the (unique) virtual redex $v^* \in vr(M)$ such that Pv is a v^*-redex. By Corollary 28, f is a bijection. It remains to use Lemma 36.

Now we strengthen the above theorem to yield the promised characterization of families via residuals of virtual redexes.

Theorem 38 Let $M \in \hat{\hat{\Lambda}}$. Redexes with histories Pv and Qu relative to M are in the same family iff there is a redex $w \subseteq M$ such that both Pv and Qu are w-redexes.

Proof. ($\Rightarrow$) Let $Pv \simeq Qu$. Since $\simeq$ is the equivalence closure of the copy relation, it is enough to consider the case then Qu is a copy of Pv, i.e., $Q \approx_L P + P'$ for some P' and $u \in v/P'$. By Corollary 29, there is a unique $w \in vr(M)$ such that Qu is an w-redex. By VCR (Theorem 22), $u \in w/(P+P')$, and by Lemma 25, $v \in w/P$.

($\Leftarrow$) Let Pv and Qu be w-redexes. Further, let $P'v'$ and $Q'u'$ be the extraction normal forms of Pv and Qu, respectively. By ($\Rightarrow$), $P'v'$ and $Q'u'$ are w-redexes, and $P'v' = Q'u'$ by Lemma 36. Hence $Pv \simeq Qu$ by Theorem 4.9 of [21].

7 Creation and Contribution Relations

In this section we show that, for any $M \in \hat{\hat{\Lambda}}$, the *contribution* relation on the set $FAM(M)$ of families relative to M is a *forest* (as a partial order). That is, the down set $\{\phi' \mid \phi' \hookrightarrow \phi\}$ of any family $\phi \in FAM(M)$, wrt $\hookrightarrow$, is a chain (i.e., is a total order).

Definition 39 (Contribution relation) Let $M \in \hat{\hat{\Lambda}}$ and let $v, v' \in vr(M)$. We define the *creation relation* $\prec$ on $vr(M)$ as the union of the following relations $\prec_1$ and $\prec_2$.

(1) $v \prec_1 v'$ iff, for some i, either v is i-th virtual redex relative to an application subterm $w \subseteq M$ with $arity(w) > i$ and v' is the $(i+1)$-th virtual redex relative to w; or v is i-th virtual redex relative to an arrow ι and v' is the $(i+1)$-th virtual redex relative to ι.

(2) $v \prec_2 v'$ iff v' is the 1st virtual redex relative to an arrow that v contracts.

If $v \prec v'$, then we say that v is a *creator* of v'. The *contribution relation* $\leq_v$ on $vr(M)$ is the transitive reflexive closure of $\prec$. Similarly, if $v \leq_v v'$, then we say that v is a *contributor* of v'.

Lemma 40 (See [10]) Every virtual redex $u \subseteq M \in \hat{\hat{\Lambda}}$ with a positive degree has a unique creator $v \in vr(M)$, and $deg(u) = deg(v) + 1$.

Lemma 41 (See [10]) Let $M \xrightarrow{u} M'$, let $v_1, v_2 \in vr(M)$ be relative to the application subterm $w = aps(u)$, and let v_1' and v_2' be the unique (by Remark 18) u-residuals of v_1 and v_2, respectively. Then $v_1 \prec_1 v_2$ (resp. $v_1 \prec_2 v_2$) iff $v_1' \prec_1 v_2'$ (resp. $v_1' \prec_2 v_2'$).

Lemma 42 (See [10]) Let $M \in \hat{\hat{\Lambda}}$, let $v_1^*, v_2^* \in vr(M)$, and let $P_1 v_1$ and $P_2 v_2$ be respectively v_1^*- and v_2^*-redexes in extraction normal form. Then $v_1^* \prec_1 v_2^*$ (resp. $v_1^* \prec_2 v_2^*$) iff $P_2 = P_1 + v_1$ and v_2 is a redex of type 1 (resp. type 2) created by v_1.

Corollary 43 A reduction $P : M_0 \xrightarrow{u_0} M_1 \xrightarrow{u_1} \ldots \xrightarrow{u_{n-1}} M_n$ is a chain iff there is a covering chain $v_0 \prec v_1 \prec \ldots \prec v_{n-1}$ of virtual redexes in M_0 such that u_i is a v_i-redex, $i = 0, \ldots, n-1$.

Let $FAM(P)$ denote the set of families (relative to the initial term of P) whose member redexes are contracted in P. The family relation induces the contribution relation on families as follows:

Definition 44 ([8]) Let ϕ_1, ϕ_2 be families relative to the same term $M \in \hat{\hat{\Lambda}}$. Then $\phi_1 \hookrightarrow \phi_2$ (read ϕ_1 *contributes* to, or is a *contributor* of, ϕ_2) iff for any redex $Pv \in \phi_2$, P contracts at least one member of ϕ_1.

It follows easily from the definition of extraction in [21] that if a reduction Q extracts to Q', then $FAM(Q') \subseteq FAM(Q)$. Hence $\phi_1 \hookrightarrow \phi_2$ iff the history of the extraction normal form of ϕ_2 contracts a member of ϕ_1. Hence we have immediately from Corollary 35 and Corollary 43 the following theorem.

Theorem 45 (1) Let $M \in \hat{\hat{\Lambda}}$, let $v_1, v_2 \in vr(M)$, and let ϕ_1 and ϕ_2 be the corresponding families relative to M. Then $v_1 \leq_v v_2$ iff $\phi_1 \hookrightarrow \phi_2$.
(2) Further, for any $M \in \hat{\hat{\Lambda}}$, $(FAM(M), \hookrightarrow)$ is a forest.

8 Conclusion

Theorem 45 implies that all application subterms (rather, their 'skeletons') can be computed *independently*, in the sense of [13], and the results can be 'combined' to yield the normal form (the combination of P and Q should yield the final term of their least upper bound $P \sqcup Q$). Note here that nesting of application subterms is not an obstacle.

At present, we do not know what constitutes the skeleton of an application subterm. The idea is that the application subterms whose relative substitution paths are isomorphic must have the same skeletons. But the above theorem shows that independent computation is possible [5]. Note that, once we compute a skeleton, we can reuse the result every time we meet the same skeleton in a hyperbalanced term. In other words, we can define 'big beta steps' (say by induction on the length of the longest relative path), that normalize skeletons in one big step. This would make computation of hyperbalanced terms efficient.

References

1. Asperti, A., Laneve, C. Paths, computations and labels in the lambda calculus. Theoretical Computer Science 142(2):277-297, 1995.
2. Asperti, A., Danos, V., Laneve, C., Regnier, L., Paths in the lambda-calculus. In *Proceedings of the Symposium on Logic in Computer Science (LICS)*, IEEE Computer Society Press, 1994.
3. Asperti, A., Mairson H.G., Parallel beta reduction is not elementary recursive. In Proc. of ACM Symposium on Principles of Programming Languages (POPL), 1998.

[5] For Recursive Program Schemes, whose contribution relation is again a forest, independent computation of redexes is very easy to implement [12].

4. Asperti A., Guerrini S. *The Optimal Implementation of Functional Programming Languages.* Cambridge Tracts in Theoretical Computer Science, Cambridge University Press, 1998.
5. Barendregt, H., *The Lambda Calculus. Its syntax and Semantics (revised edition).* North Holland, 1984.
6. Barendregt, H., *Lambda Calculi with Types.* in Handbook of Logic in Computer Science, Vol. 2, Oxford University Press, 1992.
7. H.P. Barendregt, J.A. Bergstra, J.W. Klop and H. Volken, Some notes on lambda-reduction, in: Degrees, reductions, and representability in the lambda calculus. Preprint no. 22, University of Utrecht, Department of Mathematics, 1976, 13-53.
8. Glauert, J.R.W., Khasidashvili, Relative normalization in deterministic residual structures. In: Proc. of the 19th International Colloquium on Trees in Algebra and Programming, CAAP'96, Springer LNCS, vol. 1059, H. Kirchner, ed. 1996, p. 180-195.
9. V. Kathail, Optimal interpreters for lambda-calculus based functional languages, Ph.D. Thesis, MIT, 1990.
10. Kennaway, R., Khasidashvili, Z. and Piperno, A., Static Analysis of Modularity of β-reduction in the Hyperbalanced λ-calculus (Full Version), available at `http://www.dsi.uniroma1.it/~piperno/`.
11. Khasidashvili, Z., β-reductions and β-developments of λ-terms with the least number of steps. In: Proc. of the International Conference on Computer Logic, COLOG'88, Springer LNCS, 417:105-111, 1990.
12. Khasidashvili, Z. Optimal normalization in orthogonal term rewriting systems. In: Proc. of 5th International Conference on Rewriting Techniques and Applications, RTA'93, Springer LNCS, 690:243-258, 1993.
13. Khasidashvili, Z., Glauert, J. R. W. The Geometry of orthogonal reduction spaces. In Proc. of the 24th International Colloquium on Automata, Languages, and Programming, ICALP'97, Springer LNCS, vol. 1256, P. Degano, R. Gorrieri, and A. Marchetti-Spaccamela, eds. 1997, p. 649-659.
14. Khasidashvili, Z., Ogawa, M. and van Oostrom, V. Perpetuality and uniform normalization in orthogonal rewrite systems. Information and Computation, vol.164, p.118-151, 2001.
15. Khasidashvili, Z., Piperno, A. Normalization of typable terms by superdevelopments. In Proc. of the Annual Conference of the European Association for Computer Science Logic, CSL'98, Springer LNCS, vol. 1584, 1999, p. 260-282
16. Khasidashvili, Z., Piperno, A. A syntactical analysis of normalization. Journal of Logic and Computation, Vol.10(3)381-410, 2000.
17. Klop, J.W., *Combinatory Reduction Systems.* PhD thesis, Matematisch Centrum Amsterdam, 1980.
18. J. Lamping, An algorithm for optimal lambda calculus reduction, in: *Proc. 17th ACM Symp. on Principles of Programming Languages* (1990) 6-30.
19. Lévy, J.-J., An algebraic interpretation of the λβκ-calculus and a labelled λ-calculus. *Theoretical Computer Science*, 2:97-114, 1976.
20. Lévy, J.-J., Réductions correctes et optimales dans le λ-calcul. PhD thesis, Univerité Paris 7, 1978.
21. J.-J. Lévy, Optimal reductions in the lambda-calculus, in: J.P. Seldin and J.R. Hindley, eds., *To H. B. Curry: Essays on Combinatory Logic, Lambda-calculus and Formalism* (Academic Press, 1980) 159-192.
22. van Raamsdonk, F., Confluence and superdevelopments. In: Proc. of the 5th International Conference on Rewrite Techniques and Applications, RTA'93, Springer LNCS, 690:168-182, 1993.

Exceptions in the Rewriting Calculus

Germain Faure[1] and Claude Kirchner[2]

[1] ENS Lyon & LORIA 46 allée d'Italie, 69364 Lyon Cedex 07, France
`Germain.Faure@ens-lyon.fr`
[2] LORIA & INRIA 615 rue du Jardin Botanique 54602 Villers-lès-Nancy, France
`Claude.Kirchner@loria.fr` `http://www.loria.fr/~ckirchne`

Abstract. In the context of the rewriting calculus, we introduce and study an exception mechanism that allows us to express in a simple way rewriting strategies and that is therefore also useful for expressing theorem proving tactics. The proposed exception mechanism is expressed in a confluent calculus which gives the ability to simply express the semantics of the `first` tactical and to describe in full details the expression of conditional rewriting.

The rewriting calculus, also called ρ-calculus, makes all the basic ingredients of rewriting explicit objects, in particular the notions of rule formation (*abstraction*), rule *application* and *result*. The original design of the calculus (fully described in [CK01,Cir00]) has nice properties and encompasses in a uniform setting λ-calculus and first-order rewriting. Thanks to its matching power, *i.e.* the ability to discriminate directly between patterns, *and* thanks to the first class status of rewrite rules, it also has the capability to represent in a simple and natural way object calculi [CKL01a,DK00]. It allows moreover, thanks to its abstraction mechanism, to design extremely powerful type systems generalizing the λ-cube [CKL01b].

In the ρ-calculus, a rewrite rule is represented as a ρ-term (a term of the ρ-calculus) and the application of a rule, for example $f(x,y) \to x$, at the root of a ρ-term, for example $f(a,b)$, is also represented by a ρ-term, *e.g.* $[f(x,y) \to x](f(a,b))$. Applying a rewrite rule $l \to r$ at the root of a term t consists first in computing the matching between l and t and secondly in applying the obtained substitution(s) to the term r. Since matching may fail, we can get an empty set of results. When matching is unitary, *i.e.* results in only one substitution, the evaluation of the rule application results in a singleton. So, in general the evaluation of every application is represented as a set, giving thus a first class status to the results of a rule application.

For simplicity, we consider here only syntactic matching and we have indeed in this case either zero or one substitution matching l towards t. For example, $f(x,y)$ matches $f(a,b)$ using the substitution $\sigma = \{x \mapsto a, y \mapsto b\}$. Therefore $[f(x,y) \to x](f(a,b))$ reduces by the evaluation rule of the ρ-calculus into the singleton $\{\sigma x\}$, *i.e.* $\{a\}$. Since there is no substitution matching $f(x,y)$ on $g(a,b)$, the evaluation of $[f(x,y) \to x](g(a,b))$ results in the empty set. This is denoted $[f(x,y) \to x](g(a,b)) \mapsto\!\!\!\twoheadrightarrow_\rho \emptyset$.

S. Tison (Ed.): RTA 2002, LNCS 2378, pp. 66–82, 2002.

Since sets are ρ-terms, we can go further on and represent rewriting systems. For example, the set consisting of the two rules $f(x,y) \to x$ and $f(x,b) \to b$ is a ρ-term denoted $\{f(x,y) \to x, f(x,b) \to b\}$. The application of such a rewrite system, at the root of the term $f(a,b)$ is denoted by the ρ-term $[\{f(x,y) \to x, f(x,b) \to b\}](f(a,b))$. To evaluate such a term, we need to distribute the elements of the set and thus the previous term is evaluated to $\{[f(x,y) \to x](f(a,b)), [f(x,b) \to b](f(a,b))\}$ and then to $\{a,b\}$. This shows how non-determinism is simply expressed in the rewriting calculus.

We can also see on these examples that in the ρ-calculus the failure of matching the left hand-side of a rewrite rule on a term is revealed by the evaluation of the application to the empty set. Therefore, the empty set can be used as a signal that a rule does not match a term. This is very useful information but unfortunately it could be erased latter by the evaluation process. For example, assuming a and b to be constants, we have $[x \to a]([a \to a](b)) \overset{*}{\longmapsto\!\!\!\!\to}_\rho [x \to a](\emptyset)$. Since clearly x matches the ρ-term $\emptyset$, the latter term can be reduced to $\{a\}$. This shows that in the ρ-calculus, it can happen that a sub-term of a given term leads to $\emptyset$ because of matching failure and that the final result is not the empty set. A calculus is called *strict* or equivalently it has the *strictness* property, when it allows for a *strict propagation of failure.* This means that if there is a matching failure at one step of a reduction, the only possible final result of the reduction is failure, represented here by the empty set.

One of the basic but already elaborated goals we want to reach using the ρ-calculus, is to use its matching-ability to naturally and simply express reduction and normalization strategies. Indeed, the first simple question to answer is: is there a ρ-term representing a given derivation in a rewrite theory? And the answer has been shown to be positive [CK01], leading in particular to the use of ρ-terms as proof terms of rewrite derivations [Ngu01]. Going further on, we want to represent rewrite strategies, and in particular normalization strategies. Therefore we want to answer the question: given a rewrite theory $\mathcal{R}$ is there a ρ-term $\xi_\mathcal{R}$, such that for any term u, if u normalizes to the term v in the rewrite theory $\mathcal{R}$, then $[\xi_\mathcal{R}](u)$ ρ-reduces to (a set containing) the term v.

Since the ρ-calculus embeds the λ-calculus, any computable function as the normalization one is expressible in this formalism. We want to make use of the matching power and the non-determinism of the ρ-calculus to bring an increased ease in the expression of such functions together with their expression in a uniform formalism combining standard rewrite techniques and higher-order behaviors.

In trying and using the specificities of the calculus to solve the previous questions, we encountered several difficulties, the main one being the ambivalent use of the empty set. For example, when applying on the term b the rewrite rule $a \to \emptyset$ that rewrites the constant a into the empty set, the evaluation rule of the calculus returns $\emptyset$ because *the matching against b fails*: $[a \to \emptyset](b) \longmapsto\!\!\!\!\to_\rho \emptyset$. But it is also possible to explicitly rewrite an object into the empty set like in $[a \to \emptyset](a) \overset{*}{\longmapsto\!\!\!\!\to}_\rho \emptyset$, and in this case the result is also the empty set because the rewrite rule *explicitly introduces* it.

It becomes then clear that one should avoid the ambivalent use of the empty set and introduce an *explicit distinction between failure and the empty set of results.* In fact, by making this distinction, we add to the matching power of the rewriting calculus a *catching power*, which allows to describe, in a rewriting style, exception mechanisms, and therefore to express easily convenient and elaborated strategies needed when computing or proving. Because of its fundamental role, we mainly focus on the `first` strategy combinator used in ELAN [BKK+98] that is also called IF THEN in LCF like logical frameworks.

We thus propose a confluent calculus (using a call-by-value evaluation mechanism) in which the `first` *and a simple but complete exception mechanism can be expressed.*

For that purpose, we introduce a new constant $\perp$ and an exception handler operator exn. In this new version of the ρ-calculus, like in most exception mechanisms, an exception can be either raised by the user ($\perp$ is put explicitly in the initial term like for example in $[x \rightarrow \perp](a)$) or can be caused by a "run time error" (for example a matching failure like in $[f(x,y) \rightarrow x](g(a,b))$). Either an exception can be *uncaught* ($\perp$ goes all over the term and the term is reduced to $\{\perp\}$ when using a strict evaluation strategy) or it can be stopped by exn and next *caught* thanks to matching. Thus, the calculus allows to: *(i)* catch any matching failure (and therefore to be able to detect that a term is in normal form) like in $\big[exn(\perp) \rightarrow c\big]\big(exn\big([f(x,y) \rightarrow x](g(a,b))\big)\big)$, *(ii)* ignore a raised exception like for example in the term $[x \rightarrow c](exn(M))$, *(iii)* switch the evaluation mechanism according to a possible failure in the evaluation of a term (think again of the application to normal terms). For example, a program consisting of a division x/y (encoded in the ρ-calculus by $(x/_\rho y)$), a program P and a possible switch of the evaluation to a program P_{err} if the *Div_By_Zero* exception is raised during the computation of x/y, can be expressed by the term $\big[\texttt{first}\big(exn(\perp) \rightarrow P_{err}, x \rightarrow P\big)\big]\big(exn(x/_\rho y)\big)\big)$.

Our contributions are therefore the construction, study and use of a rewriting calculus having an explicit exception mechanism. In the next section, we motivate and build the calculus. In section 2, we expose a call-by-value strategy that makes the calculus confluent. We finally show in section 3 that elaborated rewriting strategies can be described and evaluated, in particular we show that the `first` strategy combinator can be naturally expressed.

1 The ρ_ε-Calculus

We introduce here a new ρ-calculus: the ρ_ε-calculus. After the construction of the $\rho^{1^{st}}$-calculus (the ρ-calculus doped with the `first` operator, as described in [CK01]), it is quite natural to ask if the `first` operator can be *simply* expressed in the ρ-calculus. To understand the problem, we ask the question: instead of enriching the ρ-calculus with the `first` operator and its specific evaluation rules, what first class ingredients must we add to the ρ-calculus to express the `first`? The first problem is to obtain a non-ambivalent meaning for the empty set. The second problem is to obtain both the strictness and the matching failure test.

1.1 How Can the First Be *Simply* Expressed in the ρ-Calculus?

Since the ρ-calculus embeds the λ-calculus, any computable function like the **first** can be encoded with it. Using the "deep" and "shallow" terminology promoted in [BGG+92], what we are looking for is to *simply express* the **first** by avoiding the use of a deep encoding that would not fully use the matching capability of the ρ-calculus. Indeed a shallow encoding is encoding variables of the source calculus by variables of the target one, as a deep encoding typically encodes variables by constants. What we call in this paper a *simple* encoding is a shallow encoding using in a direct way the matching capabilities (*i.e.* the variable abstraction management) of the framework.

The original ρ-calculus has been built in such a way that the empty set represents both the empty set of terms and the result of the failure of a rule application. For example, we have

$$[a \to \emptyset](a) \;\stackrel{*}{\longmapsto\!\!\!\!\to}_\rho\; \emptyset,$$

where $\emptyset$ represents the empty set of terms and

$$[a \to \emptyset](b) \;\longmapsto\!\!\!\!\to_\rho\; \emptyset,$$

where $\emptyset$ encodes the failure in the matching between a and b. So, in this version of the ρ-calculus, we are not able to determine if the result of a rule application is a failure. Nevertheless, this test is useful if we want to define the **first** in the ρ-calculus, since the **first** is defined by:

$$\begin{array}{lll} \mathit{First}' & [\mathtt{first}(s_1,\ldots,s_n)](t) \leadsto \{u_k\downarrow\} & \\ & \quad \text{if } [s_i](t) \stackrel{*}{\longmapsto\!\!\!\!\to}_\rho \emptyset, 1 \leqslant i \leqslant k-1 \\ & \quad\quad [s_k](t) \stackrel{*}{\longmapsto\!\!\!\!\to}_\rho u_k\downarrow \neq \emptyset,\ \mathcal{FV}ar(u_k\downarrow) = \emptyset \end{array}$$

$$\begin{array}{lll} \mathit{First}'' & [\mathtt{first}(s_1,\ldots,s_n)](t) \leadsto \emptyset & \\ & \quad \text{if } [s_i](t) \stackrel{*}{\longmapsto\!\!\!\!\to}_\rho \emptyset, 1 \leqslant i \leqslant n \end{array}$$

where $u_k\downarrow$ denotes the normal form of u_k for the evaluation rules of the ρ-calculus (denoted $\stackrel{*}{\longmapsto\!\!\!\!\to}_\rho$) and $\mathcal{FV}ar(t)$ denotes the free variables of t.

Remark 1. The side condition in First' indicates that u_k must not contain any free variable and must be in normal form. It seems to be restrictive at first sight, but it allows for a valid definition of the **first** that preserves the confluence of the ρ-calculus.

Despite serious efforts to find it, the **first** seems not to be *simply* expressible in the ρ-calculus, and even if it were, there is a serious drawback in not distinguishing the empty set from the rule application failure. Indeed, $[\mathtt{first}(a \to \emptyset, b \to c)](a) \stackrel{*}{\longmapsto\!\!\!\!\to}_\rho \emptyset$ and $[\mathtt{first}(a \to \emptyset, b \to c)](c) \stackrel{*}{\longmapsto\!\!\!\!\to}_\rho \emptyset$ for two very different reasons.

1.2 A First Approach to Enriching the ρ-Calculus

To distinguish the empty set from rule application failure, we introduce a new symbol $\perp$ (which we assume not to belong to the signature of our calculus) to denote the latter. As a consequence, we have to adapt accordingly the definition of the operators of the calculus.

We denote by $\mathcal{S}ol(l \ll t)$ the set of solutions obtained when syntactically matching l and t. The central idea of the main rule of the calculus ($Fire$) is that the application of a rewrite rule $l \rightarrow r$ at the root position of a term t, consists in computing the solution of the matching equation $(l \ll t)$ and applying the substitution returned by the function $\mathcal{S}ol(l \ll t)$ to the term r. When there is no solution for the matching equation $(l \ll t)$ the special constant $\perp$ is used and the result is the set $\{\perp\}$. We obtain the rule described in Figure 1.

$$Fire \qquad [l \rightarrow r](t) \rightsquigarrow \begin{cases} \{\perp\} \text{ if } l \ll t \text{ has no solution} \\ \{\sigma r\} \text{ where } \sigma \in \mathcal{S}ol(l \ll t) \end{cases}$$

Fig. 1. *Fire* rule in the extended ρ-calculus

Since the symbol $\perp$ represents failure, we want it to be propagated. Therefore, we have to add the rules of the third part of Figure 2 (at this stage of the construction, the propagation of the failure is not strict). In the proposed calculus, the semantic of $\{\perp\}$ is the failure term, so we want it to be a normal form term. Moreover, we can not have $\{\perp\} \overset{*}{\longmapsto\!\!\!\rightarrow} \emptyset$ thanks to the side condition $n > 0$ in the $FlatBot$ rule, we can not have $\{\perp\} \overset{*}{\twoheadleftarrow\!\!\twoheadrightarrow} \perp$ and $\perp$ can never be a result since it is not a set (except of course, if the initial term is $\perp$).

To define the `first`, we need to express an evaluation scheme of the form: If $res = \perp$ then evaluate c. It can be expressed at this step of the construction by the term: $[\perp\rightarrow c](res)$. If $res = \perp$, then $[\perp\rightarrow c](res)$ can be rewritten in $\{c\}$. So, it seems necessary to have a new symbol function (which we assume not to belong to the signature of our calculus) to do a more precise matching, to distinguish the propagation of the failure from its catching.

Example 1. Using the `first` operator (which is discussed in Section 3.1), we can express the evaluation scheme:

if M is evaluated to the failure term
then evaluate the term P_{fail}
else evaluate the term P_{nor}.

thanks for example to the term $[\texttt{first}(\perp \rightarrow P_{fail}, x \rightarrow P_{nor})](M)$. But, this has a serious drawback: If M leads to $\{\perp\}$ then, using the rule for the $\perp$ propagation (see the third part of Figure 2) the following reduction may happen:
$[\texttt{first}(\perp \rightarrow P_{fail}, x \rightarrow P_{nor})](M)$
$\overset{*}{\longmapsto\!\!\!\rightarrow}_\rho [\texttt{first}(\perp \rightarrow P_{fail}, x \rightarrow P_{nor})](\perp) \longmapsto\!\!\!\rightarrow_{AppBotR} \{\perp\}$.

So, we need to improve the proposed calculus to obtain a confluent and strict calculus which allows failure catching and rules out reductions like $f(\emptyset, \bot) \longmapsto\!\!\!\!\rightarrow_\rho \emptyset$ and $f(\emptyset, \bot) \longmapsto\!\!\!\!\rightarrow_\rho \{\bot\}$.

1.3 An Adapted Version of the Calculus: The ρ_ε-Calculus

About the meaning of the empty set. What we propose in this new extension is to give a single meaning to the empty set: to represent *only* the empty set of terms. Consequently, the application of a term to the empty set should lead to failure (*i.e.* $[v](\emptyset) \overset{*}{\longmapsto\!\!\!\!\rightarrow}_\rho \{\bot\}$) and the application of the empty set to a term should lead too to failure (*i.e.* $[\emptyset](v) \overset{*}{\longmapsto\!\!\!\!\rightarrow}_\rho \{\bot\}$). On the other hand, provided all the t_i are in normal form, a term like $f(t_1, \ldots, \emptyset, \ldots, t_n)$ will be considered as a normal form and not re-writable to $\{\bot\}$. This is in particular useful to keep to the empty set its first class status. Moreover, we would like a ρ-term of the form $u \rightarrow \emptyset$ to be in normal form since it allows us to express a void function.

Up to the last rule (*Exn*), these design choices lead to the rewriting calculus evaluation rules described in Figure 2. Let us now explain the design choices leading to the last rule.

An extension of the calculus to deal with exceptions. In particular motivated by our goal to define the `first` operator, we need to express failure catching. Moreover, if we want to control the strictness of failure propagation, we need to encapsulate the failure symbol by a new operator to allow for reductions of the following form: $[operator(\bot) \rightarrow u](operator(\bot)) \longmapsto\!\!\!\!\rightarrow_\rho \{u\}$. This new operator is denoted *exn* and its evaluation rule is defined by:

$$\mathit{Exn} \qquad exn(t) \rightsquigarrow \{t\} \quad \text{if } \{t\}\downarrow\ \neq \{\bot\} \text{ and } \mathcal{FV}ar(t\downarrow) = \emptyset$$

Here, we can make the same comments on the side conditions of the *Exn* rule as whose of Remark 1. Moreover, thanks to a simple matching we have $[exn(\bot) \rightarrow u](exn(\bot)) \longmapsto\!\!\!\!\rightarrow_\rho \{u\}$.

Remark 2. If the evaluation of a term M leads to $\{\bot\}$, then $exn(M) \overset{*}{\longmapsto\!\!\!\!\rightarrow}_\rho exn(\{\bot\})$. It is then natural to allow for the reduction $exn(\{\bot\}) \overset{*}{\longmapsto\!\!\!\!\rightarrow}_\rho \{exn(\bot)\}$ since otherwise we prevent an evaluation switched by a possible failure in the evaluation of a term (see Example 2 in which the *OpSet* rule is essential). So, as can be seen in Figure 2, in the ρ_ε-calculus, we extend the *OpSet* rule for all ε-functional symbols (*i.e.* for all $h \in \mathcal{F}_\varepsilon \triangleq \mathcal{F} \cup \{exn, \bot\}$, where $\mathcal{F}$ is the signature of the calculus).

Remark 3. We could have replaced the *Exn* rule by a rule rewriting $exn(t)$ in $\{t\downarrow\}$ instead of $\{t\}$. But doing so will keep us away from a small step semantics, since the intrinsic operations at the object level will in this case be executed at the meta level and the operator at the object level will not contain all the evaluation information anymore.

Fire	$[l \to r](t)$	$\rightsquigarrow \begin{cases} \{\bot\} \text{ if } l \ll t \text{ has no solution} \\ \{\sigma r\} \text{ where } \sigma \in \mathcal{S}ol(l \ll t) \end{cases}$
Congr	$[f(t_1, \ldots, t_n)](f(u_1, \ldots, u_n))$	$\rightsquigarrow \{f([t_1](u_1), \ldots, [t_n](u_n))\}$
CongrFail	$[f(t_1, \ldots, t_n)](g(u_1, \ldots, u_n))$	$\rightsquigarrow \{\bot\}$
Distrib	$[\{u_1, \ldots, u_n\}](v)$	$\rightsquigarrow$ if $n > 0$, $\{[u_1](v), \ldots, [u_n](v)\}$ if $n = 0$, $\{\bot\}$
Batch	$[v](\{u_1, \ldots, u_n\})$	$\rightsquigarrow$ if $n > 0$, $\{[v](u_1), \ldots, [v](u_n)\}$ if $n = 0$, $\{\bot\}$
SwitchR	$u \to \{v_1, \ldots, v_n\}$	$\rightsquigarrow \{u \to v_1, \ldots, u \to v_n\}$ if $n > 0$
OpSet	$h(v_1, \ldots, \{u_1, \ldots, u_n\}, \ldots, v_m)$	$\rightsquigarrow \{h(v_1, \ldots, u_i, \ldots, v_m)\}_{1 \leqslant i \leqslant n}$ if $n > 0$ and $h \in \mathcal{F}_\varepsilon$
Flat	$\{u_1, \ldots, \{v_1, \ldots, v_m\}, \ldots, u_n\}$	$\rightsquigarrow \{u_1, \ldots, v_1, \ldots, v_m, \ldots, u_n\}$
AppBotR	$[v](\bot)$	$\rightsquigarrow \{\bot\}$
AppBotL	$[\bot](v)$	$\rightsquigarrow \{\bot\}$
AbsBotR	$v \to \bot$	$\rightsquigarrow \{\bot\}$
OpBot	$f(t_1, \ldots, t_k, \bot, t_{k+1}, \ldots, t_n)$	$\rightsquigarrow \{\bot\}$
FlatBot	$\{t_1, \ldots, \bot, \ldots, t_n\}$	$\rightsquigarrow \{t_1, \ldots, t_n\}$ if $n > 0$
Exn	$exn(t)$	$\rightsquigarrow \{t\}$ if $\{t\}\downarrow \neq \{\bot\}$ and $\mathcal{FV}ar(t\downarrow) = \emptyset$

Fig. 2. $\mathcal{EVAL}_\varepsilon$: the evaluation rules of the ρ_ε-calculus

Definition of the ρ_ε-calculus. The previous design decision leads us to define the ρ_ε-terms and the ε-first-order terms in the following way.

$$\rho_\varepsilon\textbf{-term} \qquad t ::= x \mid f(t, \ldots, t) \mid t \to t \mid [t](t) \mid \{t, \ldots, t\} \mid \bot \mid exn(t)$$

where $x \in \mathcal{X}$ and $f \in \mathcal{F}$.

In the ρ-calculus, first-order terms only are allowed in the left hand side of an abstraction. In spite of this restriction, we obtain an expressive calculus for which confluence and strictness are obtained under a large class of evaluation strategies [CK01,Cir00]. Introducing $\bot$ and exn, we would like to define an extension of the ρ-calculus that keeps these properties. Therefore we extend the terms which we will abstract on (*i.e.* rewrite rule left hand sides) to be as follows:

$$\varepsilon\textbf{-first-order term} \qquad t ::= x \mid f(t, \ldots, t) \mid exn(\bot)$$

Definition 1. *For $\mathcal{F}$ a set of function symbols and $\mathcal{X}$ a countable set of variables, we define the ρ_ε-calculus by:*

- *the subset $\rho_\varepsilon(\mathcal{F},\mathcal{X})$ of ρ_ε-terms whose sub-terms of the form $l \to r$ are such that l is a ε-first-order term;*
- *the higher-order application of substitutions to terms;*
- *the empty theory with the classical syntactic matching algorithm;*
- *the evaluation rules described in Figure 2;*
- *an evaluation strategy $\mathcal{S}$ that fixes the way to apply the evaluation rules.*

Fact 11 *The set $\rho_\varepsilon(\mathcal{F},\mathcal{X})$ is stable by $\mathcal{EVAL}_\varepsilon$.*

Proof. This is a simple consequence of the preservation of the set $\rho_\varepsilon(\mathcal{F},\mathcal{X})$ by the evaluation mechanism. This explains in particular why $\mathcal{EVAL}_\varepsilon$ does not contain the *SwitchL* rule of the ρ-calculus.

Example 2. If we want to express the evaluation scheme presented in Example 1, thanks to the *exn* operator the problem exposed in that example is solved. In fact, in the ρ_ε-term $[\mathtt{first}(exn(\bot) \to P_{fail}, x \to P_{nor})](exn(M))$ the propagation of failure is stopped thanks to the *exn* operator.

1. if $M \overset{*}{\longmapsto\!\!\!\!\rightarrow}_{\rho_\varepsilon} \{\bot\}$, we have the following reduction:

$$\begin{array}{ll} & \left[\mathtt{first}\left(exn(\bot) \to P_{fail}, x \to P_{nor}\right)\right]\left(exn(M)\right) \\ \overset{*}{\longmapsto\!\!\!\!\rightarrow}_{\rho_\varepsilon} & \left[\mathtt{first}\left(exn(\bot) \to P_{fail}, x \to P_{nor}\right)\right]\left(exn(\{\bot\})\right) \\ \longmapsto\!\!\!\!\rightarrow_{OpSet} & \left[\mathtt{first}\left(exn(\bot) \to P_{fail}, x \to P_{nor}\right)\right]\left(\{exn(\bot)\}\right) \\ \overset{*}{\longmapsto\!\!\!\!\rightarrow}_{\rho_\varepsilon} & \{P_{fail}\} \end{array}$$

   So, in this case, the initial term leads to P_{fail}, which is the wanted behavior.
2. if $M \overset{*}{\longmapsto\!\!\!\!\rightarrow}_{\rho_\varepsilon} \{M' \downarrow\}$ where $\{M' \downarrow\} \neq \{\bot\}$, we obtain:

$$\begin{array}{ll} & \left[\mathtt{first}\left(exn(\bot) \to P_{fail}, x \to P_{nor}\right)\right]\left(exn(M)\right) \\ \overset{*}{\longmapsto\!\!\!\!\rightarrow}_{\rho_\varepsilon} & \left[\mathtt{first}\left(exn(\bot) \to P_{fail}, x \to P_{nor}\right)\right]\left(exn(M')\right) \\ \overset{*}{\longmapsto\!\!\!\!\rightarrow}_{\rho_\varepsilon} & \{P_{nor}\} \end{array}$$

   In this case also, we have the wanted result: P_{nor}.

After the introduction of this new version of the ρ-calculus, we are going to make precise some of its properties and its expressiveness.

2 On the Confluence of the ρ_ε-Calculus. The $\rho_{\varepsilon v}$-Calculus

2.1 The Non Confluence of the ρ_ε-Calculus

In [Cir00,CK01], it is shown that the ρ-calculus is confluent under a large class of evaluation strategies. The same thing holds here. At first, it may seem easy to prove it but in fact, there are two problems that need a particular treatment: the *Exn* rule and the multiple ways to obtain failure.

Let us begin to show typical examples of confluence failure. Non-confluence is inherited from the confluence failure of the ρ-calculus (described in [Cir00, CK01]) and from specific critical overlaps due to its enrichment and on which we are focusing now.

First, if we apply a rule $l \to r$ to a term t (*i.e.* apply the *Fire* rule to $[l \to r](t)$) with t *Exn*-reducible, we can have some non confluent reductions.

Example 3 (Under-evaluated terms).

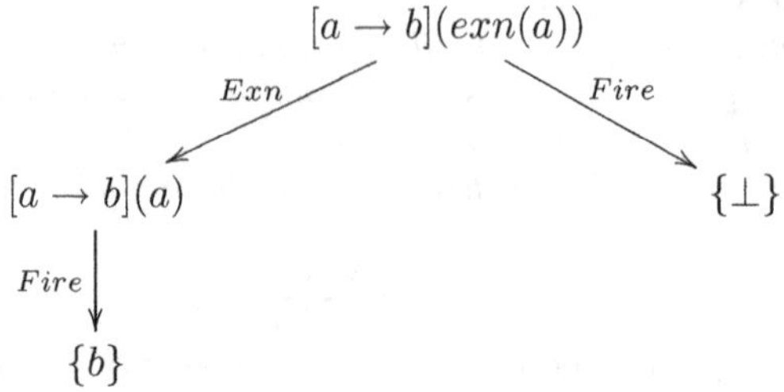

In the ρ-calculus, knowing if a term can be reduced to failure (represented in the ρ-calculus by $\emptyset$) is quite easy since we only have to guarantee that no failure is possible and that $\emptyset$ is not a sub-term of t. But in the ρ_ε-calculus, knowing if a term can be reduced to $\{\bot\}$ seems not to be so easy since for example such a term must not contain sub-terms like $[u](v)$ where u or v can be reduced to $\emptyset$ (see rule *Distrib*, *Batch*). In fact, if this condition is not verified, we have problems of confluence like the critical overlap of Example 4.

Example 4 (Operand equals to $\emptyset$).

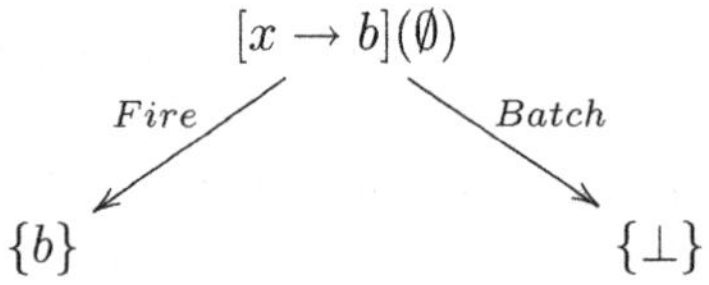

In this example, the problem is indeed more complicated to solve than it seems since we need a test to insure that a term can not be reduced to $\emptyset$. More details about this problem can be found in [Fau01].

In the ρ_ε-calculus, if we do not specify any strategy, we do not have a strict propagation of $\bot$ as shown in the following example.

Example 5 (Non-strict propagation of failure).

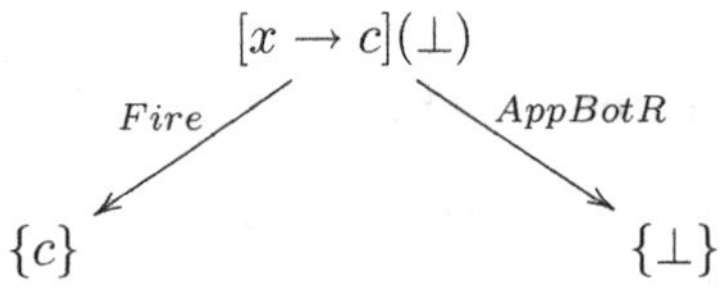

So, in addition to the difficulty to obtain, thanks to a suitable strategy, a *confluent* calculus, we are looking for a strategy which also allows for a *strict* calculus.

The $exn(\bot)$ role. The $exn(\bot)$ pattern allows to catch via matching rule application failure like in $\big[exn(\bot) \to c\big]\big(exn([f(x,y) \to x](g(a,b)))\big)$. Of course, nothing in the ρ_ε-calculus prevents from instantiating a variable by the $exn(\bot)$ pattern. This is something quite natural since it allows to propagate failure suspicion or to ignore a raised exception like in $[x \to c](exn(M))$.

2.2 How to Obtain a Confluent Calculus? The $\rho_{\varepsilon\mathrm{v}}$-Calculus

As seen in the last section, the ρ_ε-calculus is neither confluent nor strict. To obtain a confluent calculus, the evaluation mechanism described in $\mathcal{EVAL}_\varepsilon$ should be tamed by a suitable strategy. This could be described in two ways, either by an independent strategy or, and this is the choice made in [Fau01], by technical conditions directly added to the *Fire* and *Exn* rules. Here, we do not describe these conditional rules and we refer to [Fau01] for the detailed approach. We define the $\rho_{\varepsilon\mathrm{v}}$-calculus as the ρ_ε-calculus but using a call-by-value evaluation strategy.

Intuitively, in the ρ_ε-calculus, a value is a normal form term containing no free variable and no set (except the empty set as the argument of a function):

$$\textbf{value} \qquad v ::= c \mid f(v,\ldots,v) \mid f(v,\ldots,\emptyset,\ldots,v) \mid v \to v \mid exn(\bot)$$

(where $c \in \mathcal{F}_0$ and $f \in \cup_{n\geq 1}\mathcal{F}_n$)
We can define the $\rho_{\varepsilon\mathrm{v}}$-calculus either by using the classical syntactic matching algorithm or by adding a new rule to it. Actually, although it is not necessary for the confluence and the strictness, it will be judicious that during the matching no term can be instantiated by $\emptyset$. So, as one can see in Figure 3 (where $\wedge$ is supposed to be an associative and commutative operator), we add the *EmptyInstantiation* rule so as to avoid this instantiation.

Decomposition	$(f(t_1,\ldots,t_n) \ll f(t'_1,\ldots,t'_n)) \wedge P$	$\mapsto\!\!\!\!\rightarrow \bigwedge_{i=1\ldots n} t_i \ll t'_i \wedge P$
SymbolClash	$(f(t_1,\ldots,t_n) \ll g(t'_1,\ldots,t'_m)) \wedge P$	$\mapsto\!\!\!\!\rightarrow \mathbf{F}$ if $f \neq g$
MergingClash	$(x \ll t) \wedge (x \ll t') \wedge P$	$\mapsto\!\!\!\!\rightarrow \mathbf{F}$ if $t \neq t'$
EmptyInstanciation	$(x \ll \emptyset) \wedge P$	$\mapsto\!\!\!\!\rightarrow \mathbf{F}$
SymbolVariableClash	$(f(t_1,\ldots,t_n) \ll x) \wedge P$	$\mapsto\!\!\!\!\rightarrow \mathbf{F}$ if $x \in \mathcal{X}$

Fig. 3. The call-by-value syntactic matching

In the $\rho_{\varepsilon\mathrm{v}}$-calculus, the *Fire* and *Exn* rules can only be applied to values. Assume that l, r are ρ_ε-terms and v is a value, we define $Fire_v, Exn_v$ by:

$$\begin{array}{lll} Fire_v & [l \to r](v) \rightsquigarrow & \begin{cases} \{\bot\}\ if\ l \ll v\ has\ no\ solution \\ \{\sigma r\} where\ \sigma \in Sol(l \ll v) \end{cases} \\ Exn_v & exn(v) & \rightsquigarrow v \end{array}$$

So, now we can define more precisely the $\rho_{\varepsilon\mathrm{v}}$-calculus:

Definition 2. *The $\rho_{\varepsilon\mathrm{v}}$-calculus is defined as the ρ_ε-calculus except that we replace the Fire and Exn rules by the $Fire_v$ and Exn_v rules and that $Sol(l \ll v)$ is computed by the matching algorithm of Figure 3.*

In [Fau01], it is shown that this calculus is confluent and strict (since $\perp$ is not a value):

Theorem 1. *The $\rho_{\varepsilon\mathrm{v}}$-calculus is a confluent and strict calculus.*

This result is fundamental since the confluence is really necessary in our approach to express the `first`.

Thus, in the following section, we show that the $\rho_{\varepsilon\mathrm{v}}$-calculus allows us to express the `first` operator and consequently that we obtain a confluent and strict calculus in which the `first` can be expressed.

3 Expressiveness of the $\rho_{\varepsilon\mathrm{v}}$-Calculus

3.1 The `First` Operator in the $\rho_{\varepsilon\mathrm{v}}$-Calculus

Originally, the ρ_ε-calculus was introduced to obtain a calculus in which the `first` can be *simply* expressed. All over the construction of the ρ_ε-calculus, we aim at eliminating all arguments that help us to think that the `first` can not be expressed in it. We are going to show that we can indeed find a *shallow* encoding (see Section 1.1) of it in the $\rho_{\varepsilon\mathrm{v}}$-calculus. So, we define a term to express it, afterwards we prove that the given definition is valid and finally we give some examples.

As seen below, the `first` is an operator whose role is to select between its arguments the first one that, applied to a given ρ-term, does not evaluate to $\{\perp\}$. This is something quite difficult to express in the ρ-calculus. We propose to represent the term $[\mathtt{first}(t_1,\ldots,t_n)](r)$ by a set of n terms, denoted $\{u_1,\ldots,u_n\}$, with the property that every term u_i can be reduced to $\{\perp\}$ except perhaps one (say u_l). In other words, the initial set is reduced to a singleton, containing the normal form of u_l, by successively reducing each u_i ($i \neq l$) to $\{\perp\}$ and then using the *FlatBot* rule. To express the `first` in the ρ_ε-calculus, we are going to use its two main trumps: the matching and the failure catching. Each u_i is expressed according to the $i-1$ first terms: If all u_j ($j < i$) are reduced to $\{\perp\}$, then u_i leads to $[t_i](r)$ else u_i leads to $\{\perp\}$ by a rule application failure. So, we define each u_i by

$$
\begin{aligned}
u_1 &\triangleq [t_1](x)\\
u_2 &\triangleq \big[exn(\perp) \to [t_2](x)\big]\big(exn(u_1)\big)\\
u_3 &\triangleq \big[exn(\perp) \to [exn(\perp) \to [t_3](x)](exn(u_2))\big]\big(exn(u_1)\big)\\
u_4 &\triangleq \big[exn(\perp) \to [exn(\perp) \to [exn(\perp) \to [t_4](x)](exn(u_3))](exn(u_2))\big]\big(exn(u_1)\big)\\
&\vdots\\
u_n &\triangleq [exn(\perp) \to [exn(\perp) \to [\ldots \to [t_n](x)](exn(u_{n-1}))\ldots](exn(u_2))](exn(u_1))
\end{aligned}
$$

To get the right definition, one should replace in the expression of u_i, all u_j $(j < i)$ by their own definition, but such extended formulae would be rather tedious to read. We define $MyFirst$ by:

$$MyFirst(t_1, \ldots, t_n) \triangleq x \rightarrow \{u_1, \ldots, u_n\}$$

By $\texttt{first}(t_1, \ldots, t_n)$, we denote precisely the $\texttt{first}$ operator as defined in Section 1.1 by the two rules $First', First''$ and we denote by $MyFirst(t_1, \ldots, t_n)$ the operator purely defined in the ρ_ε-calculus just above.

It must be emphasized that this definition is not valid in a non-confluent calculus. In fact, if $MyFirst$ is used in this kind of calculus, it may happen that there exists i_0 such that u_{i_0} can be evaluated to a term $v_{i_0} \neq \{\bot\}$ by a first reduction and to $\{\bot\}$ by a second one. In this case, we can have for example a first reduction used to evaluate u_{i_0} that leads to $\{\bot\}$, but in one of the u_k $(k > i_0)$ we can have a different reduction of the same term u_{i_0} that leads to $v_{i_0} \neq \{\bot\}$ and so u_k is reduced as if u_{i_0} is not evaluated to $\{\bot\}$, whereas here, it is. So, we obtain a set of terms which does not necessary have the right properties. For example, it can happen that more than one of these terms can not be reduced to $\{\bot\}$. Thus, in this section, we are going to work in a confluent calculus: the $\rho_{\varepsilon v}$-calculus (although we can work in any confluent ρ_ε-calculus). Let us show the validity of the definition of $MyFirst$.

Lemma 1. *In the $\rho_{\varepsilon v}$-calculus,*

- *if for all i, $[t_i](r) \overset{*}{\longmapsto\!\!\!\!\rightarrow}_{\rho_{\varepsilon v}} \{\bot\}$, then $[MyFirst(t_1, \ldots, t_n)](r) \overset{*}{\longmapsto\!\!\!\!\rightarrow}_{\rho_{\varepsilon v}} \{\bot\}$ and $[\textbf{\textit{first}}(t_1, \ldots, t_n)](r) \overset{*}{\longmapsto\!\!\!\!\rightarrow}_{\rho_{\varepsilon v}} \{\bot\}$.*
- *if there exists l such that for all $i \leq l-1$ $[t_i](r) \overset{*}{\longmapsto\!\!\!\!\rightarrow}_{\rho_{\varepsilon v}} \{\bot\}$ and $[t_l](r) \overset{*}{\longmapsto\!\!\!\!\rightarrow}_{\rho_{\varepsilon v}} \{v_l\} \downarrow \neq \{\bot\}$ where $\mathcal{F}Var(v_l) = \emptyset$ then $[MyFirst(t_1, \ldots, t_n)](r) \overset{*}{\longmapsto\!\!\!\!\rightarrow}_{\rho_{\varepsilon v}} \{v_l\} \downarrow$ and $[\textbf{\textit{first}}(t_1, \ldots, t_n)](r) \overset{*}{\longmapsto\!\!\!\!\rightarrow}_{\rho_{\varepsilon v}} \{v_l\} \downarrow$.*

Proof. See [Fau01].

If there exists l such that for all $i \leq l-1$ $[t_i](r) \overset{*}{\longmapsto\!\!\!\!\rightarrow}_{\rho_{\varepsilon v}} \{\bot\}$ and $[t_l](r) \overset{*}{\longmapsto\!\!\!\!\rightarrow}_{\rho_{\varepsilon v}} \{v_l\}\downarrow \neq \{\bot\}$ with $\mathcal{F}Var(v_l) \neq \emptyset$, since a term with free variables is not a value, we can not "really evaluate" $[MyFirst(t_1, \ldots, t_n)](r)$ and so, the evaluation is suspended (as it is for the $\texttt{first}$). So, *w.r.t.* terms with free variables, this result is also coherent. The following example is an illustration of this fact:

Example 6.

$$\begin{array}{ll} & x \rightarrow [MyFirst(y \rightarrow [x](y), y \rightarrow c)](b) \\ \overset{*}{\longmapsto\!\!\!\!\rightarrow}_{\rho_{\varepsilon v}} & x \rightarrow \{[x](b), [exn(\bot) \rightarrow [y \rightarrow c](b)](exn([x](b)))\} \\ \overset{*}{\longmapsto\!\!\!\!\rightarrow}_{\rho_{\varepsilon v}} & \{x \rightarrow [x](b), x \rightarrow [exn(\bot) \rightarrow c](exn([x](b)))\} \end{array}$$

In the $\rho_{\varepsilon v}$-calculus this result is in normal form because we can not continue the evaluation since the variable x needs to be instantiated. Let us note that it is not a drawback since we have the same behavior if we consider the $\texttt{first}$ defined by the rules $First'$ and $First''$ given in Section 1.1.

Thanks to the above lemma, we get our main expressiveness result (see Section 1.1 for the definition of simply expressible):

Theorem 2. *The $\textbf{\textit{first}}$ is simply expressible in the $\rho_{\varepsilon v}$-calculus.*

Now, we will denote by $\mathtt{first}(t_1,\ldots,t_n)$ the term $MyFirst(t_1,\ldots,t_n)$. The above theorem is significant since this question about the expressiveness of the `first` in the ρ-calculus has arisen since the introduction of the ρ-calculus. Moreover, this result (a confluent calculus in which the `first` can be expressed) is proved here for the first time since the confluence of the ρ^{1st}-calculus (the ρ-calculus doped with the `first`) is still an open question. Last but not least, this calculus allows also a simple but complete exception scheme as illustrated below.

In the next example, we provide an example of two reductions of the same term illustrating the confluence of the proposed calculus.

Example 7. A first reduction:

$$
\begin{array}{ll}
 & [x \to [\mathtt{first}(y \to [y](x),\ y \to c)](b)](a) \\
\triangleq & [x \to \{[y \to [y](x)](b), \\
 & \quad [exn(\bot) \to [y \to c](b)](exn([y \to [y](x)](b)))\}](a) \\
\overset{*}{\longmapsto\!\!\!\!\rightarrow}_{\rho_{\varepsilon v}} & [x \to \{[b](x),\ [exn(\bot) \to [y \to c](b)](exn([b](x)))\}](a) \\
\overset{*}{\longmapsto\!\!\!\!\rightarrow}_{\rho_{\varepsilon v}} & \{[b](a)\ ,\ [exn(\bot) \to [y \to c](b)](exn([b](a)))\} \\
\overset{*}{\longmapsto\!\!\!\!\rightarrow}_{\rho_{\varepsilon v}} & \{\{\bot\},\ [y \to c](b)\} \\
\overset{*}{\longmapsto\!\!\!\!\rightarrow}_{\rho_{\varepsilon v}} & \{c\}
\end{array}
$$

An other possible reduction could be:

$$
\begin{array}{ll}
 & [x \to [\mathtt{first}(y \to [y](x),\ y \to c)](b)](a) \\
\overset{*}{\longmapsto\!\!\!\!\rightarrow}_{\rho_{\varepsilon v}} & [\mathtt{first}(y \to [y](a),\ y \to c)](b) \\
\triangleq & \{[y \to [y](a)](b),\ [exn(\bot) \to [y \to c](b)](exn([y \to [y](a)](b)))\} \\
\overset{*}{\longmapsto\!\!\!\!\rightarrow}_{\rho_{\varepsilon v}} & \{\bot,\ \{c\}\} \\
\overset{*}{\longmapsto\!\!\!\!\rightarrow}_{\rho_{\varepsilon v}} & \{c\}
\end{array}
$$

In example 2, we saw that using the `first` we can encode an evaluation scheme switched by the result of the evaluation of one term (failure or not). Let us illustrate in this example the given definition of the `first`.

Example 8.

$$
\begin{array}{ll}
 & [\mathtt{first}(exn(\bot) \to P_{fail},\ x \to P_{nor})](exn(M)) \\
\triangleq & \{[exn(\bot) \to P_{fail}](exn(M))\ , \\
 & \quad [exn(\bot) \to [x \to P_{nor}](exn(M))]([exn(\bot) \to P_{fail}](exn(M)))\ \}
\end{array}
$$

The evaluation strategy (call-by-value) forces to begin by the evaluation of M. We are going to distinguish two cases:

1. if $M \overset{*}{\longmapsto\!\!\!\!\rightarrow}_{\rho_{\varepsilon v}} \{\bot\}$, we have the following reduction:

$$
\begin{array}{ll}
\multicolumn{2}{l}{\{[exn(\bot) \to P_{fail}](exn(M))\ ,} \\
\multicolumn{2}{l}{\quad [exn(\bot) \to [x \to P_{nor}](exn(M))]([exn(\bot) \to P_{fail}](exn(M)))\ \}} \\
\overset{*}{\longmapsto\!\!\!\!\rightarrow}_{\rho_{\varepsilon v}} & \{\{P_{fail}\},\ \{\bot\}\} \\
\overset{*}{\longmapsto\!\!\!\!\rightarrow}_{\rho_{\varepsilon v}} & \{P_{fail}\}
\end{array}
$$

So, in this case , the initial term leads to P_{fail}, which is the wanted result.

2. if $M \overset{*}{\longmapsto\!\!\!\!\rightarrow}_{\rho_{\varepsilon v}} \{M' \downarrow\}$ where $\{M' \downarrow\} \neq \{\bot\}$, we obtain:

$$
\begin{array}{ll}
 & \{ [exn(\bot) \to P_{fail}](exn(M)) \;, \\
 & \quad [exn(\bot) \to [x \to P_{nor}](exn(M))]([exn(\bot) \to P_{fail}](exn(M))) \;\} \\
\overset{*}{\longmapsto\!\!\!\!\rightarrow}_{\rho_{\varepsilon v}} & \{\{\bot\},\; [x \to P_{nor}](exn(M))\} \\
\overset{*}{\longmapsto\!\!\!\!\rightarrow}_{\rho_{\varepsilon v}} & \{\{[x \to P_{nor}](M')\}\} \\
\overset{*}{\longmapsto\!\!\!\!\rightarrow}_{\rho_{\varepsilon v}} & \{P_{nor}\}
\end{array}
$$

In this last case also, we have the wanted result: P_{nor}.

3.2 Examples

In the Example 2, we have seen that the $\rho_{\varepsilon\text{v}}$-calculus allows to switch the evaluation mechanism according to a possible failure in the evaluation of a term. In the following example we are going to see that we can switch the evaluation according to a failure catching as precise as desired.

Example 9. Let us consider the evaluation scheme:

if the two arguments of f are evaluated to the failure
then evaluate $P_{fail1\&2}$
else if the first argument of f is evaluated to the failure
  then evaluate P_{fail1}
  else if the second argument of f is evaluated to the failure
    then evaluate P_{fail2}
    else evaluate P_{nor}

It can be expressed in the $\rho_{\varepsilon\text{v}}$-calculus by the following term (one should replace `first` by its given definition):

$$
\left[\ \texttt{first}\ \left(\begin{array}{ll}
f(exn(\bot), exn(\bot)) & \to P_{fail1\&2}, \\
f(exn(\bot), x) & \to P_{fail1}, \\
f(x, exn(\bot)) & \to P_{fail2}, \\
f(x, y) & \to P_{nor}
\end{array}\right)\right] \ \big(f(exn(M), exn(N))\big)
$$

So, we can switch the evaluation scheme according to the possible evaluation failure of two arguments of a function. It is clear that a similar term can be constructed for every function.

The next example shows how to express in the $\rho_{\varepsilon\text{v}}$-calculus the exception mechanism of ML [Mil78,WL96], when no named exception is considered.

Example 10. The following ML program:

```
try
 x/y;
 P1
with Div_By_Zero -> P2
```

is expressed in the $\rho_{\varepsilon\text{v}}$-calculus, as:

$$[\texttt{first}(exn(\bot) \to P2, x \to P1)](exn(x/_\rho y)))$$

where $/_\rho$ is an encoding of x/y in the rewriting calculus. Similarly, since we can ignore a raised exception (see Section 2.1), we can express the ML program

```
try
 P1
with _ -> P2
```

by the ρ-term $[x \to P2](exn(P1))$.

Example 11. One of the `first`'s strong interest is to allow for a full and explicit encoding of normalization strategies. For instance, thanks to the `first`, an innermost normalization operator can be built [Cir00,CK01] as follows:

- A term traversal operator for all operators f_i in the signature:

$$\begin{aligned}\Phi(r) \triangleq first(\ & f_1(r, id, \ldots, id), \ldots, f_1(id, \ldots, id, r),\\ & \ldots,\\ & f_m(r, id, \ldots, id), \ldots, f_m(id, \ldots, id, r))\end{aligned}$$

  Using Φ we get for example:

$$[\Phi(a \to c)](f(a,b)) \underset{\rho_\varepsilon}{\mapsto\!\!\!\to} \{f(c,b)\} \text{ and } [\Phi(a \to b)](c) \underset{\rho_\varepsilon}{\mapsto\!\!\!\to} \{\bot\}$$

- The *BottomUp* operator

$$Once_{bu}(r) \triangleq [\Theta](H_{bu}(r)) \ with \ H_{bu}(r) \triangleq f \to (x \to [first(\Phi(f), r)](x))$$

  where Θ is Turing's fixed-point combinator expressed in the ρ-calculus:

$$\Theta = [A](A) \ with \ A = x \to (y \to [y]([[x](x)](y)))$$

  For instance we have: $[Once_{bu}(a \to b)](f(a, g(a)) \mapsto\!\!\!\to_{\rho_\varepsilon} \{f(b, g(a))\})$
- A repetition operator

$$repeat*(r) \triangleq [\Theta](J(r)) \ with \ J(r) \triangleq f \to (x \to [first(r; f,\ id)](x))$$

  This can be used like in $[repeat * (\{a \to b, b \to c\})](a) \mapsto\!\!\!\to_{\rho_\varepsilon} \{c\}$.
- Finally we get the innermost normalization operator:

$$im(r) \triangleq repeat * (Once_{bu}(r))$$

  It allows us to reduce $f(a, g(a))$ to its innermost normal form according to the rewriting system $\mathcal{R} = \{a \to b, f(x, g(x)) \to x\}$. For instance, $[im(\mathcal{R})](f(a, g(a))) \mapsto\!\!\!\to_{\rho_{\varepsilon v}} \{b\}$. Note that we do not need to assume confluence of $\mathcal{R}$: considering the rewrite system $\mathcal{R}' = \{a \to b, a \to c, f(x,x) \to x\}$, we have $[im(\mathcal{R}')](f(a,a)) \mapsto\!\!\!\to_{\rho_{\varepsilon v}} \{b, f(c,b), f(b,c), c\}$.

Expressing normalization strategy becomes explicit and reasonably easy since in ρ-calculus, terms rules, rule applications and therefore strategies are considered at the object level.

So, it is quite natural to ask if we can express how to apply these rules. Since the `first` is expressible in the $\rho_{\varepsilon v}$-calculus, we can express all the operators above and we obtain as a corollary:

Corollary 1. *The operators im (and om) describing the innermost (resp. outermost) normalization can be expressed in the $\rho_{\varepsilon v}$-calculus. Moreover, given a rewriting theory and two first-order ground terms (in the sense of rewriting) such that t is normalized to $t\downarrow$ w.r.t. the set of rewrite rules $\mathcal{R}$, the term $[im(\mathcal{R})](t)$ is $\rho_{\varepsilon v}$-reduced to a set containing the term $t\downarrow$.*

Note that in this situation, a rewrite system is innermost confluent if and only if $[im(\mathcal{R})](t)$ is a singleton. Notice also that, following [CK01], normalized rewriting [DO90] using a conditional rewrite rule of the form $l \to r$ if c can be simply expressed as the ρ-term $l \to [\mathrm{True} \to r]([im(\mathcal{R})](c))$ and therefore that conditional rewriting is simply expressible in the $\rho_{\varepsilon v}$-calculus.

4 Conclusion and Further Work

By the introduction of an exception mechanism in the rewriting calculus, we have solved the *simple* expression problem of the `first` operator in a confluent calculus. Consequently, this solves the open problem of the doped calculus confluence.

Motivated by this expressiveness question, we have indeed elaborated a powerful variation of the rewriting calculus which is useful, first for theorem proving when providing general proof search strategies in a semantically well founded language, and secondly, as a useful programming paradigm for rule based languages. It has in particular the main advantage to bring the exception paradigm uniformly at the level of rewriting.

This could be extended in particular by allowing for named exceptions. Another challenging question is now to express, maybe in the $\rho_{\varepsilon v}$-calculus, strategy operators like the "don't care" one used for example in the ELAN language.

Acknowledgement. We thank Daniel Hirschkoff for his strong interest and support in this work, Horatiu Cirstea, Thérèse Hardin and Luigi Liquori for many stimulating discussions on the rewriting calculus and its applications and for their constructive remarks on the ideas developed here. We also thank the referees for their comments and suggestions.

References

[BGG+92] R. Boulton, A. Gordon, M. Gordon, J. Harrison, J. Herbert, and J. V. Tassel. Experience with embedding hardware description languages in HOL. In V. Stavridou, T. F. Melham, and R. T. Boute, editors, *Proceedings of the IFIP TC10/WG 10.2 International Conference on Theorem Provers in Circuit Design: Theory, Practice and Experience*, volume A-10 of *IFIP Transactions*, pages 129–156, Nijmegen, The Netherlands, June 1992. North-Holland/Elsevier.

[BKK+98] P. Borovanský, C. Kirchner, H. Kirchner, P.-E. Moreau, and C. Ringeissen. An overview of ELAN. In C. Kirchner and H. Kirchner, editors, *Proceedings of the second International Workshop on Rewriting Logic and Applications*, volume 15, `http://www.elsevier.nl/locate/entcs/volume15.html`, Pont-à-Mousson (France), September 1998. Electronic Notes in Theoretical Computer Science. Report LORIA 98-R-316.

[Cir00] H. Cirstea. *Calcul de réécriture : fondements et applications.* Thèse de Doctorat d'Université, Université Henri Poincaré – Nancy 1, France, October 2000.

[CK01] H. Cirstea and C. Kirchner. The rewriting calculus — Part I *and* II. *Logic Journal of the Interest Group in Pure and Applied Logics*, 9(3):427–498, May 2001.

[CKL01a] H. Cirstea, C. Kirchner, and L. Liquori. The Matching Power. In V. van Oostrom, editor, *Rewriting Techniques and Applications*, volume 2051 of *Lecture Notes in Computer Science*, Utrecht, The Netherlands, May 2001. Springer-Verlag.

[CKL01b] H. Cirstea, C. Kirchner, and L. Liquori. The Rho Cube. In F. Honsell, editor, *Foundations of Software Science and Computation Structures*, volume 2030 of *Lecture Notes in Computer Science*, pages 166–180, Genova, Italy, April 2001.

[DK00] H. Dubois and H. Kirchner. Objects, rules and strategies in ELAN. In *Proceedings of the second AMAST workshop on Algebraic Methods in Language Processing, Iowa City, Iowa, USA*, May 2000.

[DO90] N. Dershowitz and M. Okada. A rationale for conditional equational programming. *Theoretical Computer Science*, 75:111–138, 1990.

[Fau01] G. Faure. Etude des propriétés du calcul de réécriture: du $\rho_\emptyset$-calcul au ρ_ε-calcul. Rapport, LORIA and ENS-Lyon, September 2001. `http://www.loria.fr/~ckirchne/=rho/rho.html`.

[Mil78] R. Milner. A theory of type polymorphism in programming. *JCSS*, 17:348–375, 1978.

[Ngu01] Q.-H. Nguyen. Certifying Term Rewriting Proof in ELAN. In M. van den Brand and R. Verma, editors, *Proc. of RULE'01*, volume 59. Elsevier Science Publishers B. V. (North-Holland), September 2001. Available at `http://www.elsevier.com/locate/entcs/volume59.html`.

[WL96] Weiss and Leroy. *Le langage Caml.* InterEditions, 1996.

Deriving Focused Lattice Calculi

Georg Struth

Institut für Informatik, Universität Augsburg
Universitätsstr. 14, D-86135 Augsburg, Germany
Tel:+49-821-598-3109, Fax:+49-821-598-2274,
struth@informatik.uni-augsburg.de

Abstract. We derive rewrite-based ordered resolution calculi for semilattices, distributive lattices and boolean lattices. Using ordered resolution as a metaprocedure, theory axioms are first transformed into independent bases. Focused inference rules are then extracted from inference patterns in refutations. The derivation is guided by mathematical and procedural background knowledge, in particular by ordered chaining calculi for quasiorderings (forgetting the lattice structure), by ordered resolution (forgetting the clause structure) and by Knuth-Bendix completion for non-symmetric transitive relations (forgetting both structures). Conversely, all three calculi are derived and proven complete in a transparent and generic way as special cases of the lattice calculi.

1 Introduction

We propose focused ordered resolution calculi for semilattices, distributive lattices and boolean lattices as theories of order. These calculi are relevant to many interesting applications, including set-theoretic and fixed-point reasoning, program analysis and construction, substructural logics and type systems. Focusing means integrating mathematical and procedural knowledge. Here, it is achieved via domain-specific inference rules, rewriting techniques and syntactic orderings on terms, atoms and clauses. The inference rules are specific superposition rules for lattice theory. They are constrained to manipulations with maximal terms in maximal atoms. Since lattices are quite complex for automated reasoning, focusing may not only drastically improve the proof-search in comparison with an axiomatic approach, it seems even indispensable. But it is also difficult. Only very few focused ordered resolution calculi are known so far. One reason is that existing methods require guessing the inference rules and justifying them a posteriori in rather involved completeness proofs (c.f. for instance [17,15]). We therefore use an alternative method to derive the inference rules and prove refutational completeness by faithfulness of the derivation [12]. This derivation method uses ordered resolution as a metaprocedure. It has two steps. First, a specification is closed under ordered resolution with on the fly redundancy elimination. The resulting resolution basis has an independence property similar to set of support: No inference among its members is needed in a refutation. Second, the patterns arising in refutational inferences between non-theory clauses and the resolution

S. Tison (Ed.): RTA 2002, LNCS 2378, pp. 83–97, 2002.

basis are transformed into inference rules that entirely replace the resolution basis. Up to refutational completeness of the metaprocedure, the derivation method is purely constructive. It allows a fine-grained analysis of proofs and supports variations and approximations.

The concision and efficiency of focused calculi depends on the quality of the mathematical and procedural background knowledge that is integrated. Here, it is achieved mainly as follows. First, via specific axioms: for instance, via a lattice-theoretic variant of a cut rule as a surprising operational characterization of distributivity [9]. Second, extending related procedures: ordered chaining for quasiorderings [3,12], ordered resolution calculi as solutions to lattice-theoretic uniform word problems [11] and Knuth-Bendix completion for quasiorderings [13]. On the one hand, ordered resolution is (via the cut rule) a critical pair computation for quasiorderings extended to distributive lattices. On the other hand, ordered chaining rules are critical pair computations extended to clauses. Extending both to lattices and clauses motivates an ordered resolution calculus with ordered chaining rules for lattices that include ordered resolution at the lattice level. Third, via syntactic orderings: these constrain inferences, for instance to the respective critical pair computations. We extend and combine those of the background procedures. We use in particular multisets as the natural data structure for lattice-theoretic resolution to build term orderings.

Briefly, our main results are the following. We propose refutationally complete ordered chaining calculi for finite semilattices, distributive lattices and boolean lattices. These calculi yield decision procedures. We also argue how to lift the calculi to semi-decision procedures for infinite structures. The lattice calculi comprise an ordered chaining calculus for quasiorderings, a Knuth-Bendix completion procedure for quasiorderings and ordered resolution calculi as special cases. Their formal derivation and proof of refutational completeness is therefore uniform and generic. As a peculiarity, we derive propositional ordered resolution (with redundancy elimination) formally as a rewrite-based solution to the uniform word problem for distributive lattices. This yields a constructive refutational completeness proof that uses ordered resolution reflexively as a metaprocedure. Besides this, our results further demonstrate the power and applicability of the derivation method with a non-trivial example.

Our work uses some long and technical arguments. We exclude them into an extended version [14]. Here, we only give the main definitions and illustrate the main ideas of the derivation. We also restrict our attention to finite structures.

The remainder is organized as follows. Section 2 recalls some ordered resolution basics. Section 3 introduces some lattice theory and motivates our choice of the resolution basis. Section 4 introduces the syntactic orderings for our lattice calculi, section 5 the calculi themselves. Section 6 discusses the computation of the resolution basis. Section 7 sketches the derivation of the inference rules. Section 8 shows an example and discusses several extension and specializations of the calculi. Section 9 contains a conclusion.

2 Ordered Resolution and Redundancy

We first recall some well-known results about ordered resolution and redundancy elimination. Consider [2] for further reference.

Let $T_\Sigma(X)$ be a set of terms with signature Σ and variables in X, let P be a set of predicates. The set A of *atoms* consists of all expressions $p(t_1, \ldots, t_n)$, where p is an n-ary predicate and $t_1, \ldots, t_n$ are terms. A *clause* is an expression $\{\!|\phi_1, \ldots, \phi_m|\!\} \longrightarrow \{\!|\psi_1, \ldots, \psi_n|\!\}$. Its *antecedent* $\{\!|\phi_1, \ldots, \phi_m|\!\}$ and *succedent* $\{\!|\psi_1, \ldots, \psi_n|\!\}$ are finite multisets of atoms. Antecedents are schematically denoted by Γ, succedents by Δ. Brackets will usually be omitted. The above clause represents the closed universal formula $(\forall x_1 \ldots x_k)(\neg\phi_1 \vee \cdots \vee \neg\phi_m \vee \psi_1 \vee \cdots \vee \psi_n)$. A *Horn clause* contains at most one atom in its succedent. We deliberately write Δ for $\longrightarrow \Delta$ and $\neg\Gamma$ for $\Gamma \longrightarrow$ and Γ, ϕ for $\Gamma \cup \phi$.

We consider calculi with inference rules constrained by syntactic orderings. A *term ordering* is a well-founded total ordering on ground terms. An *atom ordering* is a well-founded total ordering on ground atoms. For non-ground expressions e_1 and e_2 and a term or atom ordering $\prec$ we define $e_1 \prec e_2$ iff $e_1\sigma \prec e_2\sigma$ for all ground substitutions σ. Consequently, $e_1 \not\succeq e_2$ if $e_2\sigma \succ e_1\sigma$ for some ground expressions $e_1\sigma$ and $e_2\sigma$. An atom ϕ is *maximal* with respect to a multiset Γ of atoms, if $\phi \not\prec \psi$ for all $\psi \in \Gamma$. It is *strictly maximal* with respect to Γ, if $\phi \not\preceq \psi$ for all $\psi \in \Gamma$. The non-ground orderings are still well-founded, but need no longer be total.

While atom orderings suffice to constrain the inferences of the ordered resolution calculus, clause orderings are needed for redundancy elimination. To extend atom orderings to clauses, we measure clauses as multisets of their atoms and use the multiset extension of the atom ordering. To disambiguate occurences of atoms in antecedents and succedents, we assign to those in the antecedent a greater weight than those in the succedent. See section 4 for more details. The clause ordering then inherits totality and well-foundedness from the atom ordering. Again, the non-ground extension need not be total. In unambiguous situations we denote all orderings by $\prec$.

Definition 1 (Ordered Resolution Calculus). *Let $\prec$ be an atom ordering. The* ordered resolution calculus OR *consists of the following deduction inference rules. The* ordered resolution rule

$$\frac{\Gamma \longrightarrow \Delta, \phi \qquad \Gamma', \psi \longrightarrow \Delta'}{\Gamma\sigma, \Gamma'\sigma \longrightarrow \Delta\sigma, \Delta'\sigma}, \tag{Res}$$

where σ is a most general unifier of ϕ and ψ, $\phi\sigma$ is strictly maximal with respect to $\Gamma\sigma, \Delta\sigma$ and maximal with respect to $\Gamma'\sigma, \Delta'\sigma$. The ordered factoring rule

$$\frac{\Gamma \longrightarrow \Delta, \phi, \psi}{\Gamma\sigma \longrightarrow \Delta\sigma, \phi\sigma}, \tag{Fact}$$

where σ is a most general unifier of ϕ and ψ and $\phi\sigma$ is strictly maximal with respect to $\Gamma\sigma$ and maximal with respect to $\Delta\sigma$.

In all inference rules, *side formulas* are the parts of clauses denoted by capital Greek letters. Atoms occurring explicitly in the premises are called *minor formulas*, those in the conclusion *principal formulas.*

Let S be a clause set and $\prec$ a clause ordering. A clause C is $\prec$*-redundant* or simply redundant in S, if C is a semantic consequence of instances from S which are all smaller than C with respect to $\prec$. Closing S under OR-inferences and eliminating redundant clauses on the fly transforms S into a *resolution basis* $rb(S)$. The transformation need not terminate, but we have *refutational completeness*: All fair OR-strategies derive the empty clause within finitely many steps, if S is inconsistent. We call a proof to the empty clause an OR-*refutation.*

Proposition 1. *S is inconsistent iff $rb(S)$ contains the empty clause.*

A resolution basis B is a special basis. By definition it satisfies the independence property that all conclusions of *primary B-inferences*, that is OR-inferences with both premises from B, are redundant. However, it need not be unique.

Proposition 2. *Let B be a consistent resolution basis and T a clause set. If $B \cup T$ is inconsistent, then some* OR*-refutation contains no primary B-inferences.*

In [2], variants of proposition 1 and proposition 2 have been shown for a stronger notion of redundancy. For deriving the lattice calculi, the weak notion suffices. But the strong notion can of course always be used. By proposition 2, resolution bases allow ordered resolution strategies similar to set of support. The construction of a resolution basis will constitute the first step of our derivation of focused lattice calculi. OR will be used as a metaprocedure in the derivation. Its properties, like refutational completeness and avoidance of primary theory inferences are essential ingredients.

3 Lattices

Here we are not concerned with arbitrary signatures and predicates. Let $\Sigma = \{\sqcup, \sqcap\}$ and $P = \{\leq\}$. $\sqcup$ and $\sqcap$ are operation symbols for the lattice join and meet operations. $\leq$ is a binary predicate symbol denoting a *quasiordering.* A *quasiordered set* (or *quoset*) is a set A endowed with a reflexive transitive relation $\leq$, which satisfies the clausal axioms

$$x \leq x, \qquad \text{(ref)} \qquad\qquad x \leq y, y \leq z \longrightarrow x \leq z. \qquad \text{(trans)}$$

A *join semilattice* is a quoset A closed under least upper bounds or joins for all pairs of elements. Formally, $x \leq z$ and $y \leq z$ iff $x \sqcup y \leq z$ for all $x, y, z \in A$. A *meet semilattice* is defined dually as a quoset closed under greatest lower bound or meets for all pairs of elements. The *dual* of a statement about lattices is obtained by interchanging joins and meets and taking the converse of the ordering. A *lattice* is both a join and a meet semilattice. It is *distributive*, if $x \sqcap (y \sqcup z) \leq (x \sqcap y) \sqcup (x \sqcap z)$ holds for all $x, y, z \in A$ or its dual and therewith both. The inequality $x \sqcap (y \sqcup z) \geq (x \sqcap y) \sqcup (x \sqcap z)$ and its dual hold in every lattice. In a quoset, joins and meets are unique up to the congruence $\sim = (\leq \cap \geq)$. Semantically, $\leq/\sim$ is a partial ordering, hence an antisymmetric quasiordering ($x \leq y \wedge y \leq x \Longrightarrow x = y$). Operationally, antisymmetry just splits equalities into inequalities. We can therefore disregard it. Joins and meets are associative, commutative, idempotent ($x \sqcup x = x = x \sqcap x$) and monotonic operations in the associated partial ordering. We will henceforth consider all inequalities modulo AC (associativity and commutativity) and normalize with respect to I (idempotence). See [4] for further information on lattices.

In [10] we derive the resolution bases directly from natural specifications. Here we choose a technically simpler way that uses more domain-specific knowledge. The lattice axioms allow us to transform every inequality between lattice terms into a set of simpler expressions.

Lemma 1. *Let S be a meet semilattice, L a distributive lattice and G a set of generators (that is a set of constants disjoint from Σ).*

(i) An inequality $s \leq t_1 \sqcap \ldots \sqcap t_n$ holds in S, iff $s \leq t_i$ holds for all $1 \leq i \leq n$.
(ii) For all $s, t \in T_{\Sigma \cup G}$, an inequality $s < t$ holds in L, iff there is a set of inequalities of the form $s_1 \sqcap \ldots \sqcap s_m \leq t_1 \sqcup \ldots \sqcup t_n$ that hold in L and the s_i and t_i are generators occuring in s and t.

A *reduced join (meet) semilattice inequality* is a join (meet) semilattice inequality whose left-hand (right-hand) side is a generator. A *reduced lattice inequality* is a lattice inequality whose left-hand side is a meet and whose right-hand side is a join of generators. Reduced semilattice and lattice inequalities are lattice-theoretic analogs of Horn clauses and clauses. In particular, the clausal arrow is a quasiordering. We usually write more compactly $s_1 \ldots s_m \leq t_1 \ldots t_n$ instead of $s_1 \sqcap \ldots \sqcap s_m \leq t_1 \sqcup \ldots \sqcup t_n$. By a lattice-theoretic variant of the Tseitin transformation [16], there is a linear reduction of distributive lattice terms. We speak of *reduced clausal theories*, when all inequalities in clauses are reduced.

For the remainder of this text we assume that all lattice and semilattice inequalities are reduced. We will also restrict our inference rules to reduced terms. But then, the axioms of the join and meet semilattice and also the distributive axiom can no longer be used, since they do not yield reduced clauses. However, one cannot completely dispense with their effect in the initial specification.

Here, background knowledge comes into play to find the appropriate initial specification for reduced clausal theories. Following [9] we use the cut rule

$$x_1 \leq y_1 z, x_2 z \leq y_2 \longrightarrow x_1 x_2 \leq y_1 y_2. \qquad \text{(cut)}$$

as an alternative characterization of distributivity. Written as an inference rule, this Horn clause is a lattice-theoretic variant of resolution. It combines the effect of transitivity, distributivity and monotonicity of join and meet. See [11] for a discussion and a derivation. In the context of Knuth-Bendix completion for quasiorderings, it can be restricted by ordering constraints and used as a critical pair computation to solve the uniform word problem for distributive lattices [11].

There are similar cut rules (not involving distributivity)

$$x_1 \leq y_1 z, z \leq y_2 \longrightarrow x_1 \leq y_1 y_2, \qquad \text{(jcut)}$$

$$x_1 \leq z, x_2 z \leq y_2 \longrightarrow x_1 x_2 \leq y_2, \qquad \text{(mcut)}$$

for the join and meet semilattice. These are used for solving the respective semilattice word problems. (cut), (jcut) and (mcut) work on reduced inequalities. The expression z which is cut out is necessarily a generator. Some lattice theory shows that these cut rules together with the Horn clauses

$$x \leq y \longrightarrow x \leq yz, \quad \text{(jr)} \qquad\qquad x \leq z \longrightarrow xy \leq z, \quad \text{(ml)}$$

constitute our desired initial specifications.

Lemma 2. *The following sets axiomatize their reduced clausal theories up to normalization with I and modulo AC.*

(i) $J = \{(\mathit{ref}), (\mathit{trans}), (\mathit{jr}), (\mathit{jcut})\}$ *for join semilattices,*
(ii) $M = \{(\mathit{ref}), (\mathit{trans}), (\mathit{ml}), (\mathit{mcut})\}$ *for meet semilattices,*
(iii) $D = \{(\mathit{ref}), (\mathit{trans}), (\mathit{jr}), (\mathit{ml}), (\mathit{jcut}), (\mathit{mcut}), (\mathit{cut})\}$ *for distributive lattices.*

See [14] for proofs. In implementations, normalization with respect to I and also with respect of 0 and 1, when these constants are present, must be explicit. We use the canonical equational system modulo AC

$$T = \{x \wedge x \to x, x \vee x \to x, x \vee 1 \to 1, x \wedge 1 \to x, x \vee 0 \to x, x \wedge 0 \to 0\}.$$

The orientation of the rewrite rules is compatible with the term orderings from section 2. T induces reduced inequalities modulo ACI01. AC and I01 are however handled differently: AC by a compatible ordering (c.f. section 4), not by rewrite rules (the laws are not orientable). I01 by the normalizing rewrite system T, not by a compatible ordering (which does not exist). In the non-ground case, our calculi use ACI(01)-unification or -unifiability (when working with constraints). That can be very prolific [1]. In the ground case, much cheaper operations, like string matching, suffice for comparing terms in the inference rules.

We denote the normal form of an expression e (a term, an atom, a clause) with respect to T by $(e)\downarrow_T$. In the ground case we may sort joins and meets of generators after each normalization. All consequences of a reduced clause set and J, M and D are again reduced. However, consequences of a reduced clause set in T-normal form may not be in T-normal form.

At the end of this section we again consider background knowledge to motivate the lattice calculi. As already mentioned, (cut) is used to solve the uniform word problem for distributive lattices. In this context, all non-theory clauses are atoms. We however want to admit arbitrary non-theory clauses. This is analogous to ordered chaining calculi for quasiordering, where a Knuth-Bendix completion procedure for quasiorderings, operating entirely on positive atoms, is extended to clauses via a (ground) positive chaining rule.

$$\frac{\Gamma \longrightarrow \Delta, r < s \qquad \Gamma' \longrightarrow \Delta', s < t}{\Gamma, \Gamma' \longrightarrow \Delta, \Delta', r < t} \qquad \text{(Ch+)}$$

Thereby s is the maximal term in the minor formulas and the minor formulas are strictly maximal with respect to Γ, Γ', Δ and Δ'. The rule is a clausal extension of a critical pair computation. A chaining rule for distributive lattices can then analogously be motivated as an extension of (cut) to the lattice level.

$$\frac{\Gamma \longrightarrow \Delta, s_1 < t_1 x \qquad \Gamma' \longrightarrow \Delta', s_2 x < t_2}{\Gamma, \Gamma' \longrightarrow \Delta, \Delta', s_1 s_2 < t_1 t_2}$$

But which other inference rules are needed and what are the ordering constraints? These questions require deeper considerations. They will be answered with the help of the derivation method.

Our third way of including background information is the construction of the specific syntactic orderings. In the following section, we will base them again on those of ordered resolution and ordered chaining. In section 6 we will show that J, M and D are resolution bases for these orderings and therefore independent sets. They will then be internalized into focused inference rules in section 7.

4 The Syntactic Orderings

The computation of a resolution basis, its termination and the procedural behaviour of our calculi crucially depend on the syntactic term, atom and clause orderings. In section 3 we have developed our initial axiomatizations with regard to the ordered chaining calculi for quasiorderings and ordered resolution in lattice theory. Here we build syntactic orderings for lattices that refine those of the background procedures and turn (jcut), and (cut) into lattice-theoretic variants of ordered Horn resolution and resolution.

Like in section 2, we first define a term ordering and then extend it to atoms and clauses. For the term ordering, let $\prec$ be the multiset extension of some total ordering on the set of generators. We assign minimal weight to 0 and 1, if present. $\prec$ is trivially well-founded, if the set is finite or denumerably infinite. We measure both joins and meets of generators by their multisets. This clearly suffices for the terms occurring in reduced lattice inequalities. By construction, $\prec$ is well-founded and compatible with AC: terms which are equal modulo AC are assigned the same measure. $\prec$ is natural for resolution, since multisets are a natural data structure for clauses. This choice of $\prec$ will force the calculus under construction to become resolution-like[1]. $\prec$ also appears, when resolution is modeled as a critical pair computation for distributive lattices.

We now consider the atom ordering. Let $\mathbb{B}$ be the two-element boolean algebra with ordering $<_{\mathbb{B}}$. Let $m = G \times \mathbb{B} \times \mathbb{B} \times G$, where G denotes a multiset of generators. Let A be a set of atoms occurring in some clause $C = \Gamma \longrightarrow \Delta$. The ordering $\prec_1 \subseteq m \times m$ is the lexicographic combination of $\prec$ for the first and last component of m and $<_{\mathbb{B}}$ for the others. A ground *atom measure* (for clause C) is the mapping $\mu_C : A \to m$ defined by $\mu_C : \phi \mapsto (t_\nu(\phi), p(\phi), s(\phi), t_\mu(\phi))$ for each

[1] In [10], we alternatively force tableau calculi for distributive lattices with a term ordering emphasizing the subterm ordering.

(ground) atom $\phi \in A$ occurring in C. Hereby $t_\nu(\phi)$ ($t_\mu(\phi)$) denotes the maximal (minimal) term with respect to $\prec$ in ϕ. $p(\phi) = 1$ ($p(\phi) = 0$), if ϕ occurs in Γ (in Δ). $s(\phi) = 1$ ($s(\phi) = 0$), if $\phi = s < t$ and $s \succeq t$ ($s \prec t$). The (ground) *atom ordering* $\prec_2 \subseteq A \times A$ is defined by $\phi \prec_2 \psi$ iff $\mu_C(\phi) \prec_1 \mu_C(\psi)$ for $\phi, \psi \in A$. Hence $\prec_2$ is embedded in $\prec_1$ via the atom measure. The ordering $\prec_1$ is total and well-founded by construction. Via the embedding, $\prec_2$ inherits these properties. The definition of the atom measure follows those of the ordered chaining calculi for quasiorderings and of Knuth-Bendix completion for quasiorderings. This forces the calculus to become a chaining calculus at the clausal level and a completion procedure at the lattice level. The polarity p is not needed for comparing atoms, but it is crucial for the extension to clauses, to integrate redundancy elimination.

All these orderings are extended to the non-ground level and the clause level according to section 2. In particular the polarity p assigns greater weight to an occurrence of a term in the antecedent than to an occurrence in the succedent. In unambiguous situations we denote all orderings by $\prec$.

5 The Lattice Chaining Calculi

We now restrict our attention to finite lattices. Then all non-theory clauses are ground, since existential quantification can be replaced by a finite disjunction and universal quantification by a finite conjunction. The non-ground clauses in J, M and D are completely internalized into the focused inference rules. Therefore, our entire calculi are then ground. The extension to the non-ground case is discussed in section 8.

We introduce indexed brackets to abbreviate the presentation of the calculi. A pair of brackets [.] in a clause denotes alternatively the clause without the brackets and the clause, where the brackets together with their content have been deleted. The clause $\Gamma \longrightarrow \Delta, [r]s \leq t$, for instance, denotes alternatively the clauses $\Gamma \longrightarrow \Delta, rs \leq t$ and $\Gamma \longrightarrow \Delta, s \leq t$. In inference rules, brackets with the same index are synchronized. The leftmost of the following inference rules, for instance, denotes alternatively the two others.

$$\frac{\Gamma \longrightarrow \Delta, [r]_i s \leq t}{\Gamma', [u]_i v \leq w \longrightarrow \Delta'} \qquad \frac{\Gamma \longrightarrow \Delta, rs \leq t}{\Gamma', uv \leq w \longrightarrow \Delta'} \qquad \frac{\Gamma \longrightarrow \Delta, s \leq t}{\Gamma', v \leq w \longrightarrow \Delta'}$$

Definition 2 (Distributive Lattice Chaining). *Let $\succ$ be the atom and clause ordering of section 4. Let all clauses be reduced. The* ordered chaining calculus for finite distributive lattices DC *consists of the deductive inference rules and the redundancy elimination rules of* OR[2] *and the following inference rules.*

$$\frac{\Gamma, t \leq t \longrightarrow \Delta}{\Gamma \longrightarrow \Delta} \qquad \text{(Ref)}$$

[2] Section 2 only defines a semantic *notion* of redundancy. Every set of inference rules implementing this notion is admitted.

$$\frac{\Gamma, r \wedge s \leq t \longrightarrow \Delta}{\Gamma, r \leq t \longrightarrow \Delta} \quad \text{(ML)} \qquad \frac{\Gamma, r \leq s \vee t \longrightarrow \Delta}{\Gamma, r \leq s \longrightarrow \Delta} \quad \text{(JR)}$$

Here the minor formula is maximal wrt. Γ and strictly maximal wrt. Δ.

$$\frac{\Gamma \longrightarrow \Delta, s_1 \leq [t_1]_j x \qquad \Gamma' \longrightarrow \Delta', [s_2]_m x \leq t_2}{(\Gamma, \Gamma' \longrightarrow \Delta, \Delta', s_1[s_2]_m \leq [t_1]_j t_2) \downarrow_T} \quad \text{(Cut+)}$$

Here the terms containing x are strictly maximal in the minor formulas. The minor formulas are strictly maximal wrt. the side formulas in the premises.

$$\frac{\Gamma \longrightarrow \Delta, s_1 \leq [t_1]_j x \qquad \Gamma', s_1[s_2]_m \leq [t_1]_j t_2 \longrightarrow \Delta'}{(\Gamma, \Gamma', [u]_m x \leq t_2 \longrightarrow \Delta, \Delta') \downarrow_T} \quad \text{(Cut-)}$$

Here the terms containing s_1 are strictly maximal in the minor formulas. In the first premise, the minor formula is strictly maximal wrt. the side formulas. In the second premise, the minor formula is maximal wrt. the side formulas. Moreover, $[u]_m x \neq t_2 \mod AC$. $u = s_2$ or else $u = s_1$, if s_2 is absent in the minor formula.

$$\frac{\Gamma \longrightarrow \Delta, s \leq [t_1]_m x, s \leq [t_1]_m t_2}{(\Gamma, [s]_j x \leq t_2 \longrightarrow \Delta, s \leq [t_1]_m t_2) \downarrow_T} \quad \text{(DF)}$$

Here, x is a generator, either t_1 is strictly maximal in the minor formulas or s is strictly maximal in the minor formulas and s can be set to 1 in the antecedent of the conclusion. The leftmost minor formula is strictly maximal wrt. the side formulas and the rightmost minor formula.

There are a dual (Cut-) and a dual (DF) rule, which are obtained by inverting the ordering, exchanging joins and meets and exchanging the brackets indexed by j with those indexed by m. The calculus is meant modulo AC at the lattice level.

(Ref) stands for *reflexivity*, (JR) and (ML) for *join right* and *meet left*, in analogy to the sequent calculus. (Cut+) and (Cut-) stand for *positive* and *negative cut*, (DF) for *distributivity factoring*.

Definition 3. *Under the conditions of definition 2, the calculus* DC *specializes to the following calculi.*

(i) *An* ordered chaining calculus for join semilattices JC, *removing the inference rule (ML) and discarding in the inference rules of* DC *the contents of all brackets indexed with m.*
(ii) *An* ordered chaining calculus for meet semilattices MC, *removing the inference rule (JR) and discarding in the inference rules of* DC *the contents of all brackets indexed with j.*
(iii) *An* ordered chaining calculus for quasiorderings QC, *removing the inference rules (JR) and (ML) and discarding in the inference rules of* DC *the contents of all brackets indexed with j and m.*
(iv) *An* ordered chaining calculus for transitive relations TC, *removing the inference rule (Ref) from* QC.

Definition 4 (Lattice Ordered Resolution [11]). *Under the conditions of definition 2, the deduction inference rules of the* ordered Horn resolution calculus HOR *and the* ordered resolution calculus OR *of [11][3] as rewrite-based solutions to the uniform word problem of semilattices and distributive lattices are restrictions of* DC *and* MC *to non-theory clauses consisting of positive atoms.*

In this case, the only applicable rule is (Cut+). It computes a lattice-theoretic variant of a Gröbner basis from the given presentation (the set of positive atoms). The query inequalities, which are in question as consequences of the presentation and the axioms, can then be shown to be redundant by a search method (c.f. [11]). An alternative solution to the word problem transforms the query into a set of negative atoms and then uses DC. We will see in section 8 that the resolution calculi are indeed a decision procedure.

Definition 5 (Knuth-Bendix Completion [13]). *Under the conditions of definition 2, the deduction inference rules of the* Knuth-Bendix completion procedures for quasiorderings *and* non-symmetric transitive relations *(without term structure), are restrictions either of* QC *and* TC *to non-theory clauses consisting of positive atoms or* OR, *all lattice terms being generators.*

Soundness and completeness of DC are the subject of section 6 and section 7, the other calculi are dealt with in section 8. The unordered variant of DC is an instance of theory resolution. We will derive the DC-rules as an ordered variant thereof. DC is more focused than mere reasoning with resolution bases.

6 Construction of the Resolution Bases

We now discuss the first step in the derivation of the chaining calculi: the construction of the resolution basis. A complete formal treatment can be found in [14]. We therefore close the sets J, M and D under ordered resolution and redundancy elimination. Use of lattice duality avoids repetitions. It turns out that J, M and D are already the resolution bases.

Proposition 3. *Let $\prec$ be the atom ordering of section 4 and* OR *the ordered resolution calculus of section 2. J, M and D are resolution bases for the reduced clausal theories of join semilattices, meet semilattices and distributive lattices. That is $M = rb(M)$, $J = rb(J)$ and $D = rb(D)$. We always implicitly normalize with respect to I.*

In the proof we first restrict our attention to M. The result for J then follows by duality, the result for D requires only a slight extension of the combination of J and M. We consider all possible ordered resolution inferences between members of M and show that the conclusions of these inferences are redundant, that is they are implied by smaller instances of members of M. We restrict ourselves to

[3] The ordering constraints of these calculi are weaker as those of ordered resolution given in this text in section 2.

a representative case, namely the following inference between two instances of (mcut), where y is maximal and $s_2 \succ x$.

$$\frac{s_2 \leq x, xy \leq t \longrightarrow_i \boxed{s_2 y \leq t} \qquad s_1 \leq y, \boxed{s_2 y \leq t} \longrightarrow_d s_1 s_2 \leq t}{s_1 \leq y, s_2 \leq x, xy \leq t \longrightarrow s_1 s_2 \leq t}$$

We add the indices i and d to denote that the Horn clauses increase or decrease with respect to $\prec$ from left to right. We also put the maximal atoms that are cut out into boxes. The conclusion is redundant. It semantically follows from the smaller instances

$$s_1 \leq y, xy \leq t \longrightarrow s_1 x \leq t, \qquad s_2 \leq x, s_1 x \leq t \longrightarrow s_1 s_2 \leq t$$

of (mcut). The analysis of the remaining cases is similar. It shows that J, M and D are indeed resolution bases. Proposition 2 and 3 immediately imply the following fact, which is essential for the arguments in the following section.

Corollary 1. *For every inconsistent reduced clause set containing J, M or D there exists a refutation without primary theory inferences in* OR.

By corollary 1, working with the resolution bases is already a strong improvement over plain axiomatic reasoning. We will see in the following section that the ordered chaining calculi yield even more restrictive proofs.

7 Deriving the Chaining Rules

We now turn to the second step in the derivation: the extraction of the inference rules of DC from OR-derivations with D. Formal proofs of all statements can be found in [14]. Here, we can only sketch them. Our main assumptions are refutational completeness of OC (theorem 1) and the fact that our ordering constraints rule out primary theory inferences (corollary 1). The main idea is to consider all possible interactions between non-theory clauses and the elements of D in a refutation. Disregarding the ordering constraints, the first (Cut-) inference rule of DC can be derived, for instance, as

$$\frac{\dfrac{\Gamma, s_1 s_2 \leq t_1 t_2 \longrightarrow \Delta \qquad \boxed{s_1 \leq t_1 x}, s_2 x \leq t_2 \longrightarrow \boxed{s_1 s_2 \leq t_1 t_2}}{\Gamma, s_1 \leq t_1 x, s_2 x \leq t_2 \longrightarrow \Delta} \qquad \Gamma' \longrightarrow \Delta', s_1 \leq t_1 x}{\Gamma, \Gamma', s_2 x \leq t_2 \longrightarrow \Delta, \Delta'}$$

from two non-theory clauses and an instance of (cut). Here, the atoms of (cut) that are cut out are in boxes. In (Cut-), the effect of (cut) is completely internalized. The derivation of (Cut+) is similar. There, both atoms in the antecedent of the instance of (cut) are cut out. (DF) arise from a resolution step followed by a factoring step. (JR) and (ML) arise from resolution steps with (jr) and (ml). The restrictions in DC modeled by brackets arise from derivation with (jcut), (mcut) or (trans). If it is possible to partition all refutations into these macro inferences, then the respective rules from D can be completely discarded in favor of the focused inference rules. Looking more precisely, this partition in patterns is however not straightforward. First, it must respect the ordering constraints of ordered resolution. Second, there are certain legal proof steps in refutations that may violate the pattern construction. We first briefly argue that these unwanted proof steps can be avoided. We then show how the partition of arbitrary OR-refutations into patterns corresponding to the DC-rules is obtained.

To understand the obstacles to the macro inferences, consider again the above derivation of (Cut-). First, $\Gamma' \longrightarrow \Delta', s_1 \leq t_1 x$ may be another instance of (cut). We call this situation a *secondary theory inference.* Second, Δ may contain an atom bigger than $s_1 \leq t_1 x$. Then the derivation of (Cut-) does not continue as above, but with a "bad" resolution step cutting out this bigger atom. We call this situation a *blocking inference.* In a blocking inference, only the first step uses the theory axiom. In DC, secondary theory inferences can occur in connection with (cut), (jcut), (mcut) and (trans). A OR-derivation is *regular*, when it contains neither primary and secondary theory inferences nor blocking inferences.

We first present a technical lemma that shows that we can always assume that all instances of (cut), (jcut), (mcut) and (trans) are decreasing from left to right in a refutation, that is, it can be indexed with d.

Lemma 3. *For every two-step proof in a* OR*-refutation that cuts out two atoms of (cut), (jcut), (mcut) or (trans) there is another* OR *proof from the same non-theory premises to the same conclusion, in which (another instance of) (cut) (jcut), (mcut) or (trans) decreases from left to right.*

The proof is based on a local proof transformation with the clauses in D, which can be completely hidden inside of a macro inference. We henceforth assume that all refutations are of this particular shape. This makes proofs much easier.

Lemma 4. *For every inconsistent clause set there exists a regular* OR*-refutation (possibly violating the ordering constraints).*

The proof is based on a proof transformation. In case of blocking inferences, the bad second inference must be permuted down towards the root of the proof tree, the needed second inference must be permuted up in a local way. In case of secondary theory inferences, we show that the two instances of (cut) can be replaced by one single instance and that the refutation can then be continued with only a few well-behaved modifications. It then remains to show that absence of blocking inferences and secondary theory inferences are compatible.

We are now prepared for our main theorem.

Theorem 1. *Let all clauses be reduced. The ground ordered chaining calculus* DC *is refutationally complete for finite distributive lattices: For every reduced ground clause set that is inconsistent in the first-order theory of finite distributive lattices there exists a refutation in* DC.

Again, we only discuss some representative cases. We may assume that the refutation is regular and that all instances of (cut), (jcut), (mcut) or (trans) are in *d*. We consider the (Res) and (Fact) steps of non-theory clauses with D. Terms, at which the chaining takes place, are sometimes put into boxes.

First, the resolution inference of a non-theory clause $\Gamma, s_1s_2 \leq t \longrightarrow_d \Delta$ with a ground instance of (ml) is

$$\frac{s_1 \leq t \longrightarrow_i s_1s_2 \leq t \qquad \Gamma, s_1s_2 \leq t \longrightarrow_d \Delta}{\Gamma, s_1 \leq t \longrightarrow \Delta},$$

where the minor formula is maximal with respect to the side formulas in the second premise. Internalization of (ml) yields (ML).

Second, consider the following resolution inference of the non-theory clause. Let t_1 be maximal in the following instance of (cut) and assume that there is a non-theory clause $\Gamma \longrightarrow \Delta, s_1 \leq t_1x$ in which the atom $s_1 \leq t_1x$ is strictly maximal. Then there is a possible resolution inference

$$\frac{\Gamma \longrightarrow \Delta, s_1 \leq \boxed{t_1}x \qquad s_1 \leq \boxed{t_1}x, s_2x \leq t_2 \longrightarrow s_1s_2 \leq t_1t_2}{\Gamma, s_2x \leq t_2 \longrightarrow \Delta, s_1s_2 \leq t_1t_2}.$$

In the conclusion t_1 may occur more than once, but only in Δ. Assuming a regular refutation, it can either be continued as

$$\frac{\Gamma, s_2x \leq t_2 \longrightarrow \Delta, s_1s_2 \leq \boxed{t_1}t_2 \qquad \Gamma', s_1s_2 \leq \boxed{t_1}t_2 \longrightarrow \Delta'}{\Gamma, \Gamma', s_2x \leq t_2 \longrightarrow \Delta, \Delta'},$$

which yields the first of the (Cut-) rules. Or Δ may be equal to $\Delta'', s_1s_2 \leq t_1t_2$. Then the refutation can be continued by ordered factoring as

$$\frac{\Gamma, s_2x \leq t_2 \longrightarrow \Delta'', s_1s_2 \leq \boxed{t_1}t_2, s_1s_2 \leq \boxed{t_1}t_2}{\Gamma, s_2x \leq t_2 \longrightarrow \Delta'', s_1s_2 \leq t_1t_2}$$

which yields the first of the (DF) rules. The remaining cases are similar.

Soundness of DC is trivial from the completeness proof, since all rules have been derived from OR with rules from D.

Corollary 2. *The semilattice and chaining calculi in definition 3 and the resolution calculi in definition 4 are refutationally complete.*

8 Discussion

Let us first consider an example that shows DC at work. See ([14]) for more examples. Many of them can be solved without (Cut+), (Cut-) or (DF); the effect of distributivity is sufficiently handled by the transformation to reduced lattice inequalities. Here we show that complements in distributive lattices are uniquely defined. We assume two complements b and c of an element a and show that they are identical. The corresponding reduced clauses are

$$ab \leq 0, \qquad ac \leq 0, \qquad 1 \leq ab, \qquad 1 \leq ac \qquad b \leq c, c \leq b \longrightarrow$$

Let $a \succ b \succ c \succ 1 \succ 0$. We then obtain $c \leq b$ from $1 \leq ab$ and $ac \leq 0$ as well as $b \leq c$ from $1 \leq ac$ and $ab \leq 0$ by (Cut+). The empty clause follows immediately from these two inequalities by resolution with $b \leq c, c \leq b \longrightarrow$.

We now show that we can go beyond refutational completeness.

Corollary 3. DC *decides the elementary theory of finite distributive lattices.*

This holds since finitely presented distributive lattices are finite and no DC-inference introduces a new variable or constant. Thus only finitely many inferences have irredundant conclusions. A resolution basis is finitely constructed. It contains the empty clause, iff the initial clause set was inconsistent. The result immediately specializes to the semilattice and chaining calculi in definition 3, the resolution calculi in definition 4 and the Knuth-Bendix procedure in definition 5.

We now discuss some extensions of DC.

Corollary 4. DC *is refutationally complete for the reduced clausal theory of finite boolean lattices.*

Boolean lattices are complemented distributive lattices with 1 and 0. Complements can be eliminating in reduced inequalities, just like negation in clauses. $a \sqcap b' \leq c$, for instance, is equivalent to $a \leq b \sqcup c$ and $a \leq b' \sqcup c$ to $a \sqcap b \leq c$, if b' denotes the complement of b. Thus DC suffices for the boolean case.

For infinite lattices, quantifiers can no longer be eliminated and first-order variables appear. Our calculi then become semi-decision procedures, since the corresponding first-order theories are undecidable. This extension requires a signature including free Skolem functions. Since these functions are non-monotonic, they are harmless. Then, our calculi can be easily lifted, only idempotence can no longer be used implicitly as a simplification (c.f [14]). Homomorphisms can also be treated by pushing them down through lattice terms. With monotonic functions, further complications with certain critical pairs appear that possibly cannot be handled within first-order logic. See [12,13] for a deeper discussion. Due to ACI-unification, the (Cut-) rules may become very prolific in the non-ground case. One improvement is unification constraints that use unifiability tests. With more structure, for instance in set theory, with lattice operations interpreted as union and intersection and the ordering as set inclusion, one may fortunately completely dispense with (Cut-) rules. $a \not\subseteq b$ holds iff b intersected with some singleton subset of a is empty (c.f. [7]). Thus every negative inequality may be transformed into positive ones and (Cut-) never applies.

9 Conclusion

We have derived focused ordered resolution calculi for lattices. Our method supports the integration of specific syntactic orderings as well as semantic and procedural knowledge in a smooth and fine-grained way. We consider that an important feature of the derivation method. We have also applied our method to the derivation of tableau calculi as lattice theoretic decision procedures [10], using a different syntactic ordering encoding the subformula property. In particular, this allows us to consider the cut rule from an algebraic point of view.

Our focused lattice calculi are a basis for many interesting applications. We plan in particular a consideration of (fragments of) set theory [7] and of certain extensions of semilattices and distributive lattices [8,5,6] for the construction and analysis of hard- and software systems. Proof support for these endeavors is very challenging.

References

1. F. Baader and W. Büttner. Unification in commutative idempotent monoids. *J. Theoretical Computer Science*, 56:345–352, 1988.
2. L. Bachmair and H. Ganzinger. Rewrite-based equational theorem proving with selection and simplification. *J. Logic and Comp utation*, 4(3):217–247, 1994.
3. L. Bachmair and H. Ganzinger. Rewrite techniques for transitive relations. In *Ninth Annual IEEE Symposium on Logic in Computer Science*, pages 384–393. IEEE Computer Society Press, 1994.
4. G. Birkhoff. *Lattice Theory*, volume 25 of *Colloquium Publications*. American Mathematical Society, 1984. Reprint.
5. J. H. Conway. *Regular Algebras and Finite State Machines*. Chapman and Hall, 1971.
6. P. J. Freyd and A. Scedrov. *Categories, Allegories*. North-Holland, 1990.
7. L. Hines. Str+ve⊆: The Str+ve-based Subset Prover. In M. E. Stickel, editor, *10th International Conference on Automated Deduction*, volume 449 of *LNAI*, pages 193–206. Springer-Verlag, 1990.
8. D. Kozen. Kleene algebra with tests. *Transactions on Programming Languages and Systems*, 19(3):427–443, 1997.
9. P. Lorenzen. Algebraische und logistische Untersuchungen über freie Verbände. *The Journal of Symbolic Logic*, 16(2):81–106, 1951.
10. G. Struth. *Canonical Transformations in Algebra, Universal Algebra and Logic*. PhD thesis, Institut für Informatik, Universität des Saarlandes, 1998.
11. G. Struth. An algebra of resolution. In L. Bachmair, editor, *Rewriting Techniques and Applications, 11th International Conference*, volume 1833 of *LNCS*, pages 214–228. Springer-Verlag, 2000.
12. G. Struth. Deriving focused calculi for transitive relations. In A. Middeldorp, editor, *Rewriting Techniques and Applications, 12th International Conference*, volume 2051 of *LNCS*, pages 291–305. Springer-Verlag, 2001.
13. G. Struth. Knuth-Bendix completion for non-symmetric transitive relations. In M. van den Brand and R. Verma, editors, *Second International Workshop on Rule-Based Programming (RULE2001)*, volume 59 of *Electronic Notes in Theoretical Computer Science*. Elsevier Science Publishers, 2001.
14. G. Struth. Deriving focused lattice calculi. Technical Report 2002-7, Institut für Informatik, Universität Augsburg, 2002.
15. J. Stuber. *Superposition Theorem Proving for Commutative Algebraic Theories*. PhD thesis, Institut für Informatik, Universität des Saarlandes, 1999.
16. G. S. Tseitin. On the complexity of derivations in propositional calculus. In J. Siekmann and G. Wrightson, editors, *Automation of Reasoning: Classical Papers on Computational Logic*, pages 466–483. Springer-Verlag, 1983. reprint.
17. U. Waldmann. *Cancellative Abelian Moioids in Refutational Theorem Proving*. PhD thesis, Institut für Informatik, Universität des Saarlandes, 1997.

Layered Transducing Term Rewriting System and Its Recognizability Preserving Property

Hiroyuki Seki[1], Toshinori Takai[2], Youhei Fujinaka[1], and Yuichi Kaji[1]

[1] Nara Institute of Science and Technology
Takayama 8916-5, Ikoma 630-0101, Japan
{seki,youhei-f,kaji}@is.aist-nara.ac.jp
[2] National Institute of Advanced Industrial Science and Technology
Nakoji 3-11-46, Amagasaki 661-0974, Japan
t-takai@aist.go.jp

Abstract. A term rewriting system which effectively preserves recognizability (EPR-TRS) has good mathematical properties. In this paper, a new subclass of TRSs, layered transducing TRSs (LT-TRSs) is defined and its recognizability preserving property is discussed. The class of LT-TRSs contains some EPR-TRSs, e.g., $\{f(x) \to f(g(x))\}$ which do not belong to any of the known decidable subclasses of EPR-TRSs. Bottom-up linear tree transducer, which is a well-known computation model in the tree language theory, is a special case of LT-TRS. We present a sufficient condition for an LT-TRS to be an EPR-TRS. Also some properties of LT-TRSs including reachability are shown to be decidable.

1 Introduction

Tree automaton is a natural extension of finite-state automaton on strings. A set of ground terms (tree language) T is *recognizable* if there exists a tree automaton which accepts T. Tree automaton inherits good mathematical properties from finite-state automaton. For example, the class of recognizable sets is closed under boolean operations (union, intersection and complementation), and decision problems such as emptiness and membership are decidable for a recognizable set. Let $\mathcal{L}(\mathcal{A})$ denote the language accepted by a tree automaton $\mathcal{A}$. For a TRS $\mathcal{R}$ and a tree language T, define $(\to_{\mathcal{R}}^*)(T) = \{t \mid \exists s \in T \text{ s.t. } s \to_{\mathcal{R}}^* t\}$. A TRS $\mathcal{R}$ *effectively preserves recognizability* (abbreviated as EPR) if for any tree automaton $\mathcal{A}$, $(\to_{\mathcal{R}}^*)(\mathcal{L}(\mathcal{A}))$ is also recognizable and a tree automaton $\mathcal{A}_*$ such that $\mathcal{L}(\mathcal{A}_*) = (\to_{\mathcal{R}}^*)(\mathcal{L}(\mathcal{A}))$ can be effectively constructed. Due to the above mentioned properties of recognizable sets, some important problems, e.g., reachability, joinability and local confluence are decidable for EPR-TRSs [8,9]. Furthermore, with additional conditions, strong normalization property, neededness and unifiability become decidable for EPR-TRSs [4,12,15]. Another interesting application of EPR-TRS is in automatic termination proving [11].

The problem to decide whether a given TRS is EPR is undecidable [7], and decidable subclasses of EPR-TRSs have been proposed in a series of works [14, 3,10,9,12,15]. These subclasses put a rather strong constraint on the syntax of

S. Tison (Ed.): RTA 2002, LNCS 2378, pp. 98–113, 2002.

the right-hand side of a rewrite rule. For example, the right-hand side of a rewrite rule in a linear semi-monadic TRS (L-SM-TRS) [3] is either a variable or $f(t_1, t_2, \ldots, t_n)$ where each t_i ($1 \leq i \leq n$) is either a variable or a ground term. Linear generalized semi-monadic TRS (L-GSM-TRS) [9] and right-linear finite path-overlapping TRS (RL-FPO-TRS) [15] weaken this constraint, but some simple EPR-TRSs such as $\{f(x) \to f(g(x))\}$ still do not belong to any of the known decidable subclasses of EPR-TRSs. To show that a given TRS $\mathcal{R}$ is EPR, for a given tree automaton $\mathcal{A}$, a tree automaton $\mathcal{A}_*$ such that $\mathcal{L}(\mathcal{A}_*) = (\to^*_{\mathcal{R}})(\mathcal{L}(\mathcal{A}))$ should be constructed. The above mentioned restrictions on the right-hand side of a rewrite rule are sufficient conditions for a procedure of automata construction to halt.

In this paper, a new subclass of TRSs, *layered transducing TRSs* (LT-TRSs) is defined and its recognizability preserving property is discussed. Intuitively, an LT-TRS is a TRS such that certain unary function symbols are specified as *markers* and a marker moves from leaf to root in each rewrite step. Bottom-up linear tree transducer [6], which is a well-known computation model in the tree language theory, can be considered as a special case of LT-TRS. We propose a procedure which, for a given tree automaton $\mathcal{A}$ and an LT-TRS $\mathcal{R}$, constructs a tree automaton $\mathcal{A}_*$ such that $\mathcal{L}(\mathcal{A}_*) = (\to^*_{\mathcal{R}})(\mathcal{L}(\mathcal{A}))$. The procedure introduces a state $[z, q]$ which is the product of a state z already belonging to $\mathcal{A}_*$ and a marker q and constructs a transition rule which is the product of a transition rule already in $\mathcal{A}_*$ and a rewrite rule in $\mathcal{R}$.

However, an LT-TRS is not always EPR and the above procedure does not always halt. We present a sufficient condition for the procedure to halt. The subclass of LT-TRSs which satisfy the sufficient condition is still incomparable with any of the known decidable subclasses of EPR-TRSs. Especially, the class contains some EPR-TRSs, such as $\{f(x) \to f(g(x))\}$ mentioned above. Finally, some properties including reachability of LT-TRSs are shown to be decidable.

The rest of the paper is organized as follows. After providing preliminary definitions in section 2, LT-TRS is defined in section 3. A procedure for automata construction is presented and the partial correctness of the procedure is proved in section 4. Sufficient conditions for the construction procedure to halt are presented in section 5. Also some properties including reachability are shown to be decidable for LT-TRS in section 5.

2 Preliminaries

2.1 Term Rewriting Systems

We use the usual notions for terms, substitutions, etc (see [1] for details). Let Σ be a *signature* and $\mathcal{V}$ be an enumerable set of *variables*. An element in Σ is called a *function symbol* and the *arity* of $f \in \Sigma$ is denoted by $a(f)$. A function symbol c with $a(c) = 0$ is called a *constant*. The set of *terms*, defined in the usual way, is denoted by $\mathcal{T}(\Sigma, \mathcal{V})$. The set of variables occurring in t is denoted by $\mathcal{V}ar(t)$. A term t is *ground* if $\mathcal{V}ar(t) = \emptyset$. The set of ground terms is denoted by $\mathcal{T}(\Sigma)$. A ground term in $\mathcal{T}(\Sigma)$ is also called a *Σ-term*. A term is *linear* if no variable occurs more than once in the term. A *substitution* σ is a mapping from $\mathcal{V}$ to $\mathcal{T}(\Sigma, \mathcal{V})$, and written as $\sigma = \{x_1 \mapsto t_1, \ldots, x_n \mapsto t_n\}$ where t_i with $1 \leq i \leq n$ is a term which substitutes for the variable x_i. The term obtained by applying a substitution σ to a term t is written as $t\sigma$. The term $t\sigma$ is called an *instance* of t. A *position* in a term t is defined as a sequence of positive integers as usual, and the set of all positions in a term t is denoted by $\mathcal{P}os(t)$. A subterm of t at a position o is denoted by $t|_o$. If a term t is obtained from a term t' by replacing the subterm of t' at position $o_1 \in \mathcal{P}os(t')$ with a term t_1 then we write $t = t'[t_1]_{o_1}$.

A *rewrite rule* over a signature Σ is an ordered pair of terms in $\mathcal{T}(\Sigma, \mathcal{V})$, written as $l \rightarrow r$. A *term rewriting system* (*TRS*) over Σ is a finite set of rewrite rules over Σ. For terms t, t' and a TRS $\mathcal{R}$, we write $t \rightarrow_{\mathcal{R}} t'$ if there exists a position $o \in \mathcal{P}os(t)$, a substitution σ and a rewrite rule $l \rightarrow r \in \mathcal{R}$ such that $t|_o = l\sigma$ and $t' = t[r\sigma]_o$. Define $\rightarrow^*_{\mathcal{R}}$ to be the reflexive and transitive closure of $\rightarrow_{\mathcal{R}}$. Also the transitive closure of $\rightarrow_{\mathcal{R}}$ is denoted by $\rightarrow^+_{\mathcal{R}}$. The subscript $\mathcal{R}$ of $\rightarrow_{\mathcal{R}}$ is omitted if $\mathcal{R}$ is clear from the context. A *redex* (*in* $\mathcal{R}$) is an instance of l for some $l \rightarrow r \in \mathcal{R}$. A *normal form* (in $\mathcal{R}$) is a term which has no redex as its subterm. Let $\mathrm{NF}_{\mathcal{R}}$ denote the set of all ground normal forms in $\mathcal{R}$. A rewrite rule $l \rightarrow r$ is *left-linear*(resp. *right-linear*) if l is linear (resp. r is linear). A rewrite rule is *linear* if it is left-linear and right-linear. A TRS $\mathcal{R}$ is *left-linear* (resp. *right-linear, linear*) if every rule in $\mathcal{R}$ is left-linear (resp. right-linear, linear).

Notions such as *reachability*, *joinability*, *confluence* and *local confluence* are defined in the usual way.

2.2 Tree Automata

A *tree automaton*(*TA*) [6] is defined by a 4-tuple $\mathcal{A} = (\Sigma, \mathcal{P}, \Delta, \mathcal{P}_{final})$ where Σ is a signature, $\mathcal{P}$ is a finite set of states, $\mathcal{P}_{final} \subseteq \mathcal{P}$ is a set of final states, and Δ is a finite set of transition rules of the form $f(p_1, \ldots, p_n) \rightarrow p$ where $f \in \Sigma$, $a(f) = n$, and $p_1, \ldots, p_n, p \in \mathcal{P}$ or of the form $p' \rightarrow p$ where $p', p \in \mathcal{P}$. A rule with the former form is called a *non-ε-rule* and a rule with the latter form is called an *ε-rule*. Consider the set of ground terms $\mathcal{T}(\Sigma \cup \mathcal{P})$ where we define $a(p) = 0$ for $p \in \mathcal{P}$. A *transition* of a TA can be regarded as a rewrite relation on $\mathcal{T}(\Sigma \cup \mathcal{P})$ by regarding transition rules in Δ as rewrite rules on $\mathcal{T}(\Sigma \cup \mathcal{P})$. For terms t and t' in $\mathcal{T}(\Sigma \cup \mathcal{P})$, we write $t \vdash_{\mathcal{A}} t'$ if and only if $t \rightarrow_{\Delta} t'$. The reflexive and transitive closure of $\vdash_{\mathcal{A}}$ is denoted by $\vdash^*_{\mathcal{A}}$. If $t \vdash_{\mathcal{A}} t_1 \vdash_{\mathcal{A}} t_2 \vdash_{\mathcal{A}} \cdots \vdash_{\mathcal{A}} t_k = t'$, we write $t \vdash^k_{\mathcal{A}} t'$ and k is called the length of the transition sequence. For a TA $\mathcal{A}$ and $t \in \mathcal{T}(\Sigma)$, if $t \vdash^*_{\mathcal{A}} p_f$ for a final state $p_f \in \mathcal{P}_{final}$, then we say t is *accepted* by $\mathcal{A}$. The set of ground terms accepted by $\mathcal{A}$ is denoted by $\mathcal{L}(\mathcal{A})$. A set T of ground terms is *recognizable* if there is a TA $\mathcal{A}$ such that $T = \mathcal{L}(\mathcal{A})$. A state p is *reachable* if there exists a Σ-term t such that $t \vdash^*_{\mathcal{A}} p$. It is well-known that for any TA $\mathcal{A}$ we can construct a TA $\mathcal{A}'$ such that $\mathcal{L}(\mathcal{A}') = \mathcal{L}(\mathcal{A})$ and $\mathcal{A}'$ contains only reachable states and contains no ε-rule [6]. Recognizable sets inherit some useful properties of regular (string) languages.

Lemma 1. *[6] The class of recognizable sets is effectively closed under union, intersection and complementation. For a recognizable set T, the following problems are decidable. (1) Does a given ground term t belong to T? (2) Is T empty?* □

2.3 TRS Which Preserves Recognizability

For a TRS $\mathcal{R}$ and a set T of ground terms, define $(\to_{\mathcal{R}}^*)(T) = \{t \mid \exists s \in T \text{ s.t. } s \to_{\mathcal{R}}^* t\}$. A TRS $\mathcal{R}$ is said to *effectively preserve recognizability* if, for any tree automaton $\mathcal{A}$, the set $(\to_{\mathcal{R}}^*)(\mathcal{L}(\mathcal{A}))$ is also recognizable and we can effectively construct a tree automaton which accepts $(\to_{\mathcal{R}}^*)(\mathcal{L}(\mathcal{A}))$. In this paper, the class of TRSs which effectively preserve recognizability is written as EPR-TRS.

Theorem 1. *If a TRS $\mathcal{R}$ belongs to EPR-TRS, then the reachability relation and the joinability relation for $\mathcal{R}$ are decidable [8]. It is also decidable whether $\mathcal{R}$ is locally confluent or not [9].* □

Unfortunately it is undecidable whether a given TRS belongs to EPR-TRS or not [7]. Therefore decidable subclasses of EPR-TRS have been proposed, for example, ground TRS by Brained [2], right-linear monadic TRS (RL-M-TRS) by Salomaa [14], linear semi-monadic TRS (L-SM-TRS) by Coquidé et al. [3], linear generalized semi-monadic TRS (L-GSM-TRS) by Gyenizse and Vágvölgyi [9], and right-linear finite path overlapping TRS (RL-FPO-TRS) by Takai et al. [15].

Theorem 2. *RL-M-TRS $\subset$ RL-FPO-TRS $\subset$ EPR-TRS and ground TRS $\subset$ L-SM-TRS $\subset$ L-GSM-TRS $\subset$ RL-FPO-TRS. All inclusions are proper.* □

Réty [13] defined a subclass of TRSs and showed that the class effectively preserves recognizability for the subclass $\mathcal{C}$ of tree languages each of which is a set $\{t\sigma \mid t \text{ is a linear term and } \sigma \text{ is a substitution such that } x\sigma \text{ is a constructor term for each } x \in \mathcal{V}ar(t)\}$ (abbreviated as $\mathcal{C}$-EPR). For example, $\mathcal{R}_3$ of Example 5 in section 4 is not an EPR-TRS but it is $\mathcal{C}$-EPR.

3 Layered Transducing TRS

A new class of TRS named *layered transducing TRS* (*LT-TRS*) is proposed in this section.

Definition 1. *Let $\Sigma = \mathcal{F} \cup \mathcal{Q}$ be a signature where $\mathcal{F} \cap \mathcal{Q} = \emptyset$. Suppose that for each $q \in \mathcal{Q}$, $a(q) = 1$. A function symbol in $\mathcal{Q}$ is called a marker. A layered transducing TRS (LT-TRS) is a linear TRS over Σ in which each rewrite rule has one of the following forms:*

(i) $f(q_1(x_1), \ldots, q_n(x_n)) \to q(t)$, *or*
(ii) $q_1'(x_1) \to q'(t')$

where $f \in \mathcal{F}$, $q_i(1 \le i \le n), q, q'_1, q' \in \mathcal{Q}$, $x_1, \ldots, x_n$ *are disjoint variables,* $t \in \mathcal{T}(\mathcal{F}, \{x_1, \ldots, x_n\})$, $t' \in \mathcal{T}(\mathcal{F}, \{x_1\})$ *and* t *and* t' *are linear terms.* □

Example 1. Let $g \in \mathcal{F}$ with $a(g) = 1$ and $q \in \mathcal{Q}$. $\mathcal{R}_1 = \{q(x) \to q(g(x))\}$ is an LT-TRS. Note that $\mathcal{R}_1$ is an EPR-TRS but is not an FPO-TRS since $q(g(x))$ properly sticks out of $q(x)$ and hence the sticking out graph of $\mathcal{R}_1$ has a self-looping edge of weight one [15]. □

Example 2. Let $f, g, h \in \mathcal{F}$, $q_1, q_2, q \in \mathcal{Q}$. $\mathcal{R}_2 = \{f(q_1(x_1), q_2(x_2)) \to q(g(h(x_2), x_1)), q_1(x_1) \to q(h(x_1))\}$ is an LT-TRS. □

In this paper, we use a, b, c to denote constants, f, g, h to denote non-marker symbols, $q, q', q_1, q_2, \ldots$ to denote markers, $p, p', p_1, p_2, \ldots$ to denote states and $s, t, t_1, t_2, \ldots$ to denote terms.

4 Construction of Tree Automata

In this section, we will present a procedure which takes an LT-TRS $\mathcal{R}$ and a tree automaton (TA) $\mathcal{A}$ as an input and constructs a TA $\mathcal{A}_*$ such that $\mathcal{L}(\mathcal{A}_*) = (\to^*_{\mathcal{R}})(\mathcal{L}(\mathcal{A}))$ if the procedure halts. Let $\mathcal{A} = (\Sigma, \mathcal{P}, \Delta, \mathcal{P}_{final})$ be a TA. By the definition of $(\to^*_{\mathcal{R}})(\mathcal{L}(\mathcal{A}))$, if $t \vdash^*_{\mathcal{A}} p$ and $t \to^*_{\mathcal{R}} s$ then $s \vdash^*_{\mathcal{A}_*} p$ should hold. To satisfy this property, the proposed procedure starts with $\mathcal{A}_0 = \mathcal{A}$ and constructs a series of TAs $\mathcal{A}_0, \mathcal{A}_1, \ldots$. We define $\mathcal{A}_*$ as the limit of this chain of TAs. For example, let $f(p_1, p_2) \to p \in \Delta$ and $f(q_1(x_1), q_2(x_2)) \to q(g(h(x_2), x_1)) \in \mathcal{R}$ and assume that

$$t = f(q_1(t_1), q_2(t_2)) \vdash^*_{\mathcal{A}} f(q_1(p'_1), q_2(p'_2)) \vdash^*_{\mathcal{A}} f(p_1, p_2) \vdash_{\mathcal{A}} p. \tag{4.1}$$

Note that $f(q_1(t_1), q_2(t_2)) \to_{\mathcal{R}} q(g(h(t_2), t_1))(= t')$ and hence $\mathcal{A}_*$ is required to satisfy $q(g(h(t_2), t_1)) \vdash^*_{\mathcal{A}_*} p$. The procedure constructs a 'product' rule of $f(p_1, p_2) \to p$ and $f(q_1(x_1), q_2(x_2)) \to q(g(h(x_2), x_1))$ and some auxiliary rules so that $\mathcal{A}_*$ can simulate the transition sequence (4.1) when $\mathcal{A}_*$ reads $q(g(h(t_2), t_1))$. More precisely, new states $[p, q]$ and $\langle h(p'_2)\rangle$ are introduced and rules

$$g(\langle h(p'_2)\rangle, p'_1) \to [p, q], h(p'_2) \to \langle h(p'_2)\rangle \tag{4.2}$$

are constructed. The following transition rule is also added so that $s \vdash^*_{\mathcal{A}_*} [p, q]$ if and only if $q(s) \vdash^*_{\mathcal{A}_*} p$.

$$q([p, q]) \to p. \tag{4.3}$$

When $\mathcal{A}_*$ reads $q(g(h(t_2), t_1))$, we can see by (4.1) that

$$t' = q(g(h(t_2), t_1)) \vdash^*_{\mathcal{A}} q(g(h(p'_2), p'_1)). \tag{4.4}$$

$\mathcal{A}_*$ guesses that in a term t such that $t \to_{\mathcal{R}} t'$, the markers q_1 and q_2 were placed above the subterms t_1 and t_2, respectively, and $\mathcal{A}_*$ behaves as if it reads q_1 and

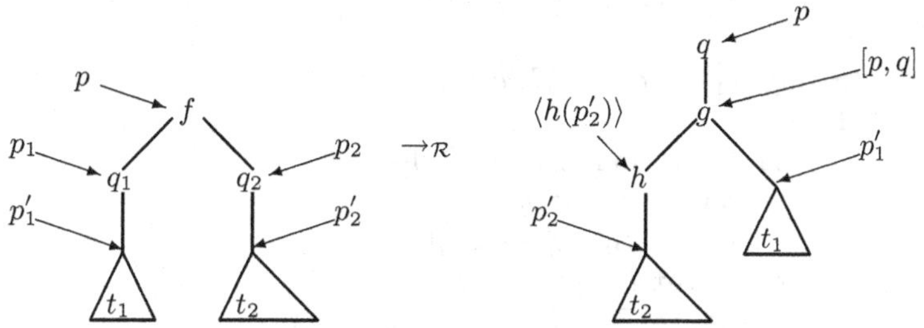

Fig. 1. Illustration of automata construction.

q_2 at p'_1 and p'_2. That is, $\mathcal{A}_*$ simulates the transition $f(p_1, p_2) \vdash_{\mathcal{A}} p$ by rules (4.2). Also see Fig. 1.

$$t' \vdash^*_{\mathcal{A}} q(g(h(p'_2), p'_1)) \vdash_{\mathcal{A}} q(g(\langle h(p'_2)\rangle, p'_1)) \vdash_{\mathcal{A}} q([p, q]) \vdash_{\mathcal{A}} p$$

The last transition is by (4.3); $\mathcal{A}_*$ encounters the marker q at the state $[p, q]$, which means that the guess was correct, and $\mathcal{A}_*$ changes its state to p by forgetting the guess q. The construction of new rules and states is repeated until $\mathcal{A}_i$ saturates. Hence, states with more than one nesting such as $[[[p, q_1], q_2], q_3]$ may be defined in general and the procedure does not always halt as illustrated in Example 5. There are two approaches to ensuring that the automata construction always halts. One of them is to redefine the procedure so that it generates a TA $\mathcal{A}_*$ such that $\mathcal{L}(\mathcal{A}_*) \supseteq (\to^*_{\mathcal{R}})(\mathcal{L}(\mathcal{A}))$ and always halts [5]. The other one is to provide a sufficient condition on a TRS and/or a TA so that the procedure halts for any TRS $\mathcal{R}$ and TA $\mathcal{A}$ which satisfy the condition [15]. This paper takes the latter approach. A major difference between [15] and this paper is that we use a 'product' state $[z, q]$, which in effect is the folding of infinitely many states and may avoid infinite state construction as illustrated in Example 4. Below we will present the construction procedure. For simplicity, if we write a signature Σ as $\Sigma = \mathcal{F} \cap \mathcal{Q}$, then we implicitly assume that $\mathcal{F} \cap \mathcal{Q} = \emptyset$ and $\mathcal{Q}$ is a set of markers.

Procedure 1 Suppose $\Sigma = \mathcal{F} \cup \mathcal{Q}$ ($\Sigma \cap \mathcal{Q} = \emptyset$) and $\forall q \in \mathcal{Q}$: $a(q) = 1$.

Input: a tree automaton $\mathcal{A} = (\Sigma, \mathcal{P}, \Delta, \mathcal{P}_{final})$ such that every state in $\mathcal{P}$ is reachable and no ε-rule is in Δ, an LT-TRS $\mathcal{R}$ over Σ.
Output: a tree automaton $\mathcal{A}_*$ s.t. $\mathcal{L}(\mathcal{A}_*) = (\to^*_{\mathcal{R}})(\mathcal{L}(\mathcal{A}))$.
Step1. Let $i := 0$, $\mathcal{A}_0 = (\Sigma, \mathcal{Z}_0, \Delta_0, \mathcal{P}_{final}) := \mathcal{A}$. In **Step2–4**, this procedure constructs $\mathcal{A}_1, \mathcal{A}_2, \cdots$ by adding new states and transition rules to $\mathcal{A}_0$.
Step2. Let $i := i + 1$, $\mathcal{A}_i = (\Sigma, \mathcal{Z}_i, \Delta_i, \mathcal{P}_{final}) := \mathcal{A}_{i-1}$.
Step3. Do the following **(S)** and **(T)** until $\mathcal{A}_i$ does not change.

(S) `if`
(S1) $f(q_1(x_1), \ldots, q_n(x_n)) \to q(t) \in \mathcal{R}$,
$f(z_1, \ldots, z_n) \to z \in \Delta_{i-1}$ $(f \in \mathcal{F})$,
$q_j(z'_j) \to z_j \in \Delta_{i-1} (1 \leq j \leq n)$

or
(**S2**) $q_1(x_1) \to q(t) \in \mathcal{R}$, $q_1(z'_1) \to z \in \Delta_{i-1}$
then
  if $t = x_l$ and $[z, q] \neq z'_l$ then
    add to $\mathcal{Z}_i$ $[z, q]$;
    add to Δ_i $z'_l \to [z, q]$;
  else let $t = g(t_1, \ldots, t_m)$
    add to $\mathcal{Z}_i$ $\langle t_1\rho\rangle, \ldots, \langle t_m\rho\rangle, [z, q]$
    add to Δ_i $g(\langle t_1\rho\rangle, \ldots, \langle t_m\rho\rangle) \to [z, q]$;
      where $\rho = \{x_j \mapsto z'_j \mid 1 \leq j \leq m\}$[1];
    ADDREC(t_j, ρ) $(1 \leq j \leq m)$.
(**T**) $\forall [z, q] \in \mathcal{Z}_i, \forall q \in \mathcal{Q}$ add to Δ_i $q([z, q]) \to z$.

Step4. If $\mathcal{A}_{i-1} = \mathcal{A}_i$, then let $\mathcal{A}_* := \mathcal{A}_i$ and output $\mathcal{A}_*$, else go to **Step2**. □

Procedure 2 (**ADDREC**) This procedure takes a term $t \in \mathcal{T}(\mathcal{F}, \mathcal{V})$ and a substitution $\rho\colon \mathcal{V} \to \mathcal{Z}_{i-1}$ as an input, and adds new states and transition rules to $\mathcal{A}_i$ so that $t\sigma \vdash^*_{\mathcal{A}_i} \langle t\rho\rangle$ holds for every substitution σ such that $\sigma = \{x_j \mapsto s_j \in \mathcal{T}(\Sigma) \mid s_j \vdash^*_{\mathcal{A}_i} x_j\rho, 1 \leq j \leq m\}$ (see Lemma 4).

ADDREC(t, ρ) =
  if $t = x$ then return;
  else let $t = h(t_1, \ldots, t_n)$
    add to Δ_i $h(\langle t_1\rho\rangle, \ldots, \langle t_n\rho\rangle) \to \langle t\rho\rangle$;
    add to $\mathcal{Z}_i$ $\langle t_1\rho\rangle, \ldots, \langle t_n\rho\rangle$;
    ADDREC(t_j, ρ) $(1 \leq j \leq n)$.

□

Example 3. Let $\mathcal{A} = (\Sigma, \mathcal{P}, \Delta, \mathcal{P}_{\mathit{final}})$ be a TA where $\Sigma = \mathcal{F} \cup \mathcal{Q}$, $\mathcal{F} = \{f, g, h, c\}$, $\mathcal{Q} = \{q_1, q_2, q\}$, $\mathcal{P} = \{p_1, p'_1, p_2, p_c, p_f\}$, $\mathcal{P}_{\mathit{final}} = \{p_f\}$ and $\Delta = \{c \to p_c, q_1(p_c) \to p'_1, q_1(p'_1) \to p_1, q_2(p_c) \to p_2, f(p_1, p_2) \to p_f\}$. It can be easily verified that $\mathcal{L}(\mathcal{A}) = \{f(q_1(q_1(c)), q_2(c))\}$. We apply Procedure 1 to $\mathcal{A}$ and LT-TRS $\mathcal{R}_2$ of Example 2. For $i = 1$ of the procedure, $f(q_1(x_1), q_2(x_2)) \to q(g(h(x_2), x_1)) \in \mathcal{R}_2$, $f(p_1, p_2) \to p_f \in \Delta_0 = \Delta$, $q_1(p'_1) \to p_1 \in \Delta_0$ and $q_2(p_c) \to p_2 \in \Delta_0$ satisfy condition (**S1**) and rules $g(\langle h(x_2)\rho\rangle, \langle x_1\rho\rangle) = g(\langle h(p_c)\rangle, p'_1) \to [p_f, q]$ and $q([p_f, q]) \to p_f$ are added to Δ_1 where $\rho = \{x_1 \mapsto p'_1, x_2 \mapsto p_c\}$. Also a rule $h(p_c) \to \langle h(p_c)\rangle$ is constructed by **ADDREC**$(h(x_2), \rho)$. Next, $q_1(x_1) \to q(h(x_1)) \in \mathcal{R}_2$ and $q_1(p'_1) \to p_1 \in \Delta_0$ satisfy condition (**S2**) and a rule $h(p'_1) \to [p_1, q]$ is constructed. The transition rules constructed in Procedure 1 are listed in Table 1. Since no rule is added when $i = 2$, the procedure halts and we obtain $\mathcal{A}_* = \mathcal{A}_1$ as the output. We can verify that $\mathcal{L}(\mathcal{A}_*) = (\to^*_{\mathcal{R}_2})(\mathcal{L}(\mathcal{A}))$. □

---

[1] For simplicity, for a state $z \in \mathcal{Z}_i$, we identify z with $\langle z\rangle$.

Table 1. The transition rules constructed by Procedure 1.

	(S)	(T)
$\mathcal{A}_1$	$g(\langle h(p_c)\rangle, p'_1) \to [p_f, q]$	$q([p_f, q]) \to p_f$
	$h(p_c) \to \langle h(p_c)\rangle$	
	$h(p'_1) \to [p_1, q]$	$q([p_1, q]) \to p_1$
	$h(p_c) \to [p'_1, q]$	$q([p'_1, q]) \to p'_1$

Example 4. Let $\mathcal{A} = (\Sigma, \mathcal{P}, \Delta, \mathcal{P}_{\mathit{final}})$, $\Sigma = \mathcal{F} \cup \mathcal{Q}$, $\mathcal{F} = \{c, g\}$, $\mathcal{Q} = \{q\}$, $\mathcal{P} = \{p_c, p_f\}$, $\mathcal{P}_{\mathit{final}} = \{p_f\}$ and $\Delta = \{c \to p_c, q(p_c) \to p_f\}$. Clearly, $\mathcal{L}(\mathcal{A}) = \{q(c)\}$. If we apply Procedure 1 to $\mathcal{A}$ and $\mathcal{R}_1$ of Example 1, then for $i = 1$ of the procedure, $q(x) \to q(g(x)) \in \mathcal{R}_1$ and $q(p_c) \to p_f \in \Delta_0$ are considered and $g(p_c) \to [p_f, q]$ and $q([p_f, q]) \to p_f$ are added. For $i = 2$, $q(x) \to q(g(x)) \in \mathcal{R}_1$ and $q([p_f, q]) \to p_f \in \Delta_1$ are considered and $g([p_f, q]) \to [p_f, q]$ is added. Since no rule is added when $i = 3$, the procedure halts with $\mathcal{A}_* = \mathcal{A}_2$. Clearly, $\mathcal{L}(\mathcal{A}_*) = \{q(g^n(c)) \mid n \geq 0\}$. Note that the transition rule $g([p_f, q]) \to [p_f, q]$ simulates infinitely many rewrite steps caused by the rule $q(x) \to q(g(x))$. In the methods proposed in [15] and [5] (without approximation), infinitely many states such as $\langle g^n(p_c)\rangle$ and $\langle q(g^n(p_c))\rangle$ $(n \geq 0)$ are introduced to simulate each rewrite step $q(g^{n-1}(c)) \to_{\mathcal{R}_1} q(g^n(c))$ by a different transition $g(\langle g^{n-1}(p_c)\rangle) \to \langle g^n(p_c)\rangle$, and thus the construction does not halt. □

Example 5. Let $\mathcal{A} = (\Sigma, \mathcal{P}, \Delta, \mathcal{P}_{\mathit{final}})$, $\Sigma = \mathcal{F} \cup \mathcal{Q}$, $\mathcal{F} = \{c, f\}$, $\mathcal{Q} = \{q\}$, $\mathcal{P} = \mathcal{P}_{\mathit{final}} = \{p\}$, $\Delta = \{c \to p, f(p) \to p, q(p) \to p\}$ and $\mathcal{R}_3 = \{f(q(x)) \to q(f(x))\}$. $\mathcal{R}_3$ is an LT-TRS. Assume that Procedure 1 is executed for $\mathcal{A}$ and $\mathcal{R}_3$. Since $f(q(x)) \to q(f(x)) \in \mathcal{R}_3$, $f(p) \to p \in \Delta_0$ and $q(p) \to p \in \Delta_0$, rules $f(p) \to [p, q]$ and $q([p, q]) \to p$ are added to Δ_1. When $i = 2$, the procedure considers $f(q(x)) \to q(f(x)) \in \mathcal{R}_3$, $f(p) \to [p, q] \in \Delta_1$ and $q(p) \to p \in \Delta_0$ and it adds $f(p) \to [[p, q], q]$ and $q([[p, q], q]) \to [p, q]$ to Δ_2. The procedure repeats a similar construction and does not halt. Note that $\mathcal{R}_3$ is not an EPR-TRS since for a recognizable set $T_1 = \{(fq)^n(c) \mid n \geq 0\}$, $(\to^*_{\mathcal{R}_3})(T_1) \cap \mathrm{NF}_{\mathcal{R}_3} = \{q^n(f^n(c)) \mid n \geq 0\}$ is not recognizable. □

In the following, we will prove the soundness ($t \vdash^*_{\mathcal{A}_i} p \in \mathcal{P} \Rightarrow \exists s\colon s \vdash^*_{\mathcal{A}} p$ and $s \to^*_{\mathcal{R}} t$) and completeness ($s \vdash^*_{\mathcal{A}} p$ and $s \to^*_{\mathcal{R}} t \Rightarrow \exists i \geq 0\colon t \vdash^*_{\mathcal{A}_i} p$) of Procedure 1. By the definition of Procedure 1, it is not difficult to prove the following three lemmas. Note that by Procedure 1, if $z \in \mathcal{Z}_i$ then either $z \in \mathcal{P}$, z is of the form $[z', q]$ for some $z' \in \mathcal{Z}_{i-1}$ and $q \in \mathcal{Q}$ or z is of the form $\langle t\rho\rangle$ for some term t and substitution $\rho\colon \mathcal{V} \to \mathcal{Z}_{i-1}$.

Lemma 2. *If $z_1 \to z \in \Delta_i$ $(i \geq 1)$, then z is of the form $[z', q]$ and z_1 is either in $\mathcal{P}$ or of the form $[z'_1, q_1]$.* □

Lemma 3. *Let $\mathcal{A} = (\Sigma, \mathcal{P}, \Delta, \mathcal{P}_{\mathit{final}})$ be a TA. Assume that every state $p \in \mathcal{P}$ is reachable. If Procedure 1 is executed for $\mathcal{A}$ and an LT-TRS $\mathcal{R}$, then every state $z \in \mathcal{Z}_i$ constructed during the execution of Procedure 1 is reachable.* □

Lemma 4. *For a transition rule $g(\langle t_1\rho\rangle, \ldots, \langle t_m\rho\rangle) \to [z, q]$ constructed in* (**S**) *of Procedure 1 and a Σ-term s, $s \vdash^*_{\mathcal{A}_i} g(\langle t_1\rho\rangle, \ldots, \langle t_m\rho\rangle) \vdash_{\mathcal{A}_i} [z, q]$ if and only if there exists a substitution $\sigma: \mathcal{V} \to \mathcal{T}(\Sigma)$ such that $s = g(t_1, \ldots, t_m)\sigma$ and for each $x \in \mathcal{V}ar(g(t_1, \ldots, t_m))$, $x\sigma \vdash^*_{\mathcal{A}_i} x\rho$.* □

For example, consider a rule $g(\langle h(p_c)\rangle, p'_1) \to [p_f, q]$ constructed in Example 3. For a Σ-term s, $s \vdash^*_{\mathcal{A}_1} g(\langle h(p_c)\rangle, p'_1) \vdash_{\mathcal{A}_1} [p_f, q]$ if and only if s can be written as $g(h(s_2), s_1)$ where s_1 and s_2 are Σ-terms such that $s_1 \vdash^*_{\mathcal{A}_1} p'_1$ and $s_2 \vdash^*_{\mathcal{A}_1} p_c$.

Lemma 5. *(Soundness) Let $t \in \mathcal{T}(\Sigma)$, $p \in \mathcal{P}$, $i \geq 1$ and $[z, q] \in \mathcal{Z}_i$.*

(**A**) $t \vdash^*_{\mathcal{A}_i} p$ $\Rightarrow$ *there exists a Σ-term s such that $s \vdash^*_{\mathcal{A}_{i-1}} p$ and $s \to^*_{\mathcal{R}} t$.*
(**B**) $t \vdash^*_{\mathcal{A}_i} [z, q] \Rightarrow$ *there exists a Σ-term s such that $s \vdash^*_{\mathcal{A}_{i-1}} z$ and $s \to^*_{\mathcal{R}} q(t)$.*

Proof. We will prove (**A**) and (**B**) simultaneously by induction on the length of the transition sequences $t \vdash^*_{\mathcal{A}_i} p$ and $t \vdash^*_{\mathcal{A}_i} [z, q]$. We will only describe the proof of (**B**) due to space limitation.

Assume $t \vdash^*_{\mathcal{A}_i} [z, q]$ and we perform the following case analysis (i)–(iii) by considering the rule applied in the last step of the transition sequence $t \vdash^*_{\mathcal{A}_i} [z, q]$.

(i) Assume $t \vdash^*_{\mathcal{A}_i} z' \vdash_{\mathcal{A}_i} [z, q]$ $(z' \in \mathcal{Z}_i)$. The rule $z' \to [z, q]$ is constructed in (**S**). Assume that condition (**S1**) holds. (The case that (**S2**) holds can be treated in a similar way.) Then there exists $f(q_1(x_1), \ldots, q_n(x_n)) \to q(x_l) \in \mathcal{R}$ with $1 \leq l \leq n$, $f(z_1, \ldots, z_n) \to z \in \Delta_{i-1}$ $(f \in \mathcal{F})$ and $q_j(z'_j) \to z_j \in \Delta_{i-1}$ $(1 \leq j \leq n)$ where $z' = z'_l$. By Lemma 3, there exists a Σ-term s_j such that $s_j \vdash^*_{\mathcal{A}_{i-1}} z'_j$ for $1 \leq j \leq n$. We further consider two subcases, namely, (i-a) $z' \in \mathcal{P}$, and (i-b) $z' = [z'', q']$ $(z'' \in \mathcal{Z}_i, q' \in \mathcal{Q})$ by Lemma 2.

(i-a) By the inductive hypothesis (**A**), there exists a Σ-term s' such that $s' \vdash^*_{\mathcal{A}_{i-1}} z'$ and $s' \to^*_{\mathcal{R}} t$. Let $s = f(q_1(s_1), \ldots, q_l(s'), \ldots, q_n(s_n))$. Then $s \vdash^*_{\mathcal{A}_{i-1}} f(q_1(z'_1), \ldots, q_l(z'), \ldots, q_n(z'_n)) \vdash^*_{\mathcal{A}_{i-1}} f(z_1, \ldots, z_n) \vdash_{\mathcal{A}_{i-1}} z$ and $s \to_{\mathcal{R}} q(s') \to^*_{\mathcal{R}} q(t)$. (i-b) There exists a Σ-term s' such that $s' \vdash^*_{\mathcal{A}_{i-1}} z''$ and $s' \to^*_{\mathcal{R}} q'(t)$ by the inductive hypothesis (**B**). Since $q_l([z'', q']) \to z_l \in \Delta_{i-1}$ by the assumption, $[z'', q'] = [z_l, q_l]$. Let $s = f(q_1(s_1), \ldots, s', \ldots, q_n(s_n))$. Then $s \vdash^*_{\mathcal{A}_{i-1}} f(q_1(z'_1), \ldots, z_l, \ldots, q_n(z'_n)) \vdash^*_{\mathcal{A}_{i-1}} f(z_1, \ldots, z_n) \vdash_{\mathcal{A}_{i-1}} z$ and $s \to^*_{\mathcal{R}} f(q_1(s_1), \ldots, q_l(t), \ldots, q_n(s_n)) \to q(t)$. In either case (i-a) or (i-b), $s \vdash^*_{\mathcal{A}_{i-1}} z$ and $s \to^*_{\mathcal{R}} q(t)$ and thus the lemma holds.

(ii) Assume $t = q'(t') \vdash^*_{\mathcal{A}_i} q'(z') \vdash_{\mathcal{A}_i} [z, q] (q' \in \mathcal{Q}, z' \in \mathcal{Z}_i)$. Since the rule $q'(z') \to [z, q]$ applied in the last step of the transition sequence is constructed in (**T**), $z' = [[z, q], q'] \in \mathcal{Z}_i$. This implies $[z, q] \in \mathcal{Z}_{i-1}$ and $q([z, q]) \to z \in \Delta_{i-1}$. By the inductive hypothesis (**B**) applied to $t' \vdash^*_{\mathcal{A}_i} [[z, q], q']$, there exists a Σ-term s' such that $s' \vdash^*_{\mathcal{A}_{i-1}} [z, q]$ and $s' \to^*_{\mathcal{R}} q'(t')$ $(= t)$. Let $s = q(s')$, then $s \vdash^*_{\mathcal{A}_{i-1}} q([z, q]) \vdash_{\mathcal{A}_{i-1}} z$ and $s \to^*_{\mathcal{R}} q(q'(t')) = q(t)$. Thus, the lemma holds.

(iii) Assume $t = g(t_1, \ldots, t_m) \vdash^k_{\mathcal{A}_i} g(z'_1, \ldots, z'_m) \vdash_{\mathcal{A}_i} [z, q]$ ($g \in \mathcal{F}$, $z'_1, \ldots, z'_m \in \mathcal{Z}_i$). This is the most difficult case. The rule $g(z'_1, \ldots, z'_m) \to [z, q] \in \Delta_i$ applied in the last step is constructed in (**S**). Assume that the rule is constructed by (**S1**). (The lemma can be shown in a similar way when the rule is constructed by (**S2**).) There exist

$$f(q_1(x_1), \ldots, q_n(x_n)) \to q(g(\xi_1, \ldots, \xi_m)) \in \mathcal{R}$$
$$(g \in \mathcal{F}, \xi_j \in \mathcal{T}(\mathcal{F}, \{x_1, \ldots, x_n\})), \quad (4.5)$$
$$f(z_1, \ldots, z_n) \to z \in \Delta_{i-1}, \quad (4.6)$$

$$q_j(z''_j) \to z_j \in \Delta_{i-1} \ (1 \le j \le n). \quad (4.7)$$

By Lemma 4, the assumed transition sequence can be written as:

$$t = g(t_1, \ldots, t_m) = g(\xi_1, \ldots, \xi_m)\sigma \vdash^*_{\mathcal{A}_i} g(\xi_1, \ldots, \xi_m)\rho$$
$$\vdash^*_{\mathcal{A}_i} g(\langle \xi_1 \rho \rangle, \ldots, \langle \xi_m \rho \rangle) = g(z'_1, \ldots, z'_m) \vdash_{\mathcal{A}_i} [z, q] \quad (4.8)$$

where

$$\sigma = \{x_j \mapsto w_j \mid 1 \le j \le n\}, \rho = \{x_j \mapsto z''_j \mid 1 \le j \le n\}, \quad (4.9)$$
$$w_j \vdash^{k_j}_{\mathcal{A}_i} z''_j \text{ with } k_j \le k (1 \le j \le n). \quad (4.10)$$

Note that for $j (1 \le j \le n)$ such that $x_j \notin \mathcal{V}ar(g(\xi_1, \ldots, \xi_m))$, we use Lemma 3 to guarantee the existence of term w_j. There are two cases to consider. (iii-a) If $z''_j \in \mathcal{P}$ then by (4.10) and the inductive hypothesis (**A**), there exists a Σ-term u_j such that $u_j \vdash^*_{\mathcal{A}_{i-1}} z''_j$ and $u_j \to^*_{\mathcal{R}} w_j$. Let $s_j = q_j(u_j)$, then $s_j \vdash^*_{\mathcal{A}_{i-1}} q_j(z''_j) \vdash_{\mathcal{A}_{i-1}} z_j$ by (4.7) and $s_j \to^*_{\mathcal{R}} q_j(w_j)$. (iii-b) If $z''_j = [z_j, q_j]$ then by the inductive hypothesis (**B**) applied to $w_j \vdash^{k_j}_{\mathcal{A}_i} [z_j, q_j]$, there exists a Σ-term s_j such that $s_j \vdash^*_{\mathcal{A}_{i-1}} z_j$ and $s_j \to^*_{\mathcal{R}} q_j(w_j)$. In either case (iii-a) or (iii-b), $s_j \vdash^*_{\mathcal{A}_{i-1}} z_j$ and $s_j \to^*_{\mathcal{R}} q_j(w_j)$. Let $s = f(s_1, \ldots, s_n)$. Then, $s \vdash^*_{\mathcal{A}_{i-1}} f(z_1, \ldots, z_n) \vdash_{\mathcal{A}_{i-1}} z$ by (4.6), and $s \to^*_{\mathcal{R}} f(q_1(w_1), \ldots, q_n(w_n)) \to_{\mathcal{R}} q(g(\xi_1, \ldots, \xi_m))\sigma = q(t)$ by (4.5),(4.8) and (4.9). Therefore, the lemma holds. □

Lemma 6. *(Completeness)*

> $s \to^*_{\mathcal{R}} t$ *and* $s \vdash^*_{\mathcal{A}_0} p \in \mathcal{P} \Rightarrow$ *there exists an integer* $i \ge 0$ *such that* $t \vdash^*_{\mathcal{A}_i} p$.

Proof. Assume that $s \to^*_{\mathcal{R}} t$ and $s \vdash^*_{\mathcal{A}_0} p$. The lemma is shown by induction on the number of rewrite steps in $s \to^*_{\mathcal{R}} t$. If $s = t$ then the lemma holds clearly. Assume $s \to^*_{\mathcal{R}} t' \to_{\mathcal{R}} t$. By the inductive hypothesis, there exists $i' \ge 0$ such that $t' \vdash^*_{\mathcal{A}_{i'}} p$. For a rewrite step $t' \to_{\mathcal{R}} t$, the applied rewrite rule has the form of either (i) $f(q_1(x_1), \ldots, q_n(x_n)) \to q(u)$, or (ii) $q_1(x_1) \to q(u)$. In either case (i) or (ii), t can be written as $t = t'[q(u\sigma)]_o$ where o is the rewrite position and σ is a substitution. Assume $u = g(u_1, \ldots, u_m)$ and

$u\sigma = g(t_1, \ldots, t_m)$. (The case when u is a variable is easier and omitted.) Consider the case (i). Since $t' \vdash^*_{\mathcal{A}_{i'}} p$ and $t'|_o$ is an instance of $f(q_1(x_1), \ldots, q_n(x_n))$, there exists a transition rule $f(z_1, \ldots, z_n) \to z_0 \in \Delta_{i'}$ and $t' \vdash^*_{\mathcal{A}_{i'}} p$ can be written as $t' = t'[f(q_1(x_1\sigma), \ldots, q_n(x_n\sigma))]_o \vdash^*_{\mathcal{A}_{i'}} t'[f(q_1(z_1'), \ldots q_n(z_n'))]_o \vdash^*_{\mathcal{A}_{i'}} t'[f(z_1, \ldots, z_n)]_o \vdash_{\mathcal{A}_{i'}} t'[z_0]_o \vdash^*_{\mathcal{A}_{i'}} p$ where $f(q_1(z_1'), \ldots, q_n(z_n')) \vdash^*_{\mathcal{A}_{i'}} f(z_1, \ldots, z_n)$ includes no ε-transition. Since $q_j(z_j') \vdash_{\mathcal{A}_{i'}} z_j$ $(1 \le j \le n)$, (**S1**) is satisfied and a transition rule $g(\langle u_1\rho\rangle, \ldots, \langle u_m\rho\rangle) \to [z_0, q]$ is constructed where $\rho = \{x_j \mapsto z_j' \mid 1 \le j \le n\}$. Since $x_j\sigma \vdash^*_{\mathcal{A}_{i'}} z_j'$, by Lemma 4, $g(t_1, \ldots, t_m) \vdash^*_{\mathcal{A}_{i'+1}} [z_0, q]$. The case (ii) can be treated in a similar way to the case (i) and we can show that (**S2**) is satisfied and transition rules which enable $g(t_1, \ldots, t_m) \vdash^*_{\mathcal{A}_{i'+1}} [z_0, q]$ are constructed. In either case, $q([z_0, q]) \to z_0$ is also added in (**T**) and we can see that $t = t'[q(g(t_1, \ldots, t_m))]_o \vdash^*_{\mathcal{A}_{i'+1}} t'[q([z_0, q])]_o \vdash_{\mathcal{A}_{i'+1}} t'[z_0]_o \vdash^*_{\mathcal{A}_{i'}} p$ and the lemma holds. □

Lemma 7. *(Partial Correctness) Let $\mathcal{A} = (\Sigma, \mathcal{P}, \Delta, \mathcal{P}_{final})$ be a TA, $\mathcal{R}$ be an LT-TRS. Assume that for input $\mathcal{A}$ and $\mathcal{R}$, Procedure 1 constructs a series of tree automata $\mathcal{A}_0, \mathcal{A}_1, \mathcal{A}_2, \cdots$. For any term $t \in \mathcal{T}(\Sigma)$ and state $p \in \mathcal{P}$,*

there exists a term $s \in \mathcal{T}(\Sigma)$ such that $s \vdash^*_{\mathcal{A}} p$ and $s \to^*_{\mathcal{R}} t$
if and only if there exists $i \ge 0$ such that $t \vdash^*_{\mathcal{A}_i} p$.

Proof. ($\Rightarrow$) By Lemma 6. ($\Leftarrow$) By induction on i and Lemma 5(**A**). □

Lemma 8. *If Procedure 1 halts for a TA $\mathcal{A}$ and an LT-TRS $\mathcal{R}$, then $\mathcal{L}(\mathcal{A}_*) = (\to^*_{\mathcal{R}})(\mathcal{L}(\mathcal{A}))$ holds for the output $\mathcal{A}_*$ of the procedure.*

Proof. ($\subseteq$) Assume $t \in \mathcal{L}(\mathcal{A}_*)$. Since $\mathcal{A}_* = \mathcal{A}_i$ for some $i \ge 0$, there exists a final state p_f such that $t \vdash^*_{\mathcal{A}_i} p_f$. By Lemma 7, there exists a Σ-term s such that $s \vdash^*_{\mathcal{A}} p_f$ and $s \to^*_{\mathcal{R}} t$. Therefore, $t \in (\to^*_{\mathcal{R}})(\mathcal{L}(\mathcal{A}))$. $\mathcal{L}(\mathcal{A}_*) \supseteq (\to^*_{\mathcal{R}})(\mathcal{L}(\mathcal{A}))$ can be shown in a similar way. □

5 Recognizability Preserving Property

In this section, two sufficient conditions for Procedure 1 to halt are proposed. One condition is that the sets of non-marker function symbols occurring in the left-hand sides and the right-hand sides of rewrite rules are disjoint. The other condition is the one which in effect restricts the class of recognizable sets. An LT-TRS $\mathcal{R}$ which satisfies the former condition effectively preserves recognizability. Although the latter condition does not directly give a subclass of LT-TRSs which are EPR, we can show that some properties of LT-TRSs are decidable, using the latter condition.

5.1 I/O-Separated LT-TRS

An LT-TRS $\mathcal{R}$ is *I/O-separated* if $\mathcal{R}$ satisfies the following condition.

Condition 1 *For a signature $\Sigma = \mathcal{F} \cup \mathcal{Q}$, $\mathcal{F}$ is further divided as $\mathcal{F} = \mathcal{F}_I \cup \mathcal{F}_O$, $\mathcal{F}_I \cap \mathcal{F}_O = \emptyset$. A function symbol in $\mathcal{F}_I$ (respectively, $\mathcal{F}_O$) is called an input symbol (respectively, output symbol). For each rewrite rule*

$$\begin{cases} f(q_1(x_1), \ldots, q_n(x_n)) \to q(t), \text{ or} \\ q_1(x_1) \to q(t) \end{cases}$$

in $\mathcal{R}$, $f \in \mathcal{F}_I$ and $t \in \mathcal{T}(\mathcal{F}_O, \{x_1, \ldots, x_n\})$. □

Lemma 9. *Let $\mathcal{R}$ be an I/O-separated LT-TRS over $\Sigma = \mathcal{F}_I \cup \mathcal{F}_O \cup \mathcal{Q}$. If Procedure 1 is executed for a TA $\mathcal{A}$ and $\mathcal{R}$, then it always halts.*

Proof. Consider the TA $\mathcal{A}_1 = (\Sigma, \mathcal{Z}_1, \Delta_1, \mathcal{P}_{final})$ constructed in Procedure 1 for a TA $\mathcal{A}$ and an I/O-separated LT-TRS $\mathcal{R}$. We can easily prove the claim that if a state $[z, q]$ is added to $\mathcal{Z}_i$ in (**S**) then $z \in \mathcal{P}$ by induction on i. Now we will show the lemma. Every rule $g(\cdots) \to z'$ constructed in (**S**) or **ADDREC** satisfies $g \in \mathcal{F}_O$ and hence the number of transition rules $f(z_1, \ldots, z_n) \to z \in \Delta_{i-1}$ ($f \in \mathcal{F}_I$) which satisfy condition (**S1**) is finite. The number of transition rules $q_j(z'_j) \to z_j \in \Delta_{i-1}$ satisfying (**S1**) or (**S2**) is also finite since $z'_j \in \mathcal{P}$ or $z'_j = [z_j, q_j]$ where $z_j \in \mathcal{P}$ by the above claim. Hence, Procedure 1 always halts. □

$\mathcal{R}_1$ in Example 1 and $\mathcal{R}_2$ in Example 2 are both I/O-separated LT-TRSs.

Theorem 3. *Every I/O-separated LT-TRS effectively preserves recognizability.* □

A bottom-up tree transducer [6] is a well-known computation model in the theory of tree languages. For a linear bottom-up tree transducer $\mathcal{M}$, if we consider the set of states of $\mathcal{M}$ as the set of markers, $\mathcal{M}$ corresponds to an I/O-separated LT-TRS. Hence, the following known property of tree transducer is obtained as a corollary.

Corollary 1. *[6] Every linear bottom-up tree transducer effectively preserves recognizability.* □

5.2 Marker-Bounded Sets

Let $\Sigma' \subseteq \Sigma$ be a subset of function symbols. Consider a tree representation of a term t. Let $depth_{\Sigma'}(t)$ denote the maximum number of occurrences of function symbols in Σ' which occur in a single path from the root to a leaf in t. That is, $depth_{\Sigma'}(t)$ is defined as:

$$depth_{\Sigma'}(g(t_1, \ldots, t_n)) = \begin{cases} \max\{depth_{\Sigma'}(t_i) \mid 1 \le i \le n\} + 1 & g \in \Sigma', \\ \max\{depth_{\Sigma'}(t_i) \mid 1 \le i \le n\} & g \notin \Sigma'. \end{cases}$$

For example, for $\Sigma = \{f, g, h, c\}$ and $\Sigma' = \{f, g\}$, we have $depth_{\Sigma'}(f(g(c), g(h(g(c))))) = 3$. For a signature $\Sigma = \mathcal{F} \cup \mathcal{Q}$, a set $T \subseteq \mathcal{T}(\Sigma)$ is *marker-bounded* if the following condition holds:

Condition 2 *There exists $k \geq 0$ such that $depth_{\mathcal{Q}}(t) \leq k$ for each $t \in T$.* □

Lemma 10. *Let $\mathcal{R}$ be an LT-TRS over $\Sigma = \mathcal{F} \cup \mathcal{Q}$. If $t \to^*_{\mathcal{R}} t'$ then $depth_{\mathcal{Q}}(t) \leq depth_{\mathcal{Q}}(t')$. If $\mathcal{R}$ contains no rewrite rule whose left-hand side is a constant, then $depth_{\mathcal{Q}}(t) = depth_{\mathcal{Q}}(t')$.*

Proof. By the form of a rewrite rule of an LT-TRS. □

Definition 2. *(deg) For each state $z \in \mathcal{Z}_i$, let $deg(z)$ denote the number of nestings in z, which is defined as follows:*

$$\begin{cases} deg(p) & = 0 \ (p \in \mathcal{P}), \\ deg([z, q]) & = deg(z) + 1 \ (z \in \mathcal{Z}_i, q \in \mathcal{Q}), \\ deg(\langle f(t_1, \ldots, t_n)\rangle) & = \max\{deg(\langle t_j \rangle) \mid 1 \leq j \leq n\} \\ & \quad \textit{where } \max\emptyset = 0 \textit{ for the empty set } \emptyset. \end{cases}$$

□

By definition, $deg(\langle f(t_1, \ldots, t_n)\rangle) = \max\{deg([z, q]) \mid [z, q] \text{ occurs in } f(t_1, \ldots, t_n)\}$. The following lemma is not difficult to prove.

Lemma 11. *(i) For $q(z_1) \to z \in \Delta_i$ ($q \in \mathcal{Q}$), $deg(z_1) \leq deg(z) + 1$.*
(ii) For $f(z_1, \ldots, z_n) \to z \in \Delta_i$ ($g \in \mathcal{F}$), $deg(z_j) \leq deg(z)$ ($1 \leq j \leq n$).
(iii) For $z_1 \to z \in \Delta_i$, $deg(z_1) \leq deg(z)$. □

Let $\mathcal{A} = (\Sigma, \mathcal{P}, \Delta, \mathcal{P}_{final})$ be a TA. A state $p \in \mathcal{P}$ is *useful* if there exists a Σ-term t, a position $o \in \mathcal{P}os(t)$ and a final state $p_f \in \mathcal{P}_{final}$ such that $t \vdash^*_{\mathcal{A}} t[p]_o \vdash^*_{\mathcal{A}} p_f$. It is known that for a given TA $\mathcal{A}$, we can construct a TA $\mathcal{A}'$ which satisfies $\mathcal{L}(\mathcal{A}') = \mathcal{L}(\mathcal{A})$, has no ε-rule and has only useful states.

Lemma 12. *Procedure 1 always halts for a TA $\mathcal{A}$ and an LT-TRS $\mathcal{R}$ which satisfy the following conditions.*

(1) $\mathcal{L}(\mathcal{A})$ is marker-bounded.
(2) $\mathcal{R}$ contains no rewrite rule whose left-hand side is a constant.

Proof. Let $\mathcal{A}$ and $\mathcal{R}$ be a TA and an LT-TRS which satisfy the conditions of the lemma. Without loss of generality, assume that $\mathcal{A} = (\Sigma, \mathcal{P}, \Delta, \mathcal{P}_{final})$ has no ε-rule and has only useful states. The proof is by contradiction. Assume that Procedure 1 does not halt for $\mathcal{A}$ and $\mathcal{R}$. For an arbitrary constant l, the number of states z_0 with $deg(z_0) < l$ and the number of rules which contain only such states are both finite. Since Procedure 1 does not halt, there exists an integer $i \geq 0$ and a state $z_0 \in \mathcal{Z}_i$ with $deg(z_0) \geq l$. In particular, let k be a constant in Condition 2 for $\mathcal{L}(\mathcal{A})$, then there exists an integer $i \geq 0$ and a state $z_0 \in \mathcal{Z}_i$ with $deg(z_0) = k' \geq k + 1$. Note that $k' \leq i$ by the definition of $deg(z_0)$ and the construction in Procedure 1. By Lemma 3, there exists a Σ-term s_0 such that $s_0 \vdash^*_{\mathcal{A}_i} z_0$. Since $deg(z_0) \geq 1$, z_0 can be written as $z_0 = \langle \cdots [z'_1, q'_1] \cdots \rangle$ $(= \langle \xi \rangle$, including the case that $z_0 = [z'_1, q'_1])$ and $deg(z_0) = deg([z'_1, q'_1])$. The state $z_0 = \langle \xi \rangle$ is introduced in (**S**) of Procedure 1 or **ADDREC** when the loop counter of the procedure is $i' \leq i$. Hence there exists $f(q_{11}(x_1), \ldots, q_{1n}(x_n)) \to q_1(t) \in \mathcal{R}$, $f(z_{11}, \ldots, z_{1n}) \to z_1 \in \Delta_{i-1}$ and $q_{1j}(z'_{1j}) \to z_{1j} \in \Delta_{i-1}$ $(1 \leq j \leq n)$ or there exists $q_{11}(x_1) \to q_1(t) \in \mathcal{R}$ and $q_{11}(z'_{11}) \to z_1 \in \Delta_{i-1}$. Also, $z_0 = [z_1, q_1]$ or $z_0 = \langle \xi \rangle = \langle t'\rho \rangle$ for some subterm t' of t where $\rho = \{x_j \mapsto z'_{1j} \mid 1 \leq j \leq n\}$. If $z_0 = [z_1, q_1]$ then clearly $deg(z_1) = deg(z_0) - 1 = k' - 1$. Suppose $z_0 = \langle t'\rho \rangle$. Since the rule $z'_{1l} \to [z_1, q_1]$ or $g(\langle t_1\rho \rangle, \ldots, \langle t_n\rho \rangle) \to [z_1, q_1]$ where $t = g(t_1, \ldots, t_n)$ is constructed, $deg(z_0) \leq deg([z_1, q_1])$ by Lemma 11. Hence, $deg(z_1) \geq deg(z_0) - 1 = k' - 1$. In either case, $deg(z_1) \geq k' - 1$. Since $s_0 \vdash^*_{\mathcal{A}_i} z_0 = \langle \xi \rangle$ and ξ is a subterm of $t\rho$, there exists a Σ-term t_0 such that $t_0 \vdash^*_{\mathcal{A}_i} [z_1, q_1]$ and s_0 is a subterm of t_0. By Lemma 5(**B**), there exists a Σ-term s_1 such that $s_1 \to^*_{\mathcal{R}} q_1(t_0)$ and $s_1 \vdash^*_{\mathcal{A}_{i-1}} z_1$. Summarizing, $s_1 \vdash^*_{\mathcal{A}_{i-1}} z_1$, $s_1 \to^*_{\mathcal{R}} q_1(t_0)$, $deg(z_1) \geq k' - 1$, and s_0 is a subterm of t_0. Repeating the above argument, we can see that there exist states $z_j \in \mathcal{Z}_{i-j}$ $(1 \leq j \leq i)$, markers $q_j \in \mathcal{Q}$ $(1 \leq j \leq i)$, Σ-terms s_j $(0 \leq j \leq i)$, t_j $(0 \leq j \leq i-1)$ such that:

$$\begin{array}{l} s_j \vdash^*_{\mathcal{A}_{i-j}} z_j, s_j \to^*_{\mathcal{R}} q_j(t_{j-1}), deg(z_j) \geq k' - j \\ \text{and } s_{j-1} \text{ is a subterm of } t_{j-1} \quad (1 \leq j \leq i). \end{array} \tag{5.1}$$

Since $s_j \to^*_{\mathcal{R}} q_j(t_{j-1})$ $(1 \leq j \leq i)$, $depth_{\mathcal{Q}}(s_j) = depth_{\mathcal{Q}}(t_{j-1}) + 1$ by Lemma 10. Since s_j is a subterm of t_j $(0 \leq j \leq i-1)$, $depth_{\mathcal{Q}}(t_j) \geq depth_{\mathcal{Q}}(s_j)$. Hence, $depth_{\mathcal{Q}}(s_i) \geq depth_{\mathcal{Q}}(s_0) + i \geq i \geq k'$. Since z_i is useful, there exists a Σ-term t', a position $o \in \mathcal{P}os(t')$ and a final state $p_f \in \mathcal{P}_{final}$ such that

$$t' \vdash^*_{\mathcal{A}} t'[z_i]_o \vdash^*_{\mathcal{A}} p_f. \tag{5.2}$$

Let $t = t'[s_i]_o$, then $t \vdash^*_{\mathcal{A}} t'[z_i]_o \vdash^*_{\mathcal{A}} p_f \in \mathcal{P}_{final}$ by (5.1) and (5.2). Thus, $t \in \mathcal{L}(\mathcal{A})$ holds. Furthermore, $depth_{\mathcal{Q}}(t) \geq depth_{\mathcal{Q}}(s_i) \geq k' \geq k + 1$. This conflicts with Condition 2. Therefore, Procedure 1 halts. □

Theorem 4. *For any TA $\mathcal{A}$ and an LT-TRS $\mathcal{R}$ which satisfy condition (1) and (2) of Lemma 12, $(\to^*_{\mathcal{R}})(\mathcal{L}(\mathcal{A}))$ is recognizable.*

Proof. By Lemmas 8 and 12. □

Note that $\mathcal{R}_1$, $\mathcal{R}_2$, $\mathcal{R}_3$ in the examples satisfy condition (2) of Lemma 12. As mentioned in section 2.3, a TRS in Réty's subclass [13] is $\mathcal{C}$-EPR. Réty's subclass and the subclass of LT-TRSs satisfying condition (2) of Lemma 12 are incomparable. In fact, any non-left-linear TRS in Réty's subclass is not an LT-TRS. On the other hand, $\{f(q(x)) \to q(f(f(x)))\}$ is an LT-TRS satisfying condition (2) but does not belong to Réty's subclass. Also the class of marker-bounded sets and $\mathcal{C}$ are incomparable. For example, consider $\mathcal{R}_3$ in Example 5. $\mathcal{L}(\mathcal{A}) = \{f(q^n(c)) \mid n \geq 0\}$ does not satisfy condition (1) but $\mathcal{R}_3$ belongs to Réty's subclass and $\mathcal{L}(\mathcal{A})$ belongs to $\mathcal{C}$.

Corollary 2. *For a finite set T of ground terms and an LT-TRS $\mathcal{R}$ which contains no rewrite rule whose left-hand side is a constant, $(\rightarrow_{\mathcal{R}}^{*})(T)$ is recognizable.* □

Corollary 3. *For an LT-TRS $\mathcal{R}$ which contains no rewrite rule whose left-hand side is a constant, reachability and joinability are decidable.*

Proof. The reachability problem is to decide whether for a given TRS $\mathcal{R}$ and Σ-terms s and t, $s \rightarrow_{\mathcal{R}}^{*} t$ holds or not. It is obvious that $s \rightarrow_{\mathcal{R}}^{*} t$ if and only if $t \in (\rightarrow_{\mathcal{R}}^{*})(\{s\})$. The latter condition is decidable by Lemma 1 and Corollary 2.

Decidability of joinability can easily be verified by noting that $\exists w\colon s \rightarrow_{\mathcal{R}}^{*} w$ and $t \rightarrow_{\mathcal{R}}^{*} w$ if and only if $(\rightarrow_{\mathcal{R}}^{*})(\{s\}) \cap (\rightarrow_{\mathcal{R}}^{*})(\{t\}) \neq \emptyset$. □

6 Conclusion

In this paper, a new subclass of TRSs called LT-TRSs is defined and a sufficient condition for an LT-TRS to effectively preserve recognizability is provided. The subclass of LT-TRSs satisfying the condition contains simple EPR-TRSs which do not belong to any of the known decidable subclasses of EPR-TRSs.

Extending the proposed class is a future study. It is not difficult to extend Procedure 1 so that a ground term can occur at an argument position of the left-hand side of a rewrite rule. For example, assume that $f(z_1, z_2) \rightarrow z \in \Delta_{i-1}$ and $f(q_1(x_1), d(c)) \rightarrow q(g(x_1))$ $(c, d, f, g \in \mathcal{F}, q_1, q \in \mathcal{Q})$ is a rewrite rule. If $q_1(z_1') \rightarrow z_1 \in \Delta_{i-1}$ and $d(c) \vdash_{\mathcal{A}_{i-1}}^{*} z_2$ then construct a transition rule $g(z_1') \rightarrow [z, q]$.

Acknowledgments. The authors would like to thank Dr. Hitoshi Ohsaki for his invaluable discussions. Also they would like to thank the anonymous referees for their carefully reading the paper and giving useful comments.

References

1. Baader, F. and Nipkow, T.: *Term Rewriting and All That*, Cambridge University Press, 1998.
2. Brained, W.S.: "Tree generating regular systems," *Inform. and control*, **14**, pp.217–231, 1969.
3. Coquidé, J.L., Dauchet, M., Gilleron, R. and Vágvölgyi, S.: "Bottom-up tree push-down automata: classification and connection with rewrite systems," *Theoretical Computer Science*, **127**, pp.69–98, 1994.
4. Durand, I. and Middeldorp, A.: "Decidable call by need computations in term rewriting (extended abstract)," Proc. of CADE-14, North Queensland, Australia, LNAI **1249**, pp.4–18, 1997.
5. Genet, T.: "Decidable approximations of sets of descendants and sets of normal forms," Proc. of RTA98, Tsukuba, Japan, LNCS **1379**, pp.151–165, 1998.
6. Gécseq, F. and Steinby, M.: *Tree Automata*, Académiai Kiadó, 1984.

7. Gilleron, R.: "Decision problems for term rewriting systems and recognizable tree languages," Proc. of STACS'91, Hamburg, Germany, LNCS **480**, pp.148–159, 1991.
8. Gilleron, R. and Tison, S.: "Regular tree languages and rewrite systems," *Fundamenta Informaticae*, **24**, pp.157–175, 1995.
9. Gyenizse, P. and Vágvölgyi, S.: "Linear generalized semi-monadic rewrite systems effectively preserve recognizability," *Theoretical Computer Science*, **194**, pp.87–122, 1998.
10. Jacquemard, F.: "Decidable approximations of term rewriting systems," Proc. of RTA96, New Brunswick, NJ, LNCS **1103**, pp.362–376, 1996.
11. Middeldorp, A.: "Approximating dependency graph using tree automata techniques," Proc. of Int'l Joint Conf. on Automated Reasoning, LNAI **2083**, pp.593–610, 2001.
12. Nagaya, T. and Toyama, Y.: "Decidability for left-linear growing term rewriting systems," Proc. of RTA99, Trento, Italy, LNCS **1631**, pp.256–270, 1999.
13. Réty, P.: "Regular sets of descendants for constructor-based rewrite systems," Proc. of LPAR'99, Tbilisi, Georgia, LNCS **1705**, pp.148–160, 1999.
14. Salomaa, K.: "Deterministic tree pushdown automata and monadic tree rewriting systems," *J. Comput. system Sci.*, **37**, pp.367–394, 1988.
15. Takai, T., Kaji, Y. and Seki, H.: "Right-linear finite path overlapping term rewriting systems effectively preserve recognizability," Proc. of RTA2000, Norwich, U.K., LNCS **1833**, pp.246–260, 2000.

Decidability and Closure Properties of Equational Tree Languages

Hitoshi Ohsaki and Toshinori Takai

National Institute of Advanced Industrial Science and Technology
Nakoji 3–11–46, Amagasaki 661–0974, Japan
{ohsaki, takai}@ni.aist.go.jp

Abstract. Equational tree automata provide a powerful tree language framework that facilitates to recognize congruence closures of tree languages. In the paper we show the emptiness problem for AC-tree automata and the intersection-emptiness problem for regular AC-tree automata, each of which was open in our previous work [20], are decidable, by a straightforward reduction to the reachability problem for ground AC-term rewriting. The newly obtained results generalize decidability of so-called *reachable property problem* of Mayr and Rusinowitch [17]. We then discuss complexity issue of AC-tree automata. Moreover, in order to solve some other questions about regular A- and AC-tree automata, we recall the basic connection between word languages and tree languages.

1 Introduction

Tree automata theory has been studied with much attention as the basis of computational model dealing with terms. Recently tree automata techniques are applied in a particular domain for automatically solving security problems of *cryptographic protocols* (network protocols with some cryptographic primitives) [7,10,19]. This is a natural phenomenon resulting from the background of tree automata, in which complexity issues, especially on decidability, have been investigated with considerable efforts. In fact, applications based on tree automata theory, e.g. for program and system verification [9,22], depend deeply upon decidability results and closure properties.

Regular tree languages, which is the counterpart of regular word languages, are known to be well-behaved, in the sense that they are closed under boolean operations and their decision problems are computable in many cases. Moreover, in term rewriting there are some useful subclasses in which rewriting closures preserve regularity [24,25]. But, on the other hand, it causes the situation where questions of non-regular tree language are considered to be troublesome. For instance, because term algebra modulo equational theory is not handled with regular tree automata (e.g., AC-congruence closure of a regular tree language is not regular [3]), the standard tree automata technique can not be applied for modeling cryptographic primitives like Diffie-Hellman exponentiation and one-time pads [23]. The difficulty of system verification allowing AC-operators is also found in another algebraic approach [18].

S. Tison (Ed.): RTA 2002, LNCS 2378, pp. 114–128, 2002.

Over the last decade tree automata framework has been generalized. So far we can find several extensions, such as tree automata with constraints (e.g. [1, 13]), tree set automata [14] and timed tree automata [12]. Each of the extensions covers a wider class of regular tree languages, while keeping the benefit of some decidability results and closure properties. *Equational tree automata* proposed in our previous paper [20] is the recent extension, with which congruence closures of recognizable tree languages are recognizable. We also proved that in certain useful cases, recognizable equational tree languages are closed under union and intersection operations. However, there are several important questions remaining open in the previous work. Our goal of this paper is to find the answers to the open questions. More precisely, we show the following results:

(1) decidability of emptiness problem for AC-tree automata,
(2) decidability of intersection-emptiness, subset and universality problems for regular A- and regular AC-tree automata, and
(3) closure properties, intersection and complement, of A-regular and AC-regular tree languages.

Moreover, we show the computability of equational rewrite descendents. This result gives rise to the decidability of reachable property problem for ground AC-term rewriting.

The rest of the paper is organized as follows. In the remaining part of this section we fix our terminologies and notations on term rewriting. In Section 2 we recall equational tree automata. And we show a positive answer to the problem (1), by generalizing the decidability result of Mayr and Rusinowitch [16]. Using this result we show the computability of ground AC-rewrite descendents. In Section 3 we discuss the complexity issue on some AC-tree automata problems. The problems (2) and (3) for A-regular tree languages are (negatively) solved in Section 4. We then partially solve the same questions for AC-regular case in Section 5. Finally we conclude this paper by summarizing decidability results and closure properties of A-, C- and AC-tree languages.

We assume familiarity with the basics of term rewriting. An *equation* over the signature $\mathcal{F}$ (a finite set of function symbols f whose arity is denoted by $\mathsf{arity}(f)$) and a set $\mathcal{V}$ of variables is a pair (s,t) of terms $s,t \in \mathcal{T}(\mathcal{F},\mathcal{V})$. An equation (s,t) is denoted by $s \approx t$. An equation $l \approx r$ is called *linear* if neither l nor r contains multiple positions of the same variable. We say $l \approx r$ is *variable-preserving* if a multiset occurrence of each variable x in l is the same as a multiset occurrence of x in r. A *ground* equation is an equation whose left- and right-hand sides do not contain any variable. An *equational system* (ES for short) $\mathcal{E}$ is a set of equations. Given two sets $\mathcal{F}_\mathrm{A}$, $\mathcal{F}_\mathrm{C}$ of some binary function symbols in $\mathcal{F}$. The intersection of $\mathcal{F}_\mathrm{A}$ and $\mathcal{F}_\mathrm{C}$ is denoted by $\mathcal{F}_\mathrm{AC}$. An ES consisting of associativity axioms $f(f(x,y),z) \approx f(x,f(y,z))$ for all $f \in \mathcal{F}_\mathrm{A}$ is denoted by $\mathsf{A}(\mathcal{F}_\mathrm{A})$. An ES of commutativity axioms $f(x,y) \approx f(y,x)$ for all $f \in \mathcal{F}_\mathrm{C}$ is $\mathsf{C}(\mathcal{F}_\mathrm{C})$. We write $\mathsf{AC}(\mathcal{F}_\mathrm{AC})$ for the union of $\mathsf{A}(\mathcal{F}_\mathrm{AC})$ and $\mathsf{C}(\mathcal{F}_\mathrm{AC})$. If unnecessary to be explicit, we simply write A, C and AC. A binary relation $\to_\mathcal{E}$ induced by an ES $\mathcal{E}$ is defined as follows: $s \to_\mathcal{E} t$ if $s = C[l\sigma]$ and $t = C[r\sigma]$ for some equation $l \approx r \in \mathcal{E}$, context $C \in \mathcal{C}(\mathcal{F},\mathcal{V})$ and substitution σ. The symmetric closure of $\to_\mathcal{E}$ is denoted by

$\vdash\!\dashv_{\mathcal{E}}$. In the paper we do not assume $r \approx l \in \mathcal{E}$ if $l \approx r \in \mathcal{E}$, so $\rightarrow_{\mathcal{E}} \neq \vdash\!\dashv_{\mathcal{E}}$. The equivalence relation of $\rightarrow_{\mathcal{E}}$ (i.e., the reflexive-transitive closure of $\vdash\!\dashv_{\mathcal{E}}$) is written by $\sim_{\mathcal{E}}$.

A term rewriting system is an ES whose equations $l \approx r$ are called *rewrite rules* and are denoted by $l \rightarrow r$. An equational TRS (ETRS for short) $\mathcal{R}/\mathcal{E}$ is a combination of a TRS $\mathcal{R}$ and an ES $\mathcal{E}$ over the same signature $\mathcal{F}$. To emphasize the signature of an ETRS, it can be denoted by the pair $(\mathcal{F}, \mathcal{R}/\mathcal{E})$ of two components. An ETRS is called *ground* if the TRS consists of ground rewrite rules. Note that $\mathcal{E}$ is not necessarily ground. We write $s \rightarrow_{\mathcal{R}/\mathcal{E}} t$ if there exist terms s' and t' such that $s \sim_{\mathcal{E}} s' \rightarrow_{\mathcal{R}} t' \sim_{\mathcal{E}} t$. As a special case, if $\mathcal{E} = \mathsf{A}$ ($\mathcal{E} = \mathsf{C}$, $\mathcal{E} = \mathsf{AC}$, respectively), $\mathcal{R}/\mathcal{E}$ is called an A-TRS (C-TRS, AC-TRS).

Given an ETRS $\mathcal{R}/\mathcal{E}$. A term s reachable from a term t with respect to $\mathcal{R}/\mathcal{E}$, i.e. $t \rightarrow^*_{\mathcal{R}/\mathcal{E}} s$ is called a *descendent* of t. For a set L of terms, descendents of terms in L, i.e. $\{ t \mid \exists s \in L.\ s \rightarrow^*_{\mathcal{R}/\mathcal{E}} t \}$, are denoted by $(\rightarrow^*_{\mathcal{R}/\mathcal{E}})[L]$.

2 AC-Tree Automata and Emptiness Problem

A *tree automaton* (TA for short) $\mathcal{A} = (\mathcal{F}, \mathcal{Q}, \mathcal{Q}_{fin}, \Delta)$ consists of a signature $\mathcal{F}$, a finite set $\mathcal{Q}$ of states (special constants with $\mathcal{F} \cap \mathcal{Q} = \varnothing$), a set $\mathcal{Q}_{fin}$ ($\subseteq \mathcal{Q}$) of final states and a finite set Δ of transition rules in one of the following forms:

$$f(p_1, \ldots, p_n) \rightarrow q \qquad \text{or} \qquad f(p_1, \ldots, p_n) \rightarrow f(q_1, \ldots, q_n)$$

for some $f \in \mathcal{F}$ with $\mathsf{arity}(f) = n$ and $p_1, \ldots, p_n, q, q_1, \ldots, q_n \in \mathcal{Q}$. In the latter form, the root function symbols of the left- and right-hand sides must be the same. An *equational tree automaton* (ETA for short) $\mathcal{A}/\mathcal{E}$ is the combination of a TA $\mathcal{A}$ and an ES $\mathcal{E}$ over the same signature $\mathcal{F}$. We often denote the 5-tuple $(\mathcal{F}, \mathcal{Q}, \mathcal{Q}_{fin}, \Delta, \mathcal{E})$ for $\mathcal{A}/\mathcal{E}$. An ETA $\mathcal{A}/\mathcal{E}$ is called *regular* if Δ consists of transition rules in the shape of $f(p_1, \ldots, p_n) \rightarrow q$. We say $\mathcal{A}/\mathcal{E}$ is an A-TA (associative-tree automaton) if $\mathcal{E} = \mathsf{A}$. An ETA $\mathcal{R}/\mathcal{E}$ with $\mathcal{E} = \mathsf{C}$ is called a C-TA (commutative-tree automaton). Likewise, if $\mathcal{E} = \mathsf{AC}$, it is called an AC-TA.

We write $s \rightarrow_{\mathcal{A}/\mathcal{E}} t$ if there exist a transition rule $l \rightarrow r \in \Delta$ and a context $C \in \mathcal{C}(\mathcal{F} \cup \mathcal{Q})$ such that $s \sim_{\mathcal{E}} C[l]$ and $t \sim_{\mathcal{E}} C[r]$. The relation $\rightarrow_{\mathcal{A}/\mathcal{E}}$ on $\mathcal{T}(\mathcal{F} \cup \mathcal{Q})$ is called *move relation* of $\mathcal{A}/\mathcal{E}$. The transitive closure and reflexive-transitive closure of $\rightarrow_{\mathcal{A}/\mathcal{E}}$ are denoted by $\rightarrow^+_{\mathcal{A}/\mathcal{E}}$ and $\rightarrow^*_{\mathcal{A}/\mathcal{E}}$. For a TA $\mathcal{A}$, we simply write $\rightarrow_{\mathcal{A}}$, $\rightarrow^+_{\mathcal{A}}$ and $\rightarrow^*_{\mathcal{A}}$, instead. A state q is called *accessible* with $\mathcal{A}/\mathcal{E}$ if there exists a term t in $\mathcal{T}(\mathcal{F})$ such that $t \rightarrow^*_{\mathcal{A}/\mathcal{E}} q$. A term t is *accepted* by $\mathcal{A}/\mathcal{E}$ if $t \in \mathcal{T}(\mathcal{F})$ and $t \rightarrow^*_{\mathcal{A}/\mathcal{E}} q$ for some $q \in \mathcal{Q}_{fin}$. Elements in the set $\mathcal{L}(\mathcal{A}/\mathcal{E})$ are ground terms accepted by $\mathcal{A}/\mathcal{E}$. A *tree language* L over $\mathcal{F}$ is some subset of $\mathcal{T}(\mathcal{F})$. We can easily observe that: membership problem for an ETA $\mathcal{A}/\mathcal{E}$ is decidable if $\mathcal{E}$ is length-preserving, i.e., if $\mathcal{E}$ is variable-preserving and for every $l \approx r \in \mathcal{E}$, the number of function symbols occurring in l is the same as the number of function symbols in r.

We write $\mathcal{E}(L)$ for $\{t \mid \exists s \in L.\ s \sim_{\mathcal{E}} t\}$. The set $\mathcal{E}(L)$ is called *$\mathcal{E}$-congruence closure* of L. A tree language L is *$\mathcal{E}$-recognizable* if there exists $\mathcal{A}/\mathcal{E}$ such that

$L = \mathcal{L}(\mathcal{A}/\mathcal{E})$. In certain useful cases, $\mathcal{E}$-recognizable tree languages are closed under union and intersection:

Lemma 1. *If $\mathcal{E}$ is a variable-preserving ES, the union of $\mathcal{E}$-recognizable tree languages L_1 and L_2 is $\mathcal{E}$-recognizable. Moreover, if $\mathcal{E} = \mathsf{A}$ (or $\mathcal{E} = \mathsf{AC}$), the intersection of L_1 and L_2 is $\mathcal{E}$-recognizable.* □

A tree language L is called *$\mathcal{E}$-regular* if L is $\mathcal{E}$-recognizable with a regular ETA $\mathcal{A}/\mathcal{E}$. If $\mathcal{E} = \varnothing$, the tree language L is called regular. Every recognizable tree language is regular, i.e. every TA is transformed to a regular TA with the same expressive power. However, $\mathcal{E}$-recognizable tree languages are not $\mathcal{E}$-regular in general.

In the previous paper [20] we studied decidability on equational tree automata. Especially we focused on emptiness problems, i.e. a question of whether $\mathcal{L}(\mathcal{A}/\mathcal{E}) = \varnothing$. In case $\mathcal{E}$ is linear, this question for regular $\mathcal{E}$-tree automata can be reduced to the same question for regular tree automata by using the commutation lemma:

Lemma 2. *Every regular ETA $\mathcal{A}/\mathcal{E}$ with $\mathcal{E}$ linear satisfies $\mathcal{E}(\mathcal{L}(\mathcal{A})) = \mathcal{L}(\mathcal{A}/\mathcal{E})$.* □

Since $\mathcal{E}(\mathcal{L}(\mathcal{A})) = \varnothing$ if and only if $\mathcal{L}(\mathcal{A}) = \varnothing$, the following property is a direct consequence of Lemma 2.

Corollary 1 ([20]). *Emptiness problem for* regular *$\mathcal{E}$-tree automata with $\mathcal{E}$ linear is decidable.* □

Thus the same problem for regular A-, C- and AC-tree automata is decidable. In contrast, the non-regular A-case is undecidable (Corollary 1, [20]). The non-regular AC-case remains open in the previous paper.

In the following part, we show the positive answer to this open question. First of all, we introduce a recent work on *reachability problem* for ground AC-TRS of Mayr and Rusinowitch [16].

Proposition 1. *Reachability problem for ground AC-TRS, i.e. a question of whether $s \to^*_{\mathcal{R}/\mathsf{AC}} t$ for an arbitrary ground AC-TRS $\mathcal{R}/\mathsf{AC}$ and terms s, t, is decidable.* □

Next we state the following property.

Lemma 3. *Given two ground AC-TRSs $(\mathcal{F}, \mathcal{R}/\mathsf{AC})$ and $(\mathcal{G}, \mathcal{S}/\mathsf{AC})$, where $\mathcal{F}_{\mathsf{AC}} = \mathcal{G}_{\mathsf{AC}}$. If $\mathcal{F}_0 \cap \mathcal{G}_0 = \varnothing$, i.e. $\mathcal{F}$ and $\mathcal{G}$ do not share any constant symbol, then for all terms $s, t \in \mathcal{T}(\mathcal{F} \cup \mathcal{G})$, $s \to^*_{(\mathcal{R}\cup\mathcal{S})/\mathsf{AC}} t$ if and only if $s \to^*_{\mathcal{R}/\mathsf{AC}} s'$ and $s' \to^*_{\mathcal{S}/\mathsf{AC}} t$ for some s'.*

Proof. Since the "if" part is trivial, it suffices to show the "only if" part. Suppose $u \to_{\mathcal{S}/\mathsf{AC}} v \to_{\mathcal{R}/\mathsf{AC}} w$. Then $u \sim_{\mathsf{AC}} C_1[l_1]$ and $v \sim_{\mathsf{AC}} C_1[r_1]$ for some $l_1 \to r_1 \in \mathcal{S}$, and $v \sim_{\mathsf{AC}} C_2[l_2]$ and $w \sim_{\mathsf{AC}} C_2[r_2]$ for some $l_2 \to r_2 \in \mathcal{R}$. By assumption, r_1 is a term in $\mathcal{T}((\mathcal{F} \cup \mathcal{G}) - \mathcal{F}_0)$ and l_2 is a term in $\mathcal{T}((\mathcal{F} \cup \mathcal{G}) - \mathcal{G}_0)$. This yields a context D such that $D[r_1, l_2] \sim_{\mathsf{AC}} C_1[r_1]$ and $D[r_1, l_2] \sim_{\mathsf{AC}} C_2[l_2]$, and thus, $u \sim_{\mathsf{AC}} D[l_1, l_2] \to_{\mathcal{S}} D[r_1, l_2] \to_{\mathcal{R}} D[r_1, r_2] \sim_{\mathsf{AC}} w$. Hence $u \sim_{\mathsf{AC}} D[l_1, l_2] \to_{\mathcal{R}} D[l_1, r_2] \to_{\mathcal{S}} D[r_1, r_2] \sim_{\mathsf{AC}} w$. This implies $\to^*_{(\mathcal{R}\cup\mathcal{S})/\mathsf{AC}} \subseteq \to^*_{\mathcal{R}/\mathsf{AC}} \cdot \to^*_{\mathcal{S}/\mathsf{AC}}$. □

In the previous lemma, by taking an AC-TA $(\mathcal{F}, \mathcal{Q}, \mathcal{Q}_{fin}, \Delta, \mathsf{AC})$ as a ground AC-TRS $(\mathcal{F} \cup \mathcal{Q}, \Delta/\mathsf{AC})$, we can extend Proposition 1 as follows.

Lemma 4. *Given a ground AC-TRS $(\mathcal{F}, \mathcal{R}/\mathsf{AC})$ and tree languages L_1, L_2 over $\mathcal{F}$. If L_1 and L_2 are AC-recognizable tree languages, it is decidable whether there exist some s in L_1 and t in L_2 such that $s \to^*_{\mathcal{R}/\mathsf{AC}} t$.*

Proof. Suppose $L_1 = \mathcal{L}(\mathcal{A}/\mathsf{AC})$ and $L_2 = \mathcal{L}(\mathcal{B}/\mathsf{AC})$ where $\mathcal{A} = (\mathcal{F}, \mathcal{P}, \mathcal{P}_{fin}, \Delta_1)$ and $\mathcal{B} = (\mathcal{F}, \mathcal{Q}, \mathcal{Q}_{fin}, \Delta_2)$, such that $\mathcal{P} \cap \mathcal{Q} = \varnothing$. Here we assume without loss of generality that p, q are state symbols such that $\mathcal{P}_{fin} = \{\mathsf{p}\}$ and $\mathcal{Q}_{fin} = \{\mathsf{q}\}$. Then we obtain

$$\exists s \in L_1,\ \exists t \in L_2.\ s \to^*_{\mathcal{R}/\mathsf{AC}} t \qquad \cdots \text{(A)}$$

if and only if

$$\exists u, v \in \mathcal{T}(\mathcal{F}).\ u \to^*_{\mathcal{A}/\mathsf{AC}} \mathsf{p} \ \wedge\ v \to^*_{\mathcal{B}/\mathsf{AC}} \mathsf{q} \ \wedge\ u \to^*_{\mathcal{R}/\mathsf{AC}} v. \qquad \cdots \text{(B)}$$

Moreover, using Lemma 3 twice, (B) holds if and only if

$$\mathsf{p} \to^*_{(\mathcal{R} \cup \mathcal{A}^{-1} \cup \mathcal{B})/\mathsf{AC}} \mathsf{q}. \qquad \cdots \text{(C)}$$

Due to Proposition 1, we know (C) is decidable, and so is (A). □

Given a term t, a predicate P and a binary relation $\to$ on terms, a question of whether there exists some s in $(\to^*)[\,\{t\}\,]$ with $P(s)$ is called *reachable property problem.* Here $P(s)$ can be replaced by the membership test $s \in [\![P]\!]$. Mayr and Rusinowitch showed in [17] that if P is a so-called state formula, this problem is decidable within their framework called *process rewrite systems.* In term rewriting setting, it corresponds to the following result:

Proposition 2. *Reachable property problem is decidable for ground AC-TRSs if $[\![P]\!]$ is a regular tree language.* □

Due to Lemma 4 we know that $(\to^*_{\mathcal{R}/\mathsf{AC}})[\,L_1\,] \cap L_2 = \varnothing$ is a computable question, provided $\mathcal{R}/\mathsf{AC}$ is a ground AC-TRS and L_1, L_2 are AC-recognizable. Therefore the above proposition is generalized by our equational tree automata framework.

One should notice that even if L_1 and/or L_2 are regular, the above decidability holds. Note that regular tree languages are not AC-recognizable: Suppose $\mathcal{F}_{\mathrm{AC}} = \{\mathsf{f}\}$ and $L = \{\mathsf{f}(\mathsf{a}, \mathsf{b})\}$. The tree language L is regular, but it is not AC-recognizable (and not AC-regular), because every AC-TA $\mathcal{A}/\mathsf{AC}$ accepting $\mathsf{f}(\mathsf{a}, \mathsf{b})$ also accepts $\mathsf{f}(\mathsf{b}, \mathsf{a})$.

On the other hand, as easy instances of Lemma 4, we can show the positive answers to our open questions:

Corollary 2. *Emptiness problem for AC-TA is decidable.*

Proof. Let $\mathcal{A}/\mathsf{AC}$ be an AC-TA over the signature $\mathcal{F}$. In Lemma 4, by taking $\mathcal{R} = \varnothing$, $L_1 = \mathcal{L}(\mathcal{A}/\mathsf{AC})$ and $L_2 = \mathcal{T}(\mathcal{F})$, we know that $L_1 \neq \varnothing$ if and only if there exist $s \in L_1$ and $t \in L_2$ such that $s \to^*_{\mathcal{R}/\mathsf{AC}} t$. □

Corollary 3. *Intersection-emptiness problem for regular AC-TA is decidable.*

Proof. Along the same lines of the previous case, we take $\mathcal{R} = \varnothing$, $L_1 = \mathcal{L}(\mathcal{A}/\mathsf{AC})$ and $L_2 = \mathcal{L}(\mathcal{B}/\mathsf{AC})$ such that $\mathcal{A}/\mathsf{AC}$ and $\mathcal{B}/\mathsf{AC}$ are regular AC-TA. □

Input $\mathcal{A}$: TA $(\mathcal{F}, \mathcal{Q}, \mathcal{Q}_{fin}, \Delta)$, $\mathcal{F}_{\mathrm{AC}}$: AC-symbols.
Output $\mathcal{Q}_{acc}$: set of accessible states.

Step 1. Let $\Delta' := \{\, l \to r \in \Delta \mid \mathsf{root}(l) \in \mathcal{F}_{\mathrm{AC}} \,\}$, $\mathcal{Q}_0, \mathcal{Q}_{acc} := \varnothing$, $k := 0$.
Step 2. Compute $\mathcal{Q}'$: set of state symbols
$$\{q \in \mathcal{Q} \mid \exists t \in \mathcal{T}(\mathcal{F} \cup \mathcal{Q}_k).\ t \to^*_{\Delta-\Delta'} q\}.$$
Step 3. Compute $\mathcal{Q}_{k+1}$: the union of $\mathcal{Q}'$ and
$$\bigcup_{f \in \mathcal{F}_{\mathrm{AC}}} \mathsf{aux}(\{l \to r \in \Delta' \mid \mathsf{root}(l) = f\}, f, \mathcal{Q}', \mathcal{Q}).$$
Here aux is a function call to the auxiliary procedure.
Step 4. If $\mathcal{Q}_{k+1} \neq \mathcal{Q}_k$ then k is increased by 1 and go to Step 2; otherwise, return $\mathcal{Q}_{acc}$ by letting $\mathcal{Q}_{acc} := \mathcal{Q}_k$.

Fig. 1. Procedure: State symbols accessible with AC-TA

Input $\mathcal{R}$: ground TRS, f: binary function symbol,
$\mathcal{Q}_1, \mathcal{Q}_2$: sets of constants such that $\mathcal{Q}_1 \subseteq \mathcal{Q}_2$.
Output $\mathcal{P}$: set of state symbols
$$\{q \in \mathcal{Q}_2 \mid \exists t \in \mathcal{T}(\{f\} \cup \mathcal{Q}_1).\ t \to^*_{\mathcal{R}/\mathsf{AC}(f)} q\}.$$

Step 1. Select a fresh constant c.
Let $\mathcal{S} := \mathcal{R} \cup \{\, c \to q \mid q \in \mathcal{Q}_1 \,\} \cup \{\, c \to f(c, c) \,\}$.
Step 2. Compute $\mathcal{P} := \{\, q \in \mathcal{Q}_2 \mid c \to^*_{\mathcal{S}/\mathsf{AC}(f)} q \,\}$.
Step 3. Return $\mathcal{P}$.

Fig. 2. Auxiliary Procedure

But an important remark for the above results is that Mayr and Rusinowitch did not actually show in [16] the proof of the general case of Proposition 1, i.e. there is no proof of decidability of the reachability problem for ground AC-TRS with *arbitrary many* AC-symbols. In other words, as far as the above proofs depend upon Proposition 1, the newly obtained result (e.g. Lemma 4) works only for the single-AC case.

So in the next section, by showing the original procedure, we generalize our results (Corollaries 2 and 3) together with doing estimation of the complexity of emptiness questions.

3 Complexity of AC-Tree Language Problems

Let us introduce a special notation for AC-tree languages: Suppose f is a function symbol in $\mathcal{F}_{\mathrm{AC}}$. An *$f$-block of a term t* is a non-empty maximal context in t consisting only of f. For notational convenience, if $t = C'[C[t_1, \ldots, t_n]]$ such that C is an f-block, we write $C'[C_f[\![t_1, \ldots, t_n]\!]]$. For instance, $g(f(a, f(b, c)))$ is represented as $g(C_f[\![a, b, c]\!])$.

First we state the basic property of the procedure in Fig. 2. This procedure is defined as in the way of Garey and Johnson [5].

Lemma 5. *At the step 2 in Fig. 2, for every $q \in \mathcal{Q}_2$, if $c \to^*_{\mathcal{S}/\mathsf{AC}(f)} q$, there exists a term $t \in \mathcal{T}(\mathcal{Q}_1 \cup \{f\})$ such that $c \to^*_{\mathcal{S}/\mathsf{AC}(f)} t \to^*_{\mathcal{R}/\mathsf{AC}(f)} q$.*

Proof. Let $\mathcal{R}_1 = \{c \to f(c, c)\}$ and $\mathcal{R}_2 = \{c \to q \mid q \in \mathcal{Q}_1\}$. Since c does not appear in the right-hand sides of rewrite rules of a ground TRS $\mathcal{R} \cup \mathcal{R}_2$, we can prove that $\to_{(\mathcal{R}\cup\mathcal{R}_2)/\mathsf{AC}(f)} \cdot \to_{\mathcal{R}_1/\mathsf{AC}(f)} \subseteq \to_{\mathcal{R}_1/\mathsf{AC}(f)} \cdot \to_{(\mathcal{R}\cup\mathcal{R}_2)/\mathsf{AC}(f)}$. By the same reason, $\to_{\mathcal{R}/\mathsf{AC}(f)} \cdot \to_{\mathcal{R}_2/\mathsf{AC}(f)} \subseteq \to_{\mathcal{R}_2/\mathsf{AC}(f)} \cdot \to_{\mathcal{R}/\mathsf{AC}(f)}$. Thus we obtain a rewrite sequence $c \to^*_{\mathcal{R}_1/\mathsf{AC}(f)} t_1 \to^*_{\mathcal{R}_2/\mathsf{AC}(f)} t_2 \to^*_{\mathcal{R}/\mathsf{AC}(f)} q$. Since rewrite rules in $\mathcal{R}$ are ground and do not contain c, the number of occurrences of c does not change in $t_2 \to^*_{\mathcal{R}/\mathsf{AC}(f)} q$. This implies $|t_2|_c = 0$, because $q \neq c$. Similarly, since $\mathcal{R}_1 \cup \mathcal{R}_2$ is ground and consists of function symbols in $\mathcal{Q}_1 \cup \{f\}$, we have $t_2 \in \mathcal{T}(\mathcal{Q}_1 \cup \{f\} \cup \{c\})$. Hence $t_2 \in \mathcal{T}(\mathcal{Q}_1 \cup \{f\})$. □

Next we show soundness of the main procedure.

Lemma 6. *For every AC-TA $\mathcal{A}/\mathsf{AC}$, a state q is accessible with $\mathcal{A}/\mathsf{AC}$ if and only if q is in $\mathcal{Q}_{acc}$.*

Proof. Let $\mathcal{A} = (\mathcal{F}, \mathcal{Q}, \mathcal{Q}_{fin}, \Delta)$. Using the induction on the loop-counter k of the procedure in Fig. 1, we show that $q \in \mathcal{Q}_k$ for some k if and only if q is accessible with $\mathcal{A}/\mathsf{AC}$. First we prove the "if" part. The base case is trivial, because $\mathcal{Q}_0 = \varnothing$. For induction step, we suppose $q \in \mathcal{Q}_k$. We divide into the three cases as follows: (1) If $q \in \mathcal{Q}_{k'}$ with $k' < k$, the lemma holds by induction hypothesis. (2) If $q \notin \mathcal{Q}_{k'}$ for any $k' < k$ but $q \in \mathcal{Q}'$ at Step 2, the lemma also holds by induction hypothesis and the definition of Step 2. (3) If $q \notin \mathcal{Q}_{k'}$ for any $k' < k$ but $q \in \mathcal{Q}_k$ at Step 3 then $c \to^*_{\mathcal{S}/\mathsf{AC}(f)} q$ for some $f \in \mathcal{F}_{\mathsf{AC}}$. By Lemma 5, we have a term $t \in \mathcal{T}(\mathcal{Q}' \cup \{f\})$ such that $t \to^*_{\mathcal{R}/\mathsf{AC}(f)} q$ where $\mathcal{R} = \{l \to r \in \Delta \mid \mathsf{root}(l) = f\}$. From the previous cases, since $\mathcal{Q}'$ at Step 2 is a set of accessible states, so is q (as $t' \to^*_{\mathcal{A}/\mathsf{AC}} t$ for some $t' \in \mathcal{T}(\mathcal{F})$). For the "only if" part, we take a state $q \in \mathcal{Q}$ such that $t \to^*_{\mathcal{A}/\mathsf{AC}} q$ for some $t \in \mathcal{T}(\mathcal{F})$. By induction on the number of blocks in t, we show that q is in $\mathcal{Q}_k$ for some k. If there is no block in t, by letting $k = 0$ the lemma holds obviously. For induction step, we assume without loss of generality that $t = C'[s_1, \ldots, s_{i-1}, C_f[\![t_1, \ldots, t_n]\!], s_{i+1}, \ldots, s_m]$, where $C' \in \mathcal{C}(\mathcal{F} - \mathcal{F}_{\mathsf{AC}})$. C' is possibly the empty context. By assumption, $t \to^*_{\mathcal{A}/\mathsf{AC}} C'[p_1, \ldots, p_m] \to^*_{\mathcal{A}} q$ and $C_f[\![t_1, \ldots, t_n]\!] \to^*_{\mathcal{A}/\mathsf{AC}} C_f[\![q_1, \ldots, q_n]\!] \to^*_{\mathcal{A}/\mathsf{AC}} p_i$ for some $p_1, \ldots, p_m, q_1, \ldots, q_n \in \mathcal{Q}$. By induction hypothesis, $q_j \in \mathcal{Q}_{k_j}$ for some k_j ($1 \leqslant j \leqslant n$). Moreover, $p_j \in \mathcal{Q}_{k'_j}$ for some k'_j ($1 \leqslant j \leqslant m$) if $j \neq i$. By definition of the procedure in Fig. 1, we obtain $p_i \in \mathcal{Q}_{k_{\max}+1}$ where $k_{\max} = \max\{k_j \mid 1 \leqslant j \leqslant n\}$. Hence, by letting $k'_{\max} = \max(\{k'_j \mid 1 \leqslant j \leqslant m \text{ and } j \neq i\} \cup \{k_{\max}+1\})$, $q \in \mathcal{Q}_{k'_{\max}}$ or $q \in \mathcal{Q}'$ at the $(k'_{\max}+1)$-th loop. □

Theorem 1. *Emptiness problem for AC-TA with arbitrary many AC-symbols is decidable.* □

The procedure in Fig. 1 needs at most $(|\mathcal{Q}|^2 \times |\mathcal{F}_{\mathsf{AC}}|)$-auxiliary function calls, and each computation is reachability test of a Petri net instance. Reachability problem for Petri nets is EXPSPACE-hard [15].

In contrast, the emptiness problem for *regular* AC-TA is solved in linear time, due to Lemma 2. From the similar observation, the complexity of membership problem can also be obtained.

Corollary 4. *Membership problem for regular AC-TA is NP-complete.*

Proof. We show that this problem is in NP. Let $\mathcal{A}/\mathsf{AC}$ be a regular AC-TA. From Lemma 2, a term t is accepted by $\mathcal{A}/\mathsf{AC}$ if and only if there exists a term t' accepted by $\mathcal{A}$ and $t' \sim_{\mathsf{AC}} t$. Thus the following procedure is sound:

(1) Guess a term t' such that $t' \sim_{\mathsf{AC}} t$.
(2) Check whether or not $t' \in \mathcal{L}(\mathcal{A})$.

Since the membership problem for regular TA is solved in linear time, the above algorithm is in NP. On the other hand, membership problem for commutative context-free grammar, which is known as an NP-hard problem [4], is reduced to this problem by assuming $\mathcal{F} = \{\mathsf{f}\} \cup \Sigma$ with $\mathcal{F}_{\mathsf{AC}} = \{\mathsf{f}\}$. Here Σ is the alphabet of a grammar. Hence the membership problem for regular AC-TA is NP-complete. □

On the other hand, because AC-recognizable tree languages are closed under intersection (Lemma 1), we can extend another decidability result.

Theorem 2. *Intersection-emptiness problem for regular AC-TA with arbitrary many AC-symbols is decidable.* □

4 Yet Negative Results on A-Tree Languages

In this section we discuss decidability of A-tree languages. In the previous paper we showed that emptiness problem for A-TA is *undecidable* in general. Using this result with the same proof argument of Lemma 4, we can prove the following undecidability.

Proposition 3 ([3]). *Reachability problem for ground A-TRS is undecidable.*

Proof. We use the following fact (Corollary 1, [20]): Emptiness problem is undecidable for A-TA. We take an A-TA $\mathcal{A}/\mathsf{A}$. Here we assume without loss of generality that $\mathcal{A} = (\mathcal{F}, \mathcal{Q}, \{\mathsf{q}\}, \Delta_1)$ like in the proof of Lemma 4. Define the TA $\mathcal{B} = (\mathcal{F}, \{\mathsf{p}\}, \{\mathsf{p}\}, \Delta_2)$ where $\Delta_2 = \{f(\mathsf{p}, \dots, \mathsf{p}) \to \mathsf{p} \mid f \in \mathcal{F}\}$. Obviously, $\mathcal{L}(\mathcal{B}/\mathsf{A}) = \mathcal{T}(\mathcal{F})$. Similar to the proof of Lemma 3, we obtain $\to_{\mathcal{A}/\mathsf{A}} \cdot \to_{\mathcal{B}^{-1}/\mathsf{A}} \subseteq \to_{\mathcal{B}^{-1}/\mathsf{A}} \cdot \to_{\mathcal{A}/\mathsf{A}}$. This implies that $\mathsf{p} \to^*_{(\mathcal{A}\cup\mathcal{B}^{-1})/\mathsf{A}} \mathsf{q}$ if and only if $\mathsf{p} \to^*_{\mathcal{B}^{-1}/\mathsf{A}} t \to^*_{\mathcal{A}/\mathsf{A}} \mathsf{q}$ for some t. By definitions of $\mathcal{A}$ and $\mathcal{B}$, $t \in \mathcal{T}(\mathcal{F} \cup \{\mathsf{p}\})$, and moreover, $t \in \mathcal{T}(\mathcal{F} \cup \mathcal{Q})$. Thus t is in $\mathcal{T}(\mathcal{F})$ if there exists. Since it is undecidable whether $t \to^*_{\mathcal{A}/\mathsf{A}} \mathsf{q}$ for some $t \in \mathcal{T}(\mathcal{F})$, so is $\mathsf{p} \to^*_{(\mathcal{A}\cup\mathcal{B}^{-1})/\mathsf{A}} \mathsf{q}$. □

Other problems, such as subset and universality problems, are also undecidable for A-TA. However, decidability of the two problems for regular A-TA remains open in [20]. In the following part, we show these decidability results by reducing to the same *word* problems.

A *grammar* $\mathcal{G}$ is the 4-tuple $(\Sigma, \mathcal{Q}, \mathsf{q}_0, \Delta)$, whose components are the alphabet Σ, a finite set $\mathcal{Q}$ of state symbols such that $\Sigma \cap \mathcal{Q} = \varnothing$, an initial state q_0 $(\in \mathcal{Q})$ and a finite set Δ of string rewrite rules (called *production rules*) in the form of $l \to r$ for some $l, r \in (\Sigma \cup \mathcal{Q})^+$. A grammar is called *context-free* if $l \in \mathcal{Q}$ for all rules $l \to r \in \Delta$. A regular grammar is a context-free grammar whose right-hand sides of rules are in $(\Sigma \cup \mathcal{Q})\,\Sigma^*$. A word generated by $\mathcal{G}$ is an element w in Σ^* such that $\mathsf{q}_0 \to_{\mathcal{G}}^* w$. Note that $\mathcal{G}$ does not generate the empty word ϵ. The language generated by $\mathcal{G}$, denoted by $\mathcal{L}(\mathcal{G})$, is a set of generated words.

A relationship between word languages and tree languages can be found in the literature, e.g. in [2,6]. A study of leaf languages is one of the examples:

Proposition 4 ([6]). *The following two statements hold true:*

1. *For every CFG $\mathcal{G}$ there exists a regular TA $\mathcal{A}$ such that $\mathcal{L}(\mathcal{G}) = \mathsf{leaf}(\mathcal{L}(\mathcal{A}))$.*
2. *For every regular TA $\mathcal{A}$ there exists a CFG $\mathcal{G}$ such that $\mathsf{leaf}(\mathcal{L}(\mathcal{A})) = \mathcal{L}(\mathcal{G})$.*

□

A *leaf-word* $\mathsf{leaf}(t)$ of a term t over the signature $\mathcal{F}$ is a word over the alphabet $\mathcal{F}_0$ (a set of constants in $\mathcal{F}$), that is inductively defined as follows: $\mathsf{leaf}(t) = t$ if t is a constant; $\mathsf{leaf}(t) = \mathsf{leaf}(t_1) \ldots \mathsf{leaf}(t_n)$ if $t = f(t_1, \ldots, t_n)$ and $\mathsf{arity}(f) \geqslant 1$. We use the same notation leaf for the extension to tree languages, i.e. let L be a tree language over $\mathcal{F}$, then $\mathsf{leaf}(L) = \{\mathsf{leaf}(t) \mid t \in L\}$. We say $\mathsf{leaf}(L)$ is a *leaf-language* of L.

One should notice that the following statement is *not* true: If $\mathsf{leaf}(L)$ is CFG then L is a regular tree language. For instance, let us consider the following example. We take $\mathcal{G}_1 = (\{\mathsf{a}, \mathsf{b}\}, \{\alpha, \beta, \gamma\}, \gamma, \Delta_1)$ where Δ_1:

$$\begin{array}{lllll} \gamma \to \mathsf{a}\,\beta & \alpha \to \mathsf{a} & \alpha \to \mathsf{b}\,\alpha\,\alpha & \beta \to \mathsf{b} & \beta \to \mathsf{a}\,\beta\,\beta \\ \gamma \to \mathsf{b}\,\alpha & \alpha \to \mathsf{a}\,\gamma & & \beta \to \mathsf{b}\,\gamma & \end{array}$$

By induction on the length of words, we can prove that for every non-empty word w over the alphabet $\{\mathsf{a}, \mathsf{b}\}$,

$$\begin{array}{l} \alpha \to_{\mathcal{G}_1}^* w \text{ if and only if } |w|_{\mathsf{a}} = |w|_{\mathsf{b}} + 1, \\ \beta \to_{\mathcal{G}_1}^* w \text{ if and only if } |w|_{\mathsf{b}} = |w|_{\mathsf{a}} + 1, \\ \gamma \to_{\mathcal{G}_1}^* w \text{ if and only if } |w|_{\mathsf{a}} = |w|_{\mathsf{b}}. \end{array}$$

This implies $\mathcal{L}(\mathcal{G}_1) = \{w \mid |w|_{\mathsf{a}} = |w|_{\mathsf{b}}\}$, i.e. $\mathcal{G}_1$ generates the set of words w such that the number of occurrences of a in w is the same as the number of occurrences of b in w. On the other hand, we consider a tree language over a signature $\mathcal{F}$ such that $\mathcal{F}_0 = \{\mathsf{a}, \mathsf{b}\}$. If $\mathcal{F}$ contains at least a function symbol f with $\mathsf{arity}(f) \geqslant 2$, a tree language $L = \{t \mid |t|_{\mathsf{a}} = |t|_{\mathsf{b}}\}$ is no longer a regular tree language, due to Pumping Lemma.

Now we discuss the relationship between word languages and equational tree languages. The following properties are obtained as an easy observation of leaf-operation.

Lemma 7. *Given an arbitrary signature $\mathcal{F}$. For all terms $s, t \in \mathcal{T}(\mathcal{F})$ and a tree language L over $\mathcal{F}$,*

1. $\mathsf{leaf}(s) = \mathsf{leaf}(t)$ *if* $s \sim_{\mathsf{A}} t$. *The reverse holds if* $\mathcal{F} = \mathcal{F}_0 \cup \{\mathsf{f}\}$ *and* $\mathcal{F}_{\mathrm{AC}} = \{\mathsf{f}\}$.
2. $\mathsf{leaf}(L) = \mathsf{leaf}(\mathsf{A}(L))$.

Suppose $\mathcal{F} = \mathcal{F}_0 \cup \{f\}$ and $\mathcal{F}_{\mathrm{AC}} = \{f\}$. If L_1, L_2 are tree languages over $\mathcal{F}$, the following statements hold true:

3. $\mathsf{leaf}(L_1 \cup L_2) = \mathsf{leaf}(L_1) \cup \mathsf{leaf}(L_2)$,
4. $\mathsf{leaf}(\mathsf{A}(L_1) \cap \mathsf{A}(L_2)) = \mathsf{leaf}(L_1) \cap \mathsf{leaf}(L_2)$,
5. $\mathsf{A}(L_1) \subseteq \mathsf{A}(L_2)$ *if and only if* $\mathsf{leaf}(L_1) \subseteq \mathsf{leaf}(L_2)$.

Proof. To prove the property 1, we observe that $\mathsf{leaf}(C[\mathsf{f}(\mathsf{f}(t_1, t_2), t_3)]) = w_1\, \mathsf{leaf}(t_1)\, \mathsf{leaf}(t_2)\, \mathsf{leaf}(t_3)\, w_2 = \mathsf{leaf}(C[\mathsf{f}(t_1, \mathsf{f}(t_2, t_3))])$ for some $w_1, w_2 \in \mathcal{F}_0^*$. It can be proved by the structural induction of C. This implies that if $s \sim_{\mathsf{A}} t$ then $\mathsf{leaf}(s) = \mathsf{leaf}(t)$. For the reverse, we define the mapping tree as follows: $\mathsf{tree}(w) = a$ if $w \in \mathcal{F}_0$; $\mathsf{tree}(w) = \mathsf{f}(a, \mathsf{tree}(w'))$ if $w = a\,w'$ for some $a \in \mathcal{F}_0$ and $w' \in \mathcal{F}_0^+$. Then it suffices to show that for all u in $\mathcal{T}(\mathcal{F})$, $\mathsf{tree}(\mathsf{leaf}(u)) \sim_{\mathcal{A}} u$. It can be proved by the structural induction of u.

The properties 2 and 3 are easy. Note that union ($\cup$) is distributive over set comprehension, i.e. $\{\mathsf{leaf}(t) \mid t \in L_1 \cup L_2\} = \{\mathsf{leaf}(t) \mid t \in L_1\} \cup \{\mathsf{leaf}(t) \mid t \in L_2\}$.

Next we show the property 4. We suppose $w \in \mathsf{leaf}(\mathsf{A}(L_1) \cap \mathsf{A}(L_2))$. By definition, there exists $t \in \mathsf{A}(L_1) \cap \mathsf{A}(L_2)$ such that $w = \mathsf{leaf}(t)$. Moreover, $\mathsf{leaf}(t) \in \mathsf{leaf}(\mathsf{A}(L_1))$ and $\mathsf{leaf}(t) \in \mathsf{leaf}(\mathsf{A}(L_2))$. Thus $w \in \mathsf{leaf}(\mathsf{A}(L_1)) \cap \mathsf{leaf}(\mathsf{A}(L_2))$. Due to the property 2, $w \in \mathsf{leaf}(L_1) \cap \mathsf{leaf}(L_2)$. For the reverse inclusion, we suppose $w \in \mathsf{leaf}(\mathsf{A}(L_1)) \cap \mathsf{leaf}(\mathsf{A}(L_2))$. Note that $\mathsf{leaf}(L_1) = \mathsf{leaf}(\mathsf{A}(L_1))$ and $\mathsf{leaf}(L_2) = \mathsf{leaf}(\mathsf{A}(L_2))$. Then there exist $t_1 \in \mathsf{A}(L_1)$ and $t_2 \in \mathsf{A}(L_2)$ such that $w = \mathsf{leaf}(t_1)$ and $w = \mathsf{leaf}(t_2)$. Due to the property 1, $t_1 \sim_{\mathsf{A}} t_2$, and thus, $t_1 \in \mathsf{A}(L_2)$ (and $t_2 \in \mathsf{A}(L_1)$ as well). Hence $t_1 \in \mathsf{A}(L_1) \cap \mathsf{A}(L_2)$. This implies $w \in \mathsf{leaf}(\mathsf{A}(L_1) \cap \mathsf{A}(L_2))$.

Finally we show the property 5. The "only if" part is trivial. To show the reverse, we suppose $t \in \mathsf{A}(L_1)$. Then $\mathsf{leaf}(t) \in \mathsf{leaf}(L_1)$ (as $\mathsf{leaf}(t) \in \mathsf{leaf}(\mathsf{A}(L_1))$), and thus, $\mathsf{leaf}(t) \in \mathsf{leaf}(L_2)$. This implies that, due to the property 1, there exists $s \in L_2$ such that $s \sim_{\mathsf{A}} t$. Hence $t \in \mathsf{A}(L_2)$. □

One should note that $\mathsf{leaf}(L_1 \cap L_2) \neq \mathsf{leaf}(L_1) \cap \mathsf{leaf}(L_2)$. For example, suppose $L_1 = \{\mathsf{f}(\mathsf{f}(\mathsf{a}, \mathsf{b}), \mathsf{c})\}$ and $L_2 = \{\mathsf{f}(\mathsf{a}, \mathsf{f}(\mathsf{b}, \mathsf{c}))\}$ such that $\mathcal{F} = \{\mathsf{f}, \mathsf{a}, \mathsf{b}, \mathsf{c}\}$ and $\mathcal{F}_{\mathsf{A}} = \{\mathsf{f}\}$. Then $\mathsf{leaf}(L_1 \cap L_2) = \varnothing$, but $\mathsf{leaf}(L_1) \cap \mathsf{leaf}(L_2) = \{\mathsf{abc}\}$. Furthermore, we see that $\mathsf{leaf}(L_1) \subseteq \mathsf{leaf}(L_2)$ does not imply $L_1 \subseteq L_2$.

We show an extension of Proposition 4 (1). In the extension, *maximality* of tree languages refines the relationship to word languages.

Definition 1. *A tree languages L over $\mathcal{F}$ is* maximal *for a word language W if for all terms t in $\mathcal{T}(\mathcal{F})$, $\mathsf{leaf}(t) \in W$ if and only if $t \in L$. An ETA $\mathcal{A}/\mathcal{E}$ is maximal for W if $\mathcal{L}(\mathcal{A}/\mathcal{E})$ is maximal for W.*

Lemma 8. *Given a CFG $\mathcal{G}$ over the alphabet Σ. Then there exists the associated regular A-TA $\mathcal{A}_{\mathcal{G}}/\mathsf{A}$ that is maximal for $\mathcal{L}(\mathcal{G})$.*

Proof. Let $\mathcal{G} = (\Sigma, \mathcal{Q}, \mathsf{q}_0, \Delta)$. Suppose Δ is in Chomsky Normal Form. We take $\mathcal{F} = \Sigma \cup \{\mathsf{f}\}$ and $\mathcal{F}_{\mathsf{A}} = \{\mathsf{f}\}$. Define the regular A-TA $\mathcal{A}_{\mathcal{G}}/\mathsf{A}$ as follows: $\mathcal{A}_{\mathcal{G}} = (\mathcal{F}, \mathcal{Q}, \{\mathsf{q}_0\}, \Delta')$ where $\Delta' = \{\mathsf{f}(q_1, q_2) \to q \mid q \to q_1\, q_2 \in \Delta\} \cup \{a \to q \mid q \to a \in \Delta\}$. It is easy to show that for every (non-empty) word $w \in \Sigma^+$, $\mathsf{q}_0 \to^*_{\mathcal{G}} w$ if and only if $\mathsf{tree}(w) \to^*_{\mathcal{A}_{\mathcal{G}}/\mathsf{A}} \mathsf{q}_0$. Here tree is the mapping defined in the previous proof. For maximality we use Lemma 7 (1) and the following property: $\mathsf{leaf}(\mathsf{tree}(w)) = w$ for every $w \in \Sigma$. Then we can prove that for every term $t \in \mathcal{T}(\Sigma \cup \{\mathsf{f}\})$ and word $w \in \Sigma^*$, $\mathsf{leaf}(t) = w$ if and only if $t \sim_{\mathsf{A}} \mathsf{tree}(w)$. Thus, if $\mathsf{leaf}(t) \in \mathcal{L}(\mathcal{G})$ then $t \sim_{\mathsf{A}} \mathsf{tree}(\mathsf{leaf}(t))$ and $\mathsf{tree}(\mathsf{leaf}(t)) \to^*_{\mathcal{A}_{\mathcal{G}}/\mathsf{A}} \mathsf{q}_0$. Hence $t \in \mathcal{L}(\mathcal{A}_{\mathcal{G}}/\mathsf{A})$. □

We consider the CFG $\mathcal{G}_2 = (\{\mathsf{a}, \mathsf{b}\}, \{\alpha, \alpha', \beta, \beta', \gamma, \delta, \xi\}, \gamma, \Delta_2)$ where Δ_2:

$$\begin{array}{llllll} \gamma \to \delta\,\beta & \alpha \to \mathsf{a} & \alpha \to \xi\,\alpha' & \beta \to \mathsf{b} & \beta \to \delta\,\beta' & \delta \to \mathsf{a} \\ \gamma \to \xi\,\alpha & \alpha \to \delta\,\gamma & \alpha' \to \alpha\,\alpha & \beta \to \xi\,\gamma & \beta' \to \beta\,\beta & \xi \to \mathsf{b} \end{array}$$

Note that $\mathcal{G}_2$ is in Chomsky Normal Form, and $\mathcal{L}(\mathcal{G}_2) = \{w \mid |w|_{\mathsf{a}} = |w|_{\mathsf{b}}\}$ because $\mathcal{G}_2$ is essentially the same as $\mathcal{G}_1$. Then the associated A-TA $\mathcal{A}_{\mathcal{G}_2}/\mathsf{A}$ over the signature $\mathcal{F} = \{\mathsf{f}, \mathsf{a}, \mathsf{b}\}$ and $\mathcal{F}_{\mathsf{A}} = \{\mathsf{f}\}$ recognizes a tree language maximal for $\mathcal{L}(\mathcal{G}_2)$ and for $\mathcal{L}(\mathcal{G}_1)$.

Next we extend Proposition 4 (2) as in the previous way.

Lemma 9. *Given an regular A-TA $\mathcal{A}/\mathsf{A}$ over the signature $\mathcal{F}$. Then there exists the associated CFG $\mathcal{G}_{\mathcal{A}/\mathsf{A}}$ such that $\mathcal{L}(\mathcal{G}_{\mathcal{A}/\mathsf{A}}) = \mathsf{leaf}(\mathcal{L}(\mathcal{A}/\mathsf{A}))$.*

Proof. From Proposition 4 (2), for the language $\mathsf{leaf}(\mathcal{L}(\mathcal{A}))$ there exists a CFG $\mathcal{G}$ such that $\mathcal{L}(\mathcal{G}) = \mathsf{leaf}(\mathcal{L}(\mathcal{A}))$. On the other hand, from Lemma 7 (2), $\mathsf{leaf}(\mathcal{L}(\mathcal{A})) = \mathsf{leaf}(\mathsf{A}(\mathcal{L}(\mathcal{A})))$. Since $\mathcal{L}(\mathcal{A}/\mathsf{A}) = \mathsf{A}(\mathcal{L}(\mathcal{A}))$ by Lemma 2, we obtain $\mathsf{leaf}(\mathcal{L}(\mathcal{A})) = \mathsf{leaf}(\mathcal{L}(\mathcal{A}/\mathsf{A}))$. By letting $\mathcal{G}_{\mathcal{A}/\mathsf{A}} = \mathcal{G}$, it satisfies that $\mathcal{L}(\mathcal{G}_{\mathcal{A}/\mathsf{A}}) = \mathsf{leaf}(\mathcal{L}(\mathcal{A}/\mathsf{A}))$. □

In this case, $\mathcal{A}/\mathsf{A}$ is not always maximal. For instance, we take an A-TA $\mathcal{A}/\mathsf{A}$ such that $\mathcal{L}(\mathcal{A}/\mathsf{A}) = \{\mathsf{a}\}$ over the signature $\{\mathsf{f}, \mathsf{a}\}$. Obviously there exists a CFG $\mathcal{G}_{\mathcal{A}/\mathsf{A}}$ such that $\mathcal{L}(\mathcal{G}_{\mathcal{A}/\mathsf{A}}) = \{\mathsf{a}\}$. But then, $\mathcal{A}/\mathsf{A}$ is not maximal for $\mathcal{L}(\mathcal{G}_{\mathcal{A}/\mathsf{A}})$, because $\mathsf{f}(\mathsf{a})$ is not in $\mathcal{L}(\mathcal{A}/\mathsf{A})$.

Corollary 5. *The following two statements hold true:*

1. *For every CFG $\mathcal{G}$ there exists a regular A-TA $\mathcal{A}/\mathsf{A}$ such that $\mathcal{L}(\mathcal{A}/\mathsf{A})$ is a maximal tree language for $\mathcal{L}(\mathcal{G})$.*
2. *For every regular A-TA $\mathcal{A}/\mathsf{A}$ there exists a CFG $\mathcal{G}$ such that $\mathsf{leaf}(\mathcal{L}(\mathcal{A}/\mathsf{A})) = \mathcal{L}(\mathcal{G})$. In case the signature $\mathcal{F} = \mathcal{F}_0 \cup \{\mathsf{f}\}$ and $\mathcal{F}_{\mathsf{A}} = \{\mathsf{f}\}$, $\mathcal{L}(\mathcal{A}/\mathsf{A})$ is a maximal tree language for $\mathcal{L}(\mathcal{G})$.*

Proof. An immediate consequence of Lemmata 8 and 9, except the case that $\mathcal{F} = \mathcal{F}_0 \cup \{\mathsf{f}\}$ and $\mathcal{F}_{\mathsf{A}} = \{\mathsf{f}\}$ in the latter statement. In the particular case, it holds that $\mathsf{tree}(\mathsf{leaf}(s)) \sim_{\mathsf{A}} s$ for any s in $\mathcal{T}(\mathcal{F})$. Thus, if $\mathsf{leaf}(t) \in \mathcal{L}(\mathcal{G})$, there exists $s \in \mathcal{L}(\mathcal{A}/\mathsf{A})$ such that $\mathsf{leaf}(s) = \mathsf{leaf}(t)$. Moreover, $s \sim_{\mathsf{A}} t$ by Lemma 7 (1). Hence $t \in \mathcal{L}(\mathcal{A}/\mathsf{A})$. □

Thus A-regular tree languages inherits the negative results of context-free languages.

Theorem 3. *A-regular tree languages are not closed under intersection or complement.*

Proof. Given CFGs $\mathcal{G}_1$ and $\mathcal{G}_2$ over the alphabet Σ. Due to Lemma 8, there are regular A-TA $\mathcal{A}/\mathsf{A}$ and $\mathcal{B}/\mathsf{A}$ over $\mathcal{F}$, where $\mathcal{F} = \Sigma \cup \{\mathsf{f}\}$ and $\mathcal{F}_A = \{\mathsf{f}\}$, such that $\mathcal{L}(\mathcal{G}_1) = \mathsf{leaf}(\mathcal{L}(\mathcal{A}/\mathsf{A}))$ and $\mathcal{L}(\mathcal{G}_2) = \mathsf{leaf}(\mathcal{L}(\mathcal{B}/\mathsf{A}))$, respectively. Suppose to the contradiction that A-regular tree languages are closed under intersection. Then there exists a regular A-TA $\mathcal{C}/\mathsf{A}$ over $\mathcal{F}$ such that $\mathcal{L}(\mathcal{C}/\mathsf{A}) = \mathcal{L}(\mathcal{A}/\mathsf{A}) \cap \mathcal{L}(\mathcal{B}/\mathsf{A})$. Moreover, due to Lemma 9, there exists a CFG $\mathcal{G}_3$ such that $\mathcal{L}(\mathcal{G}_3) = \mathsf{leaf}(\mathcal{L}(\mathcal{C}/\mathsf{A}))$. From Lemma 7 (2) and (4), we obtain $\mathcal{L}(\mathcal{G}_3) = \mathsf{leaf}(\mathcal{L}(\mathcal{A}/\mathsf{A})) \cap \mathsf{leaf}(\mathcal{L}(\mathcal{B}/\mathsf{A})) = \mathcal{L}(\mathcal{G}_1) \cap \mathcal{L}(\mathcal{G}_2)$. However, it contradicts to the fact that context-free languages are not closed under intersection. Hence A-regular tree languages are not closed under intersection.

To show not being closed under complement, we use the previous fact together with De Morgan Law. □

Theorem 4. *The following questions are undecidable in A-regular tree languages: Given $\mathcal{A}/\mathsf{A}$ and $\mathcal{B}/\mathsf{A}$ are regular A-TA over the signature $\mathcal{F}$, then*

- $\mathcal{L}(\mathcal{A}/\mathsf{A}) \subseteq \mathcal{L}(\mathcal{B}/\mathsf{A})$? *(subset)*
- $\mathcal{L}(\mathcal{A}/A) = \mathcal{T}(\mathcal{F})$? *(universarity)*
- $\mathcal{L}(\mathcal{A}/\mathsf{A}) \cap \mathcal{L}(\mathcal{B}/\mathsf{A}) = \varnothing$? *(intersection-emptiness)*

Proof. Given CFGs $\mathcal{G}_1$ and $\mathcal{G}_2$ over the alphabet Σ. Due to Lemma 8, there exist regular A-TA $\mathcal{A}/\mathsf{A}$ and $\mathcal{B}/\mathsf{A}$ over $\mathcal{F}$ such that $\mathcal{L}(\mathcal{G}_1) = \mathsf{leaf}(\mathcal{L}(\mathcal{A}/\mathsf{A}))$ and $\mathcal{L}(\mathcal{G}_2) = \mathsf{leaf}(\mathcal{L}(\mathcal{B}/\mathsf{A}))$. Here the signature $\mathcal{F} = \Sigma \cup \{\mathsf{f}\}$ and $\mathcal{F}_\mathsf{A} = \{\mathsf{f}\}$. Due to Lemma 7 (2) and (5), we know that $\mathcal{L}(\mathcal{A}/\mathsf{A}) \subseteq \mathcal{L}(\mathcal{B}/\mathsf{A})$ if and only if $\mathsf{leaf}(\mathcal{L}(\mathcal{A}/\mathsf{A})) \subseteq \mathsf{leaf}(\mathcal{L}(\mathcal{B}/\mathsf{A}))$, because $\mathcal{L}(\mathcal{A}/\mathsf{A}) = \mathsf{A}(\mathcal{L}(\mathcal{A}))$ and $\mathcal{L}(\mathcal{B}/\mathsf{A}) = \mathsf{A}(\mathcal{L}(\mathcal{B}))$, respectively. This implies the question if $\mathcal{L}(\mathcal{A}/\mathsf{A}) \subseteq \mathcal{L}(\mathcal{B}/\mathsf{A})$ is undecidable, because the question if $\mathcal{L}(\mathcal{G}_1) \subseteq \mathcal{L}(\mathcal{G}_2)$ is undecidable [8]. Similarly, using Lemma 7 we can prove that

- $\mathcal{L}(\mathcal{A}/\mathsf{A}) = \mathcal{T}(\mathcal{F})$ if and only if $\mathsf{leaf}(\mathcal{L}(\mathcal{A}/\mathsf{A})) = \mathsf{leaf}(\mathcal{T}(\mathcal{F}))$,
- $\mathcal{L}(\mathcal{A}/\mathsf{A}) \cap \mathcal{L}(\mathcal{B}/\mathsf{A}) = \varnothing$ if and only if $\mathsf{leaf}(\mathcal{L}(\mathcal{A}/\mathsf{A})) \cap \mathsf{leaf}(\mathcal{L}(\mathcal{B}/\mathsf{A})) = \varnothing$.

Hence the other two questions are also undecidable, because universality and intersection-emptiness problems for context-free languages are undecidable. □

Additionally, we can prove that equivalence problem for regular A-TA is undecidable: We take two regular A-TA $\mathcal{A}/\mathsf{A}, \mathcal{B}/\mathsf{A}$. The subset problem $\mathcal{L}(\mathcal{A}/\mathsf{A}) \subseteq \mathcal{L}(\mathcal{B}/\mathsf{A})$ is represented as $\mathcal{L}(\mathcal{A}/\mathsf{A}) \cup \mathcal{L}(\mathcal{B}/\mathsf{A}) = \mathcal{L}(\mathcal{B}/\mathsf{A})$. By Lemma 2, we have $\mathcal{L}(\mathcal{A}/\mathsf{A}) \cup \mathcal{L}(\mathcal{B}/\mathsf{A}) = \mathsf{A}(\mathcal{L}(\mathcal{A})) \cup \mathsf{A}(\mathcal{L}(\mathcal{B}))$, and then, $\mathsf{A}(\mathcal{L}(\mathcal{A})) \cup \mathsf{A}(\mathcal{L}(\mathcal{B})) = \mathsf{A}(\mathcal{L}(\mathcal{A}) \cup \mathcal{L}(\mathcal{B}))$. The A-congruence closure of a regular tree language $\mathcal{L}(\mathcal{A}) \cup \mathcal{L}(\mathcal{B})$ is A-regular, i.e. there is a regular A-TA $\mathcal{C}/\mathsf{A}$ such that $\mathcal{L}(\mathcal{C}/\mathsf{A}) = \mathsf{A}(\mathcal{L}(\mathcal{A}) \cup \mathcal{L}(\mathcal{B}))$. This implies that $\mathcal{L}(\mathcal{A}/\mathsf{A}) \subseteq \mathcal{L}(\mathcal{B}/\mathsf{A})$ if and only if $\mathcal{L}(\mathcal{C}/\mathsf{A}) = \mathcal{L}(\mathcal{B}/\mathsf{A})$. Since the former question is undecidable, so is the equivalence $\mathcal{L}(\mathcal{C}/\mathsf{A}) = \mathcal{L}(\mathcal{B}/\mathsf{A})$.

		C	A	AC
$\mathcal{L}(\mathcal{A}/\mathcal{E}) = \varnothing$?	regular	✓	✓	✓
	non-regular		×	
$\mathcal{L}(\mathcal{A}/\mathcal{E}) \subseteq \mathcal{L}(\mathcal{B}/\mathcal{E})$?	regular	✓	×	(✓)
	non-regular			?
$\mathcal{L}(\mathcal{A}/\mathcal{E}) = \mathcal{T}(\mathcal{F})$?	regular	✓	×	(✓)
	non-regular			?
$\mathcal{L}(\mathcal{A}/\mathcal{E}) \cap \mathcal{L}(\mathcal{B}/\mathcal{E}) = \varnothing$?	regular	✓	×	✓
	non-regular			

		C	A	AC
closed under $\cup$	regular	✓	✓	✓
	non-regular			
closed under $\cap$	regular	✓	×	(✓)
	non-regular		✓	✓
closed under $\overline{(\,)}$	regular	✓	×	(✓)
	non-regular		?	?

Fig. 3. Some decidability results and closure properties

5 Concluding Remarks

In the paper we have shown decidability results and closure properties of A- and AC-tree language. New results on the decidability (Theorems 1, 2 and 4) and closure properties (Theorem 3) are the solutions to the questions remaining open in our previous paper [20]. Using the following Parikh's result (Theorem 2 in [21]; the same result also in [4]), we can show a partial solution to the question of whether AC-regular tree languages are closed under intersection and complement.

Lemma 10. *Permutation closures of context-free languages are closed under boolean operations (union, intersection and complement).* □

Corollary 6. *Every AC-regular tree languages over the signature* $\mathcal{F} = \{\mathsf{f}\} \cup \mathcal{F}_0$ *and* $\mathcal{F}_{\mathrm{AC}} = \{\mathsf{f}\}$ *is closed under boolean operations.*

Proof. We show closedness under union and intersection below. Let $\mathcal{A}/\mathsf{AC}$, $\mathcal{B}/\mathsf{AC}$ be regular AC-TA over $\mathcal{F}$. By Proposition 4 (2), there exists a context-free grammar $\mathcal{G}_1$ and $\mathcal{G}_2$ such that $\mathsf{leaf}(\mathcal{L}(\mathcal{A})) = \mathcal{L}(\mathcal{G}_1)$ and $\mathsf{leaf}(\mathcal{L}(\mathcal{B})) = \mathcal{L}(\mathcal{G}_2)$. Due to Lemma 2, we have $\mathsf{leaf}(\mathcal{L}(\mathcal{A}/\mathsf{AC})) = \mathsf{leaf}(\mathsf{AC}(\mathcal{L}(\mathcal{A})))$. Moreover, by assumption of $\mathcal{F}$, $\mathsf{leaf}(\mathsf{AC}(\mathcal{L}(\mathcal{A}))) = \mathsf{perm}(\mathsf{leaf}(\mathcal{L}(\mathcal{A})))$. Thus $\mathsf{leaf}(\mathcal{L}(\mathcal{A}/\mathsf{AC})) = \mathsf{perm}(\mathcal{L}(\mathcal{G}_1))$. The same property holds for $\mathcal{B}/\mathsf{AC}$. By Lemma 10, there exists a context-free grammar $\mathcal{G}_R$ such that $\mathsf{perm}(\mathcal{L}(\mathcal{G}_R)) = \mathsf{perm}(\mathcal{L}(\mathcal{G}_1))\ R\ \mathsf{perm}(\mathcal{L}(\mathcal{G}_2))$ for each operation $R \in \{\cup, \cap\}$. By Proposition 4 (1), we obtain a TA $\mathcal{C}_R$ such that $\mathsf{leaf}(\mathcal{L}(\mathcal{C}_R)) = \mathcal{L}(\mathcal{G}_R)$, and thus, $\mathsf{leaf}(\mathcal{L}(\mathcal{C}_R/\mathsf{AC})) = \mathsf{perm}(\mathcal{L}(\mathcal{G}_R))$. Therefore $\mathcal{L}(\mathcal{C}_R/\mathsf{AC}) = \mathcal{L}(\mathcal{A}/\mathsf{AC})\ R\ \mathcal{L}(\mathcal{B}/\mathsf{AC})$ over $\mathcal{F} = \{\mathsf{f}\} \cup \mathcal{F}_0$ with $\mathcal{F}_{\mathsf{AC}} = \{\mathsf{f}\}$. □

Due to the above corollary together with decidability of emptiness problem for AC-TA (Corollary 2), the subset and universality problems are decidable in this particular case.

On the other hand, from the recent study about *multiset grammars* [11], we know that regular and non-regular AC-tree automata correspond to multiset context-free grammars and multiset monotone grammars, respectively. More precisely, the relationships like in Corollary 5 hold for them. This implies the expressive power of non-regular AC-tree automata strictly subsumes the regular case, which is also the answer to an open question of our previous paper.

We summarize the decidability results and closure properties of C-, A- and AC-tree languages in Fig. 3. New results are indicated by dotted squares. All the results of C-tree languages are easily obtained, because C-TA are essentially the same as regular TA (Lemma 3, [20]). In the figure, the check mark ✓ means "positive" and the cross × is "negative". The question mark ? means "open". If the same result holds in both regular and non-regular cases, it is represented by a single mark in a large column. A positive result proved only in a special case, e.g. of $\mathcal{F} = \{\mathsf{f}\} \cup \mathcal{F}_0$ with $\mathcal{F}_{\mathsf{AC}} = \{\mathsf{f}\}$, is denoted by the check mark with round brackets.

Acknowledgements. The authors would like to thank Ralf Treinen for his continuous help. We also thank to three anonymous referees for their comments and criticism.

References

1. B. Bogaert and S. Tison: *Equality and Disequality constraints on Direct Subterms in Tree Automata*, Proc. of 9th STACS, Cachan (France), LNCS 577, pp. 161–171 1992.
2. H. Comon, M. Dauchet, R. Gilleron, F. Jacquemard, D. Lugiez, S. Tison and M. Tommasi: *Tree Automata Techniques and Applications*, draft, 1999. Available on `http://www.grappa.univ-lille3.fr/tata/`
3. A. Deruyver and R. Gilleron: *The Reachability Problem for Ground TRS and Some Extensions*, Proc. of 14th CAAP, Barcelona (Spain), LNCS 351, pp. 227–243, 1989.
4. J. Esparza: *Petri Nets, Commutative Context-Free Grammars, and Basic Parallel Processes*, Fundamenta Informaticae 31(1), pp. 13–25, 1997.
5. M.R. Garey and D.S. Johnson: *Computers and Intractability: A Guide to the Theory of NP-Completeness*, W.H. Freeman & Company, New York, 1979.
6. F. Gécseg and M. Steinby: *Tree Automata*, Akadémiai Kiadó, Budapest, 1984.
7. T. Genet and F. Klay: *Rewriting for Cryptographic Protocol Verification*, Proc. of 17th CADE, Pittsburgh (PA), LNCS 1831, pp. 271–290, 2000.

8. J.E. Hopcroft and J.D. Ullman: *Introduction to Automata Theory, Languages, and Computation*, Addison-Wesley Publishing Company, 1979.
9. H. Hosoya, J. Vouillon and B.C. Pierce: *Regular Expression Types for XML*, Proc. of 5th ICFP, Montreal (Canada), SIGPLAN Notices 35(9), pp. 11–22, 2000.
10. Y. Kaji, T. Fujiwara and T. Kasami: *Solving a Unification Problem under Constrained Substitutions Using Tree Automata*, JSC 23(1), pp. 79–117, 1997.
11. M. Kudlek and V. Mitrana: *Normal Forms of Grammars, Finite Automata, Abstract Families, and Closure Properties of Macrosets*, Multiset Processing, LNCS 2235, pp. 135–146, 2001.
12. S. La Torre and M. Napoli: *Timed Tree Automata with an Application to Temporal Logic*, Acta Informatica 38(2), pp. 89–116, 2001.
13. D. Lugiez: *A Good Class of Tree Automata and Application to Inductive Theorem Proving*, Proc. of 25th ICALP, Aalborg (Denmark), LNCS 1443, pp. 409–420, 1998.
14. D. Lugiez and J.L. Moysset: *Tree Automata Help One to Solve Equational Formulae in AC-Theories*, JSC 18(4), pp. 297–318, 1994.
15. E.W. Mayr: *An Algorithm for the General Petri Net Reachability Problem*, SIAM J. Comput. 13(3), pp. 441–460, 1984.
16. R. Mayr and M. Rusinowitch: *Reachability is Decidable for Ground AC Rewrite Systems*, Proc. of 3rd INFINITY, Aalborg (Denmark), 1998. Draft available from `http://www.informatik.uni-freiburg.de/~mayrri/ac.ps`
17. R. Mayr and M. Rusinowitch: *Process Rewrite Systems*, Information and Computation 156, pp. 264–286, 1999.
18. J. Millen and V. Shmatikov: *Constraint Solving for Bounded-Process Cryptographic Protocol Analysis*, Proc. of 8th CCS, Philadelphia (PA), pp. 166–175, 2001.
19. D. Monniaux: *Abstracting Cryptographic Protocols with Tree Automata*, Proc. of 6th SAS, Venice (Italy), LNCS 1694, pp. 149–163, 1999.
20. H. Ohsaki: *Beyond Regularity: Equational Tree Automata for Associative and Commutative Theories*, Proc. of 15th CSL, Paris (France), LNCS 2142, pp. 539–553, 2001.
21. R.J. Parikh: *On Context-Free Languages*, JACM 13(4), pp. 570–581, 1966.
22. X. Rival and J. Goubault-Larrecq: *Experiments with Finite Tree Automata in Coq*, Proc. of 14th TPHOLs, Edinburgh (Scotland), LNCS 2152, pp. 362–377, 2001.
23. B. Schneier: *Applied Cryptography: Protocols, Algorithms, and Source Code in C*, Second Edition, John Wiley & Sons, 1996.
24. H. Seki, T. Takai, Y. Fujinaka and Y. Kaji: *Layered Transducing Term Rewriting System and Its Recognizability Preserving Property*, Proc. of 13th RTA, Copenhagen (Denmark), 2002. To appear in LNCS.
25. T. Takai, Y. Kaji and H. Seki: *Right-Linear Finite-Path Overlapping Term Rewriting Systems Effectively Preserve Recognizability*, Proc. of 11th RTA, Norwich (UK), LNCS 1833, pp. 246–260, 2000.

Regular Sets of Descendants by Some Rewrite Strategies

Pierre Réty and Julie Vuotto

LIFO - Université d'Orléans, B.P. 6759, 45067 Orléans cedex 2, France
{rety, vuotto}@lifo.univ-orleans.fr
http://www.univ-orleans.fr/SCIENCES/LIFO/Members/rety/

Abstract. For a constructor-based rewrite system R, a regular set of ground terms E, and assuming some additional restrictions, we build a finite tree automaton that recognizes the descendants of E, i.e. the terms issued from E by rewriting, according to innermost, innermost-leftmost, and outermost strategies.

1 Introduction

Tree automata have already been applied to many areas of computer science, and in particular to rewrite techniques [2]. In comparison with more sophisticated refinements [4,10,9], finite tree automata are obviously less expressive, but have plenty of good properties and lead to much simpler algorithms from a practical point of view.

Because of potential applications to automated deduction and program validation (reachability, program testing), the problem of expressing by a finite tree automaton the transitive closure of a regular set E of ground terms with respect to a set of equations, as well as the related problem of expressing the set of descendants $R^*(E)$ of E with respect to a rewrite system R, have already been investigated [1,5,14,3,11,15][1]. Except [11,15], all those papers assume that the right-hand-sides (both sides when dealing with sets of equations) of rewrite rules are shallow[2], up to slight differences. On the other hand, P. Réty's work [12] does not always preserve recognizability (E is not arbitrary), but allows rewrite rules forbidden by the other papers[3].

Reduction strategies in rewriting and programming have drawn an increasing attention within the last years, and matter both from a theoretical point of view, if the computation result is not unique, and from a practical point of view, for termination and efficiency. For a strategy st, expressing by a finite tree automaton the st-descendants $R^*_{st}(E)$ of E, can help to study st: in particular it allows to decide st-reachability since $t_1 \ {}^{st}\!\!\rightarrow^* t_2 \iff t_2 \in R^*_{st}(\{t_1\})$, and

[1] [11] computes sets of normalizable terms, which amounts to compute sets of descendants by orienting the rewrite rules in the opposite direction.

[2] Shallow means that every variable appears at depth at most one.

[3] Like $f(s(x)) \rightarrow s(f(x))$.

S. Tison (Ed.): RTA 2002, LNCS 2378, pp. 129–143, 2002.

st-joinability since $t_1 \overset{st}{\downarrow} t_2 \iff R^*_{st}(\{t_1\}) \cap R^*_{st}(\{t_2\}) \neq \emptyset$. More generally, it can help with the static analysis of rewrite programs, and by extension, of functional programs.

This paper is an extension of [12] that takes some strategies into account. As far as we know, the problem of expressing sets of descendants according to some strategies had not been addressed yet. We build finite tree automata that can express the sets of descendants of E with respect to a constructed-based rewrite system, according to innermost, innermost-leftmost, outermost strategies, assuming :

1. E is the set of ground constructor-instances (also called data-instances) of a given linear term.
2. Every rewrite rule is linear (both sides).
3. In right-hand-sides, there are no nested defined-functions, and arguments of defined-functions are either variables or ground terms.

For innermost-leftmost strategy, we temporarily assume in addition that right-hand-sides contain at most one defined-function. For outermost strategy, we also assume this extra restriction, and moreover that R has no critical pairs.

It is shown in [12] that if any restriction among 1, 2, 3 is not satisfied, then the set of descendants is not regular (even if a strategy among innermost, innermost-leftmost, outermost, is respected). About restriction 2, only non-right-linearity causes non-regularity. However, to deal with strategies, we need the regularity of the set of irreducible terms, hence left-linearity.

The paper is organized as follows. Section 2 introduces preliminaries notions. The reader used to term rewriting and tree automata may skip Subsection 2.1, but not the following ones which present more specific notions. Section 3 (resp. 4, 5) gives the computation of innermost (resp. innermost-leftmost, outermost) descendants. Missing proofs are available in the full version. See Réty's web page.

2 Preliminaries

2.1 Usual Notions : Term Rewriting and Tree Automata

Let C be a finite set of *constructors* and F be a finite set of *defined-function symbols* (*functions* in a shortened form). For $c \in C \cup F$, $ar(c)$ is the arity of c. *Terms* are denoted by letters s, t. A *data-term* is a *ground* term (i.e. without variables) that contains only constructors. T_C is the set of data-terms, $T_{C \cup F}$ is the set of ground-terms. For a term t, $Var(t)$ is the set of variables appearing in t, $Pos(t)$ is the set of *positions* of t, $\overline{Pos}(t)$ is the set of non-variable positions of t, $PosF(t)$ is the set of defined-function positions of t. t is *linear* if each variable of t appears only once in t. For $p \in Pos(t)$, $t|_p$ is the subterm of t at position p, $t(p)$ is the top symbol of $t|_p$, and $t[t']_p$ denotes the subterm replacement. For positions p, p', $p \geq p'$ means that p is located below p', i.e. $p = p'.v$ for some position v, whereas $p \| p'$ means that p and p' are incomparable, i.e. $\neg(p \geq p') \wedge \neg(p' \geq p)$. The term t contains *nested functions* if there exist $p, p' \in PosF(t)$ s.t. $p > p'$. The domain $dom(\theta)$ of a substitution θ is the set of variables x s.t. $x\theta \neq x$.

A *rewrite rule* is an oriented pair of terms, written $l \to r$. A *rewrite system* R is a finite set of rewrite rules. *lhs* stands for left-hand-side, *rhs* for right-hand-side. R is *constructor-based* if every lhs l of R is of the form $l = f(t_1, \dots, t_n)$ where $f \in F$ and $t_1, \dots, t_n$ do not contain any functions. The rewrite relation $\to_R$ is defined as follows: $t \to_R t'$ if there exist $p \in Pos(t)$, a rule $l \to r \in R$, and a substitution θ s.t. $t|_p = l\theta$ and $t' = t[r\theta]_p$ (also denoted by $t \to_{[p,l\to r,\theta]} t'$). $\to_R^*$ denotes the reflexive-transitive closure of $\to_R$. t is *irreducible* if $\neg(\exists t' \mid t \to_R t')$. $t \to_{[p]} t'$ is innermost (resp. leftmost, outermost) if $\forall v > p$ (resp. $\forall v$ occurring strictly on the left of p, $\forall v < p$) $t|_v$ is irreducible. The *narrowing* relation $\leadsto_R$ is defined as follows: $t \leadsto_R t'$ if there exist $p \in \overline{Pos}(t)$, a rule $l \to r \in R$, and a substitution θ s.t. $t|_p\theta = l\theta$ and $t' = (t\theta)[r\theta]_p$ (also denoted by $t \leadsto_{[p,l\to r,\theta]} t'$).

A (bottom-up) finite tree *automaton* is a quadruple $\mathcal{A} = (C \cup F, Q, Q_f, \Delta)$ where $Q_f \subseteq Q$ are sets of states and Δ is a set of *transitions* of the form $c(q_1, \dots, q_n) \to q$ where $c \in C \cup F$ and $q_1, \dots, q_n, q \in Q$, or of the form $q_1 \to q$ (*empty transition*). Sets of *states* are denoted by letters Q, S, D, and states by q, s, d. $\to_\Delta$ (also denoted $\to_\mathcal{A}$) is the rewrite relation induced by Δ. A ground term t is *recognized* by $\mathcal{A}$ into q if $t \to_\Delta^* q$. $L(\mathcal{A})$ is the set of terms recognized by $\mathcal{A}$ into any states of Q_f. A derivation $t \to_\Delta^* q$ where $q \in Q_f$ is called a *successful run* on t. The states of Q_f are called *final states*. $\mathcal{A}$ is *deterministic* if whenever $t \to_\Delta^* q$ and $t \to_\Delta^* q'$ we have $q = q'$. A *Q-substitution* σ is a substitution s.t. $\forall x \in dom(\sigma)$, $x\sigma \in Q$. A set E of ground terms is regular if there exists a finite automaton $\mathcal{A}$ s.t. $E = L(\mathcal{A})$.

2.2 Nesting Automata

Intuitively, the automaton $\mathcal{A}$ discriminates position p into state q means that along every successful run on $t \in L(\mathcal{A})$, $t|_p$ (and only this subterm) is recognized into q. This property allows to modify the behavior of $\mathcal{A}$ below position p without modifying the other positions, by replacing $\mathcal{A}|_p$ by another automaton $\mathcal{A}'$. See [12] for missing proofs.

Definition 2.1. *The automaton* $\mathcal{A} = (C \cup F, Q, Q_f, \Delta)$ discriminates *the position p into the state q if*

- $L(\mathcal{A}) \neq \emptyset$,
- *and* $\forall t \in L(\mathcal{A})$, $p \in Pos(t)$,
- *and for each successful derivation* $t \to_\Delta^* t[q']_{p'} \to_\Delta^* q_f$ *where* $q_f \in Q_f$, *we have*
    - $q' \to_\Delta^* q$ *(i.e. by empty transitions) if* $p' = p$,
    - *not* $(q' \to_\Delta^* q)$ *otherwise.*

In this case we define the automaton $\mathcal{A}|_p = (C \cup F, Q, \{q\}, \Delta)$.

Lemma 2.2. $L(\mathcal{A}|_p) = \{t|_p \mid t \in L(\mathcal{A})\}$.

Definition 2.3. *Let $\mathcal{A} = (C \cup F, Q, Q_f, \Delta)$ be an automaton that discriminates the position p into the state q, and let $\mathcal{A}' = (C \cup F, Q', Q'_f, \Delta')$ s.t. $Q \cap Q' = \emptyset$. We define*

$$\mathcal{A}[\mathcal{A}']_p = (C \cup F,\ Q \cup Q',\ Q_f,\ \Delta \backslash \{l \to r \mid l \to r \in \Delta \wedge r = q\} \cup \Delta' \cup \{q'_f \to q \mid q'_f \in Q'_f\})$$

Lemma 2.4. *$L(\mathcal{A}[\mathcal{A}']_p) = \{t[t']_p \mid t \in L(\mathcal{A}), t' \in L(\mathcal{A}')\}$, and $\mathcal{A}[\mathcal{A}']_p$ still discriminates p into q. Moreover, if $\mathcal{A}$ discriminates another position p' s.t. $p' \not\geq p$, into the state q', then $\mathcal{A}[\mathcal{A}']_p$ still discriminates p' into q'.*

Lemma 2.5. *Let $\mathcal{A}, \mathcal{B}$ be automata, and let $\mathcal{A} \cap \mathcal{B}$ be the classical automaton used to recognize intersection, whose states are pairs of states of $\mathcal{A}$ and $\mathcal{B}$. If $\mathcal{A}$ discriminates p into $q_{\mathcal{A}}$, $\mathcal{B}$ discriminates p into $q_{\mathcal{B}}$, and $L(\mathcal{A}) \cap L(\mathcal{B}) \neq \emptyset$, then $\mathcal{A} \cap \mathcal{B}$ discriminates p into $(q_{\mathcal{A}}, q_{\mathcal{B}})$.*

Proof. Let $t \in L(\mathcal{A} \cap \mathcal{B})$.

- since $t \in L(\mathcal{A})$, $p \in Pos(t)$
- for any successful run on t,
  $t \to^*_{\Delta_{\mathcal{A} \cap \mathcal{B}}} t[(q'_{\mathcal{A}}, q'_{\mathcal{B}})]_{p'} \to^* (qf_{\mathcal{A}}, qf_{\mathcal{B}})$
  - if $p' = p$ then from discrimination of $\mathcal{A}$ and $\mathcal{B}$, $q'_{\mathcal{A}} \to^*_{\Delta} q_{\mathcal{A}}$ and $q'_{\mathcal{B}} \to^*_{\Delta} q_{\mathcal{B}}$
  - if $p' \neq p$ then from discrimination of $\mathcal{A}$ and $\mathcal{B}$, $not(q'_{\mathcal{A}} \to^*_{\Delta} q_{\mathcal{A}})$ and $not(q'_{\mathcal{B}} \to^*_{\Delta} q_{\mathcal{B}})$.

2.3 Particular Automata

Let us define the initial automaton, i.e. the automaton that recognizes the data-instances of a given linear term t.

Definition 2.6. *We define the automaton $\mathcal{A}_{data}$ that recognizes the set of data-terms T_C :*
$\mathcal{A}_{data} = (C, Q_{data}, Q_{data_f}, \Delta_{data})$ where $Q_{data} = Q_{data_f} = \{q_{data}\}$ and $\Delta_{data} = \{c(q_{data}, \ldots, q_{data}) \to q_{data} \mid c \in C\}$.

Given a linear term t, we define the automaton $\mathcal{A}_{t\theta}$ that recognizes the data-instances of t : $\mathcal{A}_{t\theta} = (C \cup F, Q_{t\theta}, Q_{t\theta_f}, \Delta_{t\theta})$ where

$$Q_{t\theta} = \{q^p \mid p \in \overline{Pos}(t)\} \cup \{q_{data}\}$$
$$Q_{t\theta_f} = \{q^\epsilon\} \ (q_{data} \text{ if } t \text{ is a variable})$$
$$\Delta_{t\theta} = \left\{ t(p)(s_1, \ldots, s_n) \to q^p \mid p \in \overline{Pos}(t),\ s_i = \left| \begin{array}{l} q_{data} \text{ if } t|_{p.i} \text{ is a variable} \\ q^{p.i} \text{ otherwise} \end{array} \right. \right\} \cup \Delta_{data}$$

Note that $\mathcal{A}_{t\theta}$ discriminates each position p $\in \overline{Pos}(t)$ into q^p. On the other hand, $\mathcal{A}_{t\theta}$ is not deterministic, as soon as there is $p \in \overline{Pos}(t)$ s.t. $t|_p$ is a constructor-term. Indeed for any data-instance $t|_p\theta$, $t|_p\theta \to^*_{[\Delta_{t\theta}]} q^p$ and $t|_p\theta \to^*_{[\Delta_{t\theta}]} q_{data}$.

Let us now define an automaton that recognizes the terms irreducible at position p.

Definition 2.7. *Let* $IRR_p(R) = \{s \in T_{C \cup F} | p \in Pos(s)$ *and* $s|_p$ *is irreducible*$\}$.

To prove the regularity of $IRR_p(R)$, we need some more definitions.

Definition 2.8. *Let* $RED(R)$ *be the language of reducible terms:*

$$RED(R) = \{s \mid \exists p' \in Pos(s) \ \ s \rightarrow_{[p', l \rightarrow r, \sigma]} s'\}$$

Lemma 2.9. *[6] If R is left-linear,* $RED(R)$ *is a regular language.*

Lemma 2.10. $IRR_\epsilon(R) = \overline{RED(R)}$. *Therefore,* $IRR_\epsilon(R)$ *is a regular language.*

Thanks to an automaton that recognizes $IRR_\epsilon(R)$, we can now build an automaton that recognizes $IRR_p(R)$.

Theorem 2.11. *Let t be a term, and* $p \in \overline{Pos}(t)$. $IRR_p(R)$ *is a regular language and is recognized by an automaton that discriminates every position* $p' \in \overline{Pos}(t)$ *s.t.* $p' \not> p$.

Proof. Let $\mathcal{A}_\epsilon = (\mathcal{C} \cup \mathcal{F}, \mathcal{Q}_\epsilon, \mathcal{Q}_{\epsilon f}, \Delta_\epsilon)$ be an automaton that recognizes $IRR_\epsilon(R)$. Let $p = p_1. \ldots .p_k$ with $p_1, \ldots, p_k \in \mathbb{N} - \{0\}$
and $\forall i \ p_i \leq Max_{f \in F \cup C}(ar(f))$

We define $\mathcal{A}_{irr}$ as follows :

$$\mathcal{A}_{irr} = (C \cup F, Q_{irr}, Q_{irrf}, \Delta_{irr}) \text{ where}$$
$$Q_{irr} = \{q_{any},\ q_{rec}\} \cup_{v<p} \{q^v\} \cup_{v \in \overline{Pos}(t) \setminus \{v' | v' \leq p\}} \{q^v_{any}\} \cup Q_\epsilon$$
$$Q_{irrf} = \{q^\epsilon\} \text{ and}$$
$$\Delta_{irr} = \{s(S_1, \ldots, S_n) \rightarrow q^j \mid s \in F \cup C, ar(s) \geq p_{long(j)+1}$$
$$q^j \in Q_{irr},\ S_i = \begin{vmatrix} q^{j.i} & \text{if } j.i < p \\ q_{rec} & \text{if j.i=p} \\ q^{j.i}_{any} & \text{otherwise} \end{vmatrix} \ \} if\, p \neq \epsilon$$
$$\cup \{q_f \rightarrow q_{rec} \mid q_f \ \in \ Q_{\epsilon f}\} if\, p \neq \epsilon$$
$$\cup \{q_f \rightarrow q^\epsilon \mid q_f \ \in \ Q_{\epsilon f}\} if\, p = \epsilon$$
$$\cup \{s(S_1, \ldots, S_n) \rightarrow q^j_{any} \mid s \in F \cup C,\ q^j_{any} \in Q_{irr},$$
$$S_i = \begin{vmatrix} q^{j.i}_{any} \text{ if } j.i \in \overline{Pos}(t) \\ q_{any} \text{ otherwise} \end{vmatrix} \}$$
$$\cup \{s(q_{any}, \ldots, q_{any}) \rightarrow q_{any} \mid s \in F \cup C\}$$
$$\cup \Delta_\epsilon$$

$\mathcal{A}_{irr}$ recognizes $IRR_p(R)$ indeed, because:
$t|_p$ reducible i.e. $\exists$ u position s.t $u \geq p$ and $t \rightarrow_{[u]} t'$.

- q_{any} recognizes any terms.
- q^w recognize $t|_w$ *for* $w < p$.

We have written $ar(s) \geq p_{long(j)+1}$ to ensure that $p \in Pos(t)$. For example, if $p = 1.2.1$ and $s(\ldots) \rightarrow q^1$, then s should have an arity ≥ 2.
Obviously, $\mathcal{A}_{irr}$ discriminates p into q_{rec} (into q^ϵ if $p = \epsilon$), and each $p' \in \overline{Pos}(t)$ s.t. $p' \not\geq p$ into $q^{p'}_{any}$ ($q^{p'}$ if $p' < p$).

2.4 Descendants

t' is a *descendant* of t if $t \to_R^* t'$. t' is a *normal-form* of t if $t \to_R^* t'$ and t' is irreducible. If E is a set of ground terms, $R^*(E)$ denotes the set of descendants of elements of E. $IRR(R)$ denotes the set of irreducible ground terms. $R^*_{in}(E)$ (resp. $R^*_{ileft}(E)$, $R^*_{out}(E)$) denotes the set of descendants of E, according to an innermost (resp. innermost-leftmost, outermost) strategy.

Definition 2.12. *$t \to^+_{[p,rhs's]} t'$ means that t' is obtained by rewriting t at position p, plus possibly at positions coming from the rhs's.*
Formally, there exist some intermediate terms $t_1, \ldots, t_n$ and some sets of positions $P(t), P(t_1), \ldots, P(t_n)$ s.t.

$$t = t_0 \to_{[p_0, l_0 \to r_0]} t_1 \to_{[p_1, l_1 \to r_1]} \cdots \to_{[p_{n-1}, l_{n-1} \to r_{n-1}]} t_n \to_{[p_n, l_n \to r_n]} t_{n+1} = t'$$

where

- $p_0 = p$ *and* $P(t) = \{p\}$,
- $\forall j,\ p_j \in P(t_j)$,
- $\forall j,\ P(t_{j+1}) = P(t_j) \backslash \{p' \mid p' \geq p_j\} \cup \{p_j.w \mid w \in PosF(r_j)\}$.

Remark : $P(t_j)$ contains only function positions. Since there are no nested functions in rhs's, $p, p' \in P(t_j)$ implies $p \| p'$.

Definition 2.13. *Given a language E and a position p, we define $R^*_p(E)$ as follows*

$$R^*_p(E) = E \cup \{t' \mid \exists t \in E, t \to^+_{[p,rhs's]} t'\}$$

Example 1. $R = \{f(x) \to s(x),\ g(x) \to s(h(x)),\ h(x) \to f(x)\}$
$R^*_1(f(h(g(a)))) = E \cup f(f(g(a))) \cup f(s(g(a))$

An insight into the algorithm underlying the following result will be given in the sequel by Example 2, and a formal description is in the full version. The resultant automata are different from the starting one only at positions below p, and in the general case, are built by nesting automata.

Theorem 2.14. *[12] Let R be a rewrite system satisfying the restrictions given in the introduction. If E is recognized by an automaton that discriminates position p into some state q, and possibly p' into q' for some $p' \in \overline{Pos}(t)$ s.t. $p' \not\geq p$, and some states q', then so is $R^*_p(E)$.*

2.5 Positions

Given a term t, we define:

Definition 2.15. *Let $p \in Pos(t)$. $Succ(p)$ are the nearest function positions below p:*

$$Succ(p) = \{p' \in PosF(t) \mid p' > p \text{ and } \forall q \in Pos(t)\ (p < q < p' \Rightarrow q \notin PosF(t))\}$$

Definition 2.16. *Let $p, p' \in Pos(t)$. $p \lhd p'$ means that p occurs strictly on the left of p', i.e. $p = u.i.v$, $p' = u.i'.v'$, where $i, i' \in \mathbb{N}$ and $i < i'$.*

3 Innermost Descendants: $R^*_{in}(E)$

Example 2. Let a, s be constructors and f be a function, s.t. a is a constant, and s, f are unary symbols. Let $t = f(s(f(s(y))))$ and $\mathcal{A}_{t\theta}$ be the automaton that recognizes the language $E = f(s(f(s(s^*(a)))))$ of the data-instances of t.
$\mathcal{A}_{t\theta}$ can be summarized by writing :

$$\overset{q^\epsilon}{f}\,(\overset{q^1}{s}\,(\overset{q^{1.1}}{f}\,(\overset{q^{1.1.1}}{s}\,(\overset{q_{data}}{s^*}\,(a)))))$$

which means that

$$\Delta_{t\theta} = \{a \to q_{data},\ s(q_{data}) \to q_{data},\ s(q_{data}) \to q^{1.1.1},\ f(q^{1.1.1}) \to q^{1.1},\ s(q^{1.1}) \to q^1,\ f(q^1) \to q^\epsilon\}$$

where q^ϵ is the accepting state.
Consider now the rewrite system $R = \{f(s(x)) \to s(x)\}$.
Obviously, $R^*_{in}(E) = E \cup f(s(s(s^*(a)))) \cup s(s(s^*(a)))$.

We can make two remarks:

- When rewriting E, some instances of rhs's of rewrite rules are introduced by rewrite steps. So, to build an automaton that can recognize $R^*_{in}(E)$, we need to recognize the instances of rhs's into some states, without making any confusion between the various potential instances of the same rhs.
- When the starting term has nested functions, according to the innermost strategy, we first have to rewrite innermost function positions.

Note that here, we can rewrite E at positions ϵ and 1.1. According to the previous remark, we start from position 1.1.

Now, we calculate $R^*_{1.1}(E)$.

$$(1) \qquad f(s(f(s(s^*(a))))) \to_{[1.1, x/s^*(a)]} f(s(s(s^*(a))))$$

The language that instantiates the rewrite rule variable x is $s^*(a)$ (recognized into q_{data}). Therefore, we encode the first version of the rhs: $\overset{d^\epsilon_{q_{data}}}{s}\ (\overset{q_{data}}{x})$ by adding state $d^\epsilon_{q_{data}}$ and the transition $s(q_{data}) \to d^\epsilon_{q_{data}}$.

We can simulate the rewrite step, by adding transitions again. This step is called saturation in the following. Consider (1) again. Since $f(s(x))$ is the rule lhs, and $f(s(q_{data})) \to^*_{\Delta_{t\theta}} q^{1.1}$, we add the transition $d^\epsilon_{q_{data}} \to q^{1.1}$ so that the instance of the rhs by q_{data} is also recognized into $q^{1.1}$, i.e. $s(s^*(a)) \to^* q^{1.1}$. So, $R^*_{1.1}(E) = E \cup f(s(s(s^*(a))))$ is recognized by the automaton.

Now, rewriting terms of $R^*_{1.1}(E)$ at position ϵ is allowed only if position 1.1 is normalized. Consider $E' = R^*_{1.1}(E) \cap IRR_{1.1}(R)$ where $IRR_{1.1}(R)$ is the ground terms irreducible at position 1.1, over the TRS R. Thus $E' = f(s(s(s^*(a))))$, and let us calculate $R^*_\epsilon(E')$.

Let $\mathcal{A}' = (\mathcal{C} \cup \mathcal{F}, Q', \{q'^\epsilon\}, \Delta')$ an automaton that recognizes the language E' where $\Delta' = \{a \to q_{data},\ s(q_{data}) \to q_{data},\ s(q_{data}) \to q'^{1.1},\ s(q'^{1.1}) \to q'^1,\ f(q'^1) \to q'^\epsilon\}$ where q'^ϵ is the accepting state.

$$(2) \qquad f(s(s(s^*(a)))) \rightarrow_{[x/s(s^*(a))]} s(s(s^*(a)))$$

The language that instantiates x is $s(s^*(a))$ (recognized into ${q'}^{1.1}$). Therefore, we encode a second version of the rhs : $\overset{d^{\epsilon}_{q'1.1}}{s}(\overset{{q'}^{1.1}}{x})$ by adding state $d^{\epsilon}_{q'1.1}$ and the transition $s({q'}^{1.1}) \rightarrow d^{\epsilon}_{q'1.1}$.
By saturation, since $f(s(x))$ is the rule lhs and $f(s({q'}^{1.1})) \rightarrow {q'}^{\epsilon}$, we add the transition $d^{\epsilon}_{q'1.1} \rightarrow {q'}^{\epsilon}$ so that $s(s(s^*(a))) \rightarrow^* {q'}^{\epsilon}$.
So, $R^*_{\epsilon}(E') = E' \cup s(s(s^*(a)))$ is recognized by the automaton.

E' contains only terms normalized at position 1.1, which is not required by the innermost strategy when no rewrite step is applied at position ϵ.
Therefore, $R^*_{in}(E) = R^*_{\epsilon}(E') \cup R^*_{1.1}(E) = R^*_{\epsilon}(R^*_{1.1}(E) \cap IRR_{1.1}(R)) \cup R^*_{1.1}(E)$.
Remark : In the previous example, the starting term has nested functions. When it is not the case, every rewrite step is innermost, because rhs's have no nested functions either.

In general t may have more than two function positions. To generalize, we need the following notion.

Definition 3.1. *Given a language L and a position p, $R^*_{in,p}(L)$ are the innermost descendants of L over the TRS R, reducing positions below (or equal to) p, i.e.*

$$R^*_{in,p}(L) = \{s' | \exists s \in L,\ s \rightarrow^*_{[u_1,\dots,u_n]} s' \text{ by an innermost strategy}, \forall i\ (u_i \geq p)\}$$

For a language L, let $L|_p = \{s|_p \mid s \in L,\ p \in Pos(s)\}$.

Lemma 3.2. *Let R be a rewrite system satisfying the restrictions given in the introduction, and E be the data-instances of a given linear term t.*
Let $p \in PosF(t)$, and L be a language s.t. $L|_p = E|_p$, and that is recognized by an automaton $\mathcal{A}$ that discriminates every position $p' \in PosF(t) \mid p' \geq p$. Then,

$$R^*_{in,p}(L) = R^*_p(L) \ if \ Succ(p) = \emptyset$$

Otherwise, let $Succ(p) = \{p_1, \dots, p_n\}$, and in this case

$$R^*_{in,p}(L) = \left| \begin{array}{l} R^*_p[R^*_{in,p_1}(\dots(R^*_{in,p_n}(L))\dots) \cap_{p_i \in Succ(p)} IRR_{p_i}(R)] \\ \ \cup\ R^*_{in,p_1}(\dots(R^*_{in,p_n}(L))\dots) \end{array} \right.$$

and $R^*_{in,p}(L)$ is recognized by an automaton $\mathcal{A}'$ s.t. if $p' \in \overline{Pos}(t)$, $p' \not> p$, and $\mathcal{A}$ discriminates p' into q', then $\mathcal{A}'$ also discriminates p' into q'.

Proof. By noetherian induction on $(PosF(t), >)$.

- If $Succ(p) = \emptyset$, then $\forall s \in L, \forall p' \in Pos(s), (p' > p \implies s(p') \in C)$. And since rhs's have no nested functions, $R^*_p(L) = R^*_{in,p}(L)$.
We get $\mathcal{A}'$ by Theorem 2.14.
- Let $Succ(p) = \{p_1, \dots, p_n\}$. Let us define:
$R^*_{in,>p}(L) = \{s' | \exists s \in L,\ s \rightarrow^*_{[u_1,\dots,u_n]} s'$ by an innermost strat., $\forall i\ (u_i > p)\}$
Let $s \in L$. Either a rewrite step is applied at position p, and the strategy is innermost only if we first normalize s below position p by an innermost

derivation, or no rewrite step is applied at position p. And since no defined-function occurs along any branches between p and p_i :

$$R^*_{in,p}(L) = R^*_p[R^*_{in,>p}(L) \cap_{p_i \in Succ(p)} IRR_{p_i}(R)] \cup R^*_{in,>p}(L)$$

Now, note that $\forall i, j \in [1..n]$, $(i \neq j \implies p_i || p_j)$. Moreover rewrite steps at incomparable positions can be commuted. Then obviously:

$$R^*_{in,>p}(L) = R^*_{in,p_1}(\ldots (R^*_{in,p_n}(L) \ldots))$$

L is recognized by an automaton $\mathcal{A}$ that discriminates every $p' \in PosF(t)$ s.t. $p' \geq p$. For each i, $p_i > p$, then $\mathcal{A}$ discriminates every $p' \in PosF(t)$ s.t. $p' \geq p_i$. By induction hypothesis, $R^*_{in,p_n}(L)$ is recognized by an automaton $\mathcal{A}'_n$ that still discriminates p and every position p' s.t. $p' \geq p_i$, $i = 1, \ldots, n-1$, $R^*_{in,p_{n-1}}(R^*_{in,p_n}(L))$ is recognized by an automaton $\mathcal{A}'_{n-1}$ that still discriminates p and every position p' s.t. $p' \geq p_i$, $i = 1, \ldots, n-2$, $R^*_{in,p_1}(\ldots (R^*_{in,p_n}(L) \ldots))$ is recognized by an automaton $\mathcal{A}'_1$ that still discriminates p.

By Theorem 2.11, $IRR_{p_i}(R)$ is recognized by an automaton that discriminates every position $p' \in PosF(t)$ s.t. $p' \not> p_i$, then necessarily p. By lemma 2.5, $\cap_{p_i \in Succ(p)} IRR_{p_i}(R)$ is recognized by an automaton that discriminates p. Therefore $R^*_{in,p_1}(\ldots (R^*_{in,p_n}(L)) \ldots) \cap_{p_i \in Succ(p)} IRR_{p_i}(R)$ is recognized by an automaton that discriminates p, and from Theorem 2.14, so is $R^*_p[R^*_{in,p_1}(\ldots (R^*_{in,p_n}(L)) \ldots) \cap_{p_i \in Succ(p)} IRR_{p_i}(R)]$. Moreover discrimination of positions $p' \not> p$ is preserved. Finally, by union, we obtain an automaton that discriminates p and preserves the discrimination of positions $p' \not> p$.

Theorem 3.3. *Let R be a rewrite system satisfying the restrictions given in the introduction, and E be the data-instances of a given linear term t.*

$$R^*_{in}(E) = \begin{cases} R^*_{in,\epsilon}(E) & \text{if } t(\epsilon) \in F \\ R^*_{in,p_1}(\ldots (R^*_{in,p_n}(E) \ldots)) \text{ with } Succ(\epsilon) = \{p_1, \ldots, p_n\} & \text{otherwise} \end{cases}$$

and $R^*_{in}(E)$ is effectively recognized by an automaton.

Proof. We have two cases:

- If $t(\epsilon) \in F$, obviously $R^*_{in}(E) = R^*_{in,\epsilon}(E)$.
- If $t(\epsilon) \notin F$, $\forall i, j \in [1..n]$, $(i \neq j \implies p_i || p_j)$, and rewrite steps at incomparable positions can be commuted. Then $R^*_{in}(E) = R^*_{in,p_1}(\ldots (R^*_{in,p_n}(E) \ldots))$.

The automaton comes from Definition 2.6 and from applying Lemma 3.2 (several times in the second case).

Example 3. Let E the data-instances of $t = f(g(x),\ h(g(y)))$ and

$$R = \{f(x, y) \to y,\ h(x) \to s(x),\ g(x) \to x\}$$

$*$ will symbolize the data-terms that instantiate t.
$t(\epsilon) \in \mathcal{F}$, we so calculate $R^*_{in,\epsilon}(E)$ where $E = f(g(*),\ h(g(*)))$.
$R^*_{in,\epsilon}(E) = R^*_\epsilon[R^*_{in,1}(R^*_{in,2}(E)) \cap IRR_1(R) \cap IRR_2(R)] \cup R^*_{in,1}(R^*_{in,2}(E))$.

We have to compute $R^*_{in,2}(E)$.

$Succ(2) = \{2.1\}$
So, $R^*_{in,2}(E) = R^*_2[R^*_{in,2.1}(E) \cap IRR_{2.1}(R)] \cup R^*_{in,2.1}(E)$
where $R^*_{in,2.1}(E) = E \cup f(g(*), h(*))$.
$R^*_{in,2}(E) = R^*_2[f(g(*),\ h(*))] \cup R^*_{in,2.1}(E)$
$= f(g(*),\ h(*)) \cup f(g(*),\ s(*)) \cup R^*_{in,2.1}(E)$ (denoted by E1).
Now, we can compute $R^*_{in,1}(R^*_{in,2}(E))$.
$Succ(1) = \emptyset$.
So, $R^*_{in,1}(E1) = R^*_1(E1)$
$= E1 \cup f(*, h(*)) \cup f(*, s(*)) \cup f(*, h(g(*)))$ (denoted by E2).
$R^*_{in,\epsilon}(E) = R^*_\epsilon[E2 \cap IRR_1(R) \cap IRR_2(R)] \cup E2$
$= R^*_\epsilon[f(*, s(*))] \cup E2$
$= f(*, s(*)) \cup s(*) \cup E2$
Finally, we obtain $R^*_{in}(E) = E \cup f(g(*), h(*)) \cup f(g(*), s(*)) \cup f(*, h(*)) \cup f(*, s(*)) \cup f(*, h(g(*))) \cup s(*)$.

4 Innermost-Leftmost Descendants: $R^*_{ileft}(E)$

Definition 4.1. *Given a language L and a position p, let us define:*

$$R^*_{ileft,p}(L) = \{t' \mid \exists t \in L,\ t \rightarrow^*_{[u_1,\dots,u_n]} t' \text{ by an innermost}-\text{leftmost strategy, and } \forall i\ (u_i \geq p)\}$$

Lemma 4.2. *Let R be a rewrite system satisfying the restrictions given in the introduction, and E be the data-instances of a given linear term t. Let $p \in PosF(t)$ and L be a language s.t. $L|_p = E|_p$, and that is recognized by an automaton $\mathcal{A}$ that discriminates every position $p' \in PosF(t) \mid p' \geq p$. Then,*

$$R^*_{ileft,p}(L) = R^*_p(L) \text{ if } Succ(p) = \emptyset$$

Otherwise, let $Succ(p) = \{p_1, \dots, p_n\}$ s.t. $p_1 \lhd \dots \lhd p_n$, and in this case

$$R^*_{ileft,p}(L) = \left| \begin{array}{l} R^*_p[R^*_{ileft,p_n}(\dots(R^*_{ileft,p_1}(L) \cap IRR_{p_1}(R))\dots) \cap IRR_{p_n}(R)] \\ \quad \cup R^*_{ileft,p_1}(L) \cup R^*_{ileft,p_2}(R^*_{ileft,p_1}(L) \cap IRR_{p_1}(R)) \cup \dots \\ \qquad \dots \cup R^*_{ileft,p_n}(\dots(R^*_{in,p_1}(L) \cap IRR_{p_1}(R))\dots) \end{array} \right.$$

and $R^*_{ileft,p}(L)$ is recognized by an automaton $\mathcal{A}'$ s.t. if $p' \in \overline{Pos}(t)$, $p' \not> p$, and $\mathcal{A}$ discriminates p' into q', then $\mathcal{A}'$ also discriminates p' into q'.

Theorem 4.3. *Let E be the data-instances of a linear term t and let R a TRS satisfying the restrictions given in the introduction.*

$$R^*_{ileft}(E) = \left| \begin{array}{ll} R^*_{ileft,\epsilon}(E) & \text{if } \epsilon \in PosF(t) \\ & \\ R^*_{ileft,p_1}(E) \cup \dots \cup & \\ R^*_{ileft,p_n}(\dots(R^*_{ileft,p_1}(E) \cap IRR_{p_1})\dots) & \text{otherwise} \\ \text{with } Succ(\epsilon) = \{p_1, \dots, p_n\} \text{ s.t. } p_1 \lhd \dots \lhd p_n & \end{array} \right.$$

and $R^*_{ileft}(E)$ is effectively recognized by an automaton.

Proof. We have two cases:

- If $\epsilon \in PosF(t)$, obviously $R^*_{ileft}(E) = R^*_{ileft,\epsilon}(E)$.

- If $\epsilon \notin PosF(t)$, $\forall i,\ j$ *s.t.* $p_i \lhd p_j$, innermost-leftmost descendants at position p_j can be computed after to have normalized those at position p_i. Then obviously, $R^*_{ileft}(E) = R^*_{ileft,p_1}(E) \cup \ldots \cup R^*_{ileft,p_n}(\ldots(R^*_{ileft,p_1}(E) \cap IRR_{p_1})\ldots)$

The automaton comes from Definition 2.6 and from applying Lemma 4.2.

Example 4. Let E the data-instances of $t = f(g(x),\ h(y))$ and

$$R = \{f(x,y) \to s(f(x,y)),\ h(x) \to s(x),\ g(x) \to x\}.$$

$*$ will symbolize the data-terms that instantiate t.
$t(\epsilon) \in \mathcal{F}$, we so calculate $R^*_{ileft,\epsilon}(E)$ where $E = f(g(*),\ h(*))$.

$$R^*_{ileft,\epsilon}(E) = R^*_\epsilon[R^*_{ileft,2}(R^*_{ileft,1}(E) \cap IRR_1(R)) \cap IRR_2(R)] \\ \cup R^*_{ileft,1}(E) \cup R^*_{ileft,2}(R^*_{ileft,1}(E) \cap IRR_1(R)).$$

We have to compute $R^*_{ileft,1}(E)$.
So, $R^*_{ileft,1}(E) = R^*_1(E)$
$= E \cup f(*, h(*))$
Now, we can compute $R^*_{ileft,2}(R^*_{ileft,1}(E) \cap IRR_1(R))$.
$R^*_{ileft,2}(f(*, h(*)) = f(*, s(*)) \cup f(*, h(*))$
So, $R^*_\epsilon(f(*, s(*))) = s^*(f(*, s(*)))$
Finally, we obtain $R^*_{ileft}(E) = s^*(f(*, s(*))) \cup E \cup f(*, h(*))$.
Remark : The current computation of R^*_p does not take any strategies into account. Since rhs's do not have nested defined-functions, R^*_p is automatically innermost, but not leftmost. This is why we assume in addition that rhs's have at most one defined-function. However it is possible to modify the computation of R^*_p, to take the leftmost strategy into account (see [13]). Then, Theorem 4.3 still holds even if right-hand-sides contain several defined-functions.

5 Outermost Descendants: $R^*_{out}(E)$

Example 5. Consider two constructors s, a, and let

$$R = \{f(s(x),\ s(y)) \to s(f(x,\ y)),\quad g(s(x)) \to s(g(x))\}$$

Let $t = f(g(x),\ g(y))$, hence the starting language is $E = f(g(s^*(a)),\ g(s^*(a)))$. Obviously the descendants of E are $R^*(E) = s^*(f(s^*(g(s^*(a))),\ s^*(g(s^*(a)))))$, whereas the outermost descendants are:

$$R^*_{out}(E) = s^*(f(s^*(g(s^*(a))),\ s^?(g(s^*(a))))) \cup s^*(f(s^?(g(s^*(a))),\ s^*(g(s^*(a)))))$$

where $s^?$ means zero or one occurrence of s.

Surprisingly, to compute $R^*_{out}(E)$, we first reduce innermost positions 1 and 2 by computing $R^*_2(R^*_1(E)) = f(s^*(g(s^*(a))),\ s^*(g(s^*(a))))$. Now we reduce f (position ϵ in E), however we keep only the descendants s.t. f cannot be reduced any more. Let $R^!_\epsilon$ denote this local normalization. We get:

$$R^!_\epsilon[R^*_2(R^*_1(E))] = s^*(f(s^*(g(s^*(a))),\ g(s^*(a)))) \cup s^*(f(g(s^*(a)),\ s^*(g(s^*(a)))))$$

which is a subset of $R^*_{out}(E)$. Some outermost descendants are missing because the normalization of f is too harsh. To define the right one, we introduce an outermost abstract rewriting, that selects exactly the outermost descendants among the elements of $R^*(E)$. It is composed of an outermost narrowing step, a subterm decomposition, and an approximation. Subterm decomposition and approximation are just for making abstract rewriting terminate. The following definitions are illustrated by Example 6.

Notations: $\rightsquigarrow$ denotes the narrowing relation. Given a TRS R, $t \rightsquigarrow_{[p]}$ means that there exists a narrowing step at position p issued from t. Because of the restrictions, every considered term is linear: variable names do not matter, and for readability, we will often replace variables by the anonymous variable $\perp$.

Definition 5.1. *Let T_2 be the set of terms (with variables) s.t. $t(\epsilon) \in F$ and every branch of t contains at most two nested defined-functions. Given a TRS R, let us define on T_2 a binary relation $\approx$, to approximate terms. Let $t, t' \in T_2$, and for t let $Succ(\epsilon) = \{p_1, \ldots, p_n\}$. For each $i \in \{1, \ldots, n\}$, let $t_i = t[\perp]_{p_j, \forall j \neq i}$, and let $t_0 = t[\perp]_{p_j, \forall j}$. Then $t \approx t'$ iff*

1. $t' = t \wedge (1)$ *where* $(1) = (t \rightsquigarrow_{[\epsilon]} \vee Succ(\epsilon) = \emptyset)$
2. *or* $t' = t_0 \wedge \neg(1) \wedge t_0 \not\rightsquigarrow_{[\epsilon]}$
3. *or* $((t' = t_1 \wedge t_1 \not\rightsquigarrow_{[\epsilon]}) \vee \ldots \vee (t' = t_n \wedge t_n \not\rightsquigarrow_{[\epsilon]})) \wedge \neg(1) \wedge t_0 \rightsquigarrow_{[\epsilon]}$

Remark: For a given t, several t' may exist in case 3, and there is at least one because $t_1 \rightsquigarrow_{[\epsilon]} \wedge \ldots \wedge t_n \rightsquigarrow_{[\epsilon]}$ implies $t \rightsquigarrow_{[\epsilon]}$ (thanks to constructor discipline), i.e. condition (1) is true.

Comments: In case 1, t can be narrowed at position ϵ, or t does not contain nested defined-functions. Therefore a step $t \rightsquigarrow_{[\epsilon]}$ (if any) is outermost: we keep t and will try to narrow it at ϵ.

Case 2 means that t cannot be narrowed at ϵ, even if inner positions are first narrowed. Therefore, any narrowing step at an inner position is outermost. Note that a term obtained by narrowing steps at inner positions is an instance of t_0. Thus, the descendants of $t\theta$ that are instances of t_0 are outermost descendants: we keep t_0 instead of t.

Case 3 means that t cannot be narrowed at ϵ, but could be narrowed at ϵ if inner positions are first modified (narrowed). Suppose $t_i \not\rightsquigarrow_{[\epsilon]}$. Then every derivation $t \rightsquigarrow^*_{[p_{j_1}, \ldots, p_{j_l}]} s$ s.t. $\forall j_k$, $p_{j_k} \neq p_i$ is outermost, s is an instance of t_i, and $s \not\rightsquigarrow_{[\epsilon]}$: we keep t_i instead of t, and will try to narrow it at p_i.

Definition 5.2. *(outermost abstract rewriting)*
Let $t, t' \in T_2$. $t \hookrightarrow_{[p, l \to r]} t'$ if there exists a term s s.t. $t \rightsquigarrow_{[p, l \to r, \theta]} s$, and

- $p = \epsilon \wedge \exists u \in PosF(r) \mid s|_u \approx t'$,
- *or* $t \not\rightsquigarrow_{[\epsilon]}$ *and* $s \approx t'$.

Example 6. Let $R = \{h(c(x, y)) \to c(h(x), i(y)),\ i(s(x)) \to p(i(x))\}$ where c, s, p are constructors.
Since $h(h(\perp)) \not\rightsquigarrow_{[\epsilon]}$ and $h(h(\perp)) \rightsquigarrow_{[1]} h(c(h(\perp), i(\perp))) \approx_{case1} itself$, then $h(h(\perp)) \hookrightarrow h(c(h(\perp), i(\perp)))$.
Since $h(c(h(\perp), i(\perp))) \rightsquigarrow_{[\epsilon]} c(h(h(\perp)), i(i(\perp)))$ and subterms $h(h(\perp)) \approx_{case3}$ $itself$ ($n = 1$ then $t_1 = t$) and $i(i(\perp)) \approx_{case3} itself$, then $h(c(h(\perp), i(\perp))) \hookrightarrow h(h(\perp))$ and $h(c(h(\perp), i(\perp))) \hookrightarrow i(i(\perp))$.
Since $i(i(\perp)) \not\rightsquigarrow_{[\epsilon]}$ and $i(i(\perp)) \rightsquigarrow_{[1]} i(p(i(\perp))) \approx_{case2} i(p(\perp))$, then $i(i(\perp)) \hookrightarrow i(p(\perp))$.
Note that $i(p(\perp))$ does not narrow.

Lemma 5.3. *Let $t \in T_2$, σ be a data-substitution, and $t\sigma \to_{[p,l\to r]} t'$ be a rewrite step. If $p=\epsilon$ let $u \in PosF(r)$, otherwise let $u=\epsilon$. Then:*

$$\begin{array}{ll} & t\sigma \to_{[p,l\to r]} t' \text{ is outermost} \\ \Longleftrightarrow & \exists s, s' \in T_2 \mid t \approx s \ \wedge\ s \hookrightarrow^{?}_{[p,l\to r]} s' \ \wedge\ t'|_u \text{ is an instance of } s' \end{array}$$

The previous lemma does not extend to several steps if rhs's may contain several defined-functions (resp. if R has critical pairs), due to the decomposition step included in abstract rewriting, which may cause a confusion between different subterms (resp. between terms obtained by applying different rewrite rules).

Example 7.

$$R = \{i(x,y) \to c(h(x),h(y)),\quad h(x) \to x,\quad g(s(x)) \to s(g(x))\}$$

Let $t = i(s(g(x)),\, g(y))$. Then $t \approx t \hookrightarrow_{[\epsilon]} h(s(g(\bot)))$. The step

$$i(s(g(s^*(a))),\, g(s^*(a))) \to_{[\epsilon]} t' = c(h(s(g(s^*(a)))),\, h(g(s^*(a))))$$

is outermost. Now $t' \to_{[2.1]} t''$ is not outermost whereas $t''|_2 = h(s(g(s^*(a))))$ is an instance of $h(s(g(\bot)))$.

Definition 5.4. *Let E be the data-instances of a given linear term t, and $p \in PosF(t)$. Assume $Succ(p) \neq \emptyset$.*
Let us define $(t|_p)_{abs} \in T_2$ by: if $Succ(Succ(p)) = \emptyset$ (i.e. $t|_p \in T_2$), let $(t|_p)_{abs} = t|_p$, otherwise $(t|_p)_{abs} = (t[\bot]_{p_j,\forall j})|_p$ where $Succ(Succ(p)) = \{p_1,\dots,p_n\}$. Note that $t|_p$ is an instance of $(t|_p)_{abs}$.
Let $R^*_{abs,p}(E)$ be the set of descendants of $\{s \in T_2 \mid (t|_p)_{abs} \approx s\}$ by outermost abstract rewriting.
Let $\mathcal{L}(R^*_{abs,p}(E))$ be the set of ground instances of elements of $R^*_{abs,p}(E)$.

Example 8. Consider Example 5 again, and let $p=\epsilon$. Then $t_{abs} = t$, and $t \approx t_1 = f(g(\bot),\bot)$, and $t \approx t_2 = f(\bot, g(\bot))$. Now:

$$t_1 = f(g(\bot),\, \bot) \hookrightarrow_{[1]} f(s(g(\bot)),\, \bot) \hookrightarrow_{[\epsilon]} f(g(\bot),\, \bot) = t_1$$
$$t_2 = f(\bot,\, g(\bot)) \hookrightarrow_{[2]} f(\bot,\, s(g(\bot))) \hookrightarrow_{[\epsilon]} f(\bot,\, g(\bot)) = t_2$$

Therefore

$$R^*_{abs,\epsilon}(E) = \{t_2,\ f(\bot,\, s(g(\bot))),\ t_1,\ f(s(g(\bot)),\, \bot)\}$$

Now, if $R^!_\epsilon$ reduces f and keeps the descendants whose subterms headed by f are instances of elements of $R^*_{abs,\epsilon}(E)$, then even the missing descendants (when $s^?$ is exactly s) are obtained, and $R^!_\epsilon[R^*_2(R^*_1(E))] = R^*_{out}(E)$.

Lemma 5.5. *$R^*_{abs,p}(E)$ is finite. Then the set $\mathcal{L}(R^*_{abs,p}(E))$ of ground instances of elements of $R^*_{abs,p}(E)$ is effectively recognized by a tree automaton.*

Definition 5.6. *Let $p \in PosF(t)$ s.t. $Succ(p) \neq \emptyset$. $t \to^!_{[p,rhs's]} t'$ means that $t \to^*_{[p,rhs's]} t'$ (like in Definition 2.12, except that zero step is allowed), provided in addition*

$$\forall q \in P(t'),\ t'|_q \in \mathcal{L}(R^*_{abs,p}(E))$$

Given a language L, let $R^!_p(L) = \{t' \mid \exists t \in L, t \to^!_{[p,rhs's]} t'\}$.

Lemma 5.7. *If L is recognized by an automaton $\mathcal{A}$ that discriminates p, then $R^!_p(L)$ is recognized by an automaton $\mathcal{A}'$ s.t. if $p' \not> p$ and $\mathcal{A}$ discriminates p' into q', then $\mathcal{A}'$ also discriminates p' into q'.*

Definition 5.8.

$$t_0 \rightarrow_{[p_0, l_0 \rightarrow r_0]} t_1 \rightarrow \ldots t_n \rightarrow_{[p_n, l_n \rightarrow r_n]} t_{n+1} = t'$$

is outermost under a position p if

$$\forall i,\ p_i \geq p \ \wedge\ \forall q_i\, (p < q_i < p_i \implies t_i \not\rightarrow_{[q_i]})$$

Given a language L, $R^*_{out,p}(L)$ are the descendants of L outermost under p, using the TRS R, i.e. $R^*_{out,p}(L) = \{t' \mid \exists s \in L, s \rightarrow^* t'$ by a derivation outermost under $p\}$.

Lemma 5.9. *Let R be a rewrite system satisfying the restrictions given in the introduction, and E be the data-instances of a given linear term t.*
Let $p \in PosF(t)$, and L be a language s.t. $L|_p = E|_p$, and that is recognized by an automaton $\mathcal{A}$ that discriminates every position $p' \in PosF(t) \mid p' \geq p$. Then,

$$R^*_{out,p}(L) = R^*_p(L) \ if \ Succ(p) = \emptyset$$

Otherwise, let $Succ(p) = \{p_1, \ldots, p_n\}$, and in this case

$$R^*_{out,p}(L) = R^!_p[R^*_{out,p_1}(\cdots(R^*_{out,p_n}(L))\ldots)]$$

and $R^*_{out,p}(L)$ is recognized by an automaton $\mathcal{A}'$ s.t. if $p' \in \overline{Pos}(t)$, $p' \not\geq p$, and $\mathcal{A}$ discriminates p' into q', then $\mathcal{A}'$ also discriminates p' into q'.

Theorem 5.10. *Let R be a rewrite system satisfying the restrictions given in the introduction, and E be the data-instances of a given linear term t.*

$$R^*_{out}(E) = \left| \begin{array}{ll} R^*_{out,\epsilon}(E) & if\ t(\epsilon) \in F \\ & \\ R^*_{out,p_1}(\cdots(R^*_{out,p_n}(E)\ldots)) & otherwise \\ with\ Succ(\epsilon) = \{p_1, \cdots, p_n\} & \end{array} \right.$$

and $R^*_{out}(E)$ is effectively recognized by an automaton.

6 Conclusion

When computing descendants, taking some strategies into account is possible, keeping the same class of tree language (the regular ones) and assuming the same restrictions (plus left-linearity). Does it also hold for any other computations of descendants ?

References

1. H. Comon. Sequentiality, second order monadic logic and tree automata. In Proc., *Tenth Annual IEEE Symposium on logic in computer science,* pages 508-517. IEEE Computer Society Press, 26-29 June 1995.
2. H. Comon, M. Dauchet, R. Gilleron, D. Lugiez, S. Tison, and M. Tommasi. *Tree Automata techniques and Applications* (TATA). http://l3ux02.univ-lille3.fr/tata.
3. J. Coquidé, M. Dauchet, R. Gilleron, and S. Vagvolgyi. Bottom-up Tree Pushdown Automata and Rewrite Systems. In R. V. Book, editor, *Proceedings 4th Conference on Rewriting Techniques and Applications, Como (Italy),* volume 488 of LNCS, pages 287-298. Springer-Verlag, April 1991.

4. M. Dauchet, A.C. Caron, and J.L. Coquidé. Reduction Properties and Automata with Constraints. In *Journal of Symbolic Computation, 20:215-233.* 1995.
5. M. Dauchet and S. Tison. The theory of ground rewrite systems is decidable. In Proc., *Fifth Annual IEEE Symposium on logic in computer science,* pages 242-248, Philadelphia, Pennsylvania, 1990. IEEE Computer Society Press.
6. J. H. Gallier and R. V. Book. Reductions in tree replacement systems. *Theoretical Computer Science,*37:123-150, 1985.
7. T. Genet. Decidable Approximations of Sets of Descendants and Sets of Normal Forms. In *Proceedings of 9th Conference on Rewriting Techniques and Applications, Tsukuba (Japan),* volume 1379 of LNCS, pages 151-165. Springer-Verlag,1998.
8. T. Genet, F. Klay. Rewriting for Cryptographic Protocol Verification. *Technical report, CNET-France Telecom, 1999.* http://www.loria.fr/genet/Publications/GenetKlay-RR99.ps.
9. V. Gouranton, P. Réty, and H. Seidl. Synchronized Tree Languages Revisited and New Applications. In *Proceedings of FoSSaCs,* volume 2030 of LNCS, Springer-Verlag, 2001.
10. M. Hermann and R. Galbavý. Unification of Infinite Sets of Terms Schematized by Primal Grammars. *Theoretical Computer Science,* 176, 1997.
11. F. Jacquemard. Decidable Approximations of Term Rewrite Systems. In H. Ganzinger, editor, *Proceedings 7th Conference RTA, New Brunswick (USA),* volume 1103 of LNCS, pages 362-376. Springer-Verlag, 1996.
12. P. Réty. Regular Sets of Descendants for Constructor-based Rewrite Systems. In *Proceedings of the 6th international conference on LPAR, Tbilisi (Republic of Georgia),* Lecture Notes in Artificial Intelligence. Springer-Verlag, 1999.
13. P. Réty. J. Vuotto. Regular Sets of Descendants by Leftmost Strategy. Research Report RR-2002-08 LIFO, 2002.
14. K. Salomaa. Deterministic Tree Pushdown Automata and Monadic Tree Rewriting Systems. *The journal of Computer and System Sciences,* 37:367-394, 1988.
15. T. Takai, Y. Kaji, and H. Seki. Right-linear Finite Path Overlapping Term Rewriting Systems Effectively Preserve Recognizability. In L. Bachmair, editor, *Proceedings 11th Conference RTA, Norwich (UK),* volume 1833 of LNCS, pages 246-260. Springer-Verlag, 2000.

Rewrite Games

Johannes Waldmann

Institut für Informatik, Universität Leipzig
Augustusplatz 10, D-04109 Leipzig, Germany
joe@informatik.uni-leipzig.de

Abstract. For a terminating rewrite system R, and a ground term t_1, two players alternate in doing R-reductions $t_1 \to_R t_2 \to_R t_3 \to_R \dots$ That is, player 1 choses the redex in $t_1, t_3, \dots$, and player 2 choses the redex in $t_2, t_4, \dots$ The player who cannot move (because t_n is a normal form), loses.

In this note, we propose some challenging problems related to certain rewrite games. In particular, we re-formulate an open problem from combinatorial game theory (do all finite octal games have an ultimately periodic Sprague-Grundy sequence?) as a question about rationality of some tree languages.

We propose to attack this question by methods from set constraint systems, and show some cases where this works directly.

Finally we present rewrite games from to combinatory logic, and their relation to algebraic tree languages.

1 Introduction

This paper presents some fascinating open problems from the theory of combinatorial two person games. On the one hand, methods from string and term rewriting (including finite automata and set constraints) just wait to be applied there; while on the other hand, this application might stimulate some interesting developments in rewriting itself.

For a terminating rewrite system R, and a ground term t_1, two players alternate in doing R-reductions $t_1 \to_R t_2 \to_R t_3 \to_R \dots$ That is, player 1 choses the redex in $t_1, t_3, \dots$, and player 2 choses the redex in $t_2, t_4, \dots$ The player who cannot move (because t_n is a normal form), loses.

In this note, we propose some challenging problems related to certain rewrite games. In particular, we re-formulate an open problem from combinatorial game theory (do all finite octal games have an ultimately periodic Sprague-Grundy sequence?) as a question about rationality of some tree languages.

We show how octal games can be described by word rewriting systems, so the natural generalization is to look at games in arbitrary *term* rewriting systems. Then, ultimate periodicity of a Sprague-Grundy sequence is translated into the Sprague-Grundy classes being *rational*, i. e. recognized by finite automata.

S. Tison (Ed.): RTA 2002, LNCS 2378, pp. 144–158, 2002.

We propose to attack the transformed problem, using set constraint methods that would have to be extended to cope with rewrite projections. For now, we show that this approach works in the simple case of subtraction games.

We then give some examples of rewrite games in Combinatory Logic, and find that under certain restrictions, their Sprague-Grundy classes are *algebraic* (a. k. a. context-free) tree languages.

Finally, we briefly review related work and emphasize directions for future research in rewrite games.

2 Notation and Preliminaries

We assume the reader is familiar with term rewriting (see for instance [BN98]) and tree languages (see [CDG+97]). To honour the "french style", we use the words *rational* for regular or recognizable, and *algebraic* instead of context-free.

Remember that an *algebraic* tree language $\subseteq \mathrm{Term}(\Sigma)$ is produced by a grammar that is in fact a rewrite system in an extended signature $\Sigma \cup \Gamma$ with non-terminals $\Gamma = \{N_1, \ldots, N_k\}$, and *monadic* rewrite rules $N_i(x_1, \ldots, x_n) \to t$, where $x_1, \ldots, x_n$ are distinct variables, and $t \in \mathrm{Term}(\Sigma \cup \Gamma \cup \{x_1, \ldots, x_n\})$. An algebraic language is called *rational* iff all non-terminals are nullary.

For the convenience of the reader, we now present, in some detail, standard notions and results from combinatorial game theory. For a compact introduction, see [Guy98c,Guy98a]; the definitive references are [Con76,BCG83]. A complete guide to the literature is [Fra00].

2.1 Impartial Games

Combinatorial game theory studies finite deterministic two-person zero-sum games. This excludes chance, and hidden information. We also disallow ties.

A game is *normal* if the player who cannot move, loses (and the player who made the last move, wins). A game ist *impartial* if in each position, the set of possible moves does not depend on which player moves next.

We may describe a game "intensionally" by giving its ruleset, and a starting position. On the other hand, we often use an "extensional" view of a game as the (directed acyclic) graph containing all reachable positions, connected according to the rules of the game. We then even forget about the concrete positions, and just look at the unlabelled digraph: An *impartial game* G is a set of impartial games, called the *options* of G. (From now on, we just write "the *game* G".)

The simplest game is $\emptyset$, which has no options. This game $\emptyset$ is also called 0 (zero).

In a combinatorial game, for each game position, exactly one of the players has a winning strategy. For impartial games, we define unary predicates N, P by

$$N(G) :\iff \exists g \in G : P(g), \qquad P(G) :\iff \forall g \in G : N(g)$$

saying that in G, the Next (resp. Previous) player has a winning strategy.

2.2 Take-and-Break Games

A *position* in a take-and-break game is a multi-set of positive numbers, and a *move*, in general, is to pick one number, decrease it, and then split it into several parts. As usual, the object is to get the last move.

Each sequence $M = (M_1, M_2, \ldots)$ of sets $M_r \subseteq \mathbb{N}$ encodes the rules for a particular such game: It is allowed to decrease a number by r, and split the remainder into s positive summands, iff $s \in M_r$. We only consider games where the sequence M is eventually constant $\emptyset$.

For example, the game *officers* is characterized by $M = (\{1,2\}, \emptyset, \ldots)$, so the valid moves are to subtract one, and split the result into two or one (but not zero) positive numbers. That is, we are not allowed to pick a number "1" (since this would leave no positive number).

An example how this game might be played is

$$\{\underline{8}\} \to_1 \{\underline{3}, 4\} \to_2 \{1, 1, \underline{4}\} \to_1 \{1, 1, \underline{3}\} \to_2 \{1, 1, 1, 1\}.$$

We underlined the numbers that were reduced (and possibly split). In the given line of play, Player 1 loses. Actually, she could have won, starting $\{\underline{8}\} \to_1 \{2, 5\}$.

If the sets M_r are finite, they can be encoded by $d_r := \sum\{2^s : s \in M_r\}$, and the ruleset of the game is given by $(d_1, d_2, \ldots)$. Adding a leading zero and a decimal point, while suppressing trailing zeroes, *officers* is conveniently described by $0{\cdot}6$.

An *octal game* is a take-and-break game where numbers are split in at most two parts. That is, for all r, we have $\max M_r \leq 2$. Therefore, for all code digits $0 \leq d_r < 8$, and this explains the name. For example, *officers* is an octal game.

A *subtraction game* is a take-and-break game, where numbers are reduced, but never split. For all code digits $d_r \in \{0, 3\}$. The game is then characterized by its *subtraction set* $\{r : d_r = 3\}$.

For instance, the game $0 \cdot 03033$ has subtraction set $S = \{2, 4, 5\}$, and a possible line of play is $14 \to_1 9 \to_2 5 \to_1 1$. Here, Player 2 lost, but he has a better move: He can win the game if he starts $14 \to_1 9 \to_2 7$.

2.3 Basics of Sprague-Grundy Theory

In the *disjoint sum* $G + H$ of two games, a valid move is either a move in G or a move in H, leaving the other summand untouched. Formally,

$$G + H := \{g + H : g \in G\} \cup \{G + h : h \in H\}$$

Again, the object is to get the last move globally. Of course $+$ is associative and commutative and $0 + G = G = G + 0$.

Then the relation $\approx$ on games given by $G \approx H :\iff P(G+H)$ is an equivalence relation. It is stable w.r.t. disjoint sum: $\forall G, H, K : \; G \approx H \Rightarrow G + K \approx H + K$. (Note that the relation $G \sim H :\iff P(G) = P(H)$ is *not* stable.)

We define a sequence of games $(*0, *1, *2, *3, \ldots)$, called *nimbers*, by $*0 := 0, *(n+1) = \{*0, *1, \ldots, *n\}$. The nimber $*n$ corresponds to a heap of n beans, from which we may remove any positive number of beans.

The Sprague-Grundy theorem [Spr35] asserts that for each game G, there is exactly one n such that $G \approx *n$. We call n the Sprague-Grundy value of G, and write $n = \text{sg}(G)$. We have $\text{sg}(G) = 0 \iff P(G)$ (previous player wins).

The essential tool in the proof of the theorem, and in the computation of Sprague-Grundy values, is the *mex rule*: We have $\{*n_1, \ldots, *n_k\} \approx *n$, where $n = \text{mex}\{n_1, \ldots, n_k\}$. Here, the mex (*minimum excludant*) of a finite set $M \subset \mathbb{N}$ is defined by $\text{mex}\, M = \min(\mathbb{N} \setminus M)$.

For all $a, b \in \mathbb{N}$, we have $*a + *b \approx *(a \oplus b)$, where $\oplus$ denotes binary addition that disregards carries. For example, $*1 + (*2 + *3) \approx *1 + *1 \approx *0$.

2.4 Sprague-Grundy Sequences

The *Sprague-Grundy sequence* of a take-and-break game is the sequence (sg(0), sg(1), ...). For example, the Sprague-Grundy sequence of the game $0 \cdot 77$ (called *Kayles*) starts $(0, 1, 2, 3, 1, 4, 3, 2, 1, 4, 2, 6, \ldots)$

It is easy to see that for each *subtraction game*, its Sprague-Grundy sequence is bounded and ultimately periodic: Due to the mex rule, the values are bounded by $|S|$, where $S \subset \mathbb{N}$ is the subtraction set; and the period length is at most $|S|^{\max S}$, since the continuation of the sequence is determined if a slice of length $\max S$ is given.

For example, the subtraction set $S = \{7, 64, 89, 96\}$ produces a period of length 5.756.171, after a pre-period of length 1061. It is open whether there are families S_n of subtraction sets for which the period length indeed grows exponentially w.r.t. $\max S_n$. [AB95,Fla97].

For *some* octal games it is known that their Sprague-Grundy sequence is ultimately periodic. (To continue the example, the Kayles sequence reaches a period of length 12, after after a pre-period of length 71.) Whether this is true in general, is a famous open problem:

Conjecture 1 ([Con76,BCG83,Guy98b]). *For each octal game, the Sprague-Grundy sequence is bounded and ultimately periodic.*

There are small games with truly astronomic periods (e. g. $0 \cdot 16$ has period length 149459), but the games

$$0 \cdot 6, 0 \cdot 06, 0 \cdot 14, 0 \cdot 36, 0 \cdot 64, 0 \cdot 74, 0 \cdot 76, 0 \cdot 004, 0 \cdot 005, 0 \cdot 006, 0 \cdot 007$$

and others remain unsolved today, despite computing some of their sequences up to length 10^{10}.

3 Rewrite Games

Definition 2 (Rewrite Game). *For a terminating rewrite system R, and a ground term t_1, two players alternate in doing R-reductions $t_1 \to_R t_2 \to_R t_3 \to_R \dots$ That is, player 1 choses the redex in $t_1, t_3, \dots$, and player 2 choses the redex in $t_2, t_4, \dots$ The player who cannot move (because t_n is a normal form), loses.*

By definition, this gives a normal impartial game, so the Sprague-Grundy theory is applicable, and each ground term $t \in \mathrm{Term}(\Sigma)$ can be assigned its Sprague-Grundy value $\mathrm{sg}_R(t)$.

Of course, such functions sg_R can be arbitrarily complex, since R can encode any computation. On the other hand it is interesting to look for such R whose sg_R behaves nicely.

3.1 Simulating Octal Games

Proposition 3. *Each octal game can be simulated by a rewrite game.*

Proof. A game position $\{h_1, h_2, \dots, h_k\}$ is encoded as a word

$$w = DU^{h_1}DU^{h_2}\dots DU^{h_k}D.$$

A move "decrease a number by r, and split the remainder into s positive numbers" is encoded, for $s = 0, 1, 2$, by the rewrite rules

$$s = 0: \ \{DU^rD \to D\}, \quad s = 1: \ \{DU^rU \to DU\}, \quad s = 2: \ \{UU^rU \to UDU\}.$$

Note that we have to guarantee the non-emptiness of the resulting heaps, so we cannot simply write $U^r \to D$ to remove r beans. Also, we use the convention that to the left of each U string, there is a D letter. □

For some octal digits it is possible to give a more compact system. We encode $d_r = 3 = 2^1 + 2^0$ by $U^r \to \epsilon$, $d_r = 6 = 2^2 + 2^1$ by $U^{r+1} \to DU$, and $d_r = 7 = 2^2 + 2^1 + 2^0$ by $U^r \to D$.

A subtraction game with difference set S is represented by $\{U^r \to \epsilon : r \in S\}$.

The unsolved games $0 \cdot 6$ (called *officers*) resp. $0 \cdot 007$ (called *James Bond*, or *treblecross*) are encoded by the rewrite systems

$$R_{0\cdot 6} = \{UU \to DU\}, \quad R_{0\cdot 007} = \{UUU \to D\}.$$

The line of play in *officers* presented in the introduction now reads $DU^3\underline{UU}U^3 \to_1 DU\underline{UU}DUUUU \to_2 DUDUD\underline{UU}UU \to_1 DUDUDDU\underline{UU} \to_2 DUDUDDUDU$, where the chosen redexes are underlined.

We remark that the encoding of the game positions (as strings) works for all take-and-break games, but the encoding of the moves (as rewrite rules) needs the octality restriction. There seems to be no easy way to simulate the splitting of a number into three (or more) summands. Even "tricky" representations (binary strings, trees) do not seem to help.

3.2 Periodicity and Rationality

What is the correct translation of "ultimate periodicity of a Sprague-Grundy sequence" into the language of rewrite games? Note that a Sprague-Grundy sequence is in fact a mapping from sizes of heaps, that is, strings on a one-letter alphabet, to Sprague-Grundy values. For rewrite games, we need to extend this, since we deal not with heaps, but trees, in arbitrary signatures. We suggest therefore to look at the number, and complexity, of Sprague-Grundy classes of rewrite games.

Definition 4. *The Sprague-Grundy class $SG_R(n)$ of the rewrite game of R is $SG_R(n) = \{t : \mathrm{sg}_R(t) = n\}$.*

Note $P = \mathrm{SG}_R(0)$ and $N = \bigcup_{n>0} \mathrm{SG}_R(n)$.

Theorem 5. *The Sprague-Grundy sequence of a take-and-break game is eventually periodic iff it has only finitely many non-empty Sprague-Grundy classes, all of which are rational languages (using the above encoding).*

Proof. Let $l :=$ length of pre-period plus length of period of the Sprague-Grundy sequence, and $m :=$ the maximal Sprague-Grundy value (which must exist — the sequence is eventually periodic, thus bounded).

To show that each $\mathrm{SG}_R(n)$ is rational, a finite automaton has to be constructed that reads a word $w = DU^{h_1}DU^{h_2}\dots DU^{h_k}D$ and checks whether $\mathrm{sg}(h_1) \oplus \mathrm{sg}(h_2) \oplus \dots = n$.

While reading U^{h_i}, the automaton needs at most l states to compute $y := \mathrm{sg}(h_i)$. It also needs to remember what it read before, namely $x := \mathrm{sg}(h_0)\oplus \dots \oplus \mathrm{sg}(h_{i-1})$, and this can ce done with at most m states.

In all, the machine needs $l \times m$ states. The computation of $x \oplus y$ can be done without additional states since the range of x and y is bounded (for y, by the precondition of the theorem, for x, because of $a \oplus b \le a + b \le 2m$.).

On the other hand, if a Sprague-Grundy $\mathrm{SG}_R(n)$ class is rational, then so is $\mathrm{SG}_R(n) \cap DU^*D$. This is (essentially) a one-letter rational language, and thus a sum of linear sets $\{DU^{ak+b}D : k \ge 0\}$. Now the period of the Sprague-Grundy sequence is bounded by the least common multiple of all the differences a of these (finitely many) sets. □

The class SG(0) (= P) already contains all information on the other classes:

Theorem 6. *For all take-and-break games, using the above encoding: If the Sprague-Grundy class $SG(0)$ is rational, then there are only finitely many non-empty $SG(n)$, and all of them are rational.*

Proof. We write C_n for SG(n), and we let $\partial L/\partial w = \{u : w \cdot u \in L\}$ denote the left quotient of the language L by the word w.

Let C_n be a non-empty Sprague-Grundy class, and $v \in C_n$. Since $vDw \approx v + w$, we have $vDw \in C_0 \iff w \in C_n$. So $C_n = \partial C_0/\partial vD$. Therefore rationality of C_0 implies that of C_n.

Assume now that there are infinitely many non-empty Sprague-Grundy classes. By definition, these classes are mutually disjoint. But since C_0 is rational, it can only have a finite number of distinct left quotients. □

The Conjecture 1 on octal games naturally leads to the following

Question 7. *What rewrite games R have finitely many Sprague-Grundy classes, all of which are rational tree languages?*

4 Sprague-Grundy Classes and Set Constraints

Now we show that for any rewrite game, Sprague-Grundy classes are solutions of a generalized set constraint system.
We let $R^{-1}(L) := \{s : s \to_R t \in L\}$ denote the set of one-step predecessors of the set L w.r.t. the rewrite relation R.

Proposition 8. *Writing C_n for $SG_R(n)$, we have, for each $n \geq 0$,*

$$C_n = \left(R^{-1}(C_0) \cap R^{-1}(C_1) \cap \ldots \cap R^{-1}(C_{n-1})\right) \setminus R^{-1}(C_n),$$

where, for $n = 0$, the empty intersection is understood as $Term(\Sigma)$.

Proof. By the mex rule, we have $\mathrm{sg}_R(t) = n$ iff t has successors of Sprague-Grundy values $0, 1, \ldots, n-1$, but not n. □

Remeber that by Theorem 6, to "solve" an octal game it would be enough to determine C_0. (This special relation bewteen C_0 and the other classes does not hold for all rewrite games, since it is an effect of our encoding of octal games.) In other words, C_0 alone is as difficult as all the equations together. On the other hand, C_0 solves the (seemingly simple) equation $C_0 = \{D, U\}^* \setminus R^{-1}(C_0)$. For the James Bond game, corresponding to $R = \{UUU \to D\}$, it is even open whether $C_0 \cap DU^*D$ is finite.

For *subtraction games*, the constraint system can be solved easily. We only need to consider words from U^* (i. e. single heaps), since the corresponding rewrite rules cannot break D boundaries. But then the alphabet is unary, and $R^{-1}(C_n)$ is just $U^r \cdot C_n$. Now the equations from Proposition 8 form a system of positive set constraints. It is known [AW92] that such a system, if solvable, admits rational solutions. In our case, the solution is clearly unique, and therefore must be rational. (This again shows that subtraction games have ultimately periodic Sprague-Grundy sequences.)

Note that we did effectively consider a string rewrite system with rewriting only at one end of the strings. Using results for set constraint systems with projection [CP94], this idea can be extended to *top-rewriting games* (only top reductions are allowed) for rules $l \to r$, where both l and r contain at most one variable. (Details will be explained in a forthcoming technical report.)

This method (of tree set automata [GTT99]) would need to be extended to work for (certain) games with rewriting allowed anywhere, for instance, string rewriting systems encoding octal games. The known huge period lengths seem to imply that this is not straightforward, but the absence (so far) of aperiodic octal games gives hope. Astronomic periods are perhaps not that surprising, in light of the known complexities of set constraints (NEXPTIME-completeness, if projections are involved).

5 Some Rewrite Game Examples

5.1 Linear, Inverse Monadic, Non-overlapping Systems

*The Combinator **I*** has the rewrite rule $(\mathbf{I}\;x) \to x$.
This game is trivial — the number of moves is exactly one less than the number of **I** in the start term (regardless of the reduction strategy), so there are only two Sprague-Grundy classes C_0, C_1, as follows from these propositions:

Proposition 9. *If a rewrite system R is linear and has no critical pairs, then all reduction chains from t_1 to t_2 have common length.* □

Note that we use *linear* in the strong sense: a rule $l \to r$ is linear iff each variable occurs exactly twice: once in l, once in r.

Proposition 10. *If a game has the property that for all positions p, all lines of play from p to a final position p' have common length, then there are just two Sprague-Grundy classes, corresponding to values 0 and 1.*

Proof. The class C_0 contains all positions that have an even distance to a stopping position, and C_1 contains all positions with odd distance. □

Therefore, in the **I** game, the classes are given by the rational grammars

$$C_0 \to \mathbf{I} \cup (C_0\;C_1) \cup (C_1\;C_0), \quad C_1 \to (C_0\;C_0) \cup (C_1\;C_1),$$

producing the sets of all terms with an odd (C_0) resp. even (C_1) number of **I**.

*The combinator **F*** is the rewrite system $((\mathbf{F}\;x)\;y) \to (y\;x)$.
Its normal forms are given by $F_1 := \mathbf{F}$, $F_{n+1} = (\mathbf{F}\;F_n)$.
Since the preconditions of Proposition 9 hold, we can apply Proposition 10. So the game has exactly two Sprague-Grundy classes. We will prove that they are not rational, but algebraic tree languages. The construction generalizes to similar systems.

Proposition 11. *For each term t, dist(t) = size(t) − size(nf(t)), where size(t) is the number of **F** in t, and dist(t) denote the length of a reduction chain from t to its normal form nf(t).*

Proof. In each reduction step, the size decreases by one. □

Proposition 12. *$nf(F_a \overset{\bullet}{\wedge} F_b) =$ if $a > b$ then F_{a+1-b} else F_{b+2-a}.*

Proof. By induction on $a + b$:

1. $a = 1$: $F_1 \overset{\bullet}{\wedge} F_b = \mathbf{F} \overset{\bullet}{\wedge} F_b = F_{b+1} = F_{b+2-a}$.
2. $a > 1$: $F_a \overset{\bullet}{\wedge} F_b = (\mathbf{F} \overset{\bullet}{\wedge} F_{a-1}) \overset{\bullet}{\wedge} F_b \to F_b \overset{\bullet}{\wedge} F_{a-1}$
   a) $b > a - 1$ ($\iff a \leq b$): $F_b \overset{\bullet}{\wedge} F_{a-1} \to^* F_{b+1-(a-1)} = F_{b+2-a}$
   b) $b \leq a - 1$ ($\iff a > b$): $F_b \overset{\bullet}{\wedge} F_{a-1} \to^* F_{a-1+2-b} = F_{a+1-b}$ □

Proposition 13. *$dist(F_a \overset{\bullet}{\wedge} F_b) \equiv$ (if $a > b$ then 1 else 0) $\pmod 2$.*

Proof. If $a > b$, then $\mathrm{dist}(F_a \overset{\bullet}{\wedge} F_b) \equiv \mathrm{size}(F_a \overset{\bullet}{\wedge} F_b) - \mathrm{size}(\mathrm{nf}(F_a \overset{\bullet}{\wedge} F_b)) \equiv (a + b) - (a + b - 1) \equiv 1 \pmod 2$. If $a \leq b$, then $\mathrm{dist}(F_a \overset{\bullet}{\wedge} F_b) \equiv \mathrm{size}(F_a \overset{\bullet}{\wedge} F_b) - \mathrm{size}(\mathrm{nf}(F_a \overset{\bullet}{\wedge} F_b)) \equiv (a + b) - (b + 2 - a) \equiv 0 \pmod 2$. □

Proposition 14. *Neither of C_0, C_1 is rational.*

Proof. Consider the intersection of C_0 with the rational language $F_* \overset{\bullet}{\wedge} F_*$. By Proposition 13, this is $\{F_a \overset{\bullet}{\wedge} F_b : a \leq b\}$, which is not rational. Since $C_1 = \mathrm{Term}(\Sigma) \setminus C_0$, the class C_1 is not rational as well. □

Proposition 15. *Both C_0 and C_1 are algebraic.*

Proof. We use an enriched version of the inverse image $y \overset{\bullet}{\wedge} x \to (\mathbf{F} \overset{\bullet}{\wedge} x) \overset{\bullet}{\wedge} y$ of the rewrite rule to generate the set of all even-step (resp. odd-step) predecessors of normal forms. The number of reductions is counted in a "distributed" manner, as follows.

We replace the signature $\Sigma = \{\mathbf{F}, \overset{\bullet}{\wedge}\}$ by $\Sigma' = \{\mathbf{F}, \overset{0}{\wedge}, \overset{1}{\wedge}\}$. The morphism

$$\mathrm{expo} : \mathrm{Term}(\Sigma') \to \{0, 1\} : \mathbf{F} \mapsto 0, j \overset{i}{\wedge} k \mapsto (i + j + k \bmod 2)$$

adds up the exponents modulo 2, and

$$\mathrm{drop} : \mathrm{Term}(\Sigma') \to \mathrm{Term}(\Sigma) : \mathbf{F} \mapsto \mathbf{F}, x \overset{k}{\wedge} y \mapsto x \overset{\bullet}{\wedge} y$$

is the (alphabetic) morphism that just removes exponents.

Now we construct the language

$$H = \{t \in \mathrm{Term}(\Sigma') : \mathrm{dist}(\mathrm{drop}(t)) \equiv \mathrm{expo}(t) \pmod 2\}.$$

Each $t \in H$ contains, in its exponents, information about the parity of the length of the reduction of $\mathrm{drop}(t)$ to normal form. H is given by the algebraic grammar

$$H \to \mathbf{F} \cup \mathbf{F} \overset{0}{\wedge} H, \quad x \overset{i}{\wedge} y \to (\mathbf{F} \overset{1}{\wedge} y) \overset{i}{\wedge} x.$$

The languages C_0, C_1 are algebraic since $C_i = \mathrm{drop}(H \cap \mathrm{expo}^{-1}(i))$. Note that $\mathrm{expo}^{-1}(i)$ is rational since it is an inverse image of a morphism to a finite set; then we use closure of algebraic languages w.r.t. intersection with rational ones, and finally closure of algebraic languages under (alphabetic) morphisms. □

This construction easily generalizes:

Proposition 16. *If the rewrite system R is linear, non-overlapping, and inverse monadic (i. e. R^{-1} is monadic), then its rewrite game has just two Sprague-Grundy classes, both of which are algebraic tree languages.* □

Note that the rewrite systems for octal games with all digits $d_r = 7$, e. g. $U^r \to D$, are linear and inverse monadic as well. However they are overlapping (e. g. for $r = 2$, there is the critical pair $DU \leftarrow UUU \to UD$), and indeed non-confluent.

Still we know that in some cases, their Sprague-Grundy sequence is eventually periodic (equivalently, the Sprague-Grundy classes are rational): $R_{0\cdot 07} = \{U^2 \to D\}$ has period length 34, $R_{0\cdot 77} = \{U \to D, U^2 \to D\}$ has period 12, but periodicity is open for $R_{0\cdot 007} = \{U^3 \to D\}$.

So the question is, how can we lift the "non-overlapping" restriction in Proposition 16? It seems that the number and complexity of the Sprague-Grundy classes somehow measure the degree of "inconfluence" of the rewrite system.

5.2 Some More Combinators

Finally we mention two games that are unsolved, in the sense that we don't know a better way of computing Sprague-Grundy values than to expand the complete game tree.

The combinator $\mathbf{K}$ is the system $((\mathbf{K}\,x)\,y) \to x$.

For the Sprague-Grundy function $\mathrm{sg} = \mathrm{sg}_{\mathbf{K}}$ of this game, we have

Proposition 17. $sg((\mathbf{K}\,t_1)\,t_2) = sg(t_1) \oplus (1 + sg(t_2))$.

Proof. Instead of computing Sprague-Grundy values by the mex rule, we give a "strategic" proof. We write $*n$ for any term t with $\mathrm{sg}(t) = n$.

We show that $((\mathbf{K}\,*m)\,*n)$ is equivalent to the game $*m + *(n+1)$, by giving a winning strategy for the second player in the sum $((\mathbf{K}\,*m)\,*n) + *m + *(n+1)$. If the first player moves in $*m$, then the second player mirrors his move in the other copy of $*m$, and wins by induction. If the first player moves in $*n$, to $*n'$, then the other one moves from $*(n+1)$ to $*(n'+1)$, and vice versa. If the first player moves from $*(n+1)$ to some $*0$, then the second player moves from $((\mathbf{K}\,*m)\,*n)$ to $*m$. The resulting game is $*m + *m \approx 0$, a second-player win. □

Proposition 18. $sg(((\mathbf{K}\,\mathbf{K})\,t_1)\,t_2) = (1 + sg(t_1)) \oplus sg(t_2)$.

Proof. By proposition 17, it is enough to show $(((\mathbf{K}\,\mathbf{K})\,*m)\,*n) \approx ((\mathbf{K}\,*n)\,*m)$.

Moves in $*m$ or $*n$ can be mirrored, and a move $(((\mathbf{K}\,\mathbf{K})\,*m)\,*n) \to (\mathbf{K}\,*n)$ will be answered by $((\mathbf{K}\ {*n})\ {*m}) \to {*n}$, and vice versa. Of course $(\mathbf{K}\ {*n}) \approx {*n}$, since no reduction can use the outer $\mathbf{K}$. □

All numbers do occur as Sprague-Grundy values:

Proposition 19. *There is a sequence* $c_0, c_1, \ldots$ *of* $\mathbf{K}$ *terms with* $sg(c_n) = n$.

Proof. By proposition 17, we may take $c_0 = \mathbf{K}, c_{n+1} = ((\mathbf{K}\ \mathbf{K})\ c_n)$. □

Proposition 20. $sg(((\mathbf{K}\ (\mathbf{K}\ \mathbf{K}))\ t_1)\ t_1) = 0.$

Proof. We show that this game is a win for the second player. Moves in t_1 will be mirrored, the only other move goes $\to ((\mathbf{K}\ \mathbf{K})\ t_1)$, which will be answered by $((\mathbf{K}\ \mathbf{K})\ t_1) \to \mathbf{K} \approx 0$. □

The $\mathbf{K}$ game seems rather difficult, although reductions are short. The difficulty comes from the fact that sg is *not* a morphism:

Proposition 21. *There is a context* $C[]$ *and terms* t, t' *with* $sg(t) = sg(t')$, *such that* $sg(C[t]) \neq sg(C[t'])$.

Proof. Take $C[] = ((\mathbf{K}\ [])\ {*n})\ {*n}$ and $t = \mathbf{K}, t' = (\mathbf{K}\ \mathbf{K})$. We have $sg(\mathbf{K}) = sg(\mathbf{K}\,\mathbf{K}) = 0$ since both are normal forms. By Proposition 18, $sg(C[t]) = sg(((\mathbf{K}\ \mathbf{K})\ {*n})\ {*n}) = (1+n) \oplus n \neq 0$, and $sg(C[t']) = sg(((\mathbf{K}\ (\mathbf{K}\ \mathbf{K}))\ {*n})\ {*n}) = 0$ by Proposition 20. □

The combinator $\mathbf{D}$ has the reduction rule $((\mathbf{D}\ x)\ y) \to (x\ (x\ y))$.

Definition 22. Normal forms *are right spines:* $D_0 = \mathbf{D}, D_{n+1} = (\mathbf{D}\ D_n)$.

The game is difficult to analyze since it may take tower-of-exponentionals time. This can be seen from:

Lemma 23. *The normal form of* $(D_a\ D_b)$ *is* D_{2^a+b}, *and each reduction to that form takes at least* $2^a - 1$ *steps.* □

So terms may get huge, but they contain a lot of duplicated subtrees. If these are on the right spine, their Sprague-Grundy values cancel:

Proposition 24. *For all* $\mathbf{D}$ *terms* t_1, t_2, $sg(t_1\ (t_1\ t_2)) = sg(t_2)$.

Proof. We show that $(t_1\ (t_1\ t_2)) \approx t_2$. Any move inside t_i will just be mirrored. If the inner expression $(t_1\ t_2)$ is a redex, then $t_1 = (\mathbf{D}\ t_3)$. Reduction leads to $(t_1\ (t_3\ (t_3\ t_2)))$, but then the root is a redex as well, so we now have in fact $((\mathbf{D}\ t_3)\ (t_3\ (t_3\ t_2)))$, and the second player moves to $(t_3\ (t_3\ (t_3\ (t_3\ t_2))))$. By induction (twice), this is equivalent to t_2. □

Proposition 25. *sg*(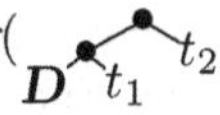) $= (1 + sg(t_1)) \oplus sg(t_2)$.

Proof. Similar to Proposition 17. □

Proposition 26. *There is a sequence* $c_0, c_1, \ldots$ *of* **D** *terms with* $sg(c_n) = n$.

Proof. Take $c_0 = \mathbf{D}$, $c_{n+1} =$ 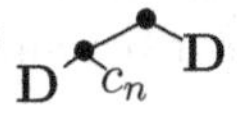, and apply Proposition 25. □

Proposition 24 does *not* extend easily to duplicate subtrees elsewhere, since again the Sprague-Grundy function is not a morphism.

For both **K** and **D**, the rewrite system is non-linear (deleting resp. duplicating). Remember that we solved the games for **I** and **F**, with the rules being linear. Now the rewrite systems associated to octal games are linear as well (since they are word rewrite systems) but linearity alone *does not* imply boundedness of Sprague-Grundy values — there is even a word rewrite game that disproves it (see the forthcoming technical report).

6 Discussion

Summary. In this paper, we have introduced the field of (impartial) rewrite games, by giving definitions, examples, a few basic results, and some open questions. Further, we discussed how rewrite games might help to solve a long standing open problem on combinatorial games. To carry out this plan, we would need to handle set constraints with rewrite projections.

Outlook. While the present paper dealt with impartial games (where both players always have identical options), game theory starts to get *really* interesting with *partizan* games (where option sets might differ). Most "natural" games (Chess, Go, ...) belong to this class.

It is straightforward to define partizan rewrite games: each of the two players player P_1, P_2 has his own rewrite system R_1, R_2, and they construct a reduction $t_1 \to_{R_1} t_2 \to_{R_2} t_3 \to_{R_1} \ldots$ There should be relations to modularity properties of rewrite systems.

Related work. Games do occur in various branches of logics and computer science. Two-person games are naturally related to formulas in predicate logic, because the sequence of alternating moves "me–you–me–you–..." corresponds to a sequence of quantifiers $\exists\forall\exists\forall\ldots$ (There *exists* a move such that for *all* your moves there *exists* an answer move such that ...) This is applied in Ehrenfeucht-Fraïssé games to characterize expressiveness of certain logics (see, e. g. chapters 2 and 3 in [EF99]). In turn, this leads to characterization of complexity classes (for instance, QBF (quantified Boolean formula) is PSPACE-complete). Another application of games in logic is the proof of Rabin's theorem (decidability of existential monadic second-order logic on trees) via the forgetful determinacy theorem [GH82], that relates an accepting computation of an ω automaten to a winning strategy in a game.

In complexity theory, one uses Pebble Games to investigate time/space tradeoffs. (These are one-person games.) Then there are the more recent developments of Arthur-Merlin games [BM88] and Interactive Proof systems [GMR89]. They characterize complexity classes, and provide means to validate knowledge, without actually revealing it. These games typically have a pre-defined (short) number of moves, and one is interested in probabilities of success. This aspect differs from combinatorial games (and rewrite games).

A lot of "natural" combinatorial games (parametrized variants of Chess, Go, etc.) are PSPACE-complete by reduction from QBF [GJ79]. On the other hand, there are complexity results on one-player games (puzzles like Sokoban or Atomix [Hol01]) that typically use a reduction from the Finite Automata Intersection Emptiness problem. This again is connected to the complexity of solving certain set constraints, which are EXPTIME- or NEXPTIME-hard (depending on the admitted operators).

We are not aware of previous work on rewrite games as defined here.

Vágvölgyi [Vág98] deals with the Ground Tree Transducer Game, which is a kind of rewrite game (since a GTT is a rewrite system) but the emphasis is on the Ehrenfeucht-Fraïssé method (one player wants to construct t_1, t_2 that are related by the GTT, the opponent wants to spoil this) while in our definition, the game is normal impartial (both players have identical options, and both want to get the last move).

In the combinatorial game theory literature, there is one topic that suggests a special form of rewrite games: Stromquist and Ullman [SU93] investigate *sequential compounds* of games, and Guy [Guy98b] suggests to investigate compounds in arbitrary posets. This idea is again mentioned by Albert and Nowakowski [AN01] when they discuss how to generalize the game of *End-Nim* onto trees. In our language, this would correspond to a rewrite game that is played at, or near to, the leaves of a tree, gradually pruning it in the process. Note that this could be seen as *ground* rewriting, which is, in a sense, complementary to the analysis of *top* rewrite games mentioned in the present paper.

Relations between games and formal languages have been studied for the case of Peg Solitaire by Ravikumar [Rav97] (when played on strings, and strips of fixed width, winning positions form a rational language) and its impartial two-player variant by Eppstein and Moore [ME00].

Conclusion. It is not all surprising that combinatorial games and term rewriting are closely connected in both directions: on the one hand, a lot of games have "natural" presentations as rewrite games, on the other hand, game trees (and their canonical forms) just *are* terms, (and Conway's "one line proofs" [Con76] *are* rewrite sequences.) I think this is a fascinating area of research where much more waits to be discovered. You are invited to play the rewrite games from this paper against my computer program here: `http://theopc.informatik.uni-leipzig.de/~joe/trs/`.

Acknowledgements. I am grateful to Ingo Althöfer, Achim Flammenkamp, Dieter Hofbauer, Jörg Rothe, and Georg Snatzke, for explanations and discussions on games, rewriting, and complexity, as well as to Mirko Rahn, Christian Vogt, and the anonymous referees for careful reading and suggesting improvements of the presentation of this paper.

References

[AB95] Ingo Althöfer and Jörg Bültermann. Superlinear period lengths in some subtraction games. *Theoretical Computer Science*, 148:111–119, 1995.

[AN01] Michael H. Albert and Richard J. Nowakowski. The game of end-nim. *Electronic Journal of Combinatorics*, 8(2), 2001.

[AW92] Alexander Aiken and Edward L. Wimmers. Solving systems of set constraints. In *Seventh Annual IEEE Symposium on Logic in Computer Science*, pages 329–340, 1992.

[BCG83] Elwyn R. Berlekamp, John H. Conway, and Richard K. Guy. *Winning Ways for your Mathematical Plays.* Academic Press, 1983. (A K Peters, 2001).

[BM88] L. Babai and S. Moran. Arthur-Merlin games: A randomized proof system, and a hierarchy of complexity classes. *Journal of Computer and System Sciences*, 36(2):254–276, 1988.

[BN98] Franz Baader and Tobias Nipkow. *Term Rewriting and All That.* Cambridge University Press, 1998.

[CDG+97] Hubert Comon, Max Dauchet, Rémi Gilleron, Denis Lugiez, Sophie Tison, and Marc Tommasi. *Tree Automata Techniques and Applications.* `http://www.grappa.univ-lille3.fr/tata/`, 1997.

[Con76] John H Conway. *On Numbers and Games.* Academic Press, 1976. (A K Peters, 2001).

[CP94] Witold Charatonik and Leszek Pacholski. Set constraints with projections are in NEXPTIME. In *IEEE Symposium on Foundations of Computer Science*, pages 642–653, 1994.

[EF99] Heinz-Dieter Ebbinghaus and Jörg Flum. *Finite Model Theory.* Springer, 1999.

[Fla97] Achim Flammenkamp. *Lange Perioden in Subtraktions-Spielen.* Hans-Jacobs Verlag, Lage, Germany, 1997. `http://wwwhomes.uni-bielefeld.de/achim/`.

[Fra00] Aviezri Fraenkel. Combinatorial games: Selected bibliography with a succinct gourmet introduction. `http://www.combinatorics.org/Surveys/index.html`, 1994–2000. Electronic Journal of Combinatorics, Dynamic Surveys.

[GH82] Y. Gurevich and L. Harrington. Trees, automata, and games. In *Proc. Symp. Theory of Computing*, pages 60–65. ACM Press, 1982.

[GJ79] M. R. Garey and D. B. Johnson. *Computers and Intractability.* W. H. Freeman, 1979.

[GMR89] S. Goldwasser, S. Micali, and C. Rackoff. The knowledge complexity of interactive proof systems. *SIAM Journal on Computing*, 18(1):186–208, February 1989.

[GTT99] Remi Gilleron, Sophie Tison, and Marc Tommasi. Set constraints and automata. *Information and Computation*, 149(1):1–41, 1999.

[Guy98a] Richard K. Guy. *Impartial Games*, pages 61–78. In Nowakowski [Now98], 1998.

[Guy98b] Richard K. Guy. *Unsolved problems in combinatorial games*, pages 475–492. In Nowakowski [Now98], 1998.

[Guy98c] Richard K. Guy. *What is a game*, pages 43–60. In Nowakowski [Now98], 1998.

[Hol01] Markus Holzer. Assembling molecules in atomix is hard. Technical Report Technical Report TUM-I0101, Technische Universität München, Institut für Informatik, 2001.

[ME00] Cristopher Moore and David Eppstein. One-dimensional peg solitaire, and duotaire. In *MSRI Workshop on Combinatorial Games*, page (to appear), 2000. `http://www.santafe.edu/~moore/pubs/peg.html`.

[Now98] Richard J. Nowakowski, editor. *Games of No Chance*. Cambridge University Press, 1998.

[Rav97] B. Ravikumar. Peg-solitaire, string rewriting systems and finite automata. In *International Symposium on Algorithms and Computation, Singapore*, 1997. `http://homepage.cs.uri.edu/faculty/ravikumar/index.html`.

[Spr35] Richard Sprague. Über mathematische kampfspiele. *Tohoku Math. J.*, 41:438–444, 1935.

[SU93] Walter Stromquist and Daniel Ullman. Sequential compounds of combinatorial games. *Theoretical Computer Science*, 119:311–321, 1993.

[Vág98] S. Vágvölgyi. The ground tree transducer game. *Fundamenta Informaticae*, (34):175–201, 1998.

An Extensional Böhm Model

Paula Severi[1]* and Fer-Jan de Vries[2]

[1] Departimento di Informatica, Università di Torino.
Corso Svizzera 185, 10149 Torino, Italy - severi@di.unito.it
[2] Department of Mathematics and Computer Science, University of Leicester
University Road, Leicester, LE1 7RH, UK - fdv1@mcs.le.ac.uk

Abstract. We show the existence of an infinitary confluent and normalising extension of the finite extensional lambda calculus with beta and eta. Besides infinite beta reductions also infinite eta reductions are possible in this extension, and terms without head normal form can be reduced to bottom. As corollaries we obtain a simple, syntax based construction of an extensional Böhm model of the finite lambda calculus; and a simple, syntax based proof that two lambda terms have the same semantics in this model if and only if they have the same eta-Böhm tree if and only if they are observationally equivalent wrt to beta normal forms. The confluence proof reduces confluence of beta, bottom and eta via infinitary commutation and postponement arguments to confluence of beta and bottom and confluence of eta.
We give counterexamples against confluence of similar extensions based on the identification of the terms without weak head normal form and the terms without top normal form (rootactive terms) respectively.

1 Introduction

In this paper we present a confluent infinitary extension $\lambda^{\mathsf{h}\infty}_{\beta\bot\eta}$ of the extensional lambda calculus $\lambda_{\beta\eta}$. In earlier work confluent infinitary extensions of the lambda calculus λ_β without the eta rule have been studied. Typically, confluence of such infinitary extensions cannot be obtained unless we add a form of bottom rule that identifies computationally insignificant terms with some added symbol $\bot$. Different choices of computationally insignificant terms may lead to different confluent and normalising extensions. The main three choices, the set of terms without head normal form, the set of terms without weak head normal form and the set of terms without top normal form (rootactive terms) result in three different confluent and normalising calculi in which the normal forms are respectively known as Böhm trees, Lévy-Longo trees and Berarducci trees.
In contrast, here for extensional lambda calculus we have no choice: only identification of all terms without head normal form with $\bot$ results in the confluent, normalising calculus $\lambda^{\mathsf{h}\infty}_{\beta\bot\eta}$. The normal forms of this calculus are known as eta-Böhm trees [2]. As corollaries of confluence and normalisation we obtain a simple,

* Partially supported by IST-2001-322222 MIKADO; IST-2001-33477 DART.

S. Tison (Ed.): RTA 2002, LNCS 2378, pp. 159–173, 2002.

syntax based, construction of an extensional Böhm model $\mathfrak{B}_\eta$ of the finite lambda calculus; plus a new and simple syntax based proof that two lambda terms have the same semantics in this model if and only if they have the same eta-Böhm tree if and only if they are observationally equivalent wrt to beta normal forms. Hence this extensional Böhm model $\mathfrak{B}_\eta$ equates more terms than Barendregt's Böhm model $\mathfrak{B}$ in [3]). It induces the same equality relation as Park's model D^*_∞ of [16,6,17].

The eta rule has not been considered before in infinitary lambda calculus mainly because of a counterexample [12,11] showing that arbitrary transfinite $\beta\eta$-reductions can not be compressed into reductions of at most ω length, as in the case without eta, since infinite β-reduction can create an eta redex.[1] The confluence proof in [12] heavily depended on compression. The recent approach of [11] uses transfinite induction and postponement of $\bot$-reduction over β-reduction.

In this paper we will follow the postponement proof technique. Roughly speaking we will show that any transfinite mixed $\beta\bot\eta$-reduction factors into a $\beta\bot$-reduction followed by an η-reduction. Confluence of $\beta\bot\eta$-reduction then follows from confluence of $\beta\bot$-reduction, commutation of β-reduction and $\bot$-reduction with η-reduction, and confluence of η-reduction. The commutation of η-reduction and $\bot$-reduction requires care: in general only outermost (or maximal [3]) $\bot$-reduction commutes with η-reduction, which can be easily overlooked, already in finite lambda calculus.

The calculus $\lambda^{\mathsf{h}\infty}_{\beta\bot\eta}$ is normalising: given a term one first reduces it via a leftmost outermost $\beta\bot$-reduction to its Böhm tree, and then via a leftmost outermost η-reduction to its eta-Böhm tree. This two step process cannot really be improved upon. We will show that at most $\omega{+}\omega$ steps are needed to compute the eta-Böhm tree of a lambda term. Using the notation of the counterexample in the footnote against ω-compression, the term $z(\lambda x.Ex)(\lambda x.Ex)(\lambda x.Ex)\ldots$ is an example of a term that needs at least $\omega{+}\omega$ steps to reduce to its eta-Böhm tree $zy^\omega y^\omega y^\omega\ldots$. This contrasts with the three infinitary extensions of λ_β, where the reduction to normal form needs at most ω steps.

2 Infinite Lambda Calculus

Infinite lambda calculus houses besides the usual finite terms and reductions also infinite terms and infinite, converging reduction. It incorporates in a natural way the open-ended view on computation that, as the computation of a program proceeds, more and more information is read from the input, and more and more of the output is produced.

We will now recall some notions and facts of infinite lambda calculus presented in [12,11]. We assume familiarity with basic notions and notations from [3].

[1] Consider a term E with the property that $Ex \to^*_\beta y(Ex)$ (e.g. the term $E = \Omega_{\lambda zw.y(zw)}$ in the notation of Definition 1) and the term $y^\omega = y(y(y(\ldots)))$. Then $\lambda x.Ex \to^*_\beta \lambda x.(y(Ex))x \to^*_\beta \lambda x.y(y(Ex))x \twoheadrightarrow^\omega_\beta \lambda x.y^\omega x \to_\eta y^\omega$. This example is related to the example of 10.1.22 in [3].

Let $\Lambda_\perp$ be the set of finite λ-terms given by the inductive grammar:

$$M ::= \perp \mid x \mid MM \mid \lambda x.M$$

Let u be any finite sequence of 0, 1 and 2's. The subterm $M|_u$ of a term $M \in \Lambda_\perp$ at occurrence u (if there is one) is defined by induction as usual:

$$M|_{\langle\rangle} = M \quad (\lambda x.M)|_{0u} = M|_u \quad (MN)|_{1u} = M|_u \quad (MN)|_{2u} = N|_u$$

We will define three length related measures for occurrences: $\mathsf{length_h}(u)$ is the number of 2's in u, $\mathsf{length_w}(u)$ is the number of 0's and 2's in u and finally $\mathsf{length_t}(u)$ is the number of 0's, 1's and 2's in u. The depth at which a subterm N in M occurs can now be measured by the length of the occurrence u of N in M. This leads to three different metrics d_x on $\Lambda_\perp$ for $\mathsf{x} \in \{\mathsf{h}, \mathsf{w}, \mathsf{t}\}$: x-metric $\mathsf{d_x}(M, N) = 0$, if $M = N$ and $\mathsf{d_x}(M, N) = 2^{-\mathsf{length_x}(u)}$, where u is a common occurrence of minimal length such that $M|_u \neq N|_u$. Now in the spirit of Arnold and Nivat [1] we define the sets $\Lambda_\perp^{\mathsf{h}\infty}$, $\Lambda_\perp^{\mathsf{w}\infty}$ and $\Lambda_\perp^{\mathsf{t}\infty}$ as the metric completions of the set of finite lambda terms $\Lambda_\perp$ over the respective metrics $\mathsf{d_h}$, $\mathsf{d_w}$ and $\mathsf{d_t}$. The indices h, w and t stand for head normal form, weak head normal form and top normal form respectively.
It may be illustrative to draw pictures and to think of terms as trees: draw the edges corresponding to the counted occurrences vertically and all other edges horizontally. Then trees in the three metric completions don't have infinite horizontal branches. In case of $\Lambda_\perp^{\mathsf{h}\infty}$ this implies, when paths of branches are coded by sequences of $0, 1$ and 2's, that its trees are characterised by the fact that their branches don't have infinite "tails" consisting of 0 and 1's only.

Definition 1. *Some abbreviations for useful finite and infinite λ-terms:*

$$\begin{array}{lll} I = \lambda x.x & K_\infty = \lambda x.\lambda x.\ldots & \Omega = (\lambda x.xx)\lambda x.xx \\ 1 = \lambda xy.xy & M^\omega = M(M(M\ldots)) & \Omega_M = (\lambda x.M(xx))\lambda x.M(xx) \\ K = \lambda xy.x & {}^\omega M = ((\ldots M)M)M & \Omega_\eta = \lambda x_0.(\lambda x_1.(\ldots)x_1)x_0 \end{array}$$

The inclusions $\Lambda_\perp^{\mathsf{h}\infty} \subset \Lambda_\perp^{\mathsf{w}\infty} \subset \Lambda_\perp^{\mathsf{t}\infty}$ are strict. For instance, $x^\omega \in \Lambda_\perp^{\mathsf{h}\infty}$, $K_\infty \in \Lambda_\perp^{\mathsf{w}\infty} - \Lambda_\perp^{\mathsf{h}\infty}$ and ${}^\omega x \in \Lambda_\perp^{\mathsf{t}\infty} - \Lambda_\perp^{\mathsf{w}\infty}$. As shown in [12,11] the sets $\Lambda_\perp^{\mathsf{h}\infty}, \Lambda_\perp^{\mathsf{w}\infty}$ and $\Lambda_\perp^{\mathsf{t}\infty}$ are the three minimal infinitary extensions of the finite lambda calculus λ_β containing respectively the Böhm trees, Lévy-Longo trees and the Berarducci trees and each closed under its respective notion of convergent reduction to be defined next. The set $\Lambda_\perp^{\mathsf{h}\infty}$ will function as the underlying set of finite and infinite lambda terms for our extensional infinitary extension $\lambda_{\beta\perp\eta}^{\mathsf{h}\infty}$.
Many notions of finite lambda calculus apply and/or extend more or less straightforwardly to the infinitary setting. The main idea which goes back to Dershowitz e.a. in [8] is that reduction sequences can be of any transfinite ordinal length α:

$$M_0 \to M_1 \to M_2 \to \ldots M_\omega \to M_{\omega+1} \to \ldots M_{\omega+\omega} \to M_{\omega+\omega+1} \to \ldots M_\alpha$$

This makes sense if the limit terms $M_\omega, M_{\omega+\omega}, \ldots$ in such sequence are all equal to the corresponding Cauchy limits $\lim_{\beta \to \lambda} M_\beta$ in the underlying metric

space for any limit ordinal $\lambda \leq \alpha$. If this is the case, the reduction $M_0 \to_\alpha M_\alpha$ is called Cauchy converging. We need the stronger concept of a strongly converging reduction that in addition satisfies that the depth of the reduced redexes goes to infinity at each limit term: $\lim_{\beta \to \lambda} d_\beta = \infty$ for each limit ordinal $\lambda \leq \alpha$, where d_β is the depth in M_β of the reduced redex in $M_\beta \to M_{\beta+1}$. We will denote strongly converging reduction by $\twoheadrightarrow$. Any finite reduction $M_0 \to^* M_n$ is strongly converging.
Finally we have to introduce the basic reduction relations of $\lambda^{\mathsf{h}\infty}_{\beta\bot\eta}$. Besides the familiar beta and eta rules it contains one of the bottom rules $\bot_{\mathsf{h}}$ (or $\bot$ for short). We define three bottom rules $\bot_{\mathsf{x}}$ where $\mathsf{x} \in \{\mathsf{h}, \mathsf{w}, \mathsf{t}\}$ as follows:

$$M \to \bot, \text{ provided } M[\bot := \Omega] \in U_{\mathsf{x}} \quad (\bot_{\mathsf{x}})$$

We usually omit the subscript x and we write $\bot$ instead of $\bot_{\mathsf{x}}$. Outermost bottom reduction, denoted as $\to_{\bot_{out}}$, is the restriction of bottom reduction to outermost redexes. The sets U_{x} are sets of $\bot$-free finite and infinite lambda terms and defined as follows:

1. The (for this paper most important) set U_{h} is the set of $\bot$-free terms in $\Lambda^{\mathsf{h}\infty}_{\bot}$ that don't have (a finite β-reduction to) a head normal form, where a term is a head normal form if it is of the form $\lambda x_1 \ldots \lambda x_n.yM_1 \ldots M_m$.
2. The set U_{w} is the set of $\bot$-free terms in $\Lambda^{\mathsf{w}\infty}_{\bot}$ that don't have (a finite β-reduction to) a weak head normal form, where a term is a weak head normal form if it is either a term of the form $yM_1 \ldots M_n$ or an abstraction $\lambda x.M$.
3. The set U_{t} is the set of $\bot$-free terms in $\Lambda^{\mathsf{t}\infty}_{\bot}$ that don't have (a finite β-reduction to) a top normal form (rootstable form), where a term is a top normal form, if it is either an abstraction $\lambda x.M$, or, an application MN where M cannot β-reduces (in a finite number of steps) to an abstraction.

Some examples: The term Ω does not have a top normal form and therefore belongs to all three sets. The term Ωx is a top normal form but has no (weak) head normal form. The term $\lambda x.\Omega$ is a weak head normal form, but has no head normal form. Hence all inclusions in $U_{\mathsf{h}} \supset U_{\mathsf{w}} \supset U_{\mathsf{t}}$ are strict.
The extensional infinite lambda calculus that is the main object of study in this paper is the calculus $\lambda^{\mathsf{h}\infty}_{\beta\bot\eta} = (\Lambda^{\mathsf{h}\infty}_{\bot}, \to_{\beta\bot\eta})$. We will also briefly mention $\lambda^{\mathsf{x}\infty}_{\beta\bot\eta} = (\Lambda^{\infty}_{\mathsf{x}}, \to_{\beta\bot\eta})$ for $\mathsf{x} \in \{\mathsf{w}, \mathsf{t}\}$. For any of these infinite lambda calculi $\lambda^{\mathsf{x}\infty}_{\rho}$ we say that

- a term M in $\lambda^{\mathsf{x}\infty}_{\rho}$ is in *ρ-normal form* if there is no N in $\lambda^{\mathsf{x}\infty}_{\rho}$ such that $M \to_\rho N$.
- $\lambda^{\mathsf{x}\infty}_{\rho}$ is *infinitary confluent* (or just *confluent* for short) if $(\Lambda^{\infty}_{\mathsf{x}}, \twoheadrightarrow_\rho)$ satisfies the diamond property, i.e. ${}_\rho\!\twoheadleftarrow \circ \twoheadrightarrow_\rho \subseteq \twoheadrightarrow_\rho \circ {}_\rho\!\twoheadleftarrow$.
- $\lambda^{\mathsf{x}\infty}_{\rho}$ is *(weakly) normalising* if for all $M \in \Lambda^{\infty}_{\mathsf{x}}$ there exists an N in ρ-normal form such that $M \twoheadrightarrow_\rho N$.
- Let α be an ordinal. We say that $\lambda^{\mathsf{x}\infty}_{\rho}$ is *α-compressible* if for all M, N such that $M \twoheadrightarrow_\rho N$ there exists a reduction from M to N of length at most α.

Without the bottom rule there is no chance of proving confluence. Berarducci's counterexample [5] is very short: $\Omega \; {}_{\beta}^{*}\!\!\leftarrow \Omega_I \twoheadrightarrow_\beta I^\omega$.
Crucial properties of those three infinite lambda calculi $\lambda_{\beta\bot}^{\mathsf{h}\infty}$, $\lambda_{\beta\bot}^{\mathsf{w}\infty}$ and $\lambda_{\beta\bot}^{\mathsf{t}\infty}$ are:

Theorem 1. Confluence, normalisation and compression of $\beta\bot$ [12,11]. *The infinite lambda calculi $\lambda_{\beta\bot}^{\mathsf{h}\infty}$, $\lambda_{\beta\bot}^{\mathsf{w}\infty}$ and $\lambda_{\beta\bot}^{\mathsf{t}\infty}$ are confluent, normalising and ω-compressible.*

Theorem 2. Postponement of β over $\bot$ [11]. *If $M \twoheadrightarrow_{\beta\bot} N$ then there exists Q such that $M \twoheadrightarrow_\beta Q \twoheadrightarrow_\bot N$.*

3 Two Non-confluent Extensional Infinite Lambda Calculi

Before we give the proof of confluence of $\lambda_{\beta\bot\eta}^{\mathsf{h}\infty}$ we will show that the two related extensional infinite lambda calculi $\lambda_{\beta\bot\eta}^{\mathsf{w}\infty}$ and $\lambda_{\beta\bot\eta}^{\mathsf{t}\infty}$ are not confluent. In fact what we note is that already the finite calculi $\lambda_{\beta\bot\eta}^{\mathsf{w}}$ and $\lambda_{\beta\bot\eta}^{\mathsf{t}}$ are not confluent for finite reductions.
For this we use the term $\Omega_\eta \in \Lambda_\bot^{\mathsf{w}\infty} \subset \Lambda_\bot^{\mathsf{t}\infty}$. Similar to Ω which β-reduces to itself in only one step, this term η-reduces to itself in only one step. The body of the outermost abstraction in Ω_η has no weak (top) normal form. The span $\bot \; {}_\bot\!\!\leftarrow \Omega_I \; {}_\eta\!\!\leftarrow \Omega_1 \twoheadrightarrow_\beta \Omega_\eta \to_\bot \lambda x.\bot$ can only be joined if $\lambda x.\bot \to_\bot \bot$, which does not hold for the $\bot_{\mathsf{w}}$-rule and the $\bot_{\mathsf{t}}$-rule. Hence this is a counterexample of confluence of both $\lambda_{\beta\bot\eta}^{\mathsf{w}\infty}$ and $\lambda_{\beta\bot\eta}^{\mathsf{t}\infty}$.
Remark also that there is a critical pair between the eta rule and each bottom rule: $\lambda x.\bot \; {}_\bot\!\!\leftarrow \lambda x.\bot x \to_\eta \bot$. The reverse step follows from the fact that the term $\bot x[\bot := \Omega] := \Omega x$ has no weak head normal form. This pair can be completed only if $\lambda x.\bot \to_\bot \bot$ which is true for the $\bot_{\mathsf{h}}$-rule, but not for the $\bot_{\mathsf{w}}$-rule. This gives an alternative counterexample against confluence of $\lambda_{\beta\bot\eta}^{\mathsf{w}\infty}$.

4 The Confluent Extensional Infinite Lambda Calculus $\lambda_{\beta\bot\eta}^{\mathsf{h}\infty}$

In this section we will prove the confluence and normalisation of $\lambda_{\beta\bot\eta}^{\mathsf{h}\infty}$. Thereto we will first prove some useful properties of the infinitary eta calculus $\lambda_\eta^{\mathsf{h}\infty}$, secondly the commutation of eta and beta and the commutation of eta and outermost bottom, and thirdly postponement of eta over beta and bottom.

4.1 The Infinitary Eta Calculus $\lambda_\eta^{\mathsf{h}\infty}$

The set $\Lambda_\bot^{\mathsf{h}\infty}$ has the very pleasant property that any Cauchy-converging η-reduction sequence in $\Lambda_\bot^{\mathsf{h}\infty}$ is h-strongly convergent. This property is not shared by the other two infinitary extensions $\Lambda_\bot^{\mathsf{w}\infty}$ and $\Lambda_\bot^{\mathsf{t}\infty}$. For instance there exists a Cauchy-converging and non-h-strongly converging η-reduction sequence from the term $\Omega_\eta \in \Lambda_\perp^{\mathsf{w}\infty} \subset \Lambda_\perp^{\mathsf{t}\infty}$.

Definition 2. *For $M \in \Lambda_\perp^{\mathsf{h}\infty}$ define $|M|_n$ as the number of nodes at* h*-depth n.*

The number $|M|_n$ of nodes of term M at h-depth n decreases if and only if we contract an η-redex in M at h-depth n. Hence:

Lemma 1. *Any Cauchy-converging η-reduction starting from a term in $\Lambda_\perp^{\mathsf{h}\infty}$ is* h*-strongly convergent.*

Proof. Suppose by contradiction that we have some transfinite Cauchy-converging η-reduction sequence $M_0 \to_\eta M_1 \to_\eta \ldots$, that reduces infinitely often at h-depth n. Then infinitely many of the inequalities in the next sequence are strict: $|M_0|_n \geq |M_1|_n \geq |M_2|_n \geq \ldots$ But infinite strictly decreasing sequences of natural numbers don't exist. Hence the limit of the h-depth of the contracted redexes in this sequence goes to infinity at each limit ordinal $\leq \alpha$. Therefore any η-reduction sequence starting from a term in $\Lambda_\perp^{\mathsf{h}\infty}$ is h-strongly converging. □

Lemma 2. *The infinitary lambda calculus $\lambda_\eta^{\mathsf{h}\infty}$ is ω-compressible.*

Proof. Let $M \twoheadrightarrow_\eta N$ a reduction sequence of length γ. We will prove that there exists a sequence from M to N of length at most ω. The proof proceeds by transfinite induction on γ. The argument of the limit case is standard, see [11]. If γ is a successor ordinal, it is sufficient to prove that a sequence of length $\omega+1$ can be compressed into one of length ω. Without loss of generality, we may suppose that we have a strongly η-reduction sequence of length $\omega+1$ from M_0 to M_ω as in:

$$\begin{array}{ccccccc} M_0 \xrightarrow[\eta]{*} \lambda x.M_k x & \xrightarrow[\eta]{} & \lambda x.M_{k+1}x & \xrightarrow[\eta]{} & \lambda x.M_{k+2}x & \cdots\cdots & \lambda x.M_\omega x \\ \vdots\,\eta & & \vdots\,\eta & & \vdots\,\eta & & \downarrow\eta \\ M_k & \overset{\cdots\cdots}{\underset{\eta}{\dashrightarrow}} & M_{k+1} & \overset{\cdots\cdots}{\underset{\eta}{\dashrightarrow}} & M_{k+2} & \cdots\cdots & M_\omega \end{array}$$

Then working to the right onwards from $\lambda x.M_k x$ the dotted squares can be all constructed. This results in a reduction of length ω starting at M_0 and after $k+1$ steps continuing as $M_k \to_\eta M_{k+1} \to_\eta \ldots$. The limit of this converging reduction is clearly M_ω. □

In the following lemma, as well as later for other reduction relations, we indicate with a superscript that the redex contracted in $M \xrightarrow{m}_\eta N$ is at depth m.

Lemma 3. *Let M_0 be a term in $\Lambda_\perp^{\mathsf{h}\infty}$. If $M_0 \xrightarrow{m}_\eta M_1$ and $M_0 \xrightarrow{n}_\eta M_2$ then one of the following two cases holds:*

$$\begin{array}{ccc} M_0 & \xrightarrow[\eta]{m} & M_1 \\ n\downarrow\eta & & n\,\vdots\,\eta \\ M_2 & \overset{m}{\underset{\eta}{\dashrightarrow}} & M_3 \end{array} \qquad\qquad \begin{array}{ccc} M_0 & \xrightarrow[\eta]{n} & M_1 \\ n\downarrow\eta & & \vdots \\ M_2 & \cdots\cdots & M_3 \end{array}$$

Proof. Note that the h-depth of an η-redex in a term does not change when we contract an η-redex elsewhere in the term. □

Theorem 3. *The infinitary lambda calculus $\lambda_\eta^{\mathsf{h}\infty}$ is confluent.*

Proof. Let two coinitial η-reductions be given. By compression (Lemma 2) we may assume that their length is at most ω. By simultaneous induction on their length we show that for any two coinitial η-reductions can be joined with a so called tiling diagram construction [11] in which all horizontal and vertical reductions are strongly converging.
Suppose we have two η-reductions of length ω: $M_{0,0} \twoheadrightarrow_\eta M_{0,\omega}$ and $M_{0,0} \twoheadrightarrow_\eta M_{\omega,0}$. We will not present the whole induction, but comment on the main cases.

The successor-successor case of the induction is in fact the previous lemma.

The successor-limit case: this follows if we can construct the following tiling diagram in which the notation $M \stackrel{=}{\rightarrow}_\eta N$ expresses that in the reduction from M to N in which at most one reduction step has been performed: at h-depth m.

$$\begin{array}{ccccccc}
M_{0,0} & \xrightarrow[\eta]{n_0} & M_{0,1} & \xrightarrow[\eta]{n_1} & M_{0,2} & \cdots\cdots & M_{0,\omega}\\
{\scriptstyle m_0}\downarrow{\scriptstyle\eta} & & {\scriptstyle m_0/=}\vdots{\scriptstyle\eta} & & {\scriptstyle m_0/=}\vdots{\scriptstyle\eta} & & {\scriptstyle m_0/=}\vdots{\scriptstyle\eta}\\
M_{1,0} & \overset{n_0/=}{\underset{\eta}{\cdots\rightarrow}} & M_{1,1} & \overset{n_1/0}{\underset{\eta}{\cdots\rightarrow}} & M_{1,2} & \cdots\cdots & M_{1,\omega}
\end{array}$$

By induction hypothesis this diagram can be constructed but for the right most edge. The bottom reduction inherits the (strong) convergence property from the top reduction. As we are dealing with η-reduction it is easy to see that there is at most one residual of the redex contracted in $M_{0,0} \rightarrow_\eta M_{1,0}$ in $M_{0,\omega}$. Contraction of this set of residuals gives the reduction of the rightmost edge: it will be at most one step.

The limit-limit case: Using the induction hypothesis we can construct diagrams for pair of coinitial sequences of respective lengths (n, m) where either $n < \omega$ and $m \leq \omega$, or $m < \omega$ and $n \leq \omega$. By the general tiling diagram theorem in [11] or in this particular instance simply the uniform nature of the strong convergence of all vertical and horizontal reductions it follows that their limits have to be the same: $M_{\omega,\omega}$. □

$$\begin{array}{ccccccc}
M_{0,0} & \xrightarrow[\eta]{n_0} & M_{0,1} & \xrightarrow[\eta]{n_1} & M_{0,2} & \cdots\cdots & M_{0,\omega}\\
{\scriptstyle m_0}\downarrow{\scriptstyle\eta} & & {\scriptstyle m_0/=}\vdots{\scriptstyle\eta} & & {\scriptstyle m_0/=}\vdots{\scriptstyle\eta} & & {\scriptstyle m_0/=}\vdots{\scriptstyle\eta}\\
M_{1,0} & \overset{n_0/=}{\underset{\eta}{\cdots\rightarrow}} & M_{1,1} & \overset{n_1/0}{\underset{\eta}{\cdots\rightarrow}} & M_{1,2} & \cdots\cdots & M_{1,\omega}\\
{\scriptstyle m_1}\downarrow{\scriptstyle\eta} & & {\scriptstyle m_1/=}\vdots{\scriptstyle\eta} & & {\scriptstyle m_1/=}\vdots{\scriptstyle\eta} & & {\scriptstyle m_1/=}\vdots{\scriptstyle\eta}\\
M_{2,0} & \overset{n_0/=}{\underset{\eta}{\cdots\rightarrow}} & M_{2,1} & \overset{n_1/=}{\underset{\eta}{\cdots\rightarrow}} & M_{2,2} & \cdots\cdots & M_{2,\omega}\\
\vdots & & \vdots & & \vdots & & \vdots\\
M_{\omega,0} & \overset{n_0/=}{\underset{\eta}{\cdots\rightarrow}} & M_{\omega,1} & \overset{n_1/=}{\underset{\eta}{\cdots\rightarrow}} & M_{\omega,2} & \cdots\cdots & M_{\omega,\omega}
\end{array}$$

Theorem 4. *The infinitary lambda calculus $\lambda_\eta^{\mathsf{h}\infty}$ is normalising.*

Proof. Let M_0 be given. We construct a reduction $M_0 \to_\eta M_1 \to_\eta M_2 \ldots$ recursively: suppose we have M_n then we construct M_{n+1} by contracting the leftmost η-redex of smallest h-depth. This gives us a possibly infinite reduction which by Lemma 1 can only be strongly converging. Its last term is an η-normal form. This is trivial in case of a finite reduction; in case of an infinite reduction, there is a standard *reductio ad absurdum* argument: Suppose there is a η-redex in the limit term, then this η-redex was already present at some finite stage in the reduction and apparently not reduced in the remaining reduction. But the reduction strategy is such that no redex can get overlooked. Contradiction. $\square$

4.2   Commutation

In this section we prove that the reductions $\twoheadrightarrow_\eta$ commutes with both $\twoheadrightarrow_\beta$ and $\twoheadrightarrow_\perp$. As each of these reduction relations on its own is ω-compressible we can assume that the length of the reductions is at most ω.

Theorem 5. [Commutation of $\twoheadrightarrow_\beta$ and $\twoheadrightarrow_\eta$]
If $M_{0,0} \twoheadrightarrow_\beta M_{0,\gamma}$ and $M_{0,0} \twoheadrightarrow_\eta M_{\delta,0}$ then there exists $M_{\delta,\gamma}$ such that $M_{0,\gamma} \twoheadrightarrow_\eta M_{\delta,\gamma}$ and $M_{\delta,0} \twoheadrightarrow_\beta M_{\delta,\gamma}$.

Proof. A double induction on the length of the two sequences, respectively γ and δ. The critical cases in this proof are: $(1,1)$, $(1,\omega)$, $(\omega, 1)$ and (ω,ω).

$(\gamma, \delta) = (1,1)$: This case is a careful analysis of cases based on the relative positions of the β-redex and the η-redex. Suppose M_0 can do both a β-reduction and an η-reduction at respectively h-depth n and m. The possible situations are:

1. The redexes do not interfere with each other:
   a) The redexes are not nested, i.e. $M_{0,0} = C[(\lambda x.M)N, (\lambda y.Py)]$.
   b) The β-redex is inside the η-redex, that is $M_{0,0}$ is of the form $C_1[\lambda x.C_2[(\lambda y.M)N]x]$.
   c) The η-redex is inside the body of the abstraction, that is $M_{0,0}$ is of the form $C_1[(\lambda x.C_2[\lambda y.My])N]$.
2. The η-redex is inside the argument of the application, that is $M_{0,0}$ is of the form $C_1[(\lambda x.M)C_2[\lambda y.Ny]]$.
3. The β-redex and η-redex overlap, that is $M_{0,0}$ is either $C[(\lambda x.Mx)N]$ or $C[\lambda x.(\lambda y.M)x]$.

These possibilities result in three different diagrams (the labels on the arrows refer to the h-depth of the contracted redex):

$$
\begin{array}{ccc}
M_{0,0} \xrightarrow{\ m\ }_\eta M_{0,1} & M_{0,0} \xrightarrow{\ m\ }_\eta M_{0,1} & M_{0,0} \xrightarrow{\ n\ }_\eta M_{0,1} \\
{\scriptstyle n}\downarrow{\scriptstyle \beta} \quad {\scriptstyle n}\vdots{\scriptstyle \beta} & {\scriptstyle n}\downarrow{\scriptstyle \beta} \quad {\scriptstyle n}\vdots{\scriptstyle \beta} & {\scriptstyle n}\downarrow{\scriptstyle \beta} \quad \vdots \\
M_{1,0} \dashrightarrow_\eta M_{1,1} & M_{1,0} \overset{\geq m-1}{\dashrightarrow}_\eta M_{1,1} & M_{1,0} \dashrightarrow M_{1,1}
\end{array}
$$

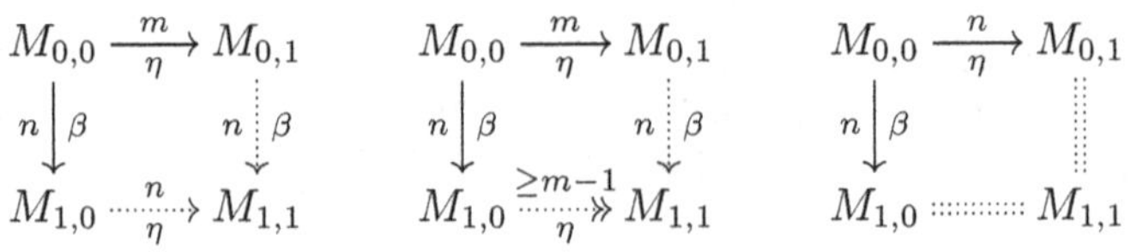

It is important to note that the depth of the eventual residual of β-redex remains the same after contraction of the η-redex.

$(\gamma, \delta) = (1, \omega)$: This case is simple. The construction of the commutation diagram comes down to an infinite horizontal chain of base case diagrams. Either the β gets cancelled against one of the η-steps or not. This implies that from a certain point onwards the vertical edges are equalities or not.

$(\gamma, \delta) = (\omega, 1)$: This case is somewhat involved. Using the previous case we construct for any natural number n the horizontal reduction $M_{n,0} \twoheadrightarrow_\eta M_{n,1}$. This way we get the vertical reduction on the right $M_{1,0} \to_\beta M_{1,1} \to_\beta M_{1,2} \to_\beta \ldots$ which is strongly converging because its reduction steps take place at the same depth as the corresponding steps on the left. Hence it has a limit, say $M_{\omega,1}$. In fact each horizontal η-reduction is the complete development of a set V_n of (occurrences of) η-redexes in $M_{n,0}$. The missing η-reduction at the bottom can now be filled in: complete development of the set of (occurrences of) η-redexes $\bigcup_{k \geq 0} \bigcap_{m \geq k} V_m$ gives us precisely a strongly convergent reduction $M_{\omega,0} \twoheadrightarrow_\eta M_{\omega,1}$.

$$
\begin{array}{ccc}
M_{0,0} & \xrightarrow{\eta} & M_{0,1} \\
\downarrow{\scriptstyle\beta} & & =\,\Big\downarrow{\scriptstyle\beta} \\
M_{1,0} & \xrightarrow{\eta} & M_{1,1} \\
=\,\Big\downarrow{\scriptstyle\beta} & & =\,\Big\downarrow{\scriptstyle\beta} \\
M_{\omega,0} & \overset{\eta}{\dashrightarrow} & M_{\omega,1}
\end{array}
$$

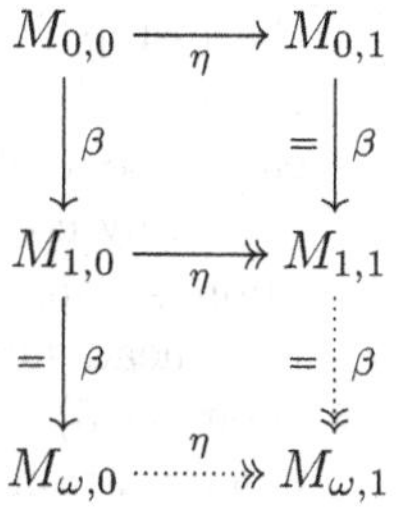

$(\gamma, \delta) = (\omega, \omega)$: Repeated application of the previous two cases allows us to construct the tiling diagram. The vertical β-reductions are all strongly converging in a uniform way: at most some β-steps can get cancelled, but the h-depth of the remaining β-residuals remains unaltered. The horizontal η-reduction sequences cannot be else than strongly converging by Lemma 1. Using the tiling diagram theorem of [12] or the uniform (nature of the) strong convergence of all vertical reductions we get that the bottom reduction and the rightmost reduction both end in the same limit term.   □

$$
\begin{array}{ccccccc}
M_{0,0} & \xrightarrow{\eta} & M_{0,1} & \xrightarrow{\eta} & M_{0,2} & \cdots & M_{0,\omega} \\
\downarrow{\scriptstyle\beta} & & =\downarrow{\scriptstyle\beta} & & =\downarrow{\scriptstyle\beta} & & =\downarrow{\scriptstyle\beta} \\
M_{1,0} & \dashrightarrow_\eta & M_{1,1} & \dashrightarrow_\eta & M_{1,2} & \cdots & M_{1,\omega} \\
\downarrow{\scriptstyle\beta} & & =\downarrow{\scriptstyle\beta} & & =\downarrow{\scriptstyle\beta} & & =\downarrow{\scriptstyle\beta} \\
M_{2,0} & \dashrightarrow_\eta & M_{2,1} & \dashrightarrow_\eta & M_{2,2} & \cdots & M_{2,\omega} \\
\vdots & & \vdots & & \vdots & & \vdots \\
M_{\omega,0} & \dashrightarrow_\eta & M_{\omega,1} & \dashrightarrow_\eta & M_{\omega,2} & \cdots & M_{\omega,\omega}
\end{array}
$$

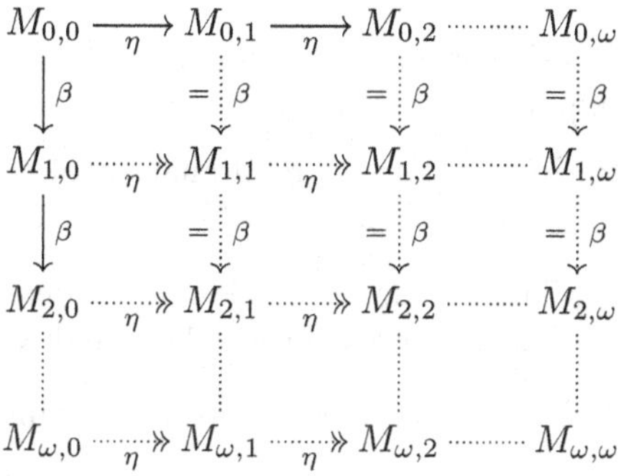

Commutation of $\twoheadrightarrow_\eta$ with $\twoheadrightarrow_\perp$ does not hold as it is shown by the counterexample $\Omega \;{}_\eta\!\leftarrow \lambda x.\Omega x \to_\perp \lambda x.\perp$. However, if we restrict ourselves to leftmost outermost $\perp$ reduction we can prove commutation of $\twoheadrightarrow_\eta$ with $\twoheadrightarrow_\perp$. First a lemma saying that $\perp$-redexes are preserved under η-reduction.

Lemma 4. Let $M \in \Lambda^{h\infty}_\perp$ be a $\perp$-free term and $M \twoheadrightarrow_\eta N$.

1. If M is a β-head normal form then so is N.
2. If N is a β-head normal form then M has a β-head normal form.
3. $M \in U_h$ if and only if $N \in U_h$.

Proof. 1. Standard inductive argument on the length of the η-reduction.

2. Without loss of generality we may assume that an η-reduction ending in a β-head normal form is finite. By induction on the length n of this reduction we will show that there is a reduction of linear β-redexes from the initial

term to a β-hnf, where a β-redex $(\lambda x.P)Q$ is linear if the bound variable x occurs free in P at most once.

The base case, when $n = 0$, is trivial. Induction step: suppose $M \to_\eta P \to_\eta^n N$ and N is a β-hnf. By induction hypothesis there exists P' in β-head normal form such that $P \to_\beta^* P'$ by contraction of linear β-redexes. By Lemma 5 (its proof does not depend on this result nor the next theorem) we can postpone the η-step from M to P over the linear β-reduction to P', so that we get $M \to_\beta^* M' \to_\eta^= P'$, where the β-reduction contracts linear β-redexes. An easy case analysis shows that either M' is in β-head normal form or reduces to one by contracting a linear β-redex.

3. Suppose that N β-reduces to a head normal form. Then M β-reduces to a head normal form by postponement of η-reduction over β-reduction (Lemma 5) and the previous part. The converse is proved similarly using commutation instead of postponement.   □

Theorem 6. Commutation of $\twoheadrightarrow_\eta$ with $\twoheadrightarrow_{\perp\text{out}}$. *If $M_{0,0} \twoheadrightarrow_{\perp\text{out}} M_{0,\gamma}$ and $M_{0,0} \twoheadrightarrow_\eta M_{\delta,0}$ then there exists $M_{\delta,\gamma}$ such that $M_{0,\gamma} \twoheadrightarrow_\eta M_{\delta,\gamma}$ and $M_{\delta,0} \twoheadrightarrow_{\perp\text{out}} M_{\delta,\gamma}$.*

Proof. The induction proof proceeds as the previous proof of Theorem 5. We skip all cases but:

$(\gamma, \delta) = (1,1)$: A careful analysis of cases based on the relative positions of the $\perp$out-redex (denoted by U below) and the η-redex leads to two basic situations:

 – The $\perp$-redex is inside the η-redex: $M_{0,0} \equiv C_1[\lambda x.C_2[U]x]$.
 – The η-redex is inside the $\perp$-redex: $M_{0,0} \equiv C_1[C_2[\lambda x.Mx]] \equiv C_1[U]$.

These two cases result in the following two diagrams:

$$
\begin{array}{ccc}
M_0 & \xrightarrow{\ m\ }_\eta & M_1 \\
{\scriptstyle n}\downarrow{\scriptstyle \perp\text{out}} & & {\scriptstyle n}\downarrow{\scriptstyle \perp\text{out}} \\
M_2 & \dashrightarrow_\eta^{\ m\ } & M_3
\end{array}
\qquad
\begin{array}{ccc}
M_0 & \xrightarrow{\ m\ }_\eta & M_1 \\
{\scriptstyle n}\downarrow{\scriptstyle \perp\text{out}} & & {\scriptstyle n}\downarrow{\scriptstyle \perp\text{out}} \\
M_2 & =\!=\!=\!=\!= & M_2
\end{array}
$$

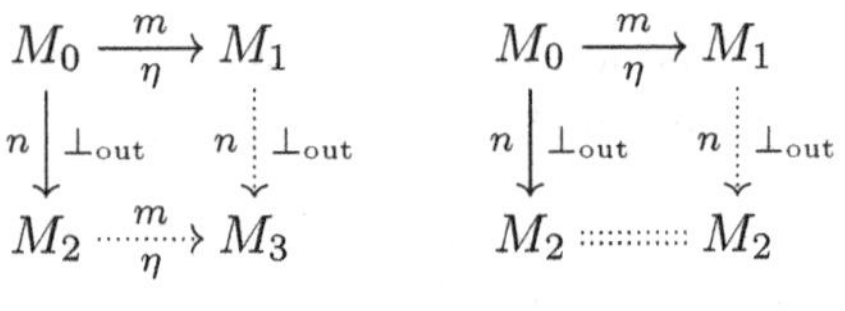

Note that the h-depth of the $\perp$out-redex and its eventual residual after contraction of the η-redex is the same. The last case follows from Lemma 4.   □

4.3   Postponement

We will prove that η-reduction can be postponed in mixed $\beta\perp\eta$-reductions and hence as as well in mixed $\beta\eta$-reductions. We first need two preparatory lemmas.

Lemma 5. *Let $\gamma, \delta \le \omega$. If $M_{0,0} \twoheadrightarrow_\eta M_{0,\gamma} \twoheadrightarrow_\beta M_{\delta,\gamma}$, then there exists an $M_{\delta,0}$ such that $M_{0,0} \twoheadrightarrow_\beta M_{\delta,0} \twoheadrightarrow_\eta M_{\delta,\gamma}$. If $M_{0,\gamma} \twoheadrightarrow_\beta M_{\delta,\gamma}$ is finite, then $M_{0,0} \twoheadrightarrow_\beta M_{\delta,0}$ will be finite as well.*

Note: in $(\lambda x.(\lambda y.M)x)N \to_\eta (\lambda y.M)N \to_\beta M[x := N]$ the resulting reduction after postponement of η-reduction requires two β-reduction steps.

Proof. The proof is again a double induction on the length of the two reductions. In the proof we try to reconstruct a tiling diagram for:

$$\begin{array}{ccc} M_{0,0} & \xrightarrow[\eta]{} \!\!\!\!\rightarrow & M_{0,\gamma} \\ \beta \downarrow\!\!\downarrow & & \downarrow\!\!\downarrow \beta \\ M_{\delta,0} & \overset{\cdots\cdots}{\underset{\eta}{\twoheadrightarrow}} & M_{\delta,\gamma} \end{array}$$

The proof by induction is an interesting variation on the proof of commutation of β and η. We skip all cases but one:

$(\gamma, \delta) = (1, \omega)$: Suppose the contracted redex in the final β-reduction $M_{0,\omega} \to_\beta M_{1,\omega}$ has h-depth n. Then because $M_{0,0} \twoheadrightarrow_\eta M_{0,\omega}$ is strongly converging, there is a k such that for $i \geq k$ the depth of all remaining η-redexes in $M_{0,k} \twoheadrightarrow_\eta M_{0,\omega}$ is at least n.

$$\begin{array}{ccccc} M_{0,0} & \xrightarrow[\eta]{*} & M_{0,k} & \overset{>n}{\underset{\eta}{\twoheadrightarrow}} & M_{0,\omega} \\ n \vdots \beta^{*} & \text{(ind hyp)} & n \vdots \beta & & n \downarrow \beta \\ M_{1,0} & \overset{\cdots\cdots}{\underset{\eta}{\twoheadrightarrow}} & M_{1,k} & \overset{\cdots\cdots}{\underset{\eta}{\twoheadrightarrow}} & M_{1,\omega} \end{array}$$

Hence all $M_{0,i}$ for $i \geq k$ contain a β-redex at the same position as the one redex contracted in $M_{0,\omega}$. Let $M_{1,i}$ be the result of contracting that β-redex in $M_{0,i}$. Since the η-reduction at $M_{0,i}$ takes place at depth lower than n, we get case 1 (a and c) and case 2 of the proof of Theorem 5 and hence $M_{1,i} \twoheadrightarrow_\eta M_{1,i+1}$. By Lemma 1 this reduction sequence is strong converging and its limit coincides with $M_{1,\omega}$. An appeal to Theorem 5 completes the proof of this case. □

The next lemma is proved in a similar way.

Lemma 6. *If $M_{0,0} \twoheadrightarrow_\eta M_{0,\gamma} \twoheadrightarrow_\bot M_{\delta,\gamma}$, then there exists an $M_{\delta,0}$ such that $M_{0,0} \twoheadrightarrow_\bot M_{\delta,0} \twoheadrightarrow_\eta M_{\delta,\gamma}$. If $M_{0,\gamma} \twoheadrightarrow_\bot M_{\delta,\gamma}$ is finite, then $M_{0,0} \twoheadrightarrow_\bot M_{\delta,0}$ will be finite as well.*

Proof. The proof is a double induction. We only do the case $(\delta, \gamma) = (1, 1)$, and skip the rest. By case analysis and using Lemma 4 we obtain the following two diagrams:

$$\begin{array}{ccc} M_{0,0} & \xrightarrow[\eta]{m} & M_{0,1} \\ n \vdots \bot & & n \downarrow \bot \\ M_{1,0} & \overset{m}{\underset{\eta}{\cdots\rightarrow}} & M_{1,1} \end{array} \qquad \begin{array}{ccc} M_{0,0} & \xrightarrow[\eta]{m} & M_{0,1} \\ n \vdots \bot & & n \downarrow \bot \\ M_{1,0} & \equiv & M_{1,1} \end{array}$$

□

Combining the previous two lemmas with Theorem 2 we get:

Corollary 1. *If* $M_{0,0} \twoheadrightarrow_\eta M_{0,\gamma} \twoheadrightarrow_{\beta\bot} M_{\delta,\gamma}$, *then there exists an* $M_{\delta,0}$ *such that* $M_{0,0} \twoheadrightarrow_{\beta\bot} M_{\delta,0} \twoheadrightarrow_\eta M_{\delta,\gamma}$. *If* $M_{0,\gamma} \twoheadrightarrow_{\beta\bot} M_{\delta,\gamma}$ *is finite, then* $M_{0,0} \twoheadrightarrow_{\beta\bot} M_{\delta,0}$ *will be finite as well.*

Theorem 7. Postponement of $\twoheadrightarrow_\eta$ **over** $\twoheadrightarrow_{\beta\bot}$. *If* $M \twoheadrightarrow_{\beta\bot\eta} N$, *then there exists an* L *such that* $M \twoheadrightarrow_{\beta\bot} L \twoheadrightarrow_\eta N$.

Proof. The proof is by (a genuine!) transfinite induction on the number of subsequences of the form $M_1 \twoheadrightarrow_\eta M_2 \twoheadrightarrow_\beta M_3$ in $M \twoheadrightarrow_{\beta\bot\eta} N$. The base case is trivial, the successor case follows directly from Corollary 1. We only show the limit case for the limit ω. The proof for arbitrary limits is similar. Consider:

$$M \equiv M_{1,0} \twoheadrightarrow_\eta M_{1,1} \twoheadrightarrow_{\beta\bot} M_{2,1} \twoheadrightarrow_\eta M_{2,2} \twoheadrightarrow_{\beta\bot} M_{3,2} \;\cdots\cdots M_{\omega,\omega} \equiv N$$

Using the induction hypothesis we construct the next diagram row by row.

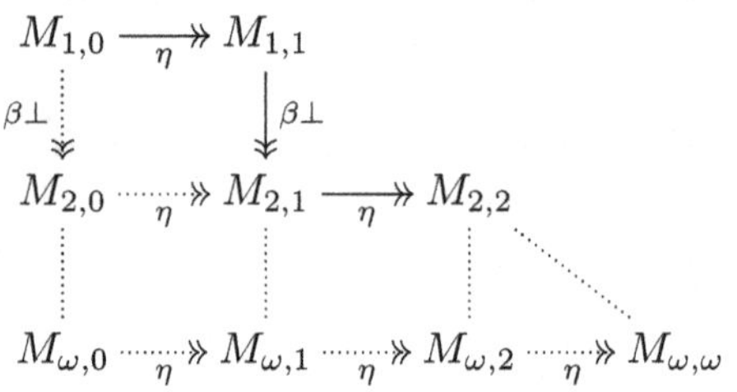

Because the diagonal is strongly converging and the horizontal η-reductions don't change the depth of the vertical $\beta\bot$-reductions, all vertical reductions are strongly convergent as well and have limits. By induction hypothesis they are connected by η-reduction steps. By Lemma 1 the combined reduction at the bottom row is strongly converging. By the uniform nature of the strong convergence of all vertical reductions the limit of reductions at the bottom row and the diagonal are the same. □

Corollary 2. *Let* $M, N \in \Lambda_\bot^{\mathsf{h}\infty}$.

1. *If* $M \twoheadrightarrow_\eta N$, *then* $\mathsf{nf}_{\beta\bot}(M) \twoheadrightarrow_\eta \mathsf{nf}_{\beta\bot}(N)$.
2. *If* $M \twoheadrightarrow_{\beta\bot\eta} N$, *then* $\mathsf{nf}_{\beta\bot}(M) \twoheadrightarrow_\eta \mathsf{nf}_{\beta\bot}(N)$.

Proof.

To prove the first item we postpone $\bot$ over β (Theorem 2). Next because the $\bot$-reduction ends in a $\beta\bot$-normal form we can remove all non-outermost $\bot$-steps. Then we can construct the diagram on the left using the two commutation theorems. The term N_2 is in $\beta\bot$-normal form, and hence by the unique $\beta\bot$-normal form property (Theorem 1) we get $N_2 = \mathsf{nf}_{\beta\bot}(N)$.

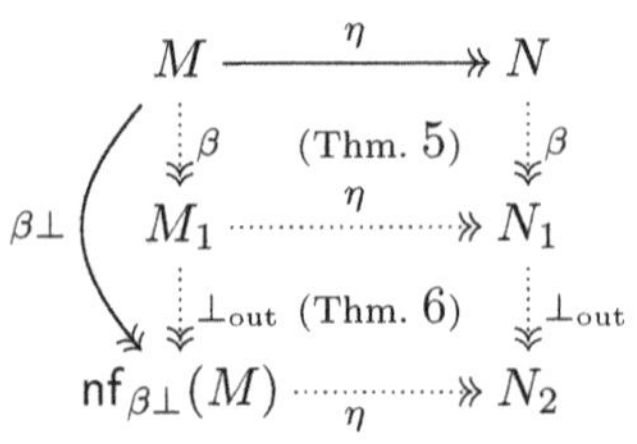

For the proof of second item we build the diagram below using postponement of η-reduction over $\beta\bot$-reduction (Theorem 7) and the previous part.

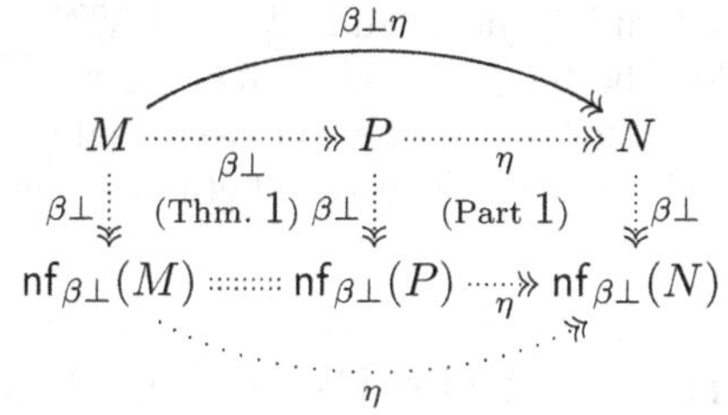

□

4.4 Confluence and Normalisation

Finally we prove that $\lambda^{\mathsf{h}\infty}_{\beta\bot\eta}$ is confluent, normalising and $\omega+\omega$-compressible.

Theorem 8. Confluence of $\beta\bot\eta$. *The extensional infinite lambda calculus* $\lambda^{\mathsf{h}\infty}_{\beta\bot\eta}$ *is confluent.*

Proof.

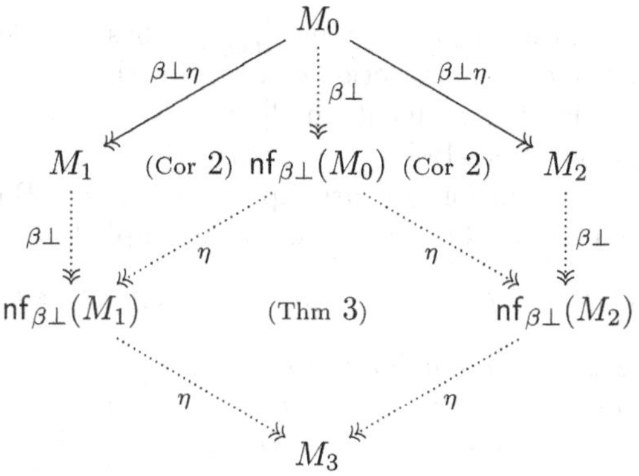

□

Our proof technique using postponement has the flavour of the confluence proof for finite $\lambda_{\beta\eta}$ of Curry and Feys [7], and may therefore seem to be related to the confluence proof of the finite $\lambda_{\beta\bot\eta}$ of 15.2.15(ii) in [3]. The latter proof makes use of η-normal forms, whereas we use the auxiliary notion of $\beta\bot$-normal forms. Note, however, that η-normal forms don't work, as can be seen by applying the proof technique of [3] to the coinitial reductions $\Omega\ {}_{\eta}\!\leftarrow \lambda x.\Omega x \rightarrow_{\bot} \lambda x.\bot$. In its compact form our proof does not restrict to a confluence proof for $\lambda_{\beta\bot\eta}$. But it is not hard to distill from the previous proofs a proof for the finite setting that make use of $\bot$-normal forms instead of $\beta\bot$-normal forms. It may be of interest to see whether other proofs of finitary confluence for finite $\lambda_{\beta\bot\eta}$ can be generalised to the infinitary setting. For instance: the older proof of Barendregt, Bergstra, Klop and Volken [4] and the more recent proof by van Oostrom [15].

Theorem 9. Normalisation and compression of $\beta\bot\eta$. *The extensional infinite lambda calculus $\lambda^{\mathsf{h}\infty}_{\beta\bot\eta}$ is normalising and is $\omega+\omega$-compressible.*

Proof. By theorems 1, 3 and 2, the calculi $\lambda^{\mathsf{h}\infty}_{\beta\bot}$ and $\lambda^{\mathsf{h}\infty}_{\eta}$ are confluent, normalising and ω-compressible. The $\beta\bot\eta$-normal form of a term can be obtained by first computing the $\beta\bot$-normal form and then the η-normal form. To prove that$\lambda^{\mathsf{h}\infty}_{\beta\bot\eta}$ is $\omega+\omega$-compressible we use postponement of eta over beta and bottom. Finally recall the last nine lines of the introduction. □

5 Eta-Böhm Trees and the Extensional Böhm Model

As a corollary of the confluence and normalisation results we see that each term in $\lambda^{\mathsf{h}\infty}_{\beta\bot\eta}$ has a unique $\beta\bot\eta$-normal form, its eta-Böhm tree. Hence we can construct an extensional Böhm model $\mathfrak{B}_\eta$ for both $\lambda_{\beta\eta}$ and $\lambda^{\mathsf{h}\infty}_{\beta\bot\eta}$ almost for free. In the notation of [3] we define the triple $\mathfrak{B}_\eta = (\mathfrak{B}_\eta, \cdot, [\![\ ]\!])$ as follows:

1. $\mathfrak{B}_\eta$ is the set of $\beta\bot\eta$-normal forms of terms in $\Lambda^{\mathsf{h}\infty}_{\bot}$.
2. $M \cdot N = \mathsf{nf}_{\beta\bot\eta}(MN)$ for all M, N in $\mathfrak{B}_\eta$.
3. $[\![M]\!]\rho = \mathsf{nf}_{\beta\bot\eta}(M^\rho)$ where M in $\mathfrak{B}_\eta$and M^ρ is the simultaneous substitution of all free variables of M for $\rho(x)$.

The unique normal form property of $\lambda^{\mathsf{h}\infty}_{\beta\bot\eta}$ implies well-definedness of this definition. The ease of this construction contrasts with the usual construction based on elaborate continuity arguments in [3] of the standard Böhm model for λ_β. An informal definition of eta-Böhm trees can be found in [2]. It is also possible to give a corecursive definition. In early approaches to eta-Böhm trees sets of finite $\beta\bot\eta$-approximants have been used, see for example [10,6].

Theorem 10. *The following statements are equivalent for terms in $\lambda^{\mathsf{h}\infty}_{\beta\bot\eta}$:*

1. *M, N have the same eta-Böhm tree.*
2. *M and N are observationally equivalent wrt finite $\beta\eta$-normal forms [13], i.e. $C[M]$ has a finite $\beta\eta$-normal form if and only if $C[N]$ has a finite $\beta\eta$-normal form for all C.*
3. *M, N have the same interpretation in Park's model D^*_∞ [16,6,17] for the extensional lambda calculus.*

($1 \Leftrightarrow 2$) has been proved by Hyland in [10] and ($2 \Leftrightarrow 3$) has been proved in [6]. Note that the direction ($1 \Rightarrow 2$) is in fact a corollary of the unique normal form property of $\lambda^{\mathsf{h}\infty}_{\beta\bot\eta}$, see [9] for an argument in a similar situation where approximants can not be used.

6 Future Research

We are currently working on another infinitary extension of $\lambda_{\beta\eta}$ that has the infinite-eta-Böhm trees as its normal forms. How to build an extension that captures the Nakajima trees [14,18,3] is still a challenge.

Acknowledgements. The first author thanks the Department of Mathematics and Computer Science of Leicester University for their hospitality. Both authors are grateful to Mariangiola Dezani-Ciancaglini and Vincent van Oostrom for discussions and Mario Severi for providing computer facilities during the preparation of this document.

References

1. A. Arnold and M. Nivat. The metric space of infinite trees. algebraic and topological properties. *Fundamenta Informaticae*, 4:445–476, 1980.
2. S. van Bakel, F. Barbanera, M. Dezani-Ciancaglini, and F.J. de Vries. Intersection types for λ-trees. *TCS*, 272(1-2):3–40, 2002.
3. H. P. Barendregt. *The Lambda Calculus Its Syntax and Semantics*. North-Holland Publishing Co., Amsterdam, Revised edition, 1984.
4. H. P. Barendregt, J. Bergstra, J. W. Klop, and H. Volken. Degrees, reductions and representability in the lambda calculus. Technical report, Department of Mathematics, Utrecht University, 1976.
5. A. Berarducci. Infinite λ-calculus and non-sensible models. In *Logic and algebra (Pontignano, 1994)*, pages 339–377. Dekker, New York, 1996.
6. M. Coppo, M. Dezani-Ciancaglini, and M. Zacchi. Type theories, normal forms, and D_∞-lambda-models. *Information and Computation*, 72(2):85–116, 1987.
7. H.B. Curry and R. Feys. *Combinatory Logic*, volume I. North-Holland Publishing Co., Amsterdam, 1958.
8. N. Dershowitz, S. Kaplan, and D.A. Plaisted. Rewrite, rewrite, rewrite, rewrite, rewrite, ... *Theoretical Computer Science*, 83(1):71–96, 1991.
9. M. Dezani-Ciancaglini, P. Severi, and F.J. de Vries. Infinitary lambda calculus and discrimination of Berarducci trees. *TCS*, 200X. Draft available on `www/mcs.le.ac.uk/~ferjan`.
10. J. M. E. Hyland. A survey of some useful partial order relations on terms of the lambda calculus. In C. Böhm, editor, *Lambda Calculus and Computer Science Theory*, volume 37 of *LNCS*, pages 83–93. Springer-Verlag, 1975.
11. J.R. Kennaway and F.J. de Vries. Infinitary rewriting. In TeReSe, editor, *Term Rewriting Systems*, Cambridge Tracts in Theoretical Computer Science. Cambridge University Press, 200X. Draft available on `www/mcs.le.ac.uk/~ferjan`.
12. J.R. Kennaway, J. W. Klop, R. Sleep, and F.J. de Vries. Infinitary lambda calculus. *TCS*, 175(1):93–125, 1997.
13. J. H. Morris Jr. *Lambda calculus models of programming languages*. PhD thesis, M.I.T., 1968.
14. R. Nakajima. Infinite normal forms for the λ-calculus. In C. Böhm, editor, *Lambda calculus and Computer Science Theory*, volume 37 of *LNCS*, pages 62–82. Springer-Verlag, 1975.
15. V. van Oostrom. Developing developments. *TCS*, 175(1):159–181, 1997.
16. D. Park. The Y-combinator in Scott's lambda-calculus models (revised version). Technical report, Department of Computer Science, University of Warwick, 1976. Copy available at `www.dcs.qmul.ac.uk/~ae/papers/others/ycslcm/`.
17. G.D. Plotkin. Set-theoretical and other elementary models of the λ-calculus. *TCS*, 121:351–409, 1993.
18. C. P. Wadsworth. The relation between computational and denotational properties for Scott's D_∞-models of the lambda-calculus. *SIAM J. Comput.*, 5(3):488–521, 1976.

A Weak Calculus with Explicit Operators for Pattern Matching and Substitution

Julien Forest

Laboratoire de Recherche en Informatique (CNRS URM 8623),
Bât 490, Université Paris-Sud,
91405 Orsay CEDEX, France.
forest@lri.fr.

Abstract. In this paper we propose a **Weak Lambda Calculus** called λP_w having explicit operators for **Pattern Matching** and **Substitution**. This formalism is able to specify functions defined by cases via pattern matching constructors as done by most modern functional programming languages such as OCAML. We show the main property enjoyed by λP_w, namely subject reduction, confluence and strong normalization.

1 Introduction

In this paper we propose a **Weak Lambda Calculus** with **Pattern Matching** and **Explicit Substitution** called λP_w. The calculus λP_w is inspired by calculi of explicit substitutions [ACCL91, BBLRD96, BR95, Kes96] and by calculi of patterns [KPT96, CK99, CK]. The weak nature of λP_w allows us to denote variables by names without requiring α-conversion to implement correctly the notion of reduction.

Theoretical study of functional programming has been enriched by the introduction of typed λ-calculi, explicit substitutions [ACCL91, BBLRD96, BR95, Kes96] and pattern matching [KPT96, CK99, CK]. These three notions are the main ingredients of the formalism we propose in this paper to model typed functional languages with function definition by cases.

In the early thirties, Church proposed the λ-*calculus* as a general theory of functions and logic. Typed versions of λ-calculus were then defined by Curry and Church, they became the standard theoretical tool for defining and implementing *typed functional programming languages*. There is however an important gap between λ-calculus and modern functional languages:

- On one hand, the operation of substitution is not incorporated in the language level but is left as a meta-operation.
- On the other hand, most popular functional languages (resp. proof assistants) allow the definition of functions (resp. proofs) by cases via pattern-matching mechanisms, while λ-calculus does not incorporate at all these constructs.

The first problem is solved by incorporating the so-called explicit substitutions into the language used to implement functional programming. To do this, one simply adds

S. Tison (Ed.): RTA 2002, LNCS 2378, pp. 174–191, 2002.

a new construction to denote substitution and new reduction rules to describe the interaction between substitution and other constructors of the language. Many calculi with explicit substitutions [ACCL91, BBLRD96, BR95, Kes96] have been proposed in the literature, operational and logic properties of these calculi were extensively studied [Les94, CHL96, Kes96].

The second problem is solved by allowing abstractions of function not only with respect to variables but also with respect to patterns. Thus, the form of the arguments of a given function can be specified in a very precise form; for instance, a term having the form $\lambda\langle x, y\rangle.M$ specify that the expected argument is a pair.

In the early nineties, Kesner, Tannen and Puel [KPT96] proposed a **Calculus of Pattern Matching** as a tool of theoretical study of pattern matching à la ML. In 1999, Kesner and Cerrito [CK99, CK] refined the ideas in [KPT96] and defined the calculus TPC_{ES} as a formalism with **explicit pattern matching** and **explicit substitution**. Other languages with explicit pattern matching, such as for example the ρ-calculus [CK98, CKL01], were recently proposed in the literature to model other programming paradigms.

The calculus presented in this paper, called λP_w, is a calculus with explicit pattern matching and explicit substitutions. This calculus is not designed as a *user level language* but as the *output calculus* of a pattern matching compilation algorithm. Such an algorithm is supposed to take a pattern matching function definition and to return an equivalent one where all the ambiguities between overlapping patterns have disappeared and where incomplete patterns definitions have been detected and completed. Such an hypothesis is not really restrictive since all the functional languages with pattern matching features ([Obj]) apply such an algorithm before evaluating programs.

The calculus λP_w is based on [CK99, CK], but have the following new features:

- λP_w is a **weak calculus** of explicit substitutions, that is, functions are lazyly evaluated. To implement this correctly, substitutions are not allowed to cross lambda constructors - so that α-conversion is no more needed to achieve correct reduction of terms - and composition of substitutions is incorporated into the substitution language in order to guarantee confluence. The syntax of λP_w is based on the weak σ-calculus with names [CHL96] in contrast to TPC_{ES}, which is a **strong** calculus based on the substitution formalism with names called x [BR95, Blo95].
- In contrast to TPC_{ES} which treats "ordinary" substitutions explicitly but the so-called "sum" substitutions implicitly [1], λP_w treats *all* the substitutions as explicit. This choice makes the formalism (typing rules and typing reduction rules) and the proofs more involved than those in [CK99], but results in a complete and self-contained formalism which is able to describe different implementations of functional languages with pattern matching.

The formalism that we present in this paper enjoys all the classical properties of typed λ-calculi.

- It is **confluent** on all terms.
- It has the **subject reduction** property.
- It is **strongly normalizing** on all well-typed terms.

[1] In fact, the first version presented in [CK99] treats sum substitutions explicitly, but the revised and corrected version in [CK] only keeps ordinary substitution as an explicit operation by moving sum substitution to the meta-level.

The paper is organized as follows. We will first give in Section 2 a formal definition of λP_w and give some basic properties such as *preservation of free variables* by reduction and confluence on all ground terms. We then introduce in Section 3 a typing system for λP_w and show that λP_w enjoys the *subject reduction* property. We finally show *strong normalization* of well-typed λP_w-terms in Section 4 before the conclusion given in Section 5.

2 Definition of λP_w

We first define the *raw* expressions of the calculus λP_w by giving three different sets to denote respectively raw terms, raw substitutions and raw sum terms. The notion of raw expression is refined by first defining the set of free variables of any raw expression which allows us to define (well-formed) expressions such as terms, substitutions and sum terms. Reduction rules of λP_w are given in Figure 1. These rules are showed to preserve free variables of expressions.

We fix two distinct infinite sets of variables: the set of *usual variables*, noted $x, y, z, \ldots$, which are used to denote ordinary terms, and the set of *sum variables*, noted $\xi, \psi, \ldots$, which are used to denote disjunction. We also fix two constants $\mathtt{L}$ and $\mathtt{R}$ and we use the notation $\mathtt{K}$ to denote indistinctly one or the other one. We will also use the notation $\mathtt{T}$ to denote indistinctly $\mathtt{L}$, $\mathtt{R}$ or a sum variable.

Types of λP_w are given by the following grammar:

$$
\begin{array}{lll}
(\textbf{Types})\ A ::= & \iota & \text{Base type} \\
 & \mid A \times A & \text{Product Type} \\
 & \mid A + A & \text{Sum Type} \\
 & \mid A \to A & \text{Functional Type}
\end{array}
$$

Patterns of λP_w are given by the following grammar:

$$
\begin{array}{lll}
(\textbf{Pattern})\ P := & \_ & \text{Wildcard} \\
 & \mid x & \text{Variable} \\
 & \mid \langle P, P \rangle & \text{Pair} \\
 & \mid @(P, P) & \text{Contraction} \\
 & \mid (P \mid_\xi P) & \text{Sum}
\end{array}
$$

The notations $\_$, x and $\langle P, Q \rangle$ are standard while the notation $@(P, Q)$ is similar (indeed more general) to the as constructor of Ocaml [Obj]. The pattern $(P \mid_\xi Q)$ is used to specify two different structures P and Q (of types A and B) corresponding to a pattern of sum type $A + B$. The sum variable ξ appearing in a pattern $(P \mid_\xi Q)$ is used to propagate the result of any matching w.r.t the pattern $(P \mid_\xi Q)$ all along the term where this variable occurs.

Raw Substitutions of λP_w are given by the following grammar:

$$
\begin{array}{lll}
(\textbf{Raw Substitution})\ s ::= & id & \text{Identity} \\
 & \mid (x/M).s & \text{Cons\_usual\_var} \\
 & \mid (\xi^{P,A}/\mathtt{K}).s & \text{Cons\_sum\_var} \\
 & \mid s \circ s & \text{Concatenation}
\end{array}
$$

In order to mark which branch of a sum pattern has been chosen we use special syntax that we called *sum terms*. A sum term is either a constant (there is one for each possible choice), a sum variable (no choice has been made), or a substituted sum variable (the full evaluation has not been made).

$$
\begin{array}{lll}
\textbf{(Raw Sum Terms)}\ \Xi ::= \xi & & \text{Sum Variable} \\
\quad \mid \mathtt{L} & & \text{Left Constant} \\
\quad \mid \mathtt{R} & & \text{Right Constant} \\
\quad \mid \xi[s] & & \text{Sum Substitution}
\end{array}
$$

We are now able to introduce λP_w-terms. The main difference between λ-calculus and pattern calculi is that the notation $\lambda x.M$ is generalized to $\lambda P.M$ where P is a pattern as given by the previous grammar. Thus, λP_w-terms are given by the following grammar:

$$
\begin{array}{lll}
\textbf{(Raw Terms)}\ M ::= & x & \text{Usual Variable} \\
 & \mid (M\ N) & \text{Application} \\
 & \mid \langle M, M\rangle & \text{Pair} \\
 & \mid \mathtt{inl}_B(M) & \text{Left injection} \\
 & \mid \mathtt{inr}_A(M) & \text{Right injection} \\
 & \mid [M \mid_\xi M] & \text{Case} \\
 & \mid [M \mid_\Xi^s M] & \text{Frozen Case} \\
 & \mid \lambda P\!:\!A.M & \text{Abstraction} \\
 & \mid M[s] & \text{Closure}
\end{array}
$$

All along the paper we may sometimes omit types from expressions in order to simplify the notation, but expressions are supposed to be as defined by this grammar. A *Case* constructor of the form $[M \mid_\xi N]$ is used to specify two different terms M and N corresponding respectively to two different patterns P and Q of a sum pattern $(P \mid_\xi Q)$ appearing somewhere in the program. The communication between the case constructor $[M \mid_\xi N]$ and its corresponding sum pattern $(P \mid_\xi Q)$ is achieved via the sum variable ξ. The introduction of the Frozen Case constructor is purely technical, the idea is to prevent reduction of the sub-term M (resp. N) inside a case constructor of the form $[M \mid_\xi N]$ where a left (resp. right) choice has been already made.

Example 1. A simple λP_w-term is $\lambda(x \mid_\xi y)\!:\!A+B.[\lambda y'\!:\!B.\langle x, y'\rangle \mid_\xi \lambda x'\!:\!A.\langle x', y\rangle]$. For a more interesting example let us suppose that we have encoded the recursive type nat as a sum type, and that $(0 \mid_\xi S\ m)$ is a pattern of type nat representing either 0 or a positive natural number of the form $S\ m$. We refer the reader to [CK99] for more examples and more details about encoding of recursive types in the formalism of pattern calculi. Indeed, the following Ocaml [Obj] term:

```
match n with
| 0      -> 0
| (S m) -> m
```

is given by the λP_w-term $\lambda(0 \mid_\xi (S\ m)) : nat.[0 \mid_\xi m]$.

Definition 21 (Raw expression) A *raw expression* is either a raw term, a raw substitution or a raw sum term.

As usually done in λ-calculus we will work modulo α-conversion. This notion must be defined with care since bound variables are not only *all* the variables appearing in *complex* patterns, but also, *all* the variables bound by substitutions.

Definition 22 (Binding Variables) The set of *Binding Variables* of a pattern (resp. a substitution) is defined as:

$$\begin{array}{ll} BVar(\_) &= \emptyset \\ BVar(x) &= \{x\} \\ BVar(\langle P, Q\rangle) &= BVar(P) \cup BVar(Q) \\ BVar((P \mid_\xi Q)) &= BVar(P) \cup BVar(Q) \cup \{\xi\} \\ BVar(@(P,Q)) &= BVar(P) \cup BVar(Q) \end{array} \qquad \begin{array}{ll} BVar(id) &= \emptyset \\ BVar((x/M).s) &= BVar(s) \cup \{x\} \\ BVar((\xi^P/\mathtt{K}).s) &= BVar(s) \cup BVar(P) \cup \{\xi\} \\ BVar(s \circ t) &= BVar(s) \cup BVar(t) \end{array}$$

We have for example, $BVar((x \mid_\xi y)) = \{\xi, x, y\}$ and $BVar((x/M).(\xi^y/\mathtt{L}).id) = \{x, y, \xi\}$.

Definition 23 (Free Variables) The set of *Free Variables* of an expression e is given by:

$$\begin{array}{ll} FV(x) &= \{x\} \\ FV(\mathtt{inl}(M)) = FV(\mathtt{inr}(M)) &= FV(M) \\ FV(M\,N) = FV(\langle M, N\rangle) &= FV(M) \cup FV(N) \\ FV(\lambda P.M) &= FV(M) \setminus BVar(P) \\ FV(M[s]) &= (FV(M) \setminus BVar(s)) \cup FV(s) \\ FV([M \mid_\xi N]) &= FV(M) \cup FV(N) \cup \{\xi\} \\ FV([M \mid_\Xi^s N]) &= ((FV(M) \cup FV(N)) \setminus BVar(s)) \cup FV(s) \cup FV(\Xi) \\ FV(id) = FV(\mathtt{K}) &= \emptyset \\ FV((x/M).s) &= FV(M) \cup FV(s) \\ FV((\xi^P/\mathtt{K}).s) &= FV(s) \\ FV(s \circ t) &= (FV(s) \setminus BVar(t)) \cup FV(t) \\ FV(\xi) &= \{\xi\} \\ FV(\xi[s]) &= (\{\xi\} \setminus BVar(s)) \cup FV(s) \end{array}$$

Thus for example, $FV(\lambda(x \mid_\xi y).[x \mid_\psi t]) = \{\psi, t\}$ and $FV([x \mid_{\xi[(x/t).id]}^{(\xi^y/\mathtt{L}).id} y]) = \{x, t, \xi\}$.

We define the set of *free sum variables* (FSV) of a raw expression e as the set of *sum variables* of e which are in $FV(e)$.

Definition 24 (Bound variables) The *Bound Variables* of a raw expression e are those variables appearing in e but not free in e.

We are now ready to define α-conversion on λP_w-expressions as simply renaming of bound variables. Thus, for example $\lambda(x \mid_\xi x).x$ and $\lambda(y \mid_\psi y).y$ are α-equivalent, but neither $\lambda(x \mid_\xi x).y$ and $\lambda(y \mid_\xi y).y$ nor $\lambda(x \mid_\xi x).x$ and $\lambda(x \mid_\xi y).x$ are α-equivalent.

We are now ready to introduce the reduction rules which are given in Figure 1.

The **Pattern matching** rules are the rules implementing the pattern matching.

The **Propagation of Substitutions, Substitutions and Variables and Constants, Substitutions and Composition** rules are a natural extension of those of the σ-calculus.

The **Case** rules explain the mechanism to distribute a substitution s with respect to a case term $[M \mid_\xi N]$ which consists in:

- We first transform the case term into a frozen case term using the rule $(Freeze)$.
- We then treat the part $\xi[s]$ of the obtained frozen case term until a result (*i.e.* the variable ξ, the constants L or R) is obtained.
- We can then distribute the substitution in the frozen case term using one of the rules $(Left)$, $(Right)$ or (Xi).

The reduction system generated by the rules Abs_id, Abs_pair, Abs_contr, Abs_left, Abs_right, Abs_var and Abs_wild is used to implement the pattern matching operation and is noted by $\longrightarrow_P$. All the other rules generate the reduction system used to implement the behavior of substitution and is noted by $\longrightarrow_{es}$. The reduction relation $\longrightarrow_{\lambda P_w}$ is generated by $\longrightarrow_{es} \cup \longrightarrow_P$. To simplify the notation we may simply note $\longrightarrow$ for $\longrightarrow_{\lambda P_w}$ in the rest of the paper.

Example 2. We show one way to propagate the substitution $s = (x/M_3).(\xi^{P.A}/\mathtt{L}).id$ inside the term $M = [M_1 \mid_\xi M_2]$.

- First of all we reduce the case term into a frozen case term by $M[s] \longrightarrow_{Freeze} [M_1 \mid^s_{\xi[s]} M_2]$
- We then "evaluate" the part $\xi[s]$: $\xi[s] \longrightarrow_{Sub\_sum\_var_4} \xi[(\xi^{P.A}/\mathtt{L}).id] \longrightarrow_{Sub\_sum\_var_2} \mathtt{L}$
- Thus we have $[M_1 \mid^s_{\xi[s]} M_2] \longrightarrow^+ [M_1 \mid^s_{\mathtt{L}} M_2]$, and thus applying the rule $(Left)$, we obtain $M[s] \longrightarrow^+ M_1[s]$.

Remark 1. Let s and s' be raw substitutions such that $s \longrightarrow s'$, then $BVar(s) = BVar(s')$.

However, the reduction system in Figure 1 is not really correct in the sense that $\longrightarrow_{\lambda P_w}$ does *not* preserve free variables. This is shown by the following example:

Example 3. Let M be a term such that $FV(M) = \emptyset$ and let $U = (\lambda(x \mid_\xi y).x\,\mathtt{inr}(M))$. Then $U \longrightarrow^*_{\lambda P_w} x[(\xi^x/\mathtt{R}).id] \longrightarrow_{\lambda P_w} x$ and $FV(U) = \emptyset$ but $FV(x) = \{x\}$.

In order to avoid this problem we restrict the set of raw expressions in order to guarantee that no new free variable does appear along reduction sequences. The notion of *acceptable* expression, or simply *expression*, is obtained via the introduction of the following concepts:

Definition 25 (Localized Free Variables) Given a sum variable ξ, a sum constant K and a raw expression e, we define the set of *localized free variables* of e w.r.t. ξ and K, written as $FV^{\mathtt{K}}_\xi(e)$, as the subset of $FV(e)$ define exactly as for $FV(e)$ except for the following cases:

$$
\begin{array}{lll}
FV^{\mathtt{L}}_\xi([M \mid_\xi N]) & = FV^{\mathtt{L}}_\xi(M) \cup \{\xi\} & \\
FV^{\mathtt{R}}_\xi([M \mid_\xi N]) & = FV^{\mathtt{R}}_\xi(N) \cup \{\xi\} & \\
FV^{\mathtt{L}}_\xi([M \mid^s_\xi N]) & = FV^{\mathtt{L}}_\xi(M[s]) \cup \{\xi\} & \\
FV^{\mathtt{R}}_\xi([M \mid^s_\xi N]) & = FV^{\mathtt{R}}_\xi(N[s]) \cup \{\xi\} & \\
FV^{\mathtt{L}}_\xi([M \mid^s_{\xi[t]} N]) & = FV^{\mathtt{L}}_\xi(M[s]) \cup \{\xi\} & \text{if } \xi \notin BVar(t) \\
FV^{\mathtt{R}}_\xi([M \mid^s_{\xi[t]} N]) & = FV^{\mathtt{R}}_\xi(M[s]) \cup \{\xi\} & \text{if } \xi \notin BVar(t)
\end{array}
$$

Start Rule			
$(\lambda P.M)\ N$	$\longrightarrow$	$(\lambda P.M)[id]\ N$	(Abs_id)
Pattern Matching			
$(\lambda\langle P_1, P_2\rangle.M)[s]\ \langle N_1, N_2\rangle$	$\longrightarrow$	$((\lambda P_1.\lambda P_2.M)[s]\ N_1)\ N_2$	(Abs_pair)
$(\lambda @(P_1, P_2).M)[s]\ N$	$\longrightarrow$	$((\lambda P_1.\lambda P_2.M)[s]N)\ N$	(Abs_contr)
$(\lambda(P_1 \mid_\xi P_2).M)[s]\ \mathtt{inl}(N)$	$\longrightarrow$	$(\lambda P_1.M)[(\xi^{P_2}/\mathtt{L}).s]\ N$	(Abs_left)
$(\lambda(P_1 \mid_\xi P_2).M)[s]\ \mathtt{inr}(N)$	$\longrightarrow$	$(\lambda P_2.M)[(\xi^{P_1}/\mathtt{R}).s]\ N$	(Abs_right)
$(\lambda x.M)[s]\ N$	$\longrightarrow$	$M[(x/N).s]$	(Abs_var)
$(\lambda\_.M)[s]\ N$	$\longrightarrow$	$M[s]$	(Abs_wild)
Case			
$[M \mid_\xi N][s]$	$\longrightarrow$	$[M \mid^{s}_{\xi[s]} N]$	$(Freeze)$
$[M \mid^{s}_{\mathtt{L}} N]$	$\longrightarrow$	$M[s]$	$(Left)$
$[M \mid^{s}_{\mathtt{R}} N]$	$\longrightarrow$	$N[s]$	$(Right)$
$[M \mid^{s}_{\xi} N]$	$\longrightarrow$	$[M[s] \mid_\xi N[s]]$	(Xi)
Propagation of Substitutions			
$(MN)[s]$	$\longrightarrow$	$M[s]N[s]$	(Sub_app)
$\mathtt{inl}(M)[s]$	$\longrightarrow$	$\mathtt{inl}(M[s])$	(Sub_left)
$\mathtt{inr}(M)[s]$	$\longrightarrow$	$\mathtt{inr}(M[s])$	(Sub_right)
$\langle M_1, M_2\rangle[s]$	$\longrightarrow$	$\langle M_1[s], M_2[s]\rangle$	(Sub_pair)
Substitutions and Variables and Constants			
$x[id]$	$\longrightarrow$	x	(Sub_var_1)
$x[(x/N).s]$	$\longrightarrow$	N	(Sub_var_2)
$y[(x/N).s]$	$\longrightarrow$	$y[s]$ if $y \neq x$	(Sub_var_3)
$x[(\xi^P/\mathtt{K}).s]$	$\longrightarrow$	$x[s]$	(Sub_var_4)
$\xi[id]$	$\longrightarrow$	ξ	$(Sub\_sum\_var_1)$
$\xi[(\xi^P/\mathtt{K}).s]$	$\longrightarrow$	$\mathtt{K}$	$(Sub\_sum\_var_2)$
$\xi[(\psi^P/\mathtt{K}).s]$	$\longrightarrow$	$\xi[s]$ if $\xi \neq \psi$	$(Sub\_sum\_var_3)$
$\xi[(x/M).s]$	$\longrightarrow$	$\xi[s]$	$(Sub\_sum\_var_4)$
Substitutions and Composition			
$M[s][t]$	$\longrightarrow$	$M[s \circ t]$	(Sub_clos)
$(s \circ t) \circ u$	$\longrightarrow$	$s \circ (t \circ u)$	(Sub_ass_env)
$((x/M).s) \circ t$	$\longrightarrow$	$(x/M[t]).(s \circ t)$	(Sub_concat_1)
$((\xi^P/K).s) \circ t$	$\longrightarrow$	$(\xi^P/K).(s \circ t)$	(Sub_concat_2)
$id \circ s$	$\longrightarrow$	s	(Sub_id)

Fig. 1. Reduction Rules for λP_w

Intuitively, $FV_\xi^{\mathtt{L}}(e)$ (resp. $FV_\xi^{\mathtt{R}}(e)$) contains all the free variables of e except those that are on the right (resp. left) part of the sum terms rooted by the sum variable ξ. Thus, for example, $FV_\xi^{\mathtt{L}}(\lambda(x \mid_\xi y).[x \mid_\xi t]) = \emptyset$, $FV_\xi^{\mathtt{L}}(\lambda y.[x \mid_\xi t]) = \{x\}$, and $FV_\xi^{\mathtt{K}}(x[(\xi^x/\mathtt{L}).id]) = \{x\}$

We are finally ready to define the notion of *acceptable expression* which will avoid the example of creation of new free variables introduced before.

Definition 26 (Acceptable Expression) The raw expression e is said to be *acceptable* (or just called an *expression*) iff $Acc(e)$, where $Acc(\;)$ is the least congruence on expressions such that every variable and every constant is acceptable and also the following requirements hold:

$$\begin{array}{lll}
Acc(\psi[s]) & \text{if } \psi \notin BVar(Q) & \forall(\xi^Q/\mathtt{K}) \in s \\
Acc(M[s]) & \text{if } FV_\xi^{\mathtt{K}}(M) \cap BVar(Q) = \emptyset & \forall(\xi^Q/\mathtt{K}) \in s \\
Acc(s \circ t) & \text{if } FV_\xi^{\mathtt{K}}(s) \cap BVar(Q) = \emptyset & \forall(\xi^Q/\mathtt{K}) \in t \\
Acc(\lambda P.M) & \text{if } (FV_\xi^{\mathtt{R}}(M) \cap BVar(Q_1)) \cup (FV_\xi^{\mathtt{L}}(M) \cap BVar(Q_2)) = \emptyset & \forall(Q_1 \mid_\xi Q_2) \in P \\
Acc([M \mid_\xi^s N]) & \text{if } Acc(M[s]) \text{ and } Acc(N[s]) & \\
Acc([M \mid_{\xi[t]}^s N]) & \text{if } Acc(M[s]) \text{ and } Acc(N[s]) \text{ and } Acc(\xi[t]) & \text{if } (\xi^Q/\mathtt{K}) \notin t \\
Acc([M \mid_{\xi[t]}^s N]) & \text{if } Acc(M[s]) \text{ and } Acc(\xi[t]) & \text{if } (\xi^Q/\mathtt{L}) \in t \\
Acc([M \mid_{\mathtt{L}}^s N]) & \text{if } Acc(M[s]) & \\
Acc([M \mid_{\xi[t]}^s N]) & \text{if } Acc(N[s]) \text{ and } Acc(\xi[t]) & \text{if } (\xi^Q/\mathtt{R}) \in t \\
Acc([M \mid_{\mathtt{R}}^s N]) & \text{if } Acc(N[s]) &
\end{array}$$

Thus for example, the term $\lambda(x \mid_\xi y).[x \mid_\xi t]$ is acceptable while $\lambda(x \mid_\xi y).[x \mid_\psi t]$ is not acceptable. Indeed $FV_\xi^{\mathtt{R}}([x \mid_\psi t]) = \{x, t, \xi\}$ and $BVar(x) = \{x\}$ and thus by definition of $Acc()$ for abstractions we have that $\lambda(x \mid_\xi y).[x \mid_\psi t]$ is not acceptable. The reader may also remark that the terms U and $x[(\xi^x/\mathtt{L}).id]$ given in Example 3 are neither acceptable.

We have to show now that this new notion of acceptable expression recently introduced is correct to prevent creation of new free variables along reduction sequences, that is, reduction preserves acceptable expressions and free variables.

Lemma 21 If e is an expression and $e \longrightarrow_{\lambda P_w} e'$ then

- e' is an expression
- $FV(e') \subseteq FV(e)$

We are now ready to state *confluence* of λP_w which guarantees that normal forms are unique.

Theorem 22 (Confluence for λP_w) λP_w is confluent.

Proof. The proof technique used to show the confluence property is the same as in [CHL96]. By lack of space we can not give here all the details of this proof but we refer the interested reader to [For02] for full technical explanations. The idea/scheme of the proof is the following.

1. We first show *strong normalization* and *confluence* of the system $\longrightarrow_{es}$. This allows us to work with es-normal forms of λP_w-expressions.
2. We then define a reduction system $\longrightarrow_{aux}$ on es-normal forms and we show that $\longrightarrow_{aux}$ is confluent.
3. We finally conclude by Hardin's Interpretation Lemma [CHL96] which relates confluence of $\longrightarrow_{\lambda P_w}$ with confluence of $\longrightarrow_{aux}$.

3 A Typing System for λP_w

As λ-calculus is strictly contained in λP_w, then λP_w is not strongly normalizing. In order to obtain strong normalization of λP_w we define a typing system which is capable of associating types to terms, sum terms and substitutions in a given environment. While typing systems have been already studied for calculi with explicit substitutions [DG01] and also for calculi with patterns [KPT96, CK], no formalism in the literature exists to correctly type explicit choice sum terms. The typing system that we present in this section is shown to have the *subject reduction* property, that is, typing is preserved under reduction. The notion of acceptable expression (Definition 26) is essential to guarantee such a property.

We restrict now our attention to a special kind of patterns called *acceptable*. For that, we say that a pattern P is *linear* if and only if every variable occurs at most once in P. We define a *type environment* to be a pair $\Phi; \Gamma$ such that Φ is a *sum environment* defined as a set of pairs of the form $\xi : \mathtt{K}$ and Γ is a *pattern environment* defined as a set of *typed patterns*, which are pairs of the form $P : A$. We say that a type environment $\Phi; \Gamma$ is *linear* if every variable occurs at most once in $\Phi; \Gamma$.

Definition 31 (Acceptable Patterns and Environments) The set of *acceptable patterns of type* A, denoted by $\mathcal{AP}(A)$, is defined to be the smallest set of *linear* patterns verifying the following properties: $\_ \in \mathcal{AP}(A)$; $x \in \mathcal{AP}(A)$ for any variable x; $@(P, Q) \in \mathcal{AP}(A)$ if $P \in \mathcal{AP}(A)$ and $Q \in \mathcal{AP}(A)$; $\langle P, Q \rangle \in \mathcal{AP}(B \times C)$ and $(P \mid_\xi Q) \in \mathcal{AP}(B + C)$ if $P \in \mathcal{AP}(B)$ and $Q \in \mathcal{AP}(C)$. The role of the notion of "acceptable patterns" is to prevent the $(wildcard)$ typing rule (corresponding to $(weakening)$ in logic) to introduce meaningless pattern expressions. This notion extends naturally to environments by defining $\Phi; \Gamma$ to be *acceptable* if and only if each pattern appearing in $\Phi; \Gamma$ is acceptable.

Thus for example the pattern $\langle x, y \rangle$ is linear but *is not* in $\mathcal{AP}(A + B)$ since a pair pattern is not compatible with a sum type.

We now introduce the typing rules for terms and substitutions (resp. for sum terms) in Figure 2 (resp. Figure 3) assuming that all the patterns and type environments appearing in these rules are acceptable.

In the rules $(Case_1)$ and $(Frozencase_1)$, ξ is a fresh sum variable and $(P \mid_\xi Q)$ is linear. In the rule $(Wildcard)$, $\Phi; P : A, \Gamma$ has to be linear. In the rule $(Proj_1)$, all the x_i's are distinct usual variables. In the rules $(Proj_2)$ and $(Nproj)$, all the ξ_i's are distinct sum variables. In the rule $(Nproj)$ ξ does not appear in Γ. In the rule (App), we require that N does not contain free sum variables. In the rules (Sub_term), $(Frozencase_1)$ and $(Frozencase_2)$ we require that s does not contain free sum variables. In the rule (Sub_cons_1) we require that M does not contain free sum variables. In the rules (Sub_cons_1), (Sub_cons_2), all contexts have to be linear.

We say that the term M (resp. the substitution s and the sum term Ξ) *has type* A (resp. *co-environment* $\Phi'; \Gamma'$ and *sum type* $\mathtt{T}$) in a type environment $\Phi; \Gamma$ if and only if there is a type derivation ending with $\Phi; \Gamma \vdash M : A$ (resp. $\Phi; \Gamma \vdash s \triangleright \Phi'; \Gamma'$ and $\Phi; \Gamma \vdash \Xi \rightsquigarrow \mathtt{T}$). We say that the term M (resp. the substitution s and the sum term Ξ) is *well-typed* in $\Phi; \Gamma$ if and only if there is a type A such that M has type A in $\Phi; \Gamma$ (resp there is an environment $\Phi'; \Gamma'$ such that s has is co-environment $\Phi'; \Gamma'$ in $\Phi; \Gamma$ and there is a sum type $\mathtt{T}$ such that Ξ has sum type $\mathtt{T}$ in $\Phi; \Gamma$). We will make an abuse of notation

$$\frac{}{\Phi; x_1 : A_1, \ldots, x_n : A_n \vdash x_i : A_i}\ (Proj_1)$$

$$\frac{\Phi; \Gamma \vdash M : A}{\Phi; \Gamma \vdash \mathtt{inl}_B(M) : A + B}\ (+Right_1) \qquad \frac{\Phi; \Gamma \vdash M : B}{\Phi; \Gamma \vdash \mathtt{inr}_A(M) : A + B}\ (+Right_2)$$

$$\frac{\Phi; \Gamma \vdash \xi \rightsquigarrow \xi \quad \Phi; P : B, \Gamma \vdash M : A \quad \Phi; Q : C, \Gamma \vdash N : A}{\Phi; (P \mid_\xi Q) : B + C, \Gamma \vdash [M \mid_\xi N] : A}\ (Case_1)$$

$$\frac{\Phi; \Gamma \vdash \xi \rightsquigarrow \mathtt{K} \quad \Phi; \Gamma \vdash M : A \quad \Phi; \Gamma \vdash N : A}{\Phi; \Gamma \vdash [M \mid_\xi N] : A}\ (Case_2)$$

$$\frac{\Phi; \Gamma \vdash \Xi \rightsquigarrow \xi \quad \Phi; P : A; \Gamma \vdash M[s] : C \quad \Phi; Q : B; \Gamma \vdash N[s] : C}{\Phi; (P \mid_\xi Q) : A + B, \Gamma \vdash [M \mid^s_\Xi N] : C}\ (Frozencase_1)$$

$$\frac{\Phi; \Gamma \vdash \Xi \rightsquigarrow \mathtt{K} \quad \Phi; \Gamma \vdash M[s] : A \quad \Phi; \Gamma \vdash N[s] : A}{\Phi; \Gamma \vdash [M \mid^s_\Xi N] : A}\ (Frozencase_2)$$

$$\frac{\Phi; P : A, Q : B, \Gamma \vdash M : C}{\Phi; \langle P, Q \rangle : A \times B, \Gamma \vdash M : C}\ (\times Left) \qquad \frac{\Phi; \Gamma \vdash M : A \quad \Phi; \Gamma \vdash N : B}{\Phi; \Gamma \vdash \langle M, N \rangle : A \times B}\ (\times Right)$$

$$\frac{\Phi; P : A, \Gamma \vdash M : B}{\Phi; \Gamma \vdash \lambda P : A.M : A \to B}\ (\to Right) \qquad \frac{\Phi; \Gamma \vdash M : A \to B \quad \Phi; \Gamma \vdash N : A}{\Phi; \Gamma \vdash (MN) : B}\ (App)$$

$$\frac{\Phi; P : A, Q : A, \Gamma \vdash M : B}{\Phi; @(P, Q) : A, \Gamma \vdash M : B}\ (Layered) \qquad \frac{\Phi; \Gamma \vdash M : B}{\Phi; P : A, \Gamma \vdash M : B}\ (Wildcard)$$

$$\frac{}{\Phi; \Gamma \vdash id \rhd \Phi; \Gamma}\ (Sub\_axiom) \qquad \frac{\Phi; \Gamma \vdash t \rhd \Phi'; \Gamma' \quad \Phi'; \Gamma' \vdash s \rhd \Phi''; \Gamma''}{\Phi; \Gamma \vdash s \circ t \rhd \Phi''; \Gamma''}\ (Sub\_concat)$$

$$\frac{\Phi; \Gamma \vdash M : A \quad \Phi; \Gamma \vdash s \rhd \Phi'; \Gamma'}{\Phi; \Gamma \vdash (x/M).s \rhd \Phi'; x : A, \Gamma'}\ (Sub\_cons_1) \qquad \frac{\Phi; \Gamma \vdash s \rhd \Phi'; \Gamma'}{\Phi; \Gamma \vdash (\xi^{P,A}/\mathtt{K}).s \rhd \xi : \mathtt{K}, \Phi'; P : A, \Gamma'}\ (Sub\_cons_2)$$

$$\frac{\Phi; \Gamma \vdash s \rhd \Phi'; \Gamma' \quad \Phi'; \Gamma' \vdash M : A}{\Phi; \Gamma \vdash M[s] : A}\ (Sub\_term)$$

Fig. 2. Typing Rules for Terms and Substitutions

by writing $\Phi; \Gamma \vdash M : A$ (resp. $\Phi; \Gamma \vdash s \rhd \Phi'; \Gamma'$ and $\Phi; \Gamma \vdash \Xi \rightsquigarrow \mathtt{T}$) to indicate that M (resp. s and Ξ) has type A in $\Phi; \Gamma$.

$$\frac{}{\xi_1 : \mathtt{K}_1, \ldots, \xi_m : \mathtt{K}_m; \Gamma \vdash \xi_j \rightsquigarrow \mathtt{K}_j} (Proj_2) \qquad \frac{\text{if } \forall i, \xi \neq \xi_i}{\xi_1 : \mathtt{K}_1, \ldots, \xi_m : \mathtt{K}_m; \Gamma \vdash \xi \rightsquigarrow \xi} (Nproj)$$

$$\frac{}{\Phi; \Gamma \vdash \mathtt{L} \rightsquigarrow \mathtt{L}} (\mathtt{L}) \qquad \frac{}{\Phi; \Gamma \vdash \mathtt{R} \rightsquigarrow \mathtt{R}} (\mathtt{R})$$

$$\frac{\Phi; \Gamma \vdash s \triangleright \Phi'; \Gamma' \qquad \Phi'; \Gamma' \vdash \xi \rightsquigarrow \mathtt{T}}{\Phi; \Gamma \vdash \xi[s] \rightsquigarrow \mathtt{T}} (Sub\_sum)$$

Fig. 3. Typing Rules for Sum Terms

First of all, we remark that for any substitution s, the co-environment of s contains its typing environment. This observation can be formalized as follows:

Remark 2. If $\Psi; \Delta \vdash s \triangleright \Phi; \Gamma$ then there exists Ψ' and Δ' such that $\Phi = \Psi\Psi'$ and $\Gamma = \Delta\Delta'$.

Also note that for any well-typed expression e in $\Phi; \Gamma$ all its free variables appear in $\Phi; \Gamma$.

An important property used in the subject reduction proof (Theorem 34) states that if an expression e is well-typed in a typing environment $\Phi; \Gamma$, then it is also well-typed in any "reasonable" typing environment containing $\Phi; \Gamma$.

Lemma 31 (Weakening for Environments)

- If $\Phi; \Gamma \vdash M : A$ then for all acceptable $\Psi; \Delta$ such that $BVar(\Psi; \Delta) \cap (BVar(\Phi; \Gamma) \cup FSV(M)) = \emptyset$, then $\Psi\Phi; \Gamma\Delta \vdash M : A$.
- If $\Phi; \Gamma \vdash s \triangleright \Phi'; \Gamma'$, then for all acceptable $\Psi; \Delta$ such that $BVar(\Psi; \Delta) \cap (BVar(\Phi; \Gamma) \cup BVar(s) \cup FSV(s)) = \emptyset$, then $\Psi\Phi; \Delta\Gamma \vdash s \triangleright \Psi\Phi'; \Delta\Gamma'$.
- If $\Phi; \Gamma \vdash \Xi \rightsquigarrow \mathtt{K}$, then for all acceptable $\Psi; \Delta$ such that $BVar(\Psi; \Delta) \cap (BVar(\Phi; \Gamma) \cap FSV(\Xi)) = \emptyset$, then $\Psi\Phi; \Delta\Gamma \vdash \Xi \rightsquigarrow \mathtt{K}$.

Proof. We prove these three statements by induction on ($|\Psi| + |\Delta|$,h) where $|\Delta|$ is the number of patterns appearing in Δ, $|\Psi|$ is the number of sum variables appearing in Ψ and h is the height of the considered proof.

The proof of subject reduction is strongly based on the possibility of deconstructing patterns into more simpler ones via the $Dec(\ )$ operation which is defined as follows:

Definition 32 Given a typed pattern $P : A$, we define its *deconstruction* as follows:

$$\begin{array}{ll} Dec(\_ : A) = \_ : A & Dec(x : A) = x : A \\ Dec((P_1 \mid_\xi P_2) : A_1 + A_2) = (P_1 \mid_\xi P_2) : A_1 + A_2 & \\ Dec(\langle P_1, P_2 \rangle : A_1 \times A_2) = Dec(P_1 : A_1), Dec(P_2 : A_2) & \\ Dec(@(P_1, P_2) : A) = Dec(P_1 : A), Dec(P_2 : A) & \end{array}$$

This notion extends naturally to a pattern environment $\Gamma = P_1 : A_1, \ldots, P_n : A_n$ by defining $Dec(\Gamma)$ as $Dec(P_1 : A_1), \ldots, Dec(P_n : A_n)$.

The typing system enjoys the property that any well-typed expression in a typing environment is also a well-typed expression in the deconstructed environment.

Lemma 32 If e is well typed in an environment $\Phi; \Gamma$ then e is also well typed in $\Phi; Dec(\Gamma)$.

Another property of typing derivations which is needed in the proof of subject reduction states that for every well typed expression e we can choose a "canonical" typing derivation ending with a particular typing rule associated to e. Thus for example, if $\Phi; \Gamma \vdash \langle M_1, M_2 \rangle : A \times B$ then there is a proof of this sequent ending with the rule $(\times Right)$. We refer the interested reader to [For02] for further details.

The deconstruction operation given in Definition 32 can be used to simplify pair and contraction patterns, but there is no operation to simplify sum patterns. There is however a property of typing derivations which allows us to decompose sum patterns appearing on the left hand side of sequents as follows:

Lemma 33 Let $\mathtt{K}$ and $\mathtt{K}'$ be $\mathtt{L}$ or $\mathtt{R}$, then

- If $\Phi; (P \mid_\xi Q) : A + B, \Gamma \vdash M : C$ then so is $\xi : \mathtt{K}, \Phi; P : A, Q : B, \Gamma \vdash M : C$.
- If $\Phi; (P \mid_\xi Q) : A + B, \Gamma \vdash s \rhd \Phi'; (P \mid_\xi Q) : A + B, \Gamma'$ then so is $\xi : \mathtt{K}, \Phi; P : A, Q : B, \Gamma \vdash s \rhd \xi : \mathtt{K}, \Phi'; P : A, Q : B, \Gamma'$.
- If $\Phi; (P \mid_\xi Q) : A + B, \Gamma \vdash \Xi \leadsto \mathtt{K}'$ then $\xi : \mathtt{K}, \Phi; P : A, Q : B, \Gamma \vdash \Xi \leadsto \mathtt{K}'$.
- If $\Phi; (P \mid_\xi Q) : A + B, \Gamma \vdash \Xi \leadsto \psi$ then $\xi : \mathtt{K}, \Phi; P : A, Q : B, \Gamma \vdash \Xi \leadsto \psi$.

We can now state the following result:

Theorem 34 (Subject reduction)

- If $\Phi; \Gamma \vdash M : A$ and $M \longrightarrow M'$ then $\Phi; \Gamma \vdash M' : A$.
- If $\Phi; \Gamma \vdash s \rhd \Phi'; \Gamma'$ and $s \longrightarrow s'$ then $\Phi; \Gamma \vdash s' \rhd \Phi'; \Gamma'$.
- If $\Phi; \Gamma \vdash \Xi \leadsto \mathtt{T}$ (with $\mathtt{T} \in \{\mathtt{L}, \mathtt{R}, \xi\}$) and $\Xi \longrightarrow \Xi'$ then $\Phi; \Gamma \vdash \Xi' \leadsto \mathtt{T}$.

Proof. By induction on expressions, using Lemmas 33, 32 and 31.

4 Strong Normalization for λP_w

This section is devoted to the proof of strong normalization of well-typed λP_w-terms. This proof is an adaptation of the one proposed by Ritter [Rit94] for a restricted version of λ_σ with de Bruijn indices where substitutions can only cross the leftmost outermost lambda. The scheme of our proof can be summarized as follows:

- We first define a calculus *modulo* an equational theory, noted $\lambda P_{w/\equiv}$.
- We then define the notion of *reducible term* and *reducible substitution* for $\lambda P_{w/\equiv}$-expressions. We show that any reducible term (resp. substitution) is strongly normalizing.

- We show that the term $M[s]$ and the substitution $t \circ s$ are reducible for any reducible substitution s, well-typed term M and well-typed substitution t.
- We use the previous point to show that any well-typed $\lambda P_{w/\equiv}$-expression is reducible and thus strongly normalizing.
- Finally, we deduce strong normalization for λP_w-expressions from strong normalization for $\lambda P_{w/\equiv}$-expressions.

We start the proof by introducing the notion of *void substitutions* which are special substitutions which do not *really* change the pattern environment part of the their typing environment (i.e. which do not bind new usual variable).

Definition 41 The set *VS* of void substitutions is defined to be the smallest set of substitutions stable by concatenation such that:

- $id \in VS$
- $(\xi^{P.A}/\mathtt{K}).s \in VS$ if and only if $s \in VS$ and $BVar(P) = \emptyset$

Void substitutions enjoy the following properties:

Remark 3.

- Let s be a void substitution such that $s \longrightarrow s'$, then s' is a void substitution.
- Any void substitution is strongly normalizing.

Lemma 41 Let s be a substitution such that $\Psi; \Delta \vdash s \rhd \Phi; \emptyset$, then we have:

- $\Delta = \emptyset$
- s is a void substitution

Proof. The first point holds by Remark 2 and the second one by contradiction.

4.1 Definition of $\lambda P_{w/\equiv}$

We can now define the notion of $\lambda P_{w/\equiv}$-reduction modulo an equational theory.

Definition 42 The congruence $\equiv$ on λP_w-terms is defined by:

$$(M\ N)[s] =_{\mathsf{Sub\_app}} M[s]\ N[s] \qquad (s \circ t) \circ u =_{\mathsf{Sub\_ass\_env}} s \circ (t \circ u)$$

$$M[s][t] =_{\mathsf{Sub\_clos}} M[s \circ t]$$

We will consider the reduction system $\lambda P_{w/\equiv}$, where $a \longrightarrow_{\lambda P_{w/\equiv}} b$ if and only if there exist a', b' such that $a \equiv a' \longrightarrow_{\mathcal{R}} b' \equiv b$, where $\mathcal{R} = \lambda P_w \setminus \{\mathsf{Sub\_app}, \mathsf{Sub\_ass\_env}, \mathsf{Sub\_clos}\}$. This definition can also be interpreted as a reduction on *equivalence classes*, i.e., $[a] \longrightarrow_{\lambda P_{w/\equiv}} [b]$ if and only if $a' \longrightarrow_{\mathcal{R}} b'$, for $a' \in [a]$ and $b' \in [b]$. We may use indistinctly both interpretations according to the context.

We remark that by subject reduction (Theorem 34), the notion of well-typed $\lambda P_{w/\equiv}$-terms is well-defined.

4.2 Strong Normalization for $\lambda P_{w/\equiv}$

We are now able to introduce the notion of reducible expressions, which makes use of the following concept.

Definition 43 (Neutral terms and substitutions)

- A term is neutral if and only if it is neither of the forms $(\lambda x.N)[s]$, $\mathtt{inr}_A(M)$, $\mathtt{inl}_B(M)$ nor $\langle M_1, M_2\rangle$.
- A substitution s is *neutral* if and only if it is not of the form $(x/M).t$.

Notation 42 Let M be a term.

- $\mathcal{P}^1_{A\times B}(M)$ denotes the term $(\lambda\langle x,\ \_\rangle : A\times B.x)[id]\ M$ where x is a fresh variable.
- $\mathcal{P}^2_{A\times B}(M)$ denotes the term $(\lambda\langle\ \_, x\rangle : A\times B.x)[id]\ M$ where x is a fresh variable.
- $\mathcal{S}_{A+B}(M)$ denotes the term $(\lambda(x \mid_\xi y) : A+B.[\langle x, w_2\rangle \mid_\xi \langle w_1, y\rangle])[id]\ M$ where x, y, w_1, w_2 and ξ are fresh variables.

Definition 44 (Reducible terms and substitutions) The set of *reducible terms* for a given type in an environment $\Phi;\Gamma$ is defined by induction on types as follows:

$[\![\iota]\!]_{\Phi;\Gamma} =_{def} \{M \mid \Phi;\Gamma \vdash M:\iota$ and M is strongly normalizing$\}$.
$[\![A \to B]\!]_{\Phi;\Gamma} =_{def} \{M \mid \Phi;\Gamma \vdash M : A \to B$ and $\forall N \in [\![A]\!]_{\Phi;\Gamma\Delta}, (M\ N) \in [\![B]\!]_{\Phi;\Gamma\Delta}\}$ where Δ satisfies the condition of the point 1 of Lemma 31.
$[\![A \times B]\!]_{\Phi;\Gamma} =_{def} \{M \mid \Phi;\Gamma \vdash M : A\times B, \mathcal{P}^1_{A\times B}(M) \in [\![A]\!]_{\Phi;\Gamma}$ and $\mathcal{P}^2_{A\times B}(M) \in [\![B]\!]_{\Phi;\Gamma}\}$
$[\![A+B]\!]_{\Phi;\Gamma} =_{def} \{M \mid \Phi;\Gamma \vdash M : A+B$ and $\forall$ fresh variables $w_1, w_2, \mathcal{S}_{A+B}(M) \in [\![A\times B]\!]_{\Phi;w_1:A,w_2:B,\Gamma}\}$

The set of *reducible substitutions for an environment $\Phi;\Gamma$ in an environment $\Psi;\Delta$* is defined as follows:

$$[\![\Phi;\Gamma]\!]_{\Psi;\Delta} =_{def} \{s \mid \Psi;\Delta \vdash s \rhd \Phi;\Gamma \text{ and } \forall (x:A) \in \Gamma, x[s] \in [\![A]\!]_{\Psi;\Delta}\}$$

Reducible terms enjoy the following expected properties:

Lemma 43 For every type C the following statements hold:

1. If $M \in [\![C]\!]_{\Phi;\Gamma}$, then M is strongly normalizing.
2. If $\Phi;\Gamma \vdash (xM_1 \dots M_n) : C$ and $M_1 \dots M_n$ are strongly normalizing, then $(xM_1 \dots M_n) \in [\![C]\!]_{\Phi;\Gamma}$.
3. If $M \in [\![C]\!]_{\Phi;\Gamma}$ and $M \longrightarrow M'$ then $M' \in [\![C]\!]_{\Phi;\Gamma}$.
4. If M is a neutral of type C and all its one-step reducts are reducible expressions, then M is reducible.

Proof. The proof is done by induction on the type C.

Lemma 44 Let M be a term in $[\![A]\!]_{\Phi;\Gamma}$. For all acceptable environment $\Psi;\Delta$ satisfying the conditions of Point 1 of Lemma 31, M is in $[\![A]\!]_{\Phi\Psi;\Gamma\Delta}$

Proof. By induction on the type A.

We can now deduce from Lemma 43 (Point 2) the following property:

Corollary 1. *All the variables are reducible.*

As for terms, reducible substitutions also enjoy the following expected properties:

Lemma 45

1. If $s \in [\![\Phi;\Gamma]\!]_{\Psi;\Delta}$, then s is strongly normalizing.
2. If $s \in [\![\Phi;\Gamma]\!]_{\Psi;\Delta}$ and $s \longrightarrow s'$ then $s' \in [\![\Phi;\Gamma]\!]_{\Psi;\Delta}$.
3. If s is a neutral substitution such that $\Psi;\Delta \vdash s \rhd \Phi;\Gamma$ and all its one-step reducts are reducible, then $s \in [\![\Phi;\Gamma]\!]_{\Psi;\Delta}$.

Proof. We prove the properties by cases on Γ. In the case of $\Gamma = \emptyset$ the proof is done by remarking that s is void. In the case of $\Gamma \neq \emptyset$, the proof is done by using Lemma 43.

Lemma 46 Let s be a substitution in $[\![\Phi;\Gamma]\!]_{\Psi;\Delta}$. For any acceptable environment $\Phi';\Gamma'$ satisfying the conditions of point 2 of Lemma 31, s is in $[\![\Phi'\Phi;\Gamma'\Gamma]\!]_{\Phi'\Psi;\Gamma'\Delta}$

Proof. By Lemmas 31 and 44.

Since id is neutral, well-typed and has no reducts, then we can conclude the following by Lemma 45.

Corollary 2. *Let $\Phi;\Gamma$ be a valid environment. Then $id \in [\![\Phi;\Gamma]\!]_{\Phi;\Gamma}$.*

We are now ready to prove the state statement of this section which allows us to prove that any well-typed expression is reducible.

Theorem 47 Let $\Psi;\Delta$ and $\Phi;\Gamma$ be valid environments and s be a substitution in $[\![\Phi;\Gamma]\!]_{\Psi;\Delta}$.

- For every *substitution* t such that $\Phi;\Gamma \vdash t \rhd \Phi';\Gamma'$, we have $t \circ s$ is in $[\![\Phi';\Gamma']\!]_{\Psi;\Delta}$.
- For every *term* M such that $\Phi;\Gamma \vdash M : A$, we have $M[s]$ is in $[\![A]\!]_{\Psi;\Delta}$.

Proof. This proof can be done by induction on the structure of the (substitution/term) e. It uses some technical lemmas which we cannot present here by lack of space but which are fully detailed in [For02]. Intuitively, these lemmas state how to deduce reducibility for a given expression from the reducibility of its sub expressions. The *equivalence relation* is used, for example, to deduce reducibility for $M[s][t]$ from reducibility of $M[s \circ t]$.

By Theorem 47 and Corollary 2 we have that the term $M[id]$ and the substitution $t \circ id$ are reducible and thus by Lemmas 43 and 45 $M[id]$ and $t \circ id$ turns out to be strongly normalizing so that the following result holds.

Theorem 48 ($\lambda P_{w/\equiv}$ strong normalization) Any well typed $\lambda P_{w/\equiv}$-expression is $\lambda P_{w/\equiv}$ strongly normalizing.

4.3 Strong Normalization for λP_w

In order to conclude with the main theorem of this section we use the following standard property [FKP99].

Lemma 49 Let $A = \langle \mathcal{O}, R_1 \cup R_2 \rangle$ be an abstract reduction system such that:

- R_2 is strongly normalizing;
- there exists a reduction system $S = \langle \mathcal{O}', R' \rangle$ and translation $\mathcal{T}$ from $\mathcal{O}$ to $\mathcal{O}'$ such that:
  - $a \longrightarrow_{R_1} b$ implies $\mathcal{T}(a) \longrightarrow_{R'} \mathcal{T}(b)$,
  - $a \longrightarrow_{R_2} b$ implies $\mathcal{T}(a) = \mathcal{T}(b)$.

Then for all term $a \in \mathcal{O}$ such that $\mathcal{T}(a)$ is R'-strongly normalizing we have that a is $(R_1 \cup R_2)$-strongly normalizing.

Theorem 410 (λP_w strong normalization) Any well typed λP_w-expression is λP_w strongly normalizing

Proof. By Lemma 48 and Lemma 49, where $R' = \longrightarrow_{\lambda P_{w/\equiv}}$, R_2 is the reduction system engendered by the three rule Sub_app, Sub_ass_env and Sub_clos, R_2 is the reduction system engendered by the remaining rules and $\mathcal{T}$ is the canonical projection from λP_w-expression to $\lambda P_{w/\equiv}$-expression.

5 Conclusion

In this paper we have presented a weak calculus λP_w with explicit operations for pattern matching and substitution. Our formalism is successively inspired by [KPT96] and [CK]. In contrast to [KPT96], which treats substitutions and pattern-matching as meta-level operations, we have incorporated them into the syntax of the language by introducing appropriate reduction and typing rules. In contrast to [CK], we have eliminated the use of α-conversion (as our calculus is weak), and we have considered here a more powerful system of substitutions having composition. However, in our opinion, the major progress w.r.t the calculus TPC_{ES} presented in [CK] is that λP_w incorporates *sum replacement* as an *explicit* operation, where the calculus TPC_{ES} uses *meta-level* substitutions for sum variables. This step was one of the main goals of the formalism we propose in this work.

We have shown that the calculus λP_w enjoys all the classical properties of typed λ-calculi, namely it is confluent on all terms, it has the subject reduction property and it is strongly normalizing on all well-typed terms.

In the future, we would like to extend λP_w with *algebraic data types*, in order to cover more realistic functional programming languages, and with more general syntax for binding structures, as for example Klop's CRS [Klo80], in order to cover not only functional programming, but also other programming paradigms.

We would also like to incorporate to our formalism some ideas of the ρ-calculus [CK98] which deals with explicit *pattern matching* in a *rewriting* formalism. In particular, the λP_w-calculus implements a fix pattern-matching algorithm in contrast to ρ-calculus which can be parametrized with different matching algorithms.

Last, but not least, we are studying different evaluation strategies for λP_w, namely *lazy* and *eager* evaluators, which represent concrete implementations of functional languages via the more theoretical notion of reduction system proposed in this paper. We expect that this future work will allow us to provide a better explanation of the interaction between the operations of pattern matching and substitution.

Acknowledgement. The author would like to thank his director Delia Kesner for the time spent to help him and for all the important remarks done on this article.

References

[ACCL91] Martín Abadi, Luca Cardelli, Pierre Louis Curien, and Jean-Jacques Lévy. Explicit substitutions. *Journal of Functional Programming*, 4(1):375–416, 1991.

[BBLRD96] Zine-El-Abidine Benaissa, Daniel Briaud, Pierre Lescanne, and Jocelyne Rouyer-Degli. λv, a calculus of explicit substitutions which preserves strong normalisation. *Journal of Functional Programming*, 6(5):699–722, 1996.

[Blo95] Roel Bloo. Preservation of strong normalization for explicit substitutions. Technical Report TR95-08, TUE Computing Science Reports, Eindhoven University of Technology, 1995.

[BR95] Roel Bloo and Kristoffer Rose. Preservation of strong normalization in named lambda calculi with explicit substitution and garbage collection. In *Computing Science in the Netherlands*, pages 62–72. Netherlands Computer Science Research Foundation, 1995.

[CHL96] Pierre-Louis Curien, Thérèse Hardin, and Jean-Jacques Lévy. Confluence properties of weak and strong calculi of explicit substitutions. *Journal of the ACM*, 43(2):362–397, March 1996.

[CK] Serenella Cerrito and Delia Kesner. Pattern matching as cut elimination. `ftp://ftp.lri.fr/LRI/articles/kesner/pm-as-ce.ps.gz`.

[CK98] Horatiu Cirstea and Claude Kirchner. ρ-calculus, the rewriting calculus. In *Fifth International Workshop on Constraints in Computational Logics*, 1998.

[CK99] Serenella Cerrito and Delia Kesner. Pattern matching as cut elimination. In Giuseppe Longo, editor, *Fourteenth Annual IEEE Symposium on Logic in Computer Science (LICS)*, pages 98–108. IEEE Computer Society Press, July 1999.

[CKL01] Horatiu Cirstea, Claude Kirchner, and Luigi Liquori. Matching Power. In Aart Middeldorp, editor, *Proceedings of RTA'2001*, Lecture Notes in Computer Science, Utrecht (The Netherlands), May 2001. Springer-Verlag.

[DG01] René David and Bruno Guillaume. A λ-calculus with explicit weakening and explicit substitution. *Mathematical Structures in Computer Science*, 11, 2001.

[FKP99] Maria C.F. Ferreira, Delia Kesner, and Laurence Puel. Lambda-calculi with explicit substitutions preserving strong normalization. *Applicable Algebra in Engineering Communication and Computing*, 9(4):333–371, 1999.

[For02] Julien Forest. A calculus of pattern matching and explicit substitution draft. Technical Report LRI-1313, 2002. `http://www.lri.fr/~forest/lpw.ps.gz`.

[Kes96] Delia Kesner. Confluence properties of extensional and non-extensional λ-calculi with explicit substitutions. In Harald Ganzinger, editor, *Seventh International Conference on Rewriting Techniques and Applications*, volume 1103 of *Lecture Notes in Computer Science*, pages 184–199. Springer-Verlag, July 1996.

[Klo80] Jan-Willem Klop. *Combinatory Reduction Systems*, volume 127 of *Mathematical Centre Tracts*. CWI, Amsterdam, 1980. PhD Thesis.

[KPT96] Delia Kesner, Laurence Puel, and Val Tannen. A Typed Pattern Calculus. *Information and Computation*, 124(1):32–61, 1996.

[Les94] Pierre Lescanne. From $\lambda\sigma$ to $\lambda\upsilon$, a journey through calculi of explicit substitutions. In *Annual ACM Symposium on Principles of Programming Languages (POPL)*, pages 60–69, Portland, Oregon, 1994.

[Obj] The Objective Caml language. `http://caml.inria.fr/`.

[Rit94] E. Ritter. Normalisation for typed lambda calculi with explicit substitution. *Lecture Notes in Computer Science*, 832:295–??, 1994.

Tradeoffs in the Intensional Representation of Lambda Terms

Chuck Liang[1] and Gopalan Nadathur[2]

[1] Department of Computer Science, Hofstra University, Hempstead, NY 11550
cscccl@hofstra.edu, Fax: 516-463-5790

[2] Department of Computer Science and Engineering, University of Minnesota, 4-192 EE/CS Building, 200 Union Street SE, Minneapolis, MN 55455
gopalan@cs.umn.edu, Fax: 612-625-0572

Abstract. Higher-order representations of objects such as programs, specifications and proofs are important to many metaprogramming and symbolic computation tasks. Systems that support such representations often depend on the implementation of an intensional view of the terms of suitable typed lambda calculi. Refined lambda calculus notations have been proposed that can be used in realizing such implementations. There are, however, choices in the actual deployment of such notations whose practical consequences are not well understood. Towards addressing this lacuna, the impact of three specific ideas is examined: the de Bruijn representation of bound variables, the explicit encoding of substitutions in terms and the annotation of terms to indicate their independence on external abstractions. Qualitative assessments are complemented by experiments over actual computations. The empirical study is based on λProlog programs executed using suitable variants of a low level, abstract machine based implementation of this language.

1 Introduction

This paper concerns the representation of lambda terms in the implementation of programming languages and systems in which it is necessary to examine the structures of such terms during execution. The best known uses of this kind of lambda terms appears within higher-order metalanguages [16,21], logical frameworks [9,19] and proof development systems [5,7,20]. Within these systems and formalisms, the terms of a chosen lambda calculus are used as data structures, with abstraction in these terms being used to encode binding notions in objects such as formulas, programs and proofs, and the attendant β-reduction operation capturing important substitution computations. Although the intensional uses of lambda terms have often pertained to this kind of "higher-order" approach to abstract syntax, they are not restricted to only this domain. Recent research on the compilation of functional programming languages has, for example, advocated the preservation of types in internal representations [23]. Typed intermediate languages that utilize this idea [22] naturally call for structural operations on lambda terms during computations.

S. Tison (Ed.): RTA 2002, LNCS 2378, pp. 192–206, 2002.

The traditional use of lambda terms as a means for computing permits each such term to be compiled into a form whose only discernible relationship to the original term is that they both reduce to the same value. Such a translation is, of course, not acceptable when lambda terms are used as data structures. Instead, a representation must be found that provides a rapid access at runtime to the *form* of a term and that also facilitates comparisons between terms based on this structure. More specifically, the relevant comparison operations usually ignore bound variable names and equality modulo α-conversion must therefore be easy to recognize. Further, comparisons of terms must factor in the β-conversion rule and, to support this, an efficient implementation of β-contraction must be provided. An essential component of β-contraction is a substitution operation over terms. Building in a fine-grained control over this operation has been thought to be useful. This control can be realized in principle by introducing a new category of terms that embody terms with 'suspended' substitutions. The detailed description of such an encoding is, however, a little complicated because the propagation of substitutions and the contraction of redexes inside the context of abstractions have, in general, to be considered when comparing terms.

The representational issues outlined above have been examined in the past and approaches to dealing with them have also been described. A well-known solution to the problem of identifying two lambda terms that differ only in the names chosen for bound variables is, for instance, to transform them into a nameless form using a scheme due to de Bruijn [4]. Similarly, several new notations for the lambda calculus have been described in recent years that have the purpose of making substitutions explicit (*e.g.*, see [1,3,10,18]). However, the actual manner in which all these devices should be deployed in a practical context is far from clear. In particular, there are tradeoffs involved with different choices and determining the precise way in which to make them requires experimentation with an actual system: the operations on lambda terms that impact performance are ones that arise dynamically and they are notoriously difficult to predict from the usual static descriptions of computations.

This paper seeks to illuminate this empirical question. The vehicle for its investigation is the *Teyjus* implementation of λProlog [17]. λProlog is a logic programming language that employs lambda terms as a representational device and that, in addition to the usual operations on such terms, uses higher-order unification as a means for probing their structures. The *Teyjus* system, therefore, implements intensional manipulations over lambda terms. This system isolates several choices in term representation, permitting them to be varied and their impact to be quantified. We employ this concrete setting to understand three different issues: the value of explicit substitutions, the benefits and drawbacks of the de Bruijn notation and the relevance of an annotation scheme that determines the dependence of (sub)terms on external abstractions. Using a mixture of qualitative characterizations and experiments, we conclude that:

1. Explicit substitutions are useful so long as they provide the ability to combine β-contraction substitutions.

2. The potential disadvantage of the de Bruijn representation, namely the need to renumber indices during β-contraction, is not significant in practice.
3. Dependency annotations can improve performance, but their effect is less pronounced when combined with the ability to merge environments.

The rest of this paper is structured as follows. In the next section, we describe a notation for lambda terms that underlies the *Teyjus* implementation. This notation embodies an explicit representation of substitutions but one that can, with suitable control strategies, be used to realize substitutions either eagerly or lazily. In Section 3, we outline the structure of λProlog computations, categorize these into conceptually distinct classes and describe specific programs in each class that we use to make measurements. The following three sections discuss, in turn, the differences in lambda term representation of present interest and provide the results of our experiments. We conclude the paper with an indication of other questions that need to be examined empirically.

2 A Notation for Lambda Terms

We use an explicit substitution notation for lambda terms in this paper that builds on the de Bruijn method for eliminating bound variable names. A notation of this kind conceptually encompasses two categories of expressions, one corresponding to terms and the other corresponding to environments that encode substitutions to be performed over terms. In a notation such as the $\lambda\sigma$-calculus [1] that use exactly these two categories, an operation must be performed on an environment expression each time it is percolated inside an abstraction towards modifying the de Bruijn indices in the terms whose substitution it represents. The notation that we have designed for use in the implementation of λProlog instead allows these adjustments to be carried out in one swoop when a substitution is actually effected rather than in an iterated manner. To support this possibility, this notation includes a third category of expressions called environment terms that encode terms together with the 'abstraction context' they come from. Our notation additionally incorporates a system for annotating terms to indicate whether or not they contain externally bound variables.

Formally, the syntax of terms, environments and environment terms of our *annotated suspension notation* are given by the following rules:

$$\begin{array}{lll}\langle T\rangle & ::= & \langle C\rangle \mid \langle V\rangle \mid \#\langle I\rangle \mid (\langle T\rangle\ \langle T\rangle)_{\langle A\rangle} \mid (\lambda_{\langle A\rangle}\ \langle T\rangle) \mid [\![\langle T\rangle, \langle N\rangle, \langle N\rangle, \langle E\rangle]\!]_{\langle A\rangle} \\ \langle E\rangle & ::= & nil \mid \langle ET\rangle :: \langle E\rangle \\ \langle ET\rangle & ::= & @\langle N\rangle \mid (\langle T\rangle, \langle N\rangle) \\ \langle A\rangle & ::= & o \mid c\end{array}$$

In these rules, $\langle C\rangle$ represents constants, $\langle V\rangle$ represent instantiatable variables (*i.e.* variables that can be substituted for by terms), $\langle I\rangle$ is the category of positive numbers and $\langle N\rangle$ is the category of natural numbers. Terms correspond to lambda terms. In keeping with the de Bruijn scheme, $\#i$ represents a variable bound by the ith abstraction looking back from the occurrence. An expression of the form $[\![t, ol, nl, e]\!]_o$ or $[\![t, ol, nl, e]\!]_c$, referred to as a *suspension*, is a new kind of term that encodes a term with a 'suspended' substitution: intuitively, such an expression represents the term t with its first ol variables being substituted for in a way determined by e and its remaining bound variables being renumbered to reflect the fact that t used to appear within ol abstractions but now appears within nl of them. Conceptually, the elements of an environment are either substitution terms generated by a contraction or are dummy substitutions corresponding to abstractions that persist in an outer context. However, renumbering of indices may have to be done during substitution, and to encode this the environment elements are annotated by a relevant abstraction level. To be deemed well-formed, suspensions must satisfy certain constraints that have a natural basis in our informal understanding of their content: in an expression of the form $[\![t, i, j, e]\!]_o$ or $[\![t, i, j, e]\!]_c$, the 'length' of the environment e must be equal to i, for each element of the form $@l$ of e it must be the case that $l < j$ and for each element of the form (t', l) of e it must be the case that $l \leq j$. A final point to note about the syntax of our expressions is that all non-atomic terms are annotated with either c or o. The former annotation indicates that the term in question does not contain any variables bound by external abstractions and the latter is used when either this is not true or when enough information is not available to determine that it is.

The expressions in our notation are complemented by a collection of rewrite rules that simulate β-contractions. These rules are presented in Figure 1. The symbols v and u that are used for annotations in these rules are schema variables that can be substituted for by either c or o. We also use the notation $e[i]$ to denote the ith element of the environment. Of the rules presented, the ones labelled (β_s) and (β_s') generate the substitutions corresponding to the β-contraction rule on de Bruijn terms and the rules (r1)-(r12), referred to as the *reading rules*, serve to actually carry out these substitutions.

The rule (r2) pertaining to 'reading' an instantiatable variable is based on a particular interpretation of such variables: substitutions that are made for them must not contain de Bruijn indices that are captured by external abstractions. This is a common understanding of such variables but not the only one. For example, treating these variables as essentially first-order ones whose instantiation can contain free de Bruijn indices provides the basis for lifting higher-order unification to an explicit substitution notation [8]. We comment briefly on this possibility at the end of the paper but do not treat it in any detail here.

The correctness of some of our reading rules, in particular, the rules (r8)-(r10), depends on consistency in the use of annotations. Our assumption is that these are correctly applied at the outset; thus, the static (compilation) process that creates initial internal representations of terms is assumed to apply the annotation c to only those terms that do not have unbound de Bruijn indices in them. It can then be seen that the rewrite rules preserve consistency while also attempting to retain the content in annotations [15].

The ultimate utility of our notation is dependent on its ability to simulate reduction in the lambda calculus. That it is capable of doing this can be seen

(β_s)	$((\lambda_u\, t_1)\; t_2)_v \longrightarrow [\![t_1, 1, 0, (t_2, 0) :: nil]\!]_v$
(β'_s)	$((\lambda_u\, [\![t_1, ol+1, nl+1, @nl :: e]\!]_o)\; t_2)_v \longrightarrow [\![t_1, ol+1, nl, (t_2, nl) :: e]\!]_v$
(r1)	$[\![c, ol, nl, e]\!]_u \longrightarrow c$ provided c is a constant
(r2)	$[\![x, ol, nl, e]\!]_u \longrightarrow x$ provided x is an instantiatable variable
(r3)	$[\![\#i, ol, nl, e]\!]_u \longrightarrow \#j$ provided $i > ol$ and $j = i - ol + nl$.
(r4)	$[\![\#i, ol, nl, e]\!]_u \longrightarrow \#j$ provided $i \leq ol$ and $e[i] = @l$ and $j = nl - l$.
(r5)	$[\![\#i, ol, nl, e]\!]_u \longrightarrow [\![t, 0, j, nil]\!]_u$ provided $i \leq ol$ and $e[i] = (t, l)$ and $j = nl - l$.
(r6)	$[\![(t_1\; t_2)_u, ol, nl, e]\!]_v \longrightarrow ([\![t_1, ol, nl, e]\!]_v\; [\![t_2, ol, nl, e]\!]_v)_v$.
(r7)	$[\![(\lambda_u\, t), ol, nl, e]\!]_v \longrightarrow (\lambda_v\, [\![t, ol+1, nl+1, @nl :: e]\!]_o)$.
(r8)	$[\![(t_1\; t_2)_c, ol, nl, e]\!]_u \longrightarrow (t_1\; t_2)_c$.
(r9)	$[\![(\lambda_c\, t), ol, nl, e]\!]_u \longrightarrow (\lambda_c\, t)$.
(r10)	$[\![[\![t, ol, nl, e]\!]_c, ol', nl', e']\!]_u \longrightarrow [\![t, ol, nl, e]\!]_c$.
(r11)	$[\![[\![t, ol, nl, e]\!]_o, 0, nl', nil]\!]_o \longrightarrow [\![t, ol, nl + nl', e]\!]_o$.
(r12)	$[\![t, 0, 0, nil]\!]_u \longrightarrow t$

Fig. 1. Rule schemata for rewriting annotated terms

in two steps [15]. First, underlying every consistently annotated term in the suspension notation is intended to be a de Bruijn term that is to be obtained by 'calculating out' the suspended substitutions. The reading rules can be shown to possess properties that support this interpretation: every sequence of rewritings using these rules terminates and any two such sequences that start at the same term ultimately produce annotated forms of the same de Bruijn term. It can then be shown that the de Bruijn term t β-reduces to s if and only if any t' that is a consistently annotated version of t can be rewritten using our rules to a term s' that is itself a consistently annotated version of s.

The (β'_s) rule is redundant to our collection if our sole purpose is to simulate β-contraction; indeed, omitting this rule yields a calculus that is similar to those in [3] and [10]. However, the (β'_s) rule is the only one in our system that permits substitutions arising from different contractions to be combined into one environment and, thereby, to be carried out in the same walk over the structure of the term being substituted into. The rule (r11) is also redundant, but it serves a similar useful purpose in that it that it allows a reduction to be combined with a renumbering walk after a term has been substituted into a new (abstraction) context. In fact, the main uses of rules (r11) and (r12) arise right after a use of rule (r5) and a reduction procedure based on our rules can actually roll these distinct rule applications into one. Rather than eliminating the effects of rules (β'_s) and (r11), it is possible to replace them with more general ones that are capable of merging the environments in *any* term of the form $[\![[\![t, ol_1, nl_1, e_1]\!]_u, ol_2, nl_2, e_2]\!]_v$ to produce an equivalent term of the form $[\![t, ol, nl, e]\!]_v$; such a collection of rules is, in fact, presented in [18]. However, embedding this larger collection of rules in an actual procedure can be difficult and an underlying assumption here is that the rules (β_s') and (r11) provide an efficient way to capture most of their useful effects. A final observation is that the rules (r8)-(r10) are redundant in a sense analogous to the (β_s') rule. However, using these rules can have practical benefits as we discuss in Section 6.

3 Computations in λProlog

The language λProlog is one that, from the perspective of this paper, extends Prolog in three important ways. First, it replaces first-order terms—the data structures of a logic programming language—with the terms of a typed lambda calculus. These lambda terms are complemented by a notion of equality given by the α-, β- and η-conversion rules. Second, λProlog uses a unification operation that builds in the extended notion of equality accompanying lambda terms. Finally, the language incorporates two new kinds of goals that provide, respectively, for scoping over names and clauses defining predicates.

The new features of λProlog endow it with useful metalanguage capabilities based essentially on the fact that the abstraction construct that is present in lambda terms provides a versatile mechanism for capturing the binding notions that appear in a variety of syntactic constructs. Suppose, for instance, that we are interested in representing a formula of the form $\forall x\ P(x)$ where $P(x)$ represents a possibly complex formula in which x appears free. Using lambda terms, this formula may be encoded as $(all\ (\lambda x\, \overline{P(x)}))$, where *all* is a constant chosen to encode the predicative force of the universal quantifier and $\overline{P(x)}$ represents $P(x)$. This kind of explicit treatment of quantifier scope has several advantages. Identity of formulas under variable renaming is implicit in the encoding; the representations of $\forall x\ P(x)$ and $\forall y\ P(y)$ are, for example, equivalent lambda terms. Quantifier instantiation can be realized uniformly through application. Thus, the instantiation of the quantifier in $(all\ Q)$, where Q itself is a complex structure possibly containing quantifiers, by the term t can be realized simply by writing down the expression $(Q\ t)$; β-contraction realizes the actual substitution with all the renamings necessary for logical correctness. Expressions with instantiatable variables can be used in conjunction with the enhanced unification notion to analyze formulas in logically sophisticated ways. Finally, many computations on formulas are based on recursion over their structure. The usual Horn clauses of a logic programming language already provide for a recursion over first-order structure. The new scoping mechanisms of λProlog complement this to realize also a recursion over *binding* structure in an elegant and logically justifiable way.

The capabilities of λProlog that we have outlined above have been exploited for a variety of purposes in the past, ranging from building theorem provers and encoding proofs to implementing experimental programming languages. A detailed discussion of these applications is beyond the scope of this paper. For our present purposes it suffices to note that *all* of them depend on the ability to perform conversion operations on lambda terms, to compare their structures and to decompose them in logical ways. Thus, λProlog provides a rich programming framework—that is not unlike others that utilize lambda terms intensionally—for studying the impact on performance of choices in the representation of these terms. In the following sections, we use this framework in quantifying some of these differences. We identify below a taxonomy of λProlog computations that is based roughly on the 'level' of the higher-order feature used and we use this to describe a collection of testing programs to be employed in our empirical study. The test suite is available from the Teyjus site at `http://teyjus.cs.umn.edu/`.

First-Order Programs. This category exercises no higher-order features and should therefore be impervious to differences in the way (lambda) terms are treated. We include two programs from this category:

- *[quicksort]* A standard Prolog implementation of the familiar sorting routine.
- *[pubkey]* An implementation of a public key security protocol described in [6]. This program uses the new scoping devices in λProlog.

First-Order Like Unification with Reduction. Genuine lambda terms may appear in programs in this category, but these figure mainly in reduction computations, most unification problems being first-order in nature. Matching the representation of a lambda term with the pattern *(all Q)*, for instance, yields a unification problem of this kind and the subsequent application *(Q t)* generates a reduction. The two programs included in this category are the following:

- *[hnorm]* A head normalization routine used to reduce a collection of randomly generated lambda terms.
- *[church]* A program that involves arithmetic computations with Church numerals and associated combinators.

L_λ Style Programs. Computation in this class proceeds by first dispensing with *all* abstractions in lambda terms using new constants, then carrying out a first-order style analysis over the remaining structure and eventually abstracting out the new constants. As an idiom, this is a popular one amongst λProlog users and it has also been adopted in other related systems such as Elf and Isabelle. Programs in this class use a controlled form of reduction—the argument of a redex is always a constant—and a restricted form of higher-order unification [12]. We test the following programs in this category:

- *[typeinf]* A program that infers type schemes for ML-like programs.
- *[compiler]* A compiler for a small imperative language [11].

Unrestricted Programs. Programs in this class make essential use of (general) higher-order unification. As such, they tend to involve significant backtracking and they also encompass β-reduction where the arguments of redexes are complex terms. The programs tested in this category are the following:

- *[hilbert]* A λProlog encoding of Hilbert's Tenth Problem [13].
- *[funtrans]* A transformer of functional programs [14].

The *Teyjus* system used in our experiments is an abstract machine and compiler based implementation of λProlog. It uses a low-level encoding of the terms in the annotated suspension notation. The use of the rewrite rules in Figure 1 within *Teyjus* are confined to a procedure that transforms terms to head normal forms, upon which all comparison operations are based. This procedure can be modified to realize, and hence to study, the effect of different choices in lambda term representation. It is this ability that we utilize in the following sections.

4 The Value of Explicit Substitutions

Reflecting an explicit substitution notation into the representation of lambda terms provides the basis for a lazy strategy in effecting reduction substitutions. There are two potential benefits to such laziness.

First, actual substitutions may be delayed till a point where it becomes evident that it is unnecessary to perform them. Thus, consider the task of determining if the two terms $((\lambda\,\lambda\,\lambda\,(\#3\ \#2\ s))\ (\lambda\,\#1))$ and $((\lambda\,\lambda\,\lambda\,(\#3\ \#1\ t))\ (\lambda\ \#1))$ are identical, assuming that s and t are some unspecified but complicated terms. We can conclude that they are not by observing that they reduce respectively to $\lambda\,\lambda\,(\#2\ s')$ and $\lambda\,\lambda\,(\#1\ t')$, where s' and t' are terms that result from s and t by appropriate substitutions. Note that, in making this determination, it is not necessary to explicitly calculate the results of the substitutions over the terms s and t. The annotated suspension notation and a suitable head-normalization procedure [15] provide the basis for such an approach.

Second, such laziness makes it possible to combine substitution walks that arise from contracting different β-redexes. Thus, suppose that we wish to instantiate the two quantifiers in the formula represented by $(all\ (\lambda x\,(all\ (\lambda y\,P))))$, where P represents an unspecified formula, with the terms t_1 and t_2. Such an instantiation is realized through two reductions, eventually requiring t_2 and t_1 to be substituted for the first and second free variables in P and the indices of all other free variables to be decremented by two. All these substitutions involves a walk over the *same* structure—the structure of P—and it would be profitable if they could all be done together. To combine walks in this manner, it is necessary to temporarily suspend substitutions generated by β-contractions. In a situation in which all the redexes are available in a single term, this kind of delaying of substitution can be built into the reduction procedure through *ad hoc* devices. However, in the case being considered, the two quantifier instantiations are ones that can only be considered incrementally and, further, intervening structure needs to be processed before the abstraction giving rise to the second redex is encountered. The structure that engenders sharing is therefore not all available within a single call to a reduction procedure and an explicit encoding of substitution over P seems to be necessary for realizing this benefit.

Towards understanding how important these factors are in practice, different strategies were experimented with in the head normalization routine in *Teyjus*. Three variations were tried. In one case, each time a β_s rule was used, the effects of the resulting substitution were immediately calculated out, mimicking the customary eager approach. The second version used lazy substitution but did not use the (β_s') rule in the rewriting process. This version gives us the benefits

Program	*Eager*		*Lazy without Merging*		*Lazy with Merging*	
	Running Time	*Reading Rule Applications*	*Running Time*	*Reading Rule Applications*	*Running Time*	*Reading Rule Applications*
quicksort	0.25 secs	0	0.25 secs	0	0.25 secs	0
pubkey	0.34 secs	0	0.34 secs	0	0.33 secs	0
church	0.36 secs	243461	0.39 secs	367314	0.27 secs	73200
hnorm	0.75 secs	266970	0.83 secs	439121	0.66 secs	89020
typeinf	14.70 secs	10291085	14.81 secs	17582708	9.58 secs	2177884
compiler	3.53 secs	2496431	3.82 secs	3391088	2.26 secs	318703
hilbert	0.48 secs	296027	0.36 secs	58894	0.34 secs	5489
funtrans	2.20 secs	44803	2.22 secs	60116	2.19 secs	24290

Fig. 2. Comparison of Reduction Substitution Strategies

of delayed substitution *without* the added advantage of combining substitution walks. The last variation used the full repertoire of rewrite rules, once again in a demand-driven mode for calculating substitutions, thereby attempting to draw on all the benefits of explicit substitutions. The overall framework in all three cases was provided by a graph-based reduction routine with an iterative control and that uses a stack as an auxiliary store. Such a procedure writes the result of each rewriting step based on the rules in Figure 1 to the heap. An alternative procedure, that was not utilized but that is applicable equally to all three variants tested and hence likely to produce similar timing behaviour, would embed most of the information written to heap within a recursion stack.

Figure 2 tabulates the running time and the number of applications of the rewrite rules that percolate the effects of substitutions over terms for each variant over the suite of programs described in Section 3. All the tests were conducted on a 400MHz UltraSparc and each figure in the table represents the average of five runs. The data in the table indicates a clear preference for a lazy approach to substitution combined with merging of substitutions in the cases where lambda terms are used intrinsically in computation. In the cases where the computation predominantly involves reduction or L_λ style processing, the time improvements range from 12% to over 35%. The measured time is over *all* computations, including backchaining over logic programming clauses. The difference attributable to the substitution strategy is pinpointed more dramatically and accurately by the counts of reading rule applications that are directly responsible for the timing differences and that are unaffected by the specifics of term representation and reduction procedure implementation. These figures indicate a reduction of structure traversal to between a third and an eigth in the cases mentioned through the use of lazy substitution with merging. A complementary observation is that while the number of β-redexes contracted remains unchanged, between 50% and 90% of these contractions are realized via the (β_s') rule.

We note that behavior degrades in some cases when lazy substitution is used without merging. There are two explanations for this. First, in this mode, terms of the form $[\![t, ol, nl, e]\!]_v$ where t is itself a suspension are often encountered. In a demand driven approach the two suspensions have to be worked inwards in tandem and this has a noticeable overhead. Second, backtracking, a possibility in a logic programming like setting, has an impact. Backtracking causes a revocation of the changes that have been made beyond the choice point that computation is retracted to. By carrying out substitution work that is deterministic late, a source of redundancy is introduced. This effect is evident in the increase in reading rule applications under a lazy substitution strategy without merging.

The data tabulated above pertain to the situation when annotations are used. Without these the differences are more dramatic as we discuss later. The conclusion from our observations thus seems to be that explicit substitutions are important for performance provided they are accompanied by merging.

5 The Treatment of Bound Variables

The common approaches to representing bound variables can be separated into two categories: those that use explicit names or, roughly equivalently, pointers to cells corresponding to binding occurrences and those that use de Bruijn indices. In comparing these approaches, it is important to distinguish two operations to which bound variables are relevant. One of these is that of checking the identity of two terms up to α-convertibility. The second is that of making substitutions generated by β-contractions into terms, in which case the treatment of bound variables is relevant to the way in which illegal capture is avoided.

The de Bruijn representation is evidently superior from the perspective of checking the identity of terms up to α-convertibility. For example, consider matching the two terms $\lambda y_1 \ldots \lambda y_n\, (y_i\; t_1\; \ldots\; t_m)$ and $\lambda z_1 \ldots \lambda z_n\, (z_i\; s_1\; \ldots\; s_m)$. The heads of these terms that are embedded under the abstractions are bound in both cases by the same abstraction. Thus, the matching problem can be translated into one over the arguments of this term. A prelude to this transformation at a formal level under a name based scheme is, however, a 'normalization' of bound variable names. This step is avoided under the de Bruijn scheme.

A further consideration of the above example indicates a more significant advantage of the de Bruijn notation. Under a name based scheme, the transformation step must produce the following set of pairs of terms to be matched:

$$\{\langle \lambda y_1 \ldots \lambda y_n\, t_1, \lambda z_1 \ldots \lambda z_n\, s_1\rangle, \ldots, \langle \lambda y_1 \ldots \lambda y_n\, t_m, \lambda z_1 \ldots \lambda z_n\, s_m\rangle\}.$$

The abstractions at the front of each of the terms are necessary: they provide the context in which the bound variables in the arguments are to be interpreted in the course of matching them. Constructing these new terms at run time is computationally costly and also a bit too complex to accommodate in a low-level, abstract machine based system such as *Teyjus*. Under the de Bruijn scheme, this context is implicitly present in the numbering of bound variables, obviating the explicit attachment of the abstractions.

From the perspective of carrying out substitutions in contrast, the de Bruijn scheme has no real benefit and may, in fact, even incur an overhead. The important observation here is that the renaming that may be needed in the substitution process in a name based scheme has a counterpart in the form of renumbering relative to the de Bruijn notation. To understand the nature of the needed mechanism, we may consider the reduction of the term $\lambda x\,((\lambda y\,\lambda z\,y\,x)\;(\lambda w\,x))$ whose de Bruijn representation is $\lambda\,((\lambda\,\lambda\,\#2\;\#3)\;(\lambda\,\#2))$. This term reduces to $\lambda x\,\lambda z\,((\lambda w\,x)\;x)$, a term whose de Bruijn representation is $\lambda\,\lambda\,((\lambda\,\#3)\;\#2)$. Comparing the two de Bruijn terms, we notice the following: When substituting the term $(\lambda\,\#2)$ inside an abstraction, the index representing the locally free variable occurrence, *i.e.*, 2, has to be incremented by 1 to avoid its inadvertent capture. Further, indices for bound variable occurrences within the scope of an abstraction that disappears on account of a β-contraction may have to be changed; here the index 3 corresponding to the variable occurrence x in the scope of the abstraction that is eliminated must be decremented by 1. The substitution operation that is used in formalizing β-contraction under the de Bruijn scheme must account for both effects.

At a detailed level, there is a difference in the renaming and renumbering devices needed in name-based and nameless representations. Given a β-redex of the form $(\lambda x\,\lambda y\,t_1)\;t_2$ whose de Bruijn version is a term of the form $(\lambda\,\lambda\,\hat{t}_1)\;\hat{t}_2$, the renaming in the first case is effected over the 'body', *i.e.*, $\lambda y\,t_1$, and in the second case over the argument, *i.e.*, $\hat{t}_2$.[1] One advantage of the name-based representation is that the renaming may be avoided altogether if there is no name clash. However determining this requires either a traversal of the term being substituted, or an explicit record of the variables that are free in it. An interesting alternative, described, for instance, in [2], is to always perform a renaming and, more significantly, to fold this into the same structure traversal as that realizing the β-contraction substitution.

The above discussion indicates that the additional cost relative to substitution that is attendant on the de Bruijn notation is bounded by the effort expended in renumbering substituted terms. A first sense of this cost can thus be obtained by measuring the proportion of substitutions that actually lead to nontrivial renumbering of compound terms. Cases of this kind can be identified as those in which rule (r5) is used where the skeletal term is non atomic and where an immediate simplification by one of the rules (r8)-(r10) or (r12) is not possible.

Figure 3 tabulates the data gathered towards this end for the programs in our test suite that use reductions. An interesting observation is that *no* renumbering is actually involved in the case of L_λ style programming. The reason for this is not hard to see—the only reductions performed are those corresponding to eliminating the binding with a new constant. Thus, for a significant set of computations carried out in λProlog and related languages, renumbering is a non-issue. In the other cases, some renumbering can occur but adopting a merging based approach to substitution can reduce this considerably. This phenomenon is also understandable; substituting a term in after more enclosing abstractions have disappeared due to contractions leaves fewer reasons to renumber.

The cases where a nontrivial renumbering needs to be done do not necessarily constitute an extra cost. In general, when a term is substituted in, it is necessary also to examine its structure and possibly reduce it to (weak) head normal form.

[1] In the de Bruijn scheme, some bound variables in $\hat{t}_1$ may also have to be renumbered, but this can be done efficiently at the same time that $\hat{t}_2$ is substituted into the term.

Program	*Eager*		*Lazy with Merging*	
	Total Substitutions	*Renumbering Substitutions*	*Total Substitutions*	*Renumbering Substitutions*
church	34999	3267	34723	12
hnorm	25497	0	32003	0
typeinf	777832	0	722291	0
compiler	154967	0	100519	0
hilbert	3840	1500	1539	411
funtrans	8587	146	7374	19

Fig. 3. Renumbering with the de Bruijn Representation

Now, the necessary renumbering can be incorporated into the same walk as the one that carries out this introspection. This structure is realized by choosing to percolate the substitution inwards first in a term of the form $[\![t, 0, nl, nil]\!]_v$, using rule (r11) to facilitate the necessary merging in the case that t is itself a suspension. The main drawback of this approach, in contrast to the scheme in [2] for instance, is that it can lead to a loss in sharing in reduction if the same term, t, has to be substituted, and reduced, in more than one place. An indication of the loss in sharing can be obtained from the differences in the number of (β_s) and (β'_s) reductions under the two strategies in those cases where renumbering is an issue. Our measurements show that, under an 'umbrella' regime of delayed substitution with merging, these numbers were *identical* in all the relevant cases. Thus, there is no loss in sharing from combining the reduction and renumbering walks and, consequently, no real renumbering overhead relative to our test suite.

Our conclusion, then, is that the de Bruijn treatment of bound variables is the preferred one in practice. It is *obviously* superior to name based schemes relative to comparing terms modulo α-conversion and, in fact, representations of the latter kind are not serious contenders from this perspective in low-level implementations. The de Bruijn scheme has the drawback of a renumbering overhead in realizing β-contraction. However, our experiments show that this overhead is either negligible or nonexistent in a large number of cases.

6 The Relevance of Annotations

Annotations that are included with terms have the potential for being useful in two different ways. First, they can lead to substitutions walks being avoided when it is known that the substitutions will not affect the term. Second, by allowing a suspension to be simplified directly to its skeleton term, they can lead to a preservation of sharing of structure, and, hence, of reduction work, in a graph based implementation. Both effects are present in the rules (r8)-(r10) in Figure 1, the only ones to actually use annotations. There is, of course, a cost associated with maintaining annotations as manifest in the structure of all the other rules. However, this cost can be considerably reduced with proper care. In the *Teyjus* implementation, for example, an otherwise unused low-end bit in the tag word corresponding to each term stores this information. The setting of this bit is generally folded into the setting of the entire tag word and a single test on the bit that ignores the structure of the term suffices in utilizing the information in the annotation.

Program	*Eager*		*Lazy with Merging*	
	Without Annotations	*With Annotations*	*Without Annotations*	*With Annotations*
quicksort	0.26	0.25	0.25	0.25
pubkey	0.33	0.34	0.31	0.33
church	1.90	0.36	0.29	0.27
hnorm	0.77	0.75	0.68	0.66
typeinf	48.56	14.70	9.56	9.58
compiler	19.51	3.53	2.29	2.26
hilbert	0.60	0.48	0.34	0.34
funtrans	2.20	2.20	2.26	2.19

Fig. 4. The Effect of Annotations

Experiments were conducted to quantify the benefits of annotations. The different versions of the head normalization routine described in Section 4 were modified to ignore annotations. The chosen programs were then executed on a 400MHz UltraSparc using the various versions of the *Teyjus* system. Figure 4 tabulates the running time in seconds for four of these versions; we have omitted the case of delayed substitutions without merging since the results are similar to the version with eager substitutions. The indication from these data is that annotations can make a significant difference in a situation where substitution is performed eagerly—for example, the running time is reduced by about 70% and 80% in the case of the two L_λ style programs—but they have little effect in the situation when lazy substitution with merging is used. Interpreting these observations, it appears that annotations can lead to the recognition of a large amount of unnecessary substitution work. In the situation where substitution walks can be merged, however, these can also be combined with walks for reducing the term. Since reduction has to be performed in any case, the redundancy when annotations are not used is minimal. Consistent with the observations from the previous section, the data for the case of lazy substitutions also indicates little benefit from shared reduction.

7 Conclusion

We have examined the tradeoffs surrounding the use of three ideas in the machine encoding of lambda terms. Our study indicates that a notation that supports a delayed application of substitution as well as a merging of substitution walks can be used to significant practical effect. Within this context the de Bruijn scheme for treating bound variables has definite advantages with little attendant costs. Finally, the benefits of annotations appear to be marginal at best when reduction and substitution are performed in tandem. Our observations have been made relative to a graph-based approach to representing terms. It is further possible to use hash-consing in the low-level encoding as has been done in the FLINT system that also employs the annotated suspension notation [22]. While we have not experimented with this technique explicitly, we believe that its use will be neutral to the other choices considered here.

The work reported here can be extended in several ways. One further question to consider is the difference between destructive and non-destructive realizations of reduction. There are 'obvious' advantages to a destructive version in a deterministic setting that become less clear with a language like λProlog that permits backtracking. This matter can be examined experimentally. Along a different direction, the implementation of higher-order unification needs to be considered more carefully. By changing the interpretation of instantiatable or meta variables, it is possible to lift higher-order unification to an explicit substitution notation. Doing so has the benefit of making the application of substitutions to meta variables very efficient. However, there are also costs: a more general mechanism for combining substitutions is needed and context information must be retained dynamically to translate metavariables prior to the presentation of output. There is a tradeoff here that can, once again, be assessed empirically. A final question concerns a more refined set of benchmarks, one that makes finer distinctions in the categories of computations on lambda terms. Such a refinement may reveal further factors that can influence the tradeoffs in representations.

Acknowledgements. This work was supported in part by NSF Grant CCR-0096322. Useful presentation suggestions were received from the referees, not all of which could be explicitly utilized due to space limitations.

References

1. M. Abadi, L. Cardelli, P.-L. Curien, and J.-J. Lévy. Explicit substitutions. *Journal of Functional Programming*, 1(4):375–416, 1991.
2. L. Aiello and G. Prini. An efficient interpreter for the lambda-calculus. *The Journal of Computer and System Sciences*, 23:383–425, 1981.
3. Z. Benaissa, D. Briaud, P. Lescanne, and J. Rouyer-Degli. $\lambda\upsilon$, a calculus of explicit substitutions which preserves strong normalization. *Journal of Functional Programming*, 6(5):699–722, 1996.
4. N. de Bruijn. Lambda calculus notation with nameless dummies, a tool for automatic formula manipulation, with application to the Church-Rosser Theorem. *Indag. Math.*, 34(5):381–392, 1972.
5. R. L. Constable et al. *Implementing Mathematics with the Nuprl Proof Development System.* Prentice-Hall, Englewood Cliffs, New Jersey, 1986.
6. G. Delzanno. Specifying and debugging security protocols via hereditary Harrop formulas and λProlog - a case-study. In *Fifth International Symposium on Functional and Logic Programming.* Springer Verlag LNCS vol. 2024, 2001.

7. G. Dowek, A. Felty, H. Herbelin, G. Huet, C. Murthy, C. Parent, C. Paulin-Mohring, and B. Werner. The Coq proof assistant user's guide. Rapport Techniques 154, INRIA, Rocquencourt, France, 1993.
8. G. Dowek, T. Hardin, and C. Kirchner. Higher-order unification via explicit substitutions. *Information and Computation*, 157:183–235, 2000.
9. R. Harper, F. Honsell, and G. Plotkin. A framework for defining logics. *Journal of the Association for Computing Machinery*, 40(1):143–184, January 1993.
10. F. Kamareddine and A. Ríos. Extending the λ-calculus with explicit substitution which preserves strong normalization into a confluent calculus on open terms. *Journal of Functional Programming*, 7(4):395–420, 1997.
11. C. Liang. Compiler construction in higher order logic programming. In *4th International Symposium on Practical Aspects of Declarative Languages*, pages 47–63. Springer Verlag LNCS No. 2257, 2002.
12. D. Miller. A logic programming language with λ-abstraction, function variables, and simple unification. *Journal of Logic and Computation*, 1(4):497–536, 1991.
13. D. Miller. Unification under a mixed prefix. *Journal of Symbolic Computation*, pages 321–358, 1992.
14. M. Mottl. Automating functional program transformation. MSc Thesis. Division of Informatics, University of Edinburgh, September 2000.
15. G. Nadathur. A fine-grained notation for lambda terms and its use in intensional operations. *Journal of Functional and Logic Programming*, 1999(2), March 1999.
16. G. Nadathur and D. Miller. An overview of λProlog. In K. A. Bowen and R. A. Kowalski, editors, *Fifth International Logic Programming Conference*, pages 810–827. MIT Press, August 1988.
17. G. Nadathur and D. J. Mitchell. System description: Teyjus—a compiler and abstract machine based implementation of λProlog. In *Automated Deduction–CADE-16*, pages 287–291. Springer-Verlag LNAI no. 1632, 1999.
18. G. Nadathur and D. S. Wilson. A notation for lambda terms: A generalization of environments. *Theoretical Computer Science*, 198(1-2):49–98, 1998.
19. C. Paulin-Mohring. Inductive definitions in the system Coq: Rules and properties. In *Proceedings of the International Conference on Typed Lambda Calculi and Applications*, pages 328–345. Springer-Verlag LNCS 664, 1993.
20. L. C. Paulson. *Isabelle: A Generic Theorem Prover*, volume 828 of *Lecture Notes in Computer Science*. Springer Verlag, 1994.
21. F. Pfenning and C. Schürmann. System description: Twelf — a meta-logical framework for deductive systems. In *Proceedings of the 16th International Conference on Automated Deduction*, pages 202–206. Springer-Verlag LNAI 1632, 1999.
22. Z. Shao. Implementing typed intermediate language. In *Proc. 1998 ACM SIGPLAN International Conference on Functional Programming (ICFP'98)*, pages 313–323. ACM Press, September 1998.
23. D. Tarditi, G. Morrisett, P. Cheng, C. Stone, R. Harper, and P. Lee. TIL: A type-directed optimizing compiler for ML. In *1996 SIGPLAN Conference on Programming Language Design and Implementation*, pages 181–192. ACM Press, 1996.

Improving Symbolic Model Checking by Rewriting Temporal Logic Formulae*

David Déharbe[1], Anamaria Martins Moreira[1], and Christophe Ringeissen[2]

[1] Universidade Federal do Rio Grande do Norte — UFRN
Natal, RN, Brazil
{david,anamaria}@dimap.ufrn.br

[2] LORIA-INRIA
Nancy, France
Christophe.Ringeissen@loria.fr

Abstract. A factor in the complexity of conventional algorithms for model checking Computation Tree Logic (CTL) is the size of the formulae, and, more precisely, the number of fixpoint operators. This paper addresses the following questions: given a CTL formula f, is there an equivalent formula with fewer fixpoint operators? and how term rewriting techniques may be used to find it? Moreover, for some sublogics of CTL, e.g. the sub-logic NF-CTL (no fixpoint computation tree logic), more efficient verification procedures are available. This paper also addresses the problem of testing whether an expression belongs or not to NF-CTL, and providing support in the choice of the most efficient amongst different available verification algorithms. In this direction, we propose a rewrite system modulo AC, and discuss its implementation in ELAN, showing how this rewriting process can be plugged in a formal verification tool.

1 Introduction

The term *model checking* is used to name formal verification techniques that show that a given property, expressed as a temporal logic formula or as an automaton, is satisfied by a given structure, such as a transition-graph, that represents a system under analysis [15,4,12]. One of the most successful approaches is symbolic model checking [13], a decision procedure for verifying that a Computation Tree Logic (CTL) formula is true in a Kripke structure. In practice, due to the state space explosion problem, symbolic model checking must often be combined with other techniques. Today, research is being carried out with the goal of potentially increasing the size of the systems that can be verified through model checking. It is well-known that the size of the structure and of the properties are the two factors in the worst-case complexity function of symbolic model checking. We can identify three possible ways to attack this problem: reducing

* Partially supported by projects CNPq–INRIA (FERUS) and CNPq–NSF (Formal Verification of Systems of Industrial Complexity).

S. Tison (Ed.): RTA 2002, LNCS 2378, pp. 207–221, 2002.

the size of the model [5] (e.g. with abstraction, symmetry, decomposition), developing more efficient verification algorithms where the size of the model and of the formula have a smaller impact on the verification activity [10,2], and reducing the size of the formulae (with a rewrite system) [6]. The research presented in this paper contributes to the last two lines of action. The main contribution of this paper is a rewrite system to test if a given CTL formula belongs to a sublogic, called NF-CTL, a sub-logic of CTL that may be verified much more efficiently than with conventional symbolic model checking algorithms [6]. This rewrite system has been implemented in the ELAN platform [3]. Further, we initiate the study of the problem of, given a CTL formula, finding an equivalent formula with fewer temporal operators. The integration of these techniques shall result in more efficient formal verification tools.

The remainder of this paper is structured as follows: Section 2 gives the required notions of symbolic model checking. Section 3 presents NF-CTL and the associated optimized verification algorithms. In Section 4, we present the main results of our study of rewriting properties of CTL and NF-CTL. The issues involved in plugging the developed rewriting systems into symbolic model checking are discussed in Section 5. Finally, Section 6 concludes and presents future work.

2 CTL Symbolic Model Checking

In this section, we provide the definitions of the main notions involved in symbolic model checking: the state-transition graph model of Kripke structures, and the syntax and semantics of the temporal logic CTL. Moreover, we sketch the symbolic model checking algorithms.

2.1 Kripke Structures

Let P be a finite set of boolean propositions. A Kripke structure over P is a quadruple $M = (S, T, I, L)$ where:

- S is a finite set of states.
- $T \subseteq S \times S$ is a transition relation, such that $\forall s \in S, \exists s' \in S, (s, s') \in T$.
- $I \subseteq S$ is the set of initial states.
- $L : S \to 2^P$ is a labeling function. L is injective and associates with each state a set of boolean propositions true in the state.

A path π in the Kripke structure M is an infinite sequence of states $s_1, s_2, \ldots$ such that $\forall i \geq 1, (s_i, s_{i+1}) \in T$. $\pi(i)$ is the ith state of π. States and transitions of a Kripke structure can be identified to their characteristic functions. The characteristic function of a state s is given by the conjunction of all propositions in the label of s, and of the negation of the propositions that are not in the label of s. The characteristic function of a transition $t = (s_1, s_2)$ is the conjunction of the functions for s_1 and s_2. These definitions are extended to sets of states and sets of transitions. The characteristic functions play an important role in symbolic model checking algorithms, in order to represent and compute the forward image (the set of states that can be reached in one transition from a given set of states $S' \subseteq S$, denoted $Forward(S')$), and the backward image (the set of states that reach a given set of states in one transition, denoted $Backward(S')$).

2.2 Computation Tree Logic

Given a set of atomic propositions P, the set of CTL formulae over P is given by the following definition.

Definition 1. *Given a finite set of boolean propositions P, $CTL(P)$ is the smallest set such that (1) $P \cup \{0,1\} \subseteq CTL(P)$ and (2) if f and g are in $CTL(P)$, then $f * g$, $f \oplus g$, $\neg f$, $f + g$, $f \Rightarrow g$, and $\mathbf{EX}f$, $\mathbf{EG}f$, $\mathbf{E}[f\mathbf{U}g]$, $\mathbf{AX}f$, $\mathbf{AG}f$, $\mathbf{AF}f$, $\mathbf{EF}f$, $\mathbf{A}[f\mathbf{U}g]$, $\mathbf{A}[f\mathbf{R}g]$, $\mathbf{E}[f\mathbf{R}g]$ are in $CTL(P)$.*

Informally, each operator is the combination of a path quantifier (**E**, for some path, or **A**, for all paths) and a state quantifier (**X**, in the next state, **F**, for some state, **G**, for all states, **U**, until, **R**, release). Formally, the following definition gives the semantics of CTL formulae in term of a given Kripke structure.

Definition 2. *Given a Kripke structure $M = (S, T, I, L)$ over a finite set of atomic propositions P, the validity of a formula $f \in CTL(P)$ at a state s of M, denoted by $M, s \models f$, is inductively defined as follows, where g and h are in $CTL(P)$, then*

1. $M, s \models p$ *iff* $p \in L(s)$.
2. $M, s \models \neg g$ *iff* $M, s \not\models g$.
3. $M, s \models g * h$ *iff* $M, s \models g$ *and* $M, s \models h$.
4. $M, s \not\models 0$ *and* $M, s \models 1$.
5. $M, s \models \mathbf{EX}g$ *iff there exists a state s' of M such that $(s, s') \in T$ and $s' \models g$. I.e. s has a successor where g is valid.*
6. $M, s \models \mathbf{EG}g$ *iff there exists a path π of M such that $\pi(1) = s$ and $\forall i \geq 1, M, \pi(i) \models g$. I.e. s is at the start of a path where g holds globally.*
7. $M, s \models \mathbf{E}[g\mathbf{U}h]$ *iff there exists a path π of M such that $\pi(1) = s$ and $\exists i \geq 1, M, \pi(i) \models h \wedge \forall j, i > j \geq 1, M, \pi(j) \models g$. I.e. s is at the start of a path where h holds eventually and g holds until h becomes valid.*

A formula f is said valid in M, *denoted by $M \models f$ if it is valid for all initial states: $M \models f$ iff $\forall s \in I, M, s \models f$.*

The remaining temporal and boolean operators can be seen as notational shortcuts. The semantics of the boolean operators is usual[1]. The remaining temporal operators obey the following set of axioms:

$$E = \begin{cases} \mathbf{AX}f &= \neg\mathbf{EX}\neg f \\ \mathbf{AG}f &= \neg\mathbf{EF}\neg f \\ \mathbf{AF}f &= \neg\mathbf{EG}\neg f \\ \mathbf{EF}f &= \mathbf{E}[1\mathbf{U}f] \\ \mathbf{A}[f\mathbf{U}g] &= \neg\mathbf{E}[\neg g\mathbf{U}(\neg f * \neg g)] * \neg\mathbf{EG}\neg g \\ \mathbf{A}[g\mathbf{R}f] &= \neg\mathbf{E}[\neg g\mathbf{U}\neg f] \\ \mathbf{E}[g\mathbf{R}f] &= \neg\mathbf{A}[\neg g\mathbf{U}\neg f] \end{cases} \qquad \begin{aligned} \mathbf{EG}0 &= 0 \\ \mathbf{E}[f\mathbf{U}0] &= 0 \\ \mathbf{EX}0 &= 0 \end{aligned}$$

[1] $+, \oplus, *$ correspond respectively to *or*, *xor*, *and*

In the definition of E, the three axioms on the right-hand side are consequences of Definition 2 and are not standard: they show how expressions with "canonical" temporal operators can be further simplified. In our setting, we consider the equational theory *opCTL* defined as the union of E and an equational theory *NEW* involving two new operators:

$$NEW = \begin{cases} \mathbf{AGE}_{\Rightarrow}(f,g) = \mathbf{AG}(f \Rightarrow \mathbf{E}[g\mathbf{R}f]) \\ \mathbf{AGA}_{\Rightarrow}(f,g) = \mathbf{AG}(f \Rightarrow \mathbf{A}[g\mathbf{R}f]) \end{cases}$$

McMillan's algorithm SMC_{ctl}, known as *symbolic model checking*, consists in verifying if a CTL formula f stands in a boolean-based representation of a Kripke structure. SMC_{ctl} recurses over the syntactic structure of the formula, returning the charateristic function of the set of states where f is valid[2]. f is considered valid in the structure when it is valid in all its initial states. The following set of equations shows the rules that direct McMillan's symbolic model checking:

1. $SMC_{\text{ctl}}(p) = p$ if $p \in P$
2. $SMC_{\text{ctl}}(\neg f) = \neg SMC_{\text{ctl}}(f)$
3. $SMC_{\text{ctl}}(f * g) = SMC_{\text{ctl}}(f) \wedge SMC_{\text{ctl}}(g)$
4. $SMC_{\text{ctl}}(\mathbf{EX}f) = Backward(SMC_{\text{ctl}}(f))$
5. $SMC_{\text{ctl}}(\mathbf{EG}f) = \mathbf{gfp}K.SMC_{\text{ctl}}(f) \wedge Backward(K)$
6. $SMC_{\text{ctl}}(\mathbf{E}[f\mathbf{U}g]) = \mathbf{lfp}K.SMC_{\text{ctl}}(g) \vee (SMC_{\text{ctl}}(f) \wedge Backward(K))$

In these equations, **gfp** and **lfp** denote the greatest and least fixpoints respectively, applied to predicate transformers for the lattice of sets of states. In practice, the computational resources used in symbolic model checking depend on the nesting and number of iterations that are required to compute these fixpoints. In the rest of this paper, we will refer to CTL operators **EG** and **E**[**U**] as *fixpoint operators*.

3 No Fixpoint Computation Tree Logic

No fixpoint computation tree logic (NF-CTL) is a sub-logic of CTL that is such that there are decision procedures with false negatives, based on induction properties of the logic, and that do not require these fixpoint computations.

3.1 Step Temporal Logic

Initially, we define Step Temporal Logic (STL), a temporal logic that does not contain fixpoint operators. The syntax of STL is given by the following definition.

Definition 3. *Given a set of propositions P, $STL(P)$ is the smallest set such that $P \subseteq STL(P)$ and, if f and g are in $STL(P)$, then $f * g, f \oplus g, \neg f, f + g, f \Rightarrow g, \mathbf{EX}f, \mathbf{AX}f$ are in $STL(P)$.*

[2] Actually, the characteristic function may also be valid for valuations of propositions that do not correspond to states of the structure, the so-called non-reachable states.

Therefore, STL is the subset of CTL containing all formulae with no fixpoint operators. The semantics of STL is the same as that of CTL, and the algorithm SMC_{stl} for symbolic model checking of STL consists of the parts of the algorithm SMC_{ctl} handling the boolean and the **EX** operators (items 1 through 4 of the definition of SMC_{ctl}).

3.2 No Fixpoint Computation Tree Logic

NF-CTL is also a subset of CTL, but additionally contains some patterns of properties including fixpoint operators. However, NF-CTL formulae may be model checked without costly fixpoint computations [6]. The following definition gives the syntax of NF-CTL.

Definition 4. *Given a set of propositions P, NFCTL(P) is the smallest set such that*

- $STL(P) \subseteq NFCTL(P)$
- *if f and g are in STL(P), then* $\mathbf{A}[f\mathbf{R}g], \mathbf{E}[f\mathbf{R}g]$, $\mathbf{AGE}_{\Rightarrow}(f,g), \mathbf{AGA}_{\Rightarrow}(f,g)$ *are in NFCTL(P)*
- *if f and g are in NFCTL(P), then* $f * g, f \oplus g, \neg f, f + g, f \Rightarrow g$ *are in NFCTL(P).*

Although at first look, NF-CTL seems to have a restricted expressive power, important classes of CTL formulae can be expressed in this subset: more than 30% of the properties found in the examples distributed with NuSMV [1] are in NF-CTL. The following formulae patterns are all in NF-CTL: $\mathbf{AG}f = \mathbf{A}[0\mathbf{R}f]$ (always), $\mathbf{AG}(f \Rightarrow \mathbf{AG}f) = \mathbf{AG}(f \Rightarrow \mathbf{A}[0\mathbf{R}f])$ (always from), $\mathbf{A}[g\mathbf{R}f]$ (release), $\mathbf{AG}(Req \Rightarrow \mathbf{A}[Ack\mathbf{R}Req])$ (handshaking), and $\mathbf{AF}f = \neg\mathbf{EG}\neg f = \neg\mathbf{E}[0\mathbf{R}\neg f]$ (eventually).

Concerning verification optimization, Table 1 presents sufficient conditions for universal NF-CTL formulae to be true of a Kripke structure [6]. For instance, if $Forward(f \wedge \neg g) \subseteq f$ holds, then we can conclude that the NF-CTL formula $\mathbf{AG}(f \Rightarrow \mathbf{A}[g\mathbf{R}f])$ is valid. If, additionally, $I \models f$ holds, then $\mathbf{A}[g\mathbf{R}f]$ is also valid. When $I \not\models f$, then $\mathbf{A}[g\mathbf{R}f]$ does not hold. Otherwise, we cannot conclude.

While standard model checking verification of these formulae requires one (in the case of $\mathbf{A}[g\mathbf{R}f]$) or two (in the case of $\mathbf{AG}(f \Rightarrow \mathbf{A}[g\mathbf{R}f])$) fixpoint computations, checking these conditions requires at most a single computation of the *Backward* function. Note that each fixpoint computation requires $O(|S|)$ calls to *Backward* (S is the number of states of the Kripke structure).

Table 2 gives sufficient conditions for existential NF-CTL formulae to hold in a Kripke structure [6]. Again, the interesting part of this table lays in the second and fourth lines. If it can be shown that $f \wedge \neg g \subseteq Backward(f)$ is true, we can conclude that the NF-CTL formula $\mathbf{AG}(f \Rightarrow \mathbf{E}[g\mathbf{R}f])$ is valid. Furthermore, if we have $I \models f$, then $\mathbf{E}[g\mathbf{R}f]$ is also valid. Again, standard model checking verification of these formulae requires one or two fixpoint computations while checking the above requirements needs no fixpoint computation at all.

3.3 Algorithms

The improved symbolic model checking algorithm is given in Figure 1. It tests if the formula to be verified is in NF-CTL, through the predicate $NFCTL_R$. In case f is not in NF-CTL, the classical algorithm is invoked. In case f does belong to NF-CTL, the algorithm SMC_{nfctl} tests the sufficient conditions. However, f is not always expressed as one of the NF-CTL template formulae and it would be operationally interesting to get an equivalent formula in one of the desired format. This equivalent formula is denoted $f\downarrow_R$ and its computation is achieved by rewriting as explained in Section 5, using the results developed in Section 4.

function $SMC(f : \mathsf{CTL}) : boolean$
    if $NFCTL_R(f)$ **then return** $SMC_{\text{nfctl}}(f\downarrow_R)$
    else return $SMC_{\text{ctl}}(f)$

Fig. 1. Improved symbolic model checking

The implementation of the predicate $NFCTL_R$ involves the detailed study of the rewriting properties of the logics CTL and NF-CTL, which is developed in Section 4. The algorithm SMC_{nfctl} is given in Figure 2 and simply tests the different sufficient conditions enumerated in Tables 1 and 2.

4 Rewriting of Temporal Logic Formulae

In this section we are interested in defining a notion of normal forms for CTL formulae that would ease the problem of finding an equivalent NF-CTL formulae. Our aim is to reach these normal forms using an appropriate term rewrite system.

4.1 Rewriting of Boolean Formulae

Since Boolean formulae are CTL formulae, we already need a rewrite system to normalize Boolean formulae. On the one hand, it is known that Boolean Algebra admits no convergent term rewrite system [17]. On the other hand, it is known that Boolean Rings admits a term rewrite system which is confluent and terminating modulo the Associativity-Commutativity of *conjunction* (denoted by $*$) and *exclusive or* (denoted by $\oplus$). The Associativity-Commutativity equational theory for a binary operator f is defined as follows:

$$AC(f) = \begin{cases} f(x, f(y, z)) = f(f(x, y), z) \\ f(x, y) \qquad\quad\; = f(y, x) \end{cases}$$

Table 1. Sufficient conditions for universal NF-CTL formulae

$I \subseteq f$	$Forward(f \wedge \neg g) \subseteq f$	$\mathbf{AG}(f \Rightarrow \mathbf{A}[g\mathbf{R}f])$	$\mathbf{A}[g\mathbf{R}f]$
0	0	?	0
0	1	1	0
1	0	?	?
1	1	1	1

Table 2. Sufficient conditions for existential NF-CTL formulae

$I \subseteq f$	$f \wedge \neg g \subseteq Backward(f)$	$\mathbf{AG}(f \Rightarrow \mathbf{E}[g\mathbf{R}f])$	$\mathbf{E}[g\mathbf{R}f]$
0	0	?	0
0	1	1	0
1	0	?	?
1	1	1	1

```
1   function SMC_nfctl(f : CTL) : boolean
2       f₁, f₂ : CTL
3       F1, F2 : BDD
4   %match(f) {
5       ¬f₁ → return not SMC_nfctl(f₁)
6       f₁ * f₂ → return SMC_nfctl(f₁) and SMC_nfctl(f₂)
7       A[f₂Rf₁] → {
8           F₁ ← SMC_stl(f₁)
9           if I⇒_bdd F₁ then
10              F₂ ← SMC_stl(f₂)
11              if Forward(F₁ ∧_bdd ¬_bdd F₂)⇒_bdd F₁ then return true
12              else return SMC_ctl(f)
13          else return false }
14      AGA_⇒(f₁, f₂) → {
15          F₁ ← SMC_stl(f₁), F₂ ← SMC_stl(f₂)
16          if Forward(F₁ ∧_bdd ¬_bdd F₂)⇒_bdd F₁ then return true
17          else return SMC_ctl(f) }
18      E[f₂Rf₁] → {
19          F₁ ← SMC_stl(f₁)
20          if I⇒_bdd F₁ then
21              F₂ ← SMC_stl(f₂)
22              if (F₁ ∧_bdd ¬_bdd F₂)⇒_bdd Backward(F₁) then return true
23              else return SMC_ctl(f)
24          else return false }
25      AGE_⇒(f₁, f₂) → {
26          F₁ ← SMC_stl(f₁), F₂ ← SMC_stl(f₂)
27          if (F₁ ∧_bdd ¬_bdd F₂)⇒_bdd Backward(F₁) then return true
28          else return SMC_ctl(f) }
29  }
```

Fig. 2. Algorithm for SMC_{nfctl}

In the rest of the paper, we consider $AC = AC(\oplus) \cup AC(*)$. We assume the reader familiar with standard notations in the field of equational reasoning. Our notations are standard in rewriting theory [7]. For each set of rules R presented in this paper (involving possibly AC operators), we consider the rewrite relation $\longrightarrow_R$ defined as follows: $s \longrightarrow_R t$ if there exist a rule $l \to r \in R$, a position ω in the term s, and a substitution σ of variables in l such that (1) the subterm of s at position ω, denoted by $s_{|\omega}$, is equal modulo AC to the σ-instance of l, and (2) t is the term obtained from s by replacing $s_{|\omega}$ with the σ-instance of r. When the rewrite rule $l \to r$, the position ω, and the substitution σ are known or necessary, we may indicate them as superscript of $\longrightarrow_R$. The rewrite relation $\longrightarrow_R$ corresponds to the classical rewrite relation modulo AC denoted by $\longrightarrow_{R,AC}$ in [11], where syntactic matching used in standard (syntactic) rewriting is replaced by AC-matching. We refer to [11] for standard definitions of confluence and termination modulo an equational theory.

Definition 5. *Given a set of propositions P, $Bool(P)$ is the smallest set such that (1) $P \cup \{0,1\} \subseteq Bool(P)$, and (2) if f and g are in $Bool(P)$, then $f * g, f \oplus g, \neg f, f + g, f \Rightarrow g$ are in $Bool(P)$.*

The equational presentation for Boolean Rings below is the classical one, where we have three axioms defining standard boolean operators, $\neg$ (not), $\Rightarrow$ (implication), $+$ (or), in terms of operators $\oplus, *, 0, 1$.

$$BR = \left\{ \begin{array}{ll} x * x & = x \\ x * 1 & = x \\ x * 0 & = 0 \\ x * (y \oplus z) & = (x * y) \oplus (x * z) \\ x \oplus x & = 0 \end{array} \right. \qquad \begin{array}{ll} x \oplus 0 & = x \\ \neg x & = x \oplus 1 \\ x \Rightarrow y & = x \oplus (x * y) \oplus 1 \\ x + y & = (x * y) \oplus x \oplus y \end{array}$$

From now on, we assume that $*$ has a greater priority than $\oplus$, and so we omit parentheses in boolean expressions. Let $BR^{\rightarrow}$ be the rewrite system obtained from the left-to-right orientation of axioms in BR.

Proposition 1. *$BR^{\rightarrow}$ is confluent and terminating modulo AC.*

Corollary 1. *AC-rewriting wrt. $BR^{\rightarrow}$ provides a decision algorithm for the equality modulo the equational theory of Boolean Rings defined by $BR \cup AC$. Every formula in $Bool(P)$ can be uniquely normalized wrt. $BR^{\rightarrow}$ to a formula involving only boolean operators $0, 1, \oplus, *$.*

4.2 Rewriting of CTL Formulae

The set of rules for Boolean Rings can be enriched with rules defining temporal operators. The resulting term rewrite system allows us to reduce formulae in $CTL(P)$. Let $opCTL^{\rightarrow}$ be the rewrite system obtained from the left-to-right orientation of axioms in $opCTL$ (cf Section 2.2).

Proposition 2. *$opCTL^{\rightarrow} \cup BR^{\rightarrow}$ is confluent and terminating modulo AC.*

Proof. It is confluent since left-hand sides of $opCTL^{\rightarrow}$ do not contain *AC*-operators, and there is no critical pairs in $opCTL^{\rightarrow}$. Moreover, any peak between a rule in $BR^{\rightarrow}$ and a rule in $opCTL^{\rightarrow}$ commutes, because left-hand sides in $opCTL^{\rightarrow}$ are built over a signature which is different from the signature of $BR^{\rightarrow}$.

Termination can be proved by using the fully syntactic AC-RPO defined in [16], with a precedence $\succ$ such that: $\Rightarrow \succ + \succ \neg \succ * \succ \oplus \succ 1 \succ 0$, $\mathbf{AG} \succ \mathbf{EF}$, $\mathbf{E}[\mathbf{R}] \succ \mathbf{A}[\mathbf{U}]$, CTL operators are greater than boolean operators, and CTL operators occurring in left-hand sides of $opCTL^{\rightarrow}$ are greater than operators in $\{\mathbf{EX}, \mathbf{EG}, \mathbf{E}[\mathbf{U}]\}$.

□

Let $\mathcal{CTL} = opCTL \cup BR \cup AC$.

Corollary 2. *AC-rewriting wrt.* $opCTL^{\rightarrow} \cup BR^{\rightarrow}$ *provides a decision algorithm for the equality modulo* $\mathcal{CTL}$. *Every formula in* $CTL(P)$ *can be uniquely normalized wrt.* $opCTL^{\rightarrow} \cup BR^{\rightarrow}$ *to a formula involving only temporal operators* $\mathbf{EX}, \mathbf{EG}, \mathbf{E}[\mathbf{U}]$ *and boolean operators* $0, 1, \oplus, *$.

4.3 NF-CTL Membership

The improved symbolic model checking function designed for NF-CTL formulae requires an algorithm for finding an equivalent formula that belongs to NF-CTL. The equivalent formula will be the normal form computed by rewriting. We have seen that the class of NF-CTL formulae is defined using operators **EX**, $\mathbf{A}[\mathbf{R}]$, $\mathbf{E}[\mathbf{R}]$, $\mathbf{AGE}_{\Rightarrow}(,)$, $\mathbf{AGA}_{\Rightarrow}(,)$. Consequently, we would like to retrieve these operators in normal forms. Hopefully, we can translate axioms in E in such way that temporal operators are expressed using **EX**, $\mathbf{A}[\mathbf{R}]$, $\mathbf{E}[\mathbf{R}]$. This leads to the following equational theory, where equalities are obtained from E (cf. Section 2.2) thanks to the principle of replacement of equals for equals (using Boolean axioms).

$$D = \begin{cases} \mathbf{AX}f &= \neg\mathbf{EX}\neg f \\ \mathbf{E}[f\mathbf{U}g] &= \neg\mathbf{A}[\neg f\mathbf{R}\neg g] \\ \mathbf{A}[f\mathbf{U}g] &= \neg\mathbf{E}[\neg f\mathbf{R}\neg g] \qquad \mathbf{E}[0\mathbf{R}0] = 0 \\ \mathbf{EF}f &= \neg\mathbf{A}[0\mathbf{R}\neg f] \qquad \mathbf{A}[f\mathbf{R}1] = 1 \\ \mathbf{AF}f &= \neg\mathbf{E}[0\mathbf{R}\neg f] \qquad \mathbf{EX}0 = 0 \\ \mathbf{EG}f &= \mathbf{E}[0\mathbf{R}f] \\ \mathbf{AG}f &= \mathbf{A}[0\mathbf{R}f] \end{cases}$$

Let $BRR = BR^{\rightarrow} \cup D^{\rightarrow}$.

Proposition 3. *BRR is confluent and terminating modulo AC.*

Proof. Similar to the proof of Proposition 2.

□

We have not yet considered the case of operators $\mathbf{AGE}_{\Rightarrow}(,)$ and $\mathbf{AGA}_{\Rightarrow}(,)$. Like other operators, $\mathbf{AGE}_{\Rightarrow}(,)$ (resp. $\mathbf{AGA}_{\Rightarrow}(,)$) can be expressed with $\mathbf{A}[\mathbf{R}]$ and $\mathbf{E}[\mathbf{R}]$:

$$\mathbf{AGE}_{\Rightarrow}(f,g) = \mathbf{AG}(f \Rightarrow \mathbf{E}[g\mathbf{R}f]) = \mathbf{A}[0\mathbf{R}(1 \oplus f \oplus f * \mathbf{E}[g\mathbf{R}f])]$$
$$\mathbf{AGA}_{\Rightarrow}(f,g) = \mathbf{A}[0\mathbf{R}(1 \oplus f \oplus f * \mathbf{A}[g\mathbf{R}f])]$$

Since we want to have occurrences of $\mathbf{AGE}_{\Rightarrow}(,)$ and $\mathbf{AGA}_{\Rightarrow}(,)$ in normal forms, we would like now to orient these two equations from right to left:

$$\mathbf{A}[0\mathbf{R}(1 \oplus f \oplus f * \mathbf{E}[g\mathbf{R}f])] \to \mathbf{AGE}_{\Rightarrow}(f,g) \quad (r)$$
$$\mathbf{A}[0\mathbf{R}(1 \oplus f \oplus f * \mathbf{A}[g\mathbf{R}f])] \to \mathbf{AGA}_{\Rightarrow}(f,g) \quad (r')$$

which means that some expressions of the form $\mathbf{A}[0\mathbf{R}?]$ will still be reduced. But considering the previous rules is too naive, if we want to obtain a confluent (modulo AC) term rewrite system. Indeed, we need to take care of the shape of the BRR-normal form of f: it can be 0, 1, a negative (in the form $u \oplus 1$), or a positive.

Definition 6. *A BRR-normal form f is* negative *if there exists a term u such that $f =_{AC} u \oplus 1$. A BRR-normal form is* positive *if it is neither negative, nor 0, nor 1.*

In the sequel, we consider four conditional variants of r and r', where the different cases are expressed by conditions. The rewrite rule r leads to the rewrite system $ARE = \{r1, \ldots, r4\}$, and similarly r' leads to $ARA = \{r1', \ldots, r4'\}$:

$$\begin{cases} \mathbf{A}[0\mathbf{R}(x \oplus y * \mathbf{E}[g\mathbf{R}f])] \to \mathbf{AGE}_{\Rightarrow}(f,g) & \\ \quad \text{if } \begin{pmatrix} x \oplus y * \mathbf{E}[g\mathbf{R}f] \text{ is in } BRR\text{-normal form} \\ x \oplus y * \mathbf{E}[g\mathbf{R}f] =_{AC} (1 \oplus f \oplus f * \mathbf{E}[g\mathbf{R}f]) \downarrow_{BRR}, \\ f \text{ is positive} \end{pmatrix} & (r1) \\ \mathbf{A}[0\mathbf{R}(x \oplus \mathbf{E}[g\mathbf{R}f \oplus 1])] \to \mathbf{AGE}_{\Rightarrow}(f \oplus 1, g) & \\ \quad \text{if } \begin{pmatrix} x \oplus \mathbf{E}[g\mathbf{R}f \oplus 1] \text{ is in } BRR\text{-normal form} \\ x \oplus \mathbf{E}[g\mathbf{R}f \oplus 1] =_{AC} (f \oplus \mathbf{E}[g\mathbf{R}f \oplus 1] \oplus f * \mathbf{E}[g\mathbf{R}f \oplus 1]) \downarrow_{BRR}, \\ f \oplus 1 \text{ is negative} \end{pmatrix} & (r2) \\ \mathbf{A}[0\mathbf{R}\ \mathbf{E}[g\mathbf{R}1]] \to \mathbf{AGE}_{\Rightarrow}(1,g) & (r3) \\ \mathbf{AGE}_{\Rightarrow}(0,g) \to \mathbf{A}[0\mathbf{R}1] & (r4) \end{cases}$$

Analogously, ARA can be defined from ARE by replacing occurrences of $\mathbf{E}[\mathbf{R}]$ (resp. $\mathbf{AGE}_{\Rightarrow}(,)$) with $\mathbf{A}[\mathbf{R}]$ (resp. $\mathbf{AGA}_{\Rightarrow}(,)$). For conditional rewrite systems ARE and ARA, it is not obvious to see that conditions of first and second rule can be satisfied, and so to be convinced that the related rewrite relation is non-empty. This is clarified by the following lemma.

Lemma 1. *Let f be a positive BRR-normal form. For each of the following terms:*

1. $r1^-(f,g) := (1 \oplus f \oplus f * \mathbf{E}[g\mathbf{R}f]) \downarrow_{BRR}$
2. $r2^+(f,g) := (f \oplus \mathbf{E}[g\mathbf{R}f \oplus 1] \oplus f * \mathbf{E}[g\mathbf{R}f \oplus 1]) \downarrow_{BRR}$

there exist two terms x, y such that

1. $r1^-(f,g) =_{AC} x \oplus y * \mathbf{E}[g\mathbf{R}f]$ *and* $r1^-(f,g)$ *is negative;*
2. $r2^+(f,g) =_{AC} x \oplus \mathbf{E}[g\mathbf{R}f \oplus 1]$ *and* $r2^+(f,g)$ *is positive.*

Note that a similar lemma can be stated for ARA. Let us now consider the following rewrite system $R = BRR \cup ARE \cup ARA$. The rewrite relation generated by R mimics the equational theory $\mathcal{CTL}$, as stated below.

Proposition 4. *For any* $f, g \in CTL(P)$, *we have* $f =_{\mathcal{CTL}} g \Longleftrightarrow f \longleftrightarrow^*_{R \cup AC} g$

At this point, a natural question arises: is R a confluent rewrite system? And if not, is it possible to turn R into an equivalent confluent rewrite system? In order to give an answer to this question, we first consider a subsystem of R.

Lemma 2. *$BRR \cup ARE$ is confluent modulo AC.*

Proof. Assume there exists a peak $t_1 \xleftarrow{\epsilon, r1}_{ARE} \mathbf{A}[0\mathbf{R}t] \xrightarrow{\epsilon, r2}_{ARE} t_2$. Lemma 1 implies that, on the one hand t is positive and on the other hand t is negative, which is impossible. Due to the use of conditions, there is no other critical pairs in ARE, and no critical pairs between BRR and ARE since the binary operator $\mathbf{A}[\mathbf{R}]$ does not occur in left-hand sides of rules in BRR.

□

The equational theory ARA is not just like ARE. Consider the first rule in ARA, say $r1'$. Its left-hand side contains two nested occurrences of $\mathbf{A}[\mathbf{R}]$, and there is a critical pair of $r1'$ on itself. This may be problematic for the confluence of $BRR \cup ARA$, when reducing complicated CTL formulae with many nested $\mathbf{A}[\mathbf{R}]$ operators. But it is no more a problem if we consider simple CTL formulae, like for instance NF-CTL formulae. To make precise what we call simple and complex CTL formulae, we stratify formulae in CTL with respect to the nesting of operators $\mathbf{A}[\mathbf{R}]$ and $\mathbf{E}[\mathbf{R}]$. The following definition is motivated by the fact that formulae in NF-CTL cannot have more than two nested *release* operators.

Definition 7. *Let* $f \in CTL(P)$. *The* depth of releases *in* f, *denoted by* $dr(f)$, *is inductively defined as follows:*

- $dr(0) = dr(1) = 0$ *and* $dr(p) = 0$ *if* $p \in P$,
- $dr(\mathbf{AX}f) = dr(\mathbf{EX}f) = dr(\neg f) = dr(f)$,
- $dr(f \oplus g) = dr(f * g) = \max(dr(f), dr(g))$,
- $dr(\mathbf{EF}f) = dr(\mathbf{AF}f) = dr(\mathbf{EG}f) = dr(\mathbf{AG}f) = 1 + dr(f)$,
- $dr(\mathbf{E}[f\mathbf{U}g]) = dr(\mathbf{A}[f\mathbf{U}g]) = 1 + \max(dr(f), dr(g))$,
- $dr(\mathbf{E}[f\mathbf{R}g]) = dr(\mathbf{A}[f\mathbf{R}g]) = 1 + \max(dr(f), dr(g))$,
- $dr(\mathbf{AGE}_{\Rightarrow}(f,g)) = dr(\mathbf{AGA}_{\Rightarrow}(f,g)) = 2 + \max(dr(f), dr(g))$

The normalization wrt. BRR does not increase the depth of releases. Given any $f \in CTL(P)$, let $f \downarrow_{NEW}$ be the normal form of f wrt. the confluent and terminating rewrite system $NEW^{\rightarrow}$, and let $\hat{f}$ be the term $(f \downarrow_{NEW}) \downarrow_{BRR}$.

Proposition 5. *For any* $f \in CTL(P)$, $dr(\hat{f}) \leq dr(f)$.

Let $CTL^n(P) = \{f \in CTL(P) \mid dr(f) \leq n\}$. In the rest of this section, we concentrate on $CTL^2(P)$ since it includes $NFCTL(P)$. By considering only formulae in $CTL^2(P)$, ARA becomes confluent, and there is no critical pair between rules in ARA and rules in $BRR \cup ARE$. Moreover, the termination of R can be easily proved using a lexicographic combination.

Proposition 6. *R is confluent and terminating modulo AC over formulae in $CTL^2(P)$.*

Proposition 7. *For any $f \in NFCTL(P)$, $dr(f) \leq 2$ and $f \downarrow_R \in NFCTL(P)$.*

We are now ready to state our main result.

Proposition 8. *The following problem:* "Given a formula f in $CTL(P)$, is there a formula g in $NFCTL(P)$ such that $f =_{CTL} g$?" *is decidable using AC-rewriting wrt. R.*

Proof. We prove the following statement: for each f in $CTL(P)$, there exists a formula g in $NFCTL(P)$ such that $f =_{CTL} g$ if and only if $dr(\hat{f}) \leq 2$ and $\hat{f} \downarrow_R \in NFCTL(P)$.

- ($\Rightarrow$) $f =_{CTL} g$ implies $\hat{f} =_{AC} \hat{g}$. According to Proposition 7, we have $dr(g) \leq 2$ and $g \downarrow_R \in NFCTL(P)$. Since $dr(\hat{g}) \leq dr(g)$ (by Proposition 5) and $\hat{g} \downarrow_R =_{AC} g \downarrow_R$, we get $dr(\hat{g}) \leq 2$ and $\hat{g} \downarrow_R \in NFCTL(P)$. Since $\hat{f} =_{AC} \hat{g}$, we can conclude that $dr(\hat{f}) \leq 2$ and $\hat{f} \downarrow_R \in NFCTL(P)$.
- ($\Leftarrow$) If $dr(\hat{f}) \leq 2$ and $\hat{f} \downarrow_R \in NFCTL(P)$, then, we can choose $g = \hat{f} \downarrow_R$ since $f =_{CTL} \hat{f} \downarrow_R$.

□

The proof of Proposition 8 motivates the definition of the following predicate used in the improved symbolic model checking function (see Figure 1):

$$NFCTL_R(f) \Leftrightarrow (dr(\hat{f}) \leq 2 \ \wedge \ \hat{f} \downarrow_R \in NFCTL(P))$$

4.4 Reducing the Number of Temporal Operators in CTL Formulæ

An interesting research direction consists now in combining the rewrite system R with other simplification rules, which are still valid in all Kripke structures. The following simplification rules aim at reducing the number of temporal operators occurring in a CTL formula. We can distinguish two sets of rules. The first set consists of rules built over temporal operators and boolean operators, whilst the second set is made of rules involving only temporal operators. Both sets of rules do not increase what we call *the depth of releases.*

$$S_1 = \begin{cases} \mathbf{AG}f * \mathbf{AG}g & \rightarrow \mathbf{AG}(f * g) \\ \mathbf{EF}f + \mathbf{EF}g & \rightarrow \mathbf{EF}(f + g) \\ \mathbf{EG}((g + \mathbf{EG}f) + (f + \mathbf{EG}g)) & \rightarrow \mathbf{EG}f + \mathbf{EG}g \\ \mathbf{AF}((g * \mathbf{AF}f) + (f * \mathbf{AF}g)) & \rightarrow \mathbf{AF}f * \mathbf{AF}g \\ \mathbf{A}[(f * f')\mathbf{U}(\mathbf{A}[f\mathbf{U}g] * g' + \mathbf{A}[f'\mathbf{U}g'] * g)] & \rightarrow \mathbf{A}[f\mathbf{U}g] * \mathbf{A}[f'\mathbf{U}g'] \end{cases}$$

$$S_2 = \left\{ \begin{array}{llll}
\mathbf{AGAG}f & \to \mathbf{AG}f & \mathbf{AFAGAF}f & \to \mathbf{AGAF}f \\
\mathbf{EFEF}f & \to \mathbf{EF}f & \mathbf{EGEFEG}f & \to \mathbf{EFEG}f \\
\mathbf{AGEFAGEF}f & \to \mathbf{EFAGEF}f & \mathbf{AGAFAG}f & \to \mathbf{AFAG}f \\
\mathbf{EFAGEFAG}f & \to \mathbf{AGEFAG}f & \mathbf{EFEGEF}f & \to \mathbf{EGEF}f \\
\mathbf{EGEG}f & \to \mathbf{EG}f & \mathbf{A}[f\mathbf{UA}[f\mathbf{U}g]] & \to \mathbf{A}[f\mathbf{U}g] \\
\mathbf{AFAF}f & \to \mathbf{AF}f & & \\
\mathbf{AFEGAF}f & \to \mathbf{EGAF}f & \mathbf{AFEGAGAF}f & \to \mathbf{EGAGAF}f \\
\mathbf{EGAFEG}f & \to \mathbf{AFEG}f & \mathbf{EGAFEFEG}f & \to \mathbf{AFEFEG}f \\
\mathbf{AGEG}f & \to \mathbf{AG}f & \mathbf{AGAFEFAGEF}f & \to \mathbf{AFEFAGEF}f \\
\mathbf{EFAF}f & \to \mathbf{EF}f & \mathbf{EFEGAGEFAG}f & \to \mathbf{EGAGEFAG}f
\end{array} \right.$$

On can remark that rules in S_2 strictly decreases the *depth of releases*, when rules are applied at top-position of terms.

Proposition 9. *$S_1 \cup S_2$ is terminating.*

Proof. Each rule of $S_1 \cup S_2$ strictly decreases the number of temporal operators. □

We have used the daTac theorem-prover [18] to check that S_1 and S_2 are confluent and terminating modulo AC. The last four rules of S_2 are critical pairs computed by daTac. However, $S_1 \cup S_2$ is not confluent since $(\mathbf{AGAG}p) * (\mathbf{AG}q)$ has two different $S_1 \cup S_2$-normal forms which are $\mathbf{AG}((\mathbf{AG}p) * q)$ and $\mathbf{AG}(p * q)$. Even if $S_1 \cup S_2$ is not confluent, it is clearly interesting to perform rewriting wrt. $S_1 \cup S_2$, and to check after each rewrite step whether or not the result term t is such that $NFCTL_R(t)$. If this is the case, we will be able to apply the improved model checking function.

5 A Rule-Based Implementation

The rewrite system R described above has been implemented using the ELAN [3] rule-based system, which provides an environment for specifying and executing rewrite programs in a language based on rewrite rules guided by strategies expressed with a strategy language. Obviously, the key feature we use from ELAN is AC-rewriting. Indeed, an ELAN program can be evaluated by two execution tools (an interpreter and a compiler) both supporting AC-rewriting. The implementation of rules in BRR is completely straightforward. Then, we declare rules in $ARE \cup ARA$ **after** rules in BRR. The order of rule declaration is significant in ELAN, since it gives implicitly the order of rule application. It means that terms involved in conditions of $ARE \cup ARA$ will be implicitly BRR-normalized. Therefore, ELAN rules corresponding to rules in $ARE \cup ARA$ can be easily implemented without an explicit call to BRR-normalization. For instance, the first rule of ARE will be declared in ELAN as the last default rule for $\mathbf{A}[0\mathbf{R}?]$:

```
[] AR[0 R x xor y*E[g R f]] => AGEimp(f,g)
  if x xor y*E[g R f] == 1 xor f xor f*E[g R f]
```

Note that terms involved in the condition are automatically *BRR*-normalized, and the equality test == corresponds to *AC*-equality. The boolean function implementing NF-CTL membership is also defined by a set of ELAN rules (as well as the function implementing the depth of releases), like the following one:

```
[] NF-CTL(E[f1 R f2]) => t if STL(f1) and STL(f2)
```

We have first tested our rewrite program with the ELAN interpreter on a set of CTL formulae [8]. We were also able to compile our rewrite program with the ELAN compiler. It generates first the related C code, and then a stand-alone executable. The communication between this executable and the symbolic model checker is possible via Unix pipes. Another more elegant solution would be to link the generated C code of the rewrite program with the C code implementing symbolic model checking. In this direction, we envision to use the TOM [14] pattern-matching compiler, which is a part of the ongoing effort to develop an ELAN toolkit. TOM can be viewed as a pre-processor for classical imperative languages such as C or Java. It allows us to combine rule-based declarations and C code, and to declare functions by pattern-matching like in functional programming. An *AC*-extension of TOM would provide us a way to compile rewrite rules in R and to link the resulting C code with, for instance, a C library for manipulating Binary Decision Diagrams.

6 Conclusion

We have shown an algorithm for testing if a CTL formula is equivalent to a formula in a subclass of CTL for which symbolic model checking can be improved thanks to less fixpoint computations. Our algorithm is based on normalization of formulae with respect to a rewrite system. This normalization is followed by a membership test, to check whether the normal form belongs to a particular subclass of CTL. In the future, we plan to generalize our approach by studying how a rewrite sytem such as R can be used to consider larger subclasses of CTL with more nested temporal operators, like the one introduced in [6]. Another interesting open problem consists in combining our rewrite system R with simplification rules dedicated to the elimination of *superfluous* temporal operators. A set of simplification rules is already identified [9], but it is not confluent and so we have to discover what would be a good winning rewriting strategy. At the implementation level, our long-term project is to combine rewriting techniques for simplifying CTL formulae with symbolic model checking of simplified CTL formulae. This future perspective opens interesting implementation problems. It is a typical application where we want to use rewriting techniques (and compilation of rewriting) in the area of symbolic model checking, in which efficient implementations are usually written with imperative programming languages.

References

1. NuSMV home page. http://nusmv.irst.itc.it, accessed on Apr. 23 2002.
2. A. Biere, A. Cimatti, E. M. Clarke, and Y. Zhu. Symbolic model checking without BDDs. In *Tools and Algorithms for Construction and Analysis of Systems*, pages 193–207, 1999.
3. P. Borovanský, C. Kirchner, H. Kirchner, P.-E. Moreau, and C. Ringeissen. An Overview of ELAN. In C. Kirchner and H. Kirchner, editors, *Proc. Second Intl. Workshop on Rewriting Logic and its Applications*, Electronic Notes in Theoretical Computer Science, Pont-à-Mousson (France), Sept. 1998. Elsevier.
4. E. Clarke and E. A. Emerson. Design and synthesis of synchronization skeletons for branching time temporal logic. In *Logics of Programs: Workshop*, volume 131 of *LNCS*, pages 52–71. Springer Verlag, 1981.
5. E. Clarke, O. Grumberg, and D. Peled. *Model Checking*. MIT Press.
6. D. Déharbe and A. M. Moreira. Symbolic model checking with fewer fixpoint computations. In *World Congress on Formal Methods and their Application(FM'99)*, volume 1708 of *LNCS*, pages 272–288, 1999.
7. N. Dershowitz and J.-P. Jouannaud. Rewrite systems. In *Handbook of Theorectical Computer Science*, chapter 15. Elsevier Science Publishers B.V., 1990.
8. M. B. Dwyer, G. S. Avrunin, and J. C. Corbett. Patterns in property specifications for finite-state verification. Technical Report UM-CS-1998-035, 1998.
9. S. Graf. *Logique du temps arborescent pour la spécification et preuve de programmes.* PhD thesis, Institut National Polytechnique de Grenoble, France, 1984.
10. H. Iwashita, T. Nakata, and F. Hirose. CTL model checking based on forward state traversal. In *ICCAD'96*, page 82, 1996.
11. J.-P. Jouannaud and H. Kirchner. Completion of a set of rules modulo a set of equations. *SIAM Journal on Computing*, 15(4):1155–1194, 1986.
12. R. P. Kurshan. *Computer-Aided Verification of Coordinating Processes.* Princeton Univ Pr, 1995.
13. K. McMillan. *Symbolic Model Checking.* Kluwer Academic Publishers, 1993.
14. P.-E. Moreau, C. Ringeissen, and M. Vittek. A pattern-matching compiler. In D. Parigot and M. van den Brand, editors, *Proceedings of the 1st International Workshop on Language Descriptions, Tools and Applications*, volume 44, Genova, april 2001. Electronic Notes in Theoretical Computer Science.
15. J.-P. Queille and J. Sifakis. Specification and verification of concurrent systems in CESAR. In *Procs.* 5th *international symposium on programming*, volume 137 of *Lecture Notes in Computer Science*, pages 244–263. Springer Verlag, 1981.
16. A. Rubio. A Fully Syntactic AC-RPO. In P. Narendran and M. Rusinowitch, editors, *Rewriting Techniques and Applications, 10th International Conference, RTA-99*, LNCS 1631, pages 133–147, Trento, Italy, July 2–4, 1999. Springer-Verlag.
17. R. Socher-Ambrosius. Boolean Algebra Admits No Convergent Term Rewriting System. In R. V. Book, editor, *Rewriting Techniques and Applications, 4th International Conference, RTA-91*, LNCS 488, pages 264–274, Como, Italy, Apr. 10–12, 1991. Springer-Verlag.
18. L. Vigneron. Automated Deduction Techniques for Studying Rough Algebras. *Fundamenta Informaticae*, 33(1):85–103, Feb. 1998.

Conditions for Efficiency Improvement by Tree Transducer Composition

Janis Voigtländer*

Department of Computer Science, Dresden University of Technology
01062 Dresden, Germany. voigt@tcs.inf.tu-dresden.de

Abstract. We study the question of efficiency improvement or deterioration for a semantic-preserving program transformation technique based on macro tree transducer composition. By annotating functional programs to reflect the internal property "computation time" explicitly in the computed output, and by manipulating such annotations, we formally prove syntactic conditions under which the composed program is guaranteed to be more efficient than the original program, with respect to call-by-need reduction to normal form. The developed criteria can be checked automatically, and thus are suitable for integration into an optimizing functional compiler.

1 Introduction

Lazy functional languages are well suited for a modular programming style, where a task is solved by combining solutions of subproblems. Unfortunately, modular programs often lack efficiency compared to other — often less understandable — programs that solve the same tasks. These inefficiencies are caused, e.g., by the production and consumption of structured intermediate results such as lists or trees. As an example, consider the following definitions in **Haskell**:

data $Nat = S\ Nat \mid Z$

$exp :: Nat \to Nat \to Nat$	$div :: Nat \to Nat$	$div' :: Nat \to Nat$
$exp\ (S\ x)\ y = exp\ x\ (exp\ x\ y)$	$div\ (S\ x) = div'\ x$	$div'\ (S\ x) = S\ (div\ x)$
$exp\ \ Z\ \ y = S\ y$	$div\ \ Z\ \ = Z$	$div'\ \ Z\ \ = Z$

The function *exp* — computing $exp\ (S^n\ Z)\ (S^m\ Z) = S^{2^n+m}\ Z$ — is defined using an accumulating parameter y, which will also be called context parameter henceforth. The functions *div* and *div'* — computing $div\ (S^n\ Z) = S^{n\ div\ 2}\ Z$ — are defined by mutual recursion. If we sequentially compose *exp* and *div* — by computing for some value t of type *Nat* the expression $e = (div\ (exp\ t\ Z))$ — an intermediate data structure is created by *exp* and consumed by the *div*- and *div'*-functions. A standard technique for eliminating intermediate results is *classical deforestation* [13], an algorithmic instance of the *unfold/fold*-technique [1]. However, classical deforestation does not succeed in optimizing e, due to its

* Research supported by the DFG under grant KU 1290/2-1.

S. Tison (Ed.): RTA 2002, LNCS 2378, pp. 222–236, 2002.

well-known problem of not reaching accumulating parameters [2]. Also, classical deforestation was only proved to be non-deteriorating for linear programs or for call-by-name evaluation without sharing [11].

Kühnemann [7,8] tackled the problem of eliminating intermediate data structures in accumulating parameters by using composition results from the theory of tree transducers [5]. The functional programs considered are extended schemes of primitive recursion — allowing mutual recursion and nesting of terms in context parameter positions — so called *macro tree transducers* (for short *mtts*). Already Engelfriet and Vogler [4] showed that the sequential composition of two mtts can be realized by a single mtt, if one of the original mtts is defined without using context parameters. For the above example, e would thus be transformed to $e' = (f\ t\ (div\ Z)\ (div'\ Z))$, with new functions f and g constructed as[1]:

$$\begin{array}{ll} f\ (S\ x)\ y_1\ y_2 &= f\ x\ (f\ x\ y_1\ y_2)\ (g\ x\ y_1\ y_2) \\ f\ \ \ Z\ \ \ y_1\ y_2 &= y_2 \\ g\ (S\ x)\ y_1\ y_2 &= g\ x\ (f\ x\ y_1\ y_2)\ (g\ x\ y_1\ y_2) \\ g\ \ \ Z\ \ \ y_1\ y_2 &= S\ y_1 \end{array}$$

The transformed program avoids the creation of an intermediate result and its eventual consumption, with obvious benefits for the efficiency. In particular, e' needs fewer lazy evaluation steps for reduction to normal form than e.

In general, the number of call-by-need reduction steps performed by the transformed program might also *increase* compared to the original one. If, for example, we consider the expression $(exp\ (div\ t)\ Z)$, tree transducer composition can again eliminate the intermediate result. However, in this case the transformed program will — except for very small inputs — perform more call-by-need reduction steps than the original program.

Clearly, in order to use tree transducer composition as a program transformation technique in an optimizing compiler, we should be able to discriminate programs for which the transformation degrades efficiency from those for which the transformation is indeed beneficial. In this paper we develop such criteria that can be checked automatically by a compiler. These criteria are sufficient to classify the various examples of the composition techniques from [4] given in [9], where the performance improvement achieved by tree transducer composition for particular programs was assured by ad hoc reasoning or by experiments. Our results improve on formal efficiency considerations by Kühnemann [8] for linear programs and by Höff [6] also for nonlinear ones. For example, our criteria can detect that the replacement of e from the introductory example by e' is safe with respect to efficiency, which was not captured by previous results. Since in the case that the first involved mtt has no context parameters the tree transducer composition technique is equivalent to classical deforestation (cf. [8]), our results also establish call-by-need performance improvements through classical deforestation for some nonlinear programs.

The remainder of this paper is organized as follows. In Sect. 2 we define basic notations and concepts of macro tree transducer units. Section 3 recalls

[1] The basic idea is to construct definitions for f and g such that for every $t = (S^n\ Z)$ and $t_1 = (S^m\ Z)$: $f\ t\ (div\ t_1)\ (div'\ t_1) = div\ (exp\ t\ t_1)$

$$g\ t\ (div\ t_1)\ (div'\ t_1) = div'\ (exp\ t\ t_1)\ .$$

program transformation by tree transducer composition. In Sect. 4 we develop our formal efficiency analysis and give the main theorems, with application to tree transducer composition and classical deforestation. Finally, Sect. 5 contains future research topics.

2 Preliminaries

We denote by $\mathbb{N}$ the set of natural numbers including 0. For $n \in \mathbb{N}$, we denote by $[n]$ the set $\{1, \ldots, n\} \subseteq \mathbb{N}$, and by X_n the finite set $\{x_1, \ldots, x_n\}$ of variables; analogously for Y_n and Z_n.

We denote simultaneous substitution of $v_1, \ldots, v_n$ for $u_1, \ldots, u_n$ in v by $v[u_1, \ldots, u_n \leftarrow v_1, \ldots, v_n]$, but will also use an alternative notation similar to set comprehensions, e.g., $v[u_i \leftarrow v_i \mid i \in [n]]$. We write substitutions left-associative.

A *ranked alphabet* is a pair $(\Sigma, rank_\Sigma)$, where Σ is a finite, nonempty set of symbols and $rank_\Sigma$ assigns to every $\sigma \in \Sigma$ a natural number k, which will also be given by writing $\sigma^{(k)}$. For every $k \in \mathbb{N}$, we define $\Sigma^{(k)} = \{\sigma \in \Sigma \mid rank_\Sigma(\sigma) = k\}$. For every set A disjoint from Σ, we define the set $T_\Sigma(A)$ of *trees over Σ indexed by A* as the smallest set T such that *(i)* $A \subseteq T$ and *(ii)* if $\sigma \in \Sigma^{(k)}$ and $t_1, \ldots, t_k \in T$, then also $(\sigma\ t_1 \cdots t_k) \in T$. If readability allows, outer brackets of trees will be omitted. We denote $T_\Sigma(\emptyset)$ by T_Σ. For every set $C \subseteq \Sigma \cup A$, we denote the *number of occurrences of symbols from C in a tree t* by $|t|_C$. For a singleton set $C = \{c\}$, we denote $|t|_{\{c\}}$ by $|t|_c$.

We consider *left-linear rewrite systems* [3] with *rewrite rules* of the form $lhs = rhs$ — with lhs and rhs being trees over ranked alphabets indexed by a set of variables, which will usually not be mentioned explicitly — where the right-hand side rhs contains only variables that also occur in lhs. A rewrite system R induces a binary nondeterministic *reduction relation* $\Rightarrow_R$ describing left-to-right application of a rule from R in some context. We use the well-known concepts of *confluence*, *termination* and *normal form*. For a confluent and terminating reduction relation $\Rightarrow_R$, we denote the unique normal form of a tree t with respect to $\Rightarrow_R$ by $nf(\Rightarrow_R, t)$. While $\Rightarrow_R$ does not fix a reduction strategy, our efficiency analysis is concerned with lazy functional programming languages. Hence, we consider *call-by-need reduction steps* [14] (leftmost-outermost reduction with sharing) and denote by $cbn(R, t)$ the number of such steps required to reach the normal form $nf(\Rightarrow_R, t)$.

We model functional programs by macro tree transducer units (for short *mtt units*). Firstly, we describe the possible shapes of right-hand sides of their rules.

Definition 1. *Let Q and Δ be ranked alphabets and $k, r \in \mathbb{N}$.*
The set $RHS(Q, \Delta, k, r)$ of right-hand sides over Q and Δ, with k recursion variables and r context variables, *is the smallest set $RHS \subseteq T_{\Delta \cup Q}(X_k \cup Y_r)$ such that (i) $Y_r \subseteq RHS$, (ii) for every $\delta \in \Delta^{(n)}$ and $\phi_1, \ldots, \phi_n \in RHS$: $(\delta\ \phi_1 \cdots \phi_n) \in RHS$, and (iii) for every $q \in Q^{(n+1)}$ with $n \in \mathbb{N}$, $x_i \in X_k$ and $\phi_1, \ldots, \phi_n \in RHS$: $(q\ x_i\ \phi_1 \cdots \phi_n) \in RHS$.*

Definition 2. *An* mtt unit *M is a tuple (Q, Σ, Δ, R) with a ranked alphabet Q of* states, *where $Q^{(0)} = \emptyset$, a ranked alphabet Σ of* input symbols, *a ranked* *alphabet* Δ *of* output symbols, *where* $Q \cap (\Sigma \cup \Delta) = \emptyset$, *and a set* R *of* rules, *such that* R *contains for every* $k, r \in \mathbb{N}$, $\sigma \in \Sigma^{(k)}$ *and* $q \in Q^{(r+1)}$, *exactly one rule of the form:* $q\ (\sigma\ x_1 \cdots x_k)\ y_1 \cdots y_r = rhs_{M,q,\sigma}$, *with* $rhs_{M,q,\sigma} \in RHS(Q, \Delta, k, r)$.

Of course, the actual variable names used in rules R of an mtt unit are not fixed to come from X_k and Y_r for some $k, r \in \mathbb{N}$; consistent renaming is allowed. The semantics of an mtt unit is given by the reduction relation induced by its rules.

We give a rather artificial example of three mtt units that will be used to illustrate important phenomena in Sect. 4. For more practical examples see [9], where it was demonstrated that also typical functions on polymorphic data types and some higher-order functions like *map* can be viewed as mtt units by choosing appropriate function and constructor symbols.

Example 1. Let $\Sigma = \Omega = \{S^{(1)}, Z^{(0)}\}$ and $\Delta = \{\delta^{(2)}, \gamma^{(1)}, \alpha^{(0)}\}$. Then $M_1 = (\{q^{(2)}\}, \Sigma, \Delta, R_1)$, $M_1' = (\{q'^{(3)}\}, \Sigma, \Delta, R_1')$ and $M_2 = (\{p_1^{(1)}, p_2^{(1)}\}, \Delta, \Omega, R_2)$ are mtt units, where R_1, R_1' and R_2 contain rules as follows:

$$
\begin{array}{l}
R_1 : \\
\quad q\ (S\ x_1)\ y_1 = \gamma\ (q\ x_1\ (\delta\ (q\ x_1\ y_1)\ y_1)) \\
\quad q\ \ Z\ \ y_1 = y_1 \\
R_1' : \\
\quad q'\ (S\ x_1)\ y_1\ y_2 = \delta\ (q'\ x_1\ y_2\ (\gamma\ y_1))\ (\gamma\ (q'\ x_1\ y_1\ y_2)) \\
\quad q'\ \ Z\ \ y_1\ y_2 = y_1
\end{array}
\qquad
\begin{array}{l}
R_2 : p_1\ (\delta\ x_1\ x_2) = p_2\ x_1 \\
\quad\ \ p_1\ \ (\gamma\ x_1) = S\ (p_2\ x_1) \\
\quad\ \ p_1\ \ \alpha = Z \\
\quad\ \ p_2\ (\delta\ x_1\ x_2) = p_1\ x_2 \\
\quad\ \ p_2\ \ (\gamma\ x_1) = p_1\ x_1 \\
\quad\ \ p_2\ \ \alpha = Z\ .
\end{array}
$$

Definition 3. *An mtt unit* $M = (Q, \Sigma, \Delta, R)$ *is:*

- *a* top-down tree transducer unit *(for short* tdtt unit*), if* $Q = Q^{(1)}$
- recursion-linear, *if for every* $q \in Q$, $\sigma \in \Sigma^{(k)}$, $i \in [k]$: $|rhs_{M,q,\sigma}|_{x_i} \leqq 1$
- context-linear, *if for every* $q \in Q^{(r+1)}$, $\sigma \in \Sigma$, $h \in [r]$: $|rhs_{M,q,\sigma}|_{y_h} \leqq 1$
- linear, *if it is recursion-linear and context-linear*
- recursion-nondeleting, *if for every* $q \in Q$, $\sigma \in \Sigma^{(k)}$ *and* $i \in [k]$: $|rhs_{M,q,\sigma}|_{x_i} \geqq 1$
- context-nondeleting, *if for every* $q \in Q^{(r+1)}$, $\sigma \in \Sigma$ *and* $h \in [r]$: $|rhs_{M,q,\sigma}|_{y_h} \geqq 1$
- basic, *if the right-hand sides of its rules do not contain nested calls, i.e., subtrees of the form* $(q\ x_i \cdots (q'\ x_{i'} \cdots) \cdots)$
- atmost, *if it is recursion-linear, and it is context-linear or basic*
- atleast, *if it is recursion-nondeleting, and it is context-nondeleting or basic.*

3 Tree Transducer Composition

In this section we recall the semantic-preserving composition of two mtt units into a single mtt unit (the meaning of "semantic-preserving" will be made precise in Lemma 1 below). Constructions for performing such a composition in the cases that the first or the second mtt unit is a tdtt unit were given already in [4]. Here, we present a single transformation that captures both cases.

Construction 1. Let $M_1 = (Q, \Sigma, \Delta, R_1)$ and $M_2 = (P, \Delta, \Omega, R_2)$ be mtt units, such that one of the two is a tdtt unit, and $Q \cap P = \emptyset$. Let $m \in \mathbb{N}$ be the number of elements of P, and fix some ordering on P, such that $P = \{p_1, \ldots, p_m\}$. The composed mtt unit will have the set of states $F = \{\overline{qp}^{(r*m+s+1)} \mid q \in Q^{(r+1)}, p \in P^{(s+1)}\}$. We use two rewrite systems:

Pre, which contains for every $h \in [r]$ with $Q^{(r+1)} \neq \emptyset$ and every $p \in P^{(1)}$, the rewrite rule: $p\ y_h = y_{h,p}$. Here the y_h and $y_{h,p}$ are treated as ordinary symbols, rather than as variables of the rewrite system.

Comp, which contains for every $q \in Q^{(r+1)}$ and $p \in P^{(s+1)}$, the rewrite rule:

$$p\ (q\ x\ y_1 \cdots y_r)\ z_1 \cdots z_s = \overline{qp}\ x\ (p_1\ y_1) \cdots (p_m\ y_1) \cdots (p_1\ y_r) \cdots (p_m\ y_r)\ z_1 \cdots z_s\ .$$

Here x, $y_1, \ldots, y_r$ and $z_1, \ldots, z_s$ are considered as variables of the rewrite system. Since M_1 or M_2 is a tdtt unit, never both the ys and zs will be present. We abbreviate the right-hand side of the above rewrite rule as $\zeta_{q,p}$.

Since the ranked alphabets $P, \Delta, \{y_1^{(0)}, y_2^{(0)}, \ldots\}$ and Q are pairwise disjoint, there are no *critical pairs* [3] in $R_2 \cup Pre \cup Comp$. Hence, the reduction relation $\Rightarrow_{R_2 \cup Pre \cup Comp}$ is confluent. It is also terminating, because for every rule the first arguments of calls to states of P in the right-hand side are proper subtrees of the first argument of the call on the left-hand side.

Now, we can construct the mtt unit $\overline{M_1 M_2} = (F, \Sigma, \Omega, \overline{R_1 R_2})$ with rules as follows. For every $q \in Q^{(r+1)}$, $\sigma \in \Sigma^{(k)}$, rule $q\ (\sigma\ x_1 \cdots x_k)\ y_1 \cdots y_r = rhs_{M_1,q,\sigma}$ in R_1, and $p \in P^{(s+1)}$, $\overline{R_1 R_2}$ contains the rule:

$$\overline{qp}\ (\sigma\ x_1 \cdots x_k)\ y_{1,p_1} \cdots y_{r,p_m}\ z_1 \cdots z_s = nf(\Rightarrow_{R_2 \cup Pre \cup Comp}, p\ rhs_{M_1,q,\sigma}\ z_1 \cdots z_s)$$

Then, $\overline{M_1 M_2}$ implements the sequential composition of M_1 and M_2, in the sense of the following lemma.

Lemma 1. *For every rewrite rule* $(p\ (q\ x\ y_1 \cdots y_r)\ z_1 \cdots z_s = \zeta_{q,p}) \in Comp$, $t \in T_\Sigma$, $t_1, \ldots, t_r \in T_\Delta$ *and* $t'_1, \ldots, t'_s \in T_\Omega$:

1. $nf(\Rightarrow_{R_1 \cup R_2}, p\ (q\ t\ y_1 \cdots y_r)\ z_1 \cdots z_s) = nf(\Rightarrow_{\overline{R_1 R_2}}, \zeta_{q,p}[x \leftarrow t])$
2. $nf(\Rightarrow_{R_1 \cup R_2}, p\ (q\ t\ t_1 \cdots t_r)\ t'_1 \cdots t'_s)$
   $= nf(\Rightarrow_{\overline{R_1 R_2} \cup R_2}, \zeta_{q,p}[x \leftarrow t][y_1, \ldots, y_r, z_1, \ldots, z_s \leftarrow t_1, \ldots, t_r, t'_1, \ldots, t'_s])$.

Proof. Both assertions follow from statement (II)(a)(i) of Lemma A.12 in [12] for $\phi = (q\ x_1\ y_1 \cdots y_r)$, respectively for $\phi = (q\ x_1\ t_1 \cdots t_r)$. □

Example 2. We compose the mtt units M_1 and M_2 from Example 1, yielding the mtt unit $\overline{M_1 M_2} = (\{\overline{qp_1}^{(3)}, \overline{qp_2}^{(3)}\}, \Sigma, \Omega, \overline{R_1 R_2})$ with the following set of rules:

$$\begin{aligned}
\overline{qp_1}\ (S\ x_1)\ y_{1,p_1}\ y_{1,p_2} &= S\ (\overline{qp_2}\ x_1\ (\overline{qp_2}\ x_1\ y_{1,p_1}\ y_{1,p_2})\ y_{1,p_1})\\
\overline{qp_1}\quad Z\quad y_{1,p_1}\ y_{1,p_2} &= y_{1,p_1}\\
\overline{qp_2}\ (S\ x_1)\ y_{1,p_1}\ y_{1,p_2} &= \overline{qp_1}\ x_1\ (\overline{qp_2}\ x_1\ y_{1,p_1}\ y_{1,p_2})\ y_{1,p_1}\\
\overline{qp_2}\quad Z\quad y_{1,p_1}\ y_{1,p_2} &= y_{1,p_2}\ .
\end{aligned}$$

Note that here we have $Comp = \{p_1\ (q\ x\ y_1) = \zeta_{q,p_1}\ ,\ p_2\ (q\ x\ y_1) = \zeta_{q,p_2}\}$, where $\zeta_{q,p_1} = \overline{qp_1}\ x\ (p_1\ y_1)\ (p_2\ y_1)$ and $\zeta_{q,p_2} = \overline{qp_2}\ x\ (p_1\ y_1)\ (p_2\ y_1)$.

Another example of the application of Construction 1 can be found in the introduction, where $f = \overline{expdiv}$ and $g = \overline{expdiv'}$.

An optimizing compiler can take advantage of the composition construction by detecting appropriate places in the program where the rewrite rules from *Comp* can be applied. While the soundness of these rewritings is guaranteed by Lemma 1, it remains to be shown under which conditions Construction 1 actually leads to performance improvements over the original program (using mtt units M_1 and M_2) by the composed program (additionally containing the mtt unit $\overline{M_1 M_2}$, and with rewritings from *Comp* having been applied). From the form of rewrite rules in *Comp* it is obvious that intermediate data structures produced by M_1 in the original program have been removed in the composed program. Thus, no memory cells for this intermediate result have to be allocated in the heap and later be deallocated by the garbage collector. This beneficial effect, however, might be rendered useless, if the composed program needs to perform more reduction steps than the original one. Hence, we would like to establish conditions under which we have for every $q \in Q^{(r+1)}$, $p \in P^{(s+1)}$, $t \in T_\Sigma$, $t_1, \ldots, t_r \in T_\Delta$ and $t'_1, \ldots, t'_s \in T_\Omega$:

$$\begin{aligned} & cbn(\overline{R_1 R_2} \cup R_2, \zeta_{q,p}[x \leftarrow t][y_1, \ldots, y_r, z_1, \ldots, z_s \leftarrow t_1, \ldots, t_r, t'_1, \ldots, t'_s]) \\ \leqq\ & cbn(R_1 \cup R_2, p\,(q\,t\,t_1 \cdots t_r)\,t'_1 \cdots t'_s) \ . \end{aligned}$$

4 Efficiency Analysis

We want to formally relate the efficiency of a composed program obtained by Construction 1 to the efficiency of the original program. As computation time is an intensional property — i.e., it cannot be extracted from the result of a computation — such a study cannot directly use the algebraic methods developed to reason about extensional properties, e.g., to establish the correctness of the composition construction in Lemma 1. A well-known strategy in this situation is to externalize the internal property, e.g., by transforming or annotating programs.

Rosendahl [10] produces for every first-order functional program a step-counting version that — when called with the same arguments — returns the number of call-by-value reduction steps performed by the original program. Sands [11] developed a call-by-name improvement theory and used it to prove the correctness of unfold/fold-transformations [1] by annotating functional programs with a special identity function that represents a single "tick" of computation time, and by stating a set of laws that can be used to derive statements about the relative efficiency of program expressions.

This use of a special symbol to indicate performed reduction steps is similar to what we will be doing in Lemmata 2 and 3 in Sect. 4.2. Note, however, that our transformation technique goes beyond unfold/fold-steps if M_1 is not a tdtt unit, hence Sands' results can neither be used to prove Lemma 1, nor to establish criteria under which Construction 1 improves the efficiency of programs.

4.1 Annotating Programs

In the following, let $M_1 = (Q, \Sigma, \Delta, R_1)$ and $M_2 = (P, \Delta, \Omega, R_2)$ be two fixed mtt units, one of which is a tdtt unit, with $\diamond, \circ, \bullet, \star \notin \Sigma \cup \Delta \cup \Omega \cup Q \cup P$. We use the notions and naming conventions from Construction 1. Based on M_1 and M_2, we define — by annotating and adding rules — several new mtt units, the relevance of which will only become clear later.

Definition 4. *An mtt unit* $M_1^{\text{left}\to\text{right}}$ *has components* $(Q, \Sigma', \Delta', R_1^{\text{left}\to\text{right}})$ *with* Σ' *and* Δ' *obtained by adding to* Σ *and* Δ *symbols from* $\{\diamond^{(1)}, \circ^{(1)}, \bullet^{(1)}, \star^{(1)}\}$ *that are mentioned in* left *and in* right, *respectively, and with* $R_1^{\text{left}\to\text{right}}$ *containing rules as given in the table below.*
Several mtt units $M_2^{\text{left}\to\text{right}} = (P, \Delta', \Omega', R_2^{\text{left}\to\text{right}})$ *are introduced analogously.*

$R_1^{\to\diamond}$	$: q\ (\sigma\ x_1 \cdots x_k)\ y_1 \cdots y_r = \diamond\ (rhs_{M_1,q,\sigma})$	$\forall q \in Q^{(r+1)}, \sigma \in \Sigma^{(k)}$
$R_2^{\diamond\to\bullet}$	$: p\ (\delta\ x_1 \cdots x_k)\ z_1 \cdots z_s = \bullet\ (rhs_{M_2,p,\delta})$ $p\ (\diamond\ x_1)\ z_1 \cdots z_s = \bullet\ (p\ x_1\ z_1 \cdots z_s)$	$\forall p \in P^{(s+1)}, \delta \in \Delta^{(k)}$ $\forall p \in P^{(s+1)}$
$R_2^{\to\bullet}$	$: p\ (\delta\ x_1 \cdots x_k)\ z_1 \cdots z_s = \bullet\ (rhs_{M_2,p,\delta})$	$\forall p \in P^{(s+1)}, \delta \in \Delta^{(k)}$
$R_2^{\diamond\to\circ}$	$: p\ (\delta\ x_1 \cdots x_k)\ z_1 \cdots z_s = rhs_{M_2,p,\delta}$ $p\ (\diamond\ x_1)\ z_1 \cdots z_s = \circ\ (p\ x_1\ z_1 \cdots z_s)$	$\forall p \in P^{(s+1)}, \delta \in \Delta^{(k)}$ $\forall p \in P^{(s+1)}$
$R_2^{\to\star}$	$: p\ (\delta\ x_1 \cdots x_k)\ z_1 \cdots z_s = \star\ (rhs_{M_2,p,\delta})$	$\forall p \in P^{(s+1)}, \delta \in \Delta^{(k)}$
$R_2^{\circ\bullet\to\circ\bullet}$	$: p\ (\delta\ x_1 \cdots x_k)\ z_1 \cdots z_s = rhs_{M_2,p,\delta}$ $p\ (\circ\ x_1)\ z_1 \cdots z_s = \circ\ (p\ x_1\ z_1 \cdots z_s)$ $p\ (\bullet\ x_1)\ z_1 \cdots z_s = \bullet\ (p\ x_1\ z_1 \cdots z_s)$	$\forall p \in P^{(s+1)}, \delta \in \Delta^{(k)}$ $\forall p \in P^{(s+1)}$ $\forall p \in P^{(s+1)}$
$R_1^{\to\circ\bullet,0}$	$: q\ (\sigma\ x_1 \cdots x_k)\ y_1 \cdots y_r =$ $\circ\ (rhs_{M_1,q,\sigma}[(\delta \cdots) \leftarrow \bullet\ (\delta \cdots) \mid \delta \in \Delta])$	$\forall q \in Q^{(r+1)}, \sigma \in \Sigma^{(k)}$
$R_1^{\to\circ\bullet,1}$	$: q\ (\sigma\ x_1 \cdots x_k)\ y_1 \cdots y_r =$ $rhs_{M_1,q,\sigma}\ [(\delta \cdots) \leftarrow \bullet\ (\delta \cdots) \mid \delta \in \Delta]$ $[(q' \cdots) \leftarrow \circ\ (q' \cdots) \mid q' \in Q]$	$\forall q \in Q^{(r+1)}, \sigma \in \Sigma^{(k)}$
$R_2^{\diamond\to\circ\bullet}$	$: p\ (\delta\ x_1 \cdots x_k)\ z_1 \cdots z_s = \bullet\ (rhs_{M_2,p,\delta})$ $p\ (\diamond\ x_1)\ z_1 \cdots z_s = \circ\ (p\ x_1\ z_1 \cdots z_s)$	$\forall p \in P^{(s+1)}, \delta \in \Delta^{(k)}$ $\forall p \in P^{(s+1)}$

Note that if M_1 fulfills one of the restrictions introduced in Definition 3, then also all the introduced $M_1^{\text{left}\to\text{right}}$ fulfill this restriction; analogously for M_2.

4.2 Ticking of Original Program

Note that $M_1^{\to\diamond}$ from Definition 4 outputs a $\diamond$-symbol in every rule application. If none of these symbols is afterwards duplicated, then the number of $\diamond$-symbols in the output produced by $M_1^{\to\diamond}$ is exactly the number of performed call-by-need reduction steps. This fact is used in the following lemma to determine the efficiency of the sequential composition of M_1 and M_2. For the rest of the paper, we fix some $q \in Q^{(r+1)}$, $p \in P^{(s+1)}$, $t \in T_\Sigma$, $t_1, \ldots, t_r \in T_\Delta$ and $t'_1, \ldots, t'_s \in T_\Omega$, and the substitution $\kappa = [y_1, \ldots, y_r, z_1, \ldots, z_s \leftarrow t_1, \ldots, t_r, t'_1, \ldots, t'_s]$.

Lemma 2. *If* M_1 *is context-linear or basic, and* M_2 *is atmost, then:*
$cbn(R_1 \cup R_2, p\,(q\,t\,t_1 \cdots t_r)\,t'_1 \cdots t'_s) = |nf(\Rightarrow_{R_1^{\to\diamond} \cup R_2^{\diamond\to\bullet}}, p\,(q\,t\,t_1 \cdots t_r)\,t'_1 \cdots t'_s)|_\bullet$

Proof. In every application of a rule from $R_1^{\to\diamond}$ (corresponding to an R_1-step), one $\diamond$-symbol is produced in the intermediate result. There is no other way how $\diamond$-symbols can be introduced, and since $M_1^{\to\diamond}$ is *(i)* context-linear or *(ii)* basic —

i.e, no $\diamond$-symbols will appear in context parameters of states from Q in case *(ii)*, or those parameters cannot be copied in case *(i)* — none of these $\diamond$-symbols will be duplicated. Only those $\diamond$-symbols are reached and reproduced as $\bullet$-symbols by $R_2^{\diamond\to\bullet}$ that correspond to an R_1-step being forced by call-by-need reduction of $(p\ (q\ t\ t_1 \cdots t_r)\ t'_1 \cdots t'_s)$. Since $M_2^{\diamond\to\bullet}$ is recursion-linear, every $\diamond$-symbol will be reproduced as $\bullet$-symbol at most once. Since $M_2^{\diamond\to\bullet}$ is context-linear or basic, those $\bullet$-symbols and also the $\bullet$-symbols produced by $M_2^{\diamond\to\bullet}$ at Δ-symbols — corresponding to the R_2-steps during the call-by-need reduction — will not be duplicated. Hence, the resulting number of $\bullet$-symbols is equal to the number of steps of R_1 and R_2 in the original program. □

Example 3. Recall the mtt units M'_1 and M_2 from Example 1, and $M_1'^{\to\diamond}$ and $M_2^{\diamond\to\bullet}$ as obtained from them according to Definition 4. Note that M'_1 is basic and M_2 is atmost, hence the preconditions of Lemma 2 are satisfied. With $t = S\ Z$ and $t_1 = t_2 = \alpha$, we have, e.g., the following reduction:

$$\begin{array}{lll}
& p_2\ (q'\ t\ t_1\ t_2) & \\
\Rightarrow_{R_1'^{\to\diamond}} & p_2\ (\diamond\ (\delta\ (q'\ Z\ t_2\ (\gamma\ t_1))\ (\gamma\ (q'\ Z\ t_1\ t_2)))) & \text{(mark step of } R'_1) \\
\Rightarrow_{R_2^{\diamond\to\bullet}} & \bullet\ (p_2\ (\delta\ (q'\ Z\ t_2\ (\gamma\ t_1))\ (\gamma\ (q'\ Z\ t_1\ t_2)))) & \text{(count marked step)} \\
\Rightarrow_{R_1'^{\to\diamond}} & \bullet\ (p_2\ (\delta\ (\diamond\ t_2)\ (\gamma\ (q'\ Z\ t_1\ t_2)))) & \text{(will } \textit{not} \text{ be counted)} \\
\Rightarrow_{R_2^{\diamond\to\bullet}} & \bullet\ (\bullet\ (p_1\ (\gamma\ (q'\ Z\ t_1\ t_2)))) & \text{(count step of } R_2) \\
\Rightarrow_{R_2^{\diamond\to\bullet}} & \bullet\ (\bullet\ (\bullet\ (S\ (p_2\ (q'\ Z\ t_1\ t_2))))) & \text{(count step of } R_2) \\
\Rightarrow_{R_1'^{\to\diamond}} & \bullet\ (\bullet\ (\bullet\ (S\ (p_2\ (\diamond\ t_1))))) & \text{(mark step of } R'_1) \\
\Rightarrow_{R_2^{\diamond\to\bullet}} & \bullet\ (\bullet\ (\bullet\ (S\ (\bullet\ (p_2\ t_1))))) & \text{(count marked step)} \\
\Rightarrow_{R_2^{\diamond\to\bullet}} & \bullet\ (\bullet\ (\bullet\ (S\ (\bullet\ (\bullet\ Z))))) & \text{(count step of } R_2)
\end{array}$$

Indeed, we have $cbn(R'_1 \cup R_2, p_2\ (q'\ t\ t_1\ t_2)) = |\bullet\ (\bullet\ (\bullet\ (S\ (\bullet\ (\bullet\ Z)))))|_\bullet = 5$. Note that a step of R'_1 that is not forced by call-by-need evaluation is not counted in the final output.

If we refrain from counting the steps of M_1 and settle for the approximation of only counting the steps of M_2, we can drop part of the restrictions in Lemma 2:

Lemma 3. *If M_2 is context-linear or basic, then:*
$cbn(R_1 \cup R_2, p\ (q\ t\ t_1 \cdots t_r)\ t'_1 \cdots t'_s) > |nf(\Rightarrow_{R_1 \cup R_2^{\to\bullet}}, p\ (q\ t\ t_1 \cdots t_r)\ t'_1 \cdots t'_s)|_\bullet$

Proof. Every step of $R_2^{\to\bullet}$ (corresponding to an R_2-step) produces one $\bullet$-symbol. Since $M_2^{\to\bullet}$ is context-linear or basic, no such symbol is duplicated. Hence, the number of $\bullet$-symbols in $nf(\Rightarrow_{R_1 \cup R_2^{\to\bullet}}, p\ (q\ t\ t_1 \cdots t_r)\ t'_1 \cdots t'_s)$ is equal to the number of R_2-steps during call-by-need reduction of $(p\ (q\ t\ t_1 \cdots t_r)\ t'_1 \cdots t'_s)$. Since during this reduction at least one step must be performed by R_1, the inequality follows. □

4.3 Ticking of Composed Program

Assume that Construction 1 composes M_1 and M_2 to $\overline{M_1 M_2} = (F, \Sigma, \Omega, \overline{R_1 R_2})$. Since $M_1^{\to\diamond}$ or $M_2^{\diamond\to\circ}$ is a tdtt unit, Construction 1 is also applicable to these mtt units, yielding $\overline{M_1^{\to\diamond} M_2^{\diamond\to\circ}} = (F, \Sigma, \Omega \cup \{\circ^{(1)}\}, \overline{R_1^{\to\diamond} R_2^{\diamond\to\circ}})$.

Example 4. For $M_1^{\rightarrow\diamond}$ and $M_2^{\diamond\rightarrow\circ}$ as obtained from the mtt units in Example 1, Construction 1 yields the mtt unit $\overline{M_1^{\rightarrow\diamond} M_2^{\diamond\rightarrow\circ}}$ with set of rules:

$$\begin{array}{llll}
\overline{qp_1} & (S\ x_1) & y_{1,p_1}\ y_{1,p_2} & = \circ\ (S\ (\overline{qp_2}\ x_1\ (\overline{qp_2}\ x_1\ y_{1,p_1}\ y_{1,p_2})\ y_{1,p_1})) \\
\overline{qp_1} & Z & y_{1,p_1}\ y_{1,p_2} & = \circ\ y_{1,p_1} \\
\overline{qp_2} & (S\ x_1) & y_{1,p_1}\ y_{1,p_2} & = \circ\ (\overline{qp_1}\ x_1\ (\overline{qp_2}\ x_1\ y_{1,p_1}\ y_{1,p_2})\ y_{1,p_1}) \\
\overline{qp_2} & Z & y_{1,p_1}\ y_{1,p_2} & = \circ\ y_{1,p_2}\ .
\end{array}$$

In the previous example, the rules obtained by the composition construction for $M_1^{\rightarrow\diamond}$ and $M_2^{\diamond\rightarrow\circ}$ correspond to the rules obtained in Example 2 by composing M_1 and M_2, except that every right-hand side has an additional $\circ$-symbol on top. This is due to the $\diamond$-symbols on top of all right-hand sides of $R_1^{\rightarrow\diamond}$ and due to the added rules $p_1\ (\diamond\ x_1) = \circ\ (p_1\ x_1)$ and $p_2\ (\diamond\ x_1) = \circ\ (p_2\ x_1)$ in $R_2^{\diamond\rightarrow\circ}$. The following lemma establishes that this observation holds in general.

Lemma 4. *For every $f \in F$ and $\sigma \in \Sigma$: $rhs_{\overline{M_1^{\rightarrow\diamond} M_2^{\diamond\rightarrow\circ}},f,\sigma} = \circ\ (rhs_{\overline{M_1 M_2},f,\sigma})$.*

Proof. Straightforward by noting that the definitions of *Pre* and *Comp* only depend on Q and P, and that for every $q \in Q$, $p \in P^{(s+1)}$ and $\sigma \in \Sigma$:

$$\begin{aligned}
& nf(\Rightarrow_{R_2^{\diamond\rightarrow\circ} \cup Pre \cup Comp}, p\ (\diamond\ (rhs_{M_1,q,\sigma}))\ z_1 \cdots z_s) \\
&= nf(\Rightarrow_{R_2^{\diamond\rightarrow\circ} \cup Pre \cup Comp}, \circ\ (p\ rhs_{M_1,q,\sigma}\ z_1 \cdots z_s)) \\
&= \circ\ (nf(\Rightarrow_{R_2 \cup Pre \cup Comp}, p\ rhs_{M_1,q,\sigma}\ z_1 \cdots z_s))
\end{aligned}$$ □

Hence, $\overline{M_1^{\rightarrow\diamond} M_2^{\diamond\rightarrow\circ}}$ can be used to approximate the number of reduction steps of $\overline{M_1 M_2}$, as exploited in the proof of the following lemma, which estimates the efficiency of the composed program without actually performing Construction 1.

Lemma 5. $$\begin{aligned} & cbn(\overline{R_1 R_2} \cup R_2, \zeta_{q,p}[x \leftarrow t]\kappa) \\ & \leqq |nf(\Rightarrow_{R_2^{\rightarrow\star}}, (nf(\Rightarrow_{R_1^{\rightarrow\diamond} \cup R_2^{\diamond\rightarrow\circ}}, p\ (q\ t\ y_1 \cdots y_r)\ z_1 \cdots z_s))\ \kappa)|_{\{\circ,\star\}} \end{aligned}$$

Proof. By Lemma 4, we know that for every $f \in F$ and $\sigma \in \Sigma$, the following holds: $rhs_{\overline{M_1^{\rightarrow\diamond} M_2^{\diamond\rightarrow\circ}},f,\sigma} = \circ\ (rhs_{\overline{M_1 M_2},f,\sigma})$. Since all rules simply have an additional symbol on top, we get the following equivalence:

$$cbn(\overline{R_1 R_2} \cup R_2, \zeta_{q,p}[x \leftarrow t]\kappa) = cbn(\overline{R_1^{\rightarrow\diamond} R_2^{\diamond\rightarrow\circ}} \cup R_2^{\rightarrow\star}, \zeta_{q,p}[x \leftarrow t]\kappa).$$

During call-by-need reduction of $\zeta_{q,p}[x \leftarrow t]\kappa$ with $\Rightarrow_{\overline{R_1^{\rightarrow\diamond} R_2^{\diamond\rightarrow\circ}} \cup R_2^{\rightarrow\star}}$, every step of $\overline{R_1^{\rightarrow\diamond} R_2^{\diamond\rightarrow\circ}}$, produces a $\circ$-symbol, while every step of $R_2^{\rightarrow\star}$ produces a $\star$-symbol. Hence, the overall number of steps is $\leqq |nf(\Rightarrow_{\overline{R_1^{\rightarrow\diamond} R_2^{\diamond\rightarrow\circ}} \cup R_2^{\rightarrow\star}}, \zeta_{q,p}[x \leftarrow t]\kappa)|_{\{\circ,\star\}}$. By confluence considerations, we can obtain the same normal form by first reducing $\zeta_{q,p}[x \leftarrow t]$ to its normal form with respect to $\Rightarrow_{\overline{R_1^{\rightarrow\diamond} R_2^{\diamond\rightarrow\circ}}}$, then performing the substitution κ and further reducing to normal form with $\Rightarrow_{R_2^{\rightarrow\star}}$. This gives us the equivalent expression $|nf(\Rightarrow_{R_2^{\rightarrow\star}}, (nf(\Rightarrow_{\overline{R_1^{\rightarrow\diamond} R_2^{\diamond\rightarrow\circ}}}, \zeta_{q,p}[x \leftarrow t]))\ \kappa)|_{\{\circ,\star\}}$, which by Lemma 1 from Sect. 3 is equal to:

$$|nf(\Rightarrow_{R_2^{\rightarrow\star}}, (nf(\Rightarrow_{R_1^{\rightarrow\diamond} \cup R_2^{\diamond\rightarrow\circ}}, p\ (q\ t\ y_1 \cdots y_r)\ z_1 \cdots z_s))\ \kappa)|_{\{\circ,\star\}}\ .$$ □

4.4 Combining Approximations

With Lemmata 2 and 3 we have two different ways to estimate the efficiency of the original program, while Lemma 5 approximates the efficiency of the composed program. In each case, this is done by annotating the rules of M_1 and

M_2 appropriately and counting certain symbols in the produced output. In the following, we will combine these results to determine the relative efficiency of the composed vs. the original program. Using Lemmata 2 and 5, we obtain the first of our main theorems:

Theorem 1. *If M_1 is context-linear or basic, and M_2 is atmost, then:*

$$cbn(\overline{R_1R_2} \cup R_2, \zeta_{q,p}[x \leftarrow t]\kappa) - cbn(R_1 \cup R_2, p\,(q\,t\,t_1 \cdots t_r)\,t'_1 \cdots t'_s) \leqq 0$$

Proof. By Lemmata 5 and 2, we have:

$$cbn(\overline{R_1R_2} \cup R_2, \zeta_{q,p}[x \leftarrow t]\kappa) - cbn(R_1 \cup R_2, p\,(q\,t\,t_1 \cdots t_r)\,t'_1 \cdots t'_s)$$
$$\leqq |\chi|_{\{\circ,\star\}} - |nf(\Rightarrow_{R_1^{\to\diamond} \cup R_2^{\diamond\to\bullet}}, p\,(q\,t\,t_1 \cdots t_r)\,t'_1 \cdots t'_s)|_\bullet \;,$$

where $\chi = nf(\Rightarrow_{R_2^{\to\star}}, (nf(\Rightarrow_{R_1^{\to\diamond} \cup R_2^{\diamond\to\circ}}, p\,(q\,t\,y_1 \cdots y_r)\,z_1 \cdots z_s))\,\kappa)$.
It is clear that: $|nf(\Rightarrow_{R_1^{\to\diamond} \cup R_2^{\diamond\to\bullet}}, p\,(q\,t\,t_1 \cdots t_r)\,t'_1 \cdots t'_s)|_\bullet$

$$= |nf(\Rightarrow_{R_2^{\diamond\to\bullet}}, (nf(\Rightarrow_{R_1^{\to\diamond} \cup R_2^{\diamond\to\bullet}}, p\,(q\,t\,y_1 \cdots y_r)\,z_1 \cdots z_s))\,\kappa)|_\bullet.$$

Since we have $t_1, \ldots, t_r \in T_\Delta$, the outer reduction with $\Rightarrow_{R_2^{\diamond\to\bullet}}$ will only apply rules at symbols from Δ. Hence, the previous expression is equivalent to $|\chi'|_{\{\bullet,\star\}}$, where $\chi' = nf(\Rightarrow_{R_2^{\to\star}}, (nf(\Rightarrow_{R_1^{\to\diamond} \cup R_2^{\diamond\to\bullet}}, p\,(q\,t\,y_1 \cdots y_r)\,z_1 \cdots z_s))\,\kappa)$. By comparing the definitions of $R_2^{\diamond\to\bullet}$ and $R_2^{\diamond\to\circ}$, it should be obvious that $|\chi|_{\{\circ,\star\}} - |\chi'|_{\{\bullet,\star\}} = |\chi|_\circ - |\chi'|_\bullet \leqq 0$. □

Note that by further considering the value of $|\chi|_\circ - |\chi'|_\bullet$ in the previous proof, we could obtain the more precise statement that under the preconditions of Theorem 1 the composed program saves at least as many reduction steps as were performed in the original program by states of M_2 on the part of the intermediate result produced by rules of M_1.

Also, note that Theorem 1 generalizes the efficiency statements about the composed program compared to the original program in Corollary 21 and Theorem 23 of [8], where M_1 and M_2 were required to be linear, and where in the case that M_1 is not a tdtt unit M_2 was restricted to have only one state.

In order to relax the a priori restrictions imposed by Theorem 1 on the involved mtt units, we can use Lemma 3 (instead of Lemma 2) as starting point:

Lemma 6. *If M_2 is context-linear or basic, then for every $\lambda \in \{0,1\}$:*

$$cbn(\overline{R_1R_2} \cup R_2, \zeta_{q,p}[x \leftarrow t]\kappa) - cbn(R_1 \cup R_2, p\,(q\,t\,t_1 \cdots t_r)\,t'_1 \cdots t'_s)$$
$$< \lambda + |\chi|_\circ - |\chi|_\bullet \;,$$

where $\chi = nf(\Rightarrow_{R_2^{\circ\bullet\to\circ\bullet}}, p\,(nf(\Rightarrow_{R_1^{\to\circ\bullet,\lambda}}, q\,t\,t_1 \cdots t_r))\,t'_1 \cdots t'_s)$.

Proof. By Lemmata 5 and 3, we have:

$$cbn(\overline{R_1R_2} \cup R_2, \zeta_{q,p}[x \leftarrow t]\kappa) - cbn(R_1 \cup R_2, p\,(q\,t\,t_1 \cdots t_r)\,t'_1 \cdots t'_s)$$
$$< |\chi_1|_{\{\circ,\star\}} - |nf(\Rightarrow_{R_1 \cup R_2^{\to\bullet}}, p\,(q\,t\,t_1 \cdots t_r)\,t'_1 \cdots t'_s)|_\bullet \;,$$

where $\chi_1 = nf(\Rightarrow_{R_2^{\to\star}}, (nf(\Rightarrow_{R_1^{\to\diamond} \cup R_2^{\diamond\to\circ}}, p\,(q\,t\,y_1 \cdots y_r)\,z_1 \cdots z_s))\,\kappa)$.
Replacing $R_2^{\diamond\to\circ}$ by $R_2^{\diamond\to\circ\bullet}$ in the expression defining χ_1 does not change the number of occurrences of $\circ$ and $\star$, because only additional $\bullet$-symbols will appear in the output, hence $|\chi_1|_{\{\circ,\star\}} = |\chi_2|_{\{\circ,\star\}}$, where

$$\chi_2 = nf(\Rightarrow_{R_2^{\to\star}}, (nf(\Rightarrow_{R_1^{\to\diamond} \cup R_2^{\diamond\to\circ\bullet}}, p\,(q\,t\,y_1 \cdots y_r)\,z_1 \cdots z_s))\,\kappa).$$

Also, it is clear that:

$$\begin{aligned} & |nf(\Rightarrow_{R_1 \cup R_2^{\rightarrow\bullet}}, p\,(q\,t\,t_1 \cdots t_r)\,t'_1 \cdots t'_s)|_\bullet \\ = & |nf(\Rightarrow_{R_1^{\rightarrow\diamond} \cup R_2^{\diamond\rightarrow\circ\bullet}}, p\,(q\,t\,t_1 \cdots t_r)\,t'_1 \cdots t'_s)|_\bullet \\ = & |nf(\Rightarrow_{R_2^{\diamond\rightarrow\circ\bullet}}, (nf(\Rightarrow_{R_1^{\rightarrow\diamond} \cup R_2^{\diamond\rightarrow\circ\bullet}}, p\,(q\,t\,y_1 \cdots y_r)\,z_1 \cdots z_s))\,\kappa)|_\bullet \ . \end{aligned}$$

Since we have $t_1, \ldots, t_r \in T_\Delta$, the outer reduction with $\Rightarrow_{R_2^{\diamond\rightarrow\circ\bullet}}$ will only apply rules at symbols from Δ, producing $\bullet$-symbols. Hence, the previous expression is equivalent to $|\chi_2|_{\{\bullet,\star\}}$, and thus the right-hand side of the above inequality is equal to $|\chi_2|_{\{\circ,\star\}} - |\chi_2|_{\{\bullet,\star\}}$, respectively to $|\chi_3|_\circ - |\chi_3|_\bullet$, where

$$\chi_3 = nf(\Rightarrow_{R_2}, (nf(\Rightarrow_{R_2^{\diamond\rightarrow\circ\bullet}}, p\,(nf(\Rightarrow_{R_1^{\rightarrow\diamond}}, q\,t\,y_1 \cdots y_r))\,z_1 \cdots z_s))\,\kappa)\,.$$

The rules from $R_2^{\diamond\rightarrow\circ\bullet}$ replace every $\diamond$-symbol by one $\circ$-symbol, and produce a $\bullet$-symbol for every consumed symbol from the intermediate ranked alphabet Δ. The same effect can be achieved by firstly directly producing $\circ$-symbols instead of $\diamond$-symbols, and secondly marking every produced intermediate symbol from Δ with a $\bullet$-symbol and then just reproducing those. Hence, χ_3 is equal to:

$$nf(\Rightarrow_{R_2}, (nf(\Rightarrow_{R_2^{\circ\bullet\rightarrow\circ\bullet}}, p\,(nf(\Rightarrow_{R_1^{\rightarrow\circ\bullet,0}}, q\,t\,y_1 \cdots y_r))\,z_1 \cdots z_s))\,\kappa)\,.$$

Since, furthermore, the $t_1, \ldots, t_r \in T_\Delta$ do not contain $\circ$- or $\bullet$-symbols, and the rules of R_2 are contained in $R_2^{\circ\bullet\rightarrow\circ\bullet}$, the previous expression is by confluence considerations equal to:

$$\chi_4 = nf(\Rightarrow_{R_2^{\circ\bullet\rightarrow\circ\bullet}}, p\,(nf(\Rightarrow_{R_1^{\rightarrow\circ\bullet,0}}, q\,t\,t_1 \cdots t_r))\,t'_1 \cdots t'_s).$$

In $nf(\Rightarrow_{R_1^{\rightarrow\circ\bullet,0}}, q\,t\,t_1 \cdots t_r)$, the rules from $R_1^{\rightarrow\circ\bullet,0}$ produce one $\circ$-symbol whenever a state is applied at some input symbol. Alternatively, we could produce one $\circ$-symbol atop every state call. Hence, χ_4 is also equal to:

$$\chi_5 = nf(\Rightarrow_{R_2^{\circ\bullet\rightarrow\circ\bullet}}, p\,(nf(\Rightarrow_{R_1^{\rightarrow\circ\bullet,1}}, \circ\,(q\,t\,t_1 \cdots t_r)))\,t'_1 \cdots t'_s).$$

In the case $\lambda = 0$, the lemma now follows from $|\chi_4|_\circ - |\chi_4|_\bullet = 0 + |\chi|_\circ - |\chi|_\bullet$, while in the case $\lambda = 1$, the lemma follows from $|\chi_5|_\circ - |\chi_5|_\bullet = 1 + |\chi|_\circ - |\chi|_\bullet$ (by normal form and definition of $|\cdot|_\circ$). □

Lemma 6 is promising, because it estimates the efficiency improvement or deterioration of the composed program over the original one, without the need of actually performing the composition construction. But, in contrast to the above Theorem 1, the approximation is still input-dependent, while we would like to obtain a statement about *all* runs of the original and the composed program.

Since we are interested in the *difference* in the number of occurrences of $\circ$- and $\bullet$-symbols, we can manipulate the right-hand sides of $R_1^{\rightarrow\circ\bullet,\lambda}$, as long as we do not decrease the value of $|\chi|_\circ - |\chi|_\bullet$ in Lemma 6. A trivial way to do this is by removing $\bullet\,(\circ\cdots)$- and $\circ\,(\bullet\cdots)$-contexts, because the rules in $R_2^{\circ\bullet\rightarrow\circ\bullet}$ would just reproduce those. If $\circ$- and $\bullet$-symbols do *not* occur together, certain conditions have to be fulfilled to allow "bringing them closer" without decreasing the overall value of $|\chi|_\circ - |\chi|_\bullet$. Such conditions are established in the following definition and lemma, and are then used to prove our second main theorem.

Definition 5. *The rewrite system Elim contains the following rewrite rules (with variables $u, u_1, u_2, \ldots$):*

1. $\bullet\,(\circ\,u) = u$
2. $\circ\,(\bullet\,u) = u$

3. *if* M_2 *is atmost, then for every* $\delta \in \Delta^{(n)}$ *and* $i \in [n]$:
   $\bullet\,(\delta\, u_1 \cdots u_n) = (\delta\, u_1 \cdots (\bullet\, u_i) \cdots u_n)$
4. *if* M_2 *is atleast, then for every* $\delta \in \Delta^{(n)}$ *and* $i \in [n]$:
   $(\delta\, u_1 \cdots (\bullet\, u_i) \cdots u_n) = \bullet\,(\delta\, u_1 \cdots u_n)$
5. *if* M_1 *is context-nondeleting and* M_2 *is atleast, then for every* $q' \in Q^{(n+1)}$ *and* $i \in [n]$: $(q'\, u\, u_1 \cdots (\bullet\, u_i) \cdots u_n) = \bullet\,(q'\, u\, u_1 \cdots u_n)$

Lemma 7. *Let* $M = (Q, \Sigma, \Delta \cup \{\circ^{(1)}, \bullet^{(1)}\}, R)$ *be an mtt unit, where* Q, Σ *and* Δ *are the ranked alphabets of* M_1 *(and* R *will typically contain annotated versions of the rules from* M_1*, which however is not a technical precondition of this lemma). Assume that* M *is context-nondeleting, if* M_1 *is context-nondeleting.*
If we rewrite one right-hand side in R *with one rewrite step of Elim, yielding an mtt unit* M' *with set of rules* R'*, then* $|\chi|_\circ - |\chi|_\bullet \leqq |\chi'|_\circ - |\chi'|_\bullet$*, where:*

$$\begin{aligned}\chi &= nf(\Rightarrow_{R_2^{\circ\bullet\to\circ\bullet}}, p\,(nf(\Rightarrow_R, q\,t\,t_1 \cdots t_r))\,t'_1 \cdots t'_s)\\ \chi' &= nf(\Rightarrow_{R_2^{\circ\bullet\to\circ\bullet}}, p\,(nf(\Rightarrow_{R'}, q\,t\,t_1 \cdots t_r))\,t'_1 \cdots t'_s)\ .\end{aligned}$$

Proof sketch. The lemma can be proved by structural induction on $t \in T_\Sigma$, using a nested induction on the structures of right-hand sides from R and $R_2^{\circ\bullet\to\circ\bullet}$, respectively, and using — for every $p' \in P^{(s'+1)}$, $n \in \mathbb{N}$, $i \in [n]$, $\tau, \tau_1, \ldots, \tau_n \in T_{\Delta\cup\{\circ^{(1)},\bullet^{(1)}\}}$, $t' \in T_\Sigma$, and $\delta \in \Delta^{(n)}$ or $q' \in Q^{(n+1)}$ — one of the following properties (depending on the rule from *Elim* that was applied):

1. $|\chi_1|_\circ - |\chi_1|_\bullet \leqq |\chi'_1|_\circ - |\chi'_1|_\bullet$
2. $|\chi_2|_\circ - |\chi_2|_\bullet \leqq |\chi'_2|_\circ - |\chi'_2|_\bullet$
3. $|\chi_3|_\circ - |\chi_3|_\bullet \leqq |\chi'_3|_\circ - |\chi'_3|_\bullet$, if M_2 is atmost
4. $|\chi_4|_\circ - |\chi_4|_\bullet \leqq |\chi'_4|_\circ - |\chi'_4|_\bullet$, if M_2 is atleast
5. $|\chi_5|_\circ - |\chi_5|_\bullet \leqq |\chi'_5|_\circ - |\chi'_5|_\bullet$, if M_1 is context-nondeleting and M_2 is atleast,

where:

$$\begin{aligned}\chi_1 &= nf(\Rightarrow_{R_2^{\circ\bullet\to\circ\bullet}}, p'\,(\bullet\,(\circ\,\tau))\,z_1 \cdots z_{s'})\\ \chi_2 &= nf(\Rightarrow_{R_2^{\circ\bullet\to\circ\bullet}}, p'\,(\circ\,(\bullet\,\tau))\,z_1 \cdots z_{s'})\\ \chi'_1 = \chi'_2 &= nf(\Rightarrow_{R_2^{\circ\bullet\to\circ\bullet}}, p'\,\tau\,z_1 \cdots z_{s'})\\ \chi_3 = \chi'_4 &= nf(\Rightarrow_{R_2^{\circ\bullet\to\circ\bullet}}, p'\,(\bullet\,(\delta\,\tau_1 \cdots \tau_n))\,z_1 \cdots z_{s'})\\ \chi'_3 = \chi_4 &= nf(\Rightarrow_{R_2^{\circ\bullet\to\circ\bullet}}, p'\,(\delta\,\tau_1 \cdots (\bullet\,\tau_i) \cdots \tau_n)\,z_1 \cdots z_{s'})\\ \chi_5 &= nf(\Rightarrow_{R_2^{\circ\bullet\to\circ\bullet}}, p'\,(nf(\Rightarrow_R, q'\,t'\,\tau_1 \cdots (\bullet\,\tau_i) \cdots \tau_n))\,z_1 \cdots z_{s'})\\ \chi'_5 &= nf(\Rightarrow_{R_2^{\circ\bullet\to\circ\bullet}}, p'\,(nf(\Rightarrow_R, \bullet\,(q'\,t'\,\tau_1 \cdots \tau_n)))\,z_1 \cdots z_{s'})\ .\end{aligned}$$

Properties 1 and 2 are immediate from $\chi_1 = \bullet\,(\circ\,\chi'_1)$ and $\chi_2 = \circ\,(\bullet\,\chi'_2)$. For $j \in \{3,4,5\}$ it can easily be seen that $|\chi_j|_\circ = |\chi'_j|_\circ$. Hence, to validate properties 3–5, it remains to prove that under the appropriate restrictions: $|\chi'_3|_\bullet \leqq |\chi_3|_\bullet$, $|\chi'_4|_\bullet \leqq |\chi_4|_\bullet$, respectively, $|\chi'_5|_\bullet \leqq |\chi_5|_\bullet$. These inequations can be established by using the rule $p'\,(\bullet\,x_1)\,z_1 \cdots z_{s'} = \bullet\,(p'\,x_1\,z_1 \cdots z_{s'})$ in $R_2^{\circ\bullet\to\circ\bullet}$, and the following two facts[2]:

Fact 1: if M_2 is atmost, $\tau' \in T_{\Delta\cup\{\circ^{(1)},\bullet^{(1)}\}}$, and τ'' is obtained by inserting at most one $\bullet$-symbol into τ', then:
$|nf(\Rightarrow_{R_2^{\circ\bullet\to\circ\bullet}}, p'\,\tau''\,z_1 \cdots z_{s'})|_\bullet \leqq 1 + |nf(\Rightarrow_{R_2^{\circ\bullet\to\circ\bullet}}, p'\,\tau'\,z_1 \cdots z_{s'})|_\bullet$.

[2] For property 5 we additionally use the observation that for context-nondeleting M, $nf(\Rightarrow_R, q'\,t'\,\tau_1 \cdots \tau_n)$ is obtained from $nf(\Rightarrow_R, q'\,t'\,\tau_1 \cdots (\bullet\,\tau_i) \cdots \tau_n)$ by removing at least one $\bullet$-symbol.

Fact 2: if M_2 is atleast, $\tau' \in T_{\Delta\cup\{\circ^{(1)},\bullet^{(1)}\}}$, and τ'' is obtained by removing at least one $\bullet$-symbol from τ', then:
$1 + |nf(\Rightarrow_{R_2^{\circ\bullet\to\circ\bullet}}, p'\ \tau''\ z_1 \cdots z_{s'})|_\bullet \leqq |nf(\Rightarrow_{R_2^{\circ\bullet\to\circ\bullet}}, p'\ \tau'\ z_1 \cdots z_{s'})|_\bullet$. □

Theorem 2. *If M_2 is context-linear or basic, and there exists $\lambda \in \{0,1\}$, such that the rules of $R_1^{\to\circ\bullet,\lambda}$ can be rewritten with (finitely many applications of) Elim until no $\circ$-symbols remain, then:*

$$cbn(\overline{R_1 R_2} \cup R_2, \zeta_{q,p}[x \leftarrow t]\kappa) - cbn(R_1 \cup R_2, p\ (q\ t\ t_1 \cdots t_r)\ t'_1 \cdots t'_s) < \lambda$$

Proof. By the preconditions we know that $R_1^{\to\circ\bullet,\lambda}$ can be rewritten with *Elim* to some R^* containing no $\circ$-symbols. By Lemma 6 and repeated applications of Lemma 7, we then have:

$$cbn(\overline{R_1 R_2} \cup R_2, \zeta_{q,p}[x \leftarrow t]\kappa) - cbn(R_1 \cup R_2, p\ (q\ t\ t_1 \cdots t_r)\ t'_1 \cdots t'_s) < \lambda + |\chi'|_\circ - |\chi'|_\bullet\ ,$$

where $\chi' = nf(\Rightarrow_{R_2^{\circ\bullet\to\circ\bullet}}, p\ (nf(\Rightarrow_{R^*}, q\ t\ t_1 \cdots t_r))\ t'_1 \cdots t'_s)$. Since R^* produces no $\circ$-symbols, no such symbols can occur in χ', i.e. $|\chi'|_\circ = 0$. Hence, the theorem follows from $|\chi'|_\bullet \in \mathbb{N}$. Actually, we could obtain a more precise statement by further considering the value of $|\chi'|_\bullet$. □

Note that since Theorem 2 is based on the approximation from Lemma 3, which disregarded the steps performed by M_1 in the original program, we actually obtain that the composed program performs at most as many reduction steps as were performed in the original program by states of M_2, plus λ.

4.5 Application

In Theorems 1 and 2 we have given criteria under which the composed program obtained from Construction 1 is at least as efficient as the original program. Note that these sufficient conditions for non-deterioration can be checked on the *original* program, hence an optimizing compiler will perform the composition construction only if it has ensured beforehand that this is indeed beneficial.

As an example, consider $e = (div\ (exp\ t\ Z))$ from the introduction. Since the mtt unit defining *exp* is context-linear and the tdtt unit defining *div* and *div'* is atmost, Theorem 1 is sufficient to automatically decide that a replacement of e by e' is safe with respect to efficiency (for every possible input t).

Note that also Theorem 2 is constructive, in the sense that we can *algorithmically* decide, whether there exists an $\lambda \in \{0,1\}$ such that the rules of a given $R_1^{\to\circ\bullet,\lambda}$ can be rewritten with *Elim* until no $\circ$-symbols remain, because the right-hand sides of mtt rules are finite trees and none of the rewrite rules in *Elim* introduces new symbols.

Example 5. Recall the mtt units M_1 and M_2 from Example 1 (composed in Example 2). Since M_2 is context-linear, Theorem 2 is applicable as follows. Consider the rules in $R_1^{\to\circ\bullet,1}$ (i.e., $\lambda = 1$):

$$\begin{array}{l} q\ (S\ x_1)\ y_1 = \bullet\ (\gamma\ (\circ\ (q\ x_1\ (\bullet\ (\delta\ (\circ\ (q\ x_1\ y_1))\ y_1))))) \\ q\quad Z \quad y_1 = y_1 \end{array}$$

Since M_2 is atmost, *Elim* contains the rewrite rules of points 1–3 in Definition 5 from the previous subsection. Hence, we can rewrite as follows:

$rhs_{M_1 \to \circ \bullet, 1, q, S} \Rightarrow_{Elim(3)} \gamma\,(\bullet\,(\circ\,(q\,x_1\,(\bullet\,(\delta\,(\circ\,(q\,x_1\,y_1))\,y_1)))))$
$\Rightarrow_{Elim(1)} \gamma\,(q\,x_1\,(\bullet\,(\delta\,(\circ\,(q\,x_1\,y_1))\,y_1))) \Rightarrow_{Elim(3)}$
$\gamma\,(q\,x_1\,(\delta\,(\bullet\,(\circ\,(q\,x_1\,y_1)))\,y_1))$
$\Rightarrow_{Elim(1)} \gamma\,(q\,x_1\,(\delta\,(q\,x_1\,y_1)\,y_1))$.

From Theorem 2 it now follows that for every $t \in T_\Sigma$ and $t_1 \in T_\Delta$:

$$cbn(\overline{R_1 R_2} \cup R_2, \overline{qp_1}\,t\,(p_1\,t_1)\,(p_2\,t_1)) - cbn(R_1 \cup R_2, p_1\,(q\,t\,t_1)) < 1$$
$$cbn(\overline{R_1 R_2} \cup R_2, \overline{qp_2}\,t\,(p_1\,t_1)\,(p_2\,t_1)) - cbn(R_1 \cup R_2, p_2\,(q\,t\,t_1)) < 1 \ .$$

4.6 Results about Classical Deforestation

Kühnemann [8] compares tree transducer composition with classical deforestation [13]. From his Lemma 20 follows that if M_1 is a tdtt unit, then our composition construction essentially yields the same result as classical deforestation with implicit let-expressions. Hence, Theorems 1 and 2 can be used to establish conditions under which classical deforestation for lazy languages is guaranteed to improve efficiency even for nonlinear programs:

Corollary 1. *In the following cases, classical deforestation leads to a program at least as efficient as the original program*[3]*:*

1. *M_1 is a tdtt unit and M_2 is an atmost mtt unit.*
2. *M_1 is a tdtt unit, every rule of which has a right-hand side of the form $(\delta \cdots)$ for some $\delta \in \Delta$, and M_2 is a context-linear or basic mtt unit.*

Proof. The proposition for case 1 follows from Theorem 1. In case 2 the proposition follows from Theorem 2 with $\lambda = 0$, applying the *Elim*-rule 2 from Definition 5. □

5 Future Work

The presented efficiency analysis gives sufficient conditions for when call-by-need reduction of the composed program to normal form does not need more steps than for the original expression, independent from the input. In lazy functional programs, however, such expressions might also occur in a context where their reduction to normal form is not necessary. Hence, the analysis should be made context-independent. We conjecture that the efficiency statement from Theorem 1 remains valid also for partial call-by-need reductions, as does the statement of Theorem 2, if the *Elim*-rules under points 4 and 5 of Definition 5 are abandoned.

Voigtländer and Kühnemann [12] presented a new composition technique that generalizes the transformation considered here, by handling also cases where both involved mtt units use accumulating parameters. Our formal efficiency analysis method is also applicable to the extended transformation and then successfully classifies all examples from [7,8,9,12], but due to space constraints we could not elaborate on this more general setting in the present paper.

[3] By the remarks below the proofs of Theorems 1 and 2, we can even show that the resulting program performs strictly fewer reduction steps than the original program.

Acknowledgment. I would like to thank Armin Kühnemann and the anonymous referees for helpful comments and suggestions.

References

1. R.M. Burstall and J. Darlington. A transformation system for developing recursive programs. *J. ACM*, 24:44–67, 1977.
2. W.N. Chin. Safe fusion of functional expressions II: Further improvements. *J. Funct. Prog.*, 4:515–555, 1994.
3. N. Dershowitz and J.P. Jouannaud. Rewrite systems. In J. van Leeuwen, editor, *Handbook of Theoretical Computer Science*, volume B, chapter 6, pages 243–320. Elsevier Science Publishers B.V., 1990.
4. J. Engelfriet and H. Vogler. Macro tree transducers. *J. Comput. Syst. Sci.*, 31:71–145, 1985.
5. Z. Fülöp and H. Vogler. *Syntax-Directed Semantics—Formal Models Based on Tree Transducers.* Monographs in Theoretical Computer Science. Springer-Verlag, 1998.
6. M. Höff. Vergleich von Verfahren zur Elimination von Zwischenergebnissen bei funktionalen Programmen. Master thesis, Dresden University of Technology, 1999.
7. A. Kühnemann. Benefits of tree transducers for optimizing functional programs. In *Foundations of Software Technology & Theoretical Computer Science, Chennai, India, Proceedings*, volume 1530 of *LNCS*, pages 146–157. Springer-Verlag, 1998.
8. A. Kühnemann. Comparison of deforestation techniques for functional programs and for tree transducers. In *Functional and Logic Programming, Tsukuba, Japan, Proceedings*, volume 1722 of *LNCS*, pages 114–130. Springer-Verlag, 1999.
9. A. Kühnemann and J. Voigtländer. Tree transducer composition as deforestation method for functional programs. Technical Report TUD-FI01-07, Dresden University of Technology, 2001.
10. M. Rosendahl. Automatic complexity analysis. In *Functional Programming Languages and Computer Architecture, London, England, Proceedings*, pages 144–156. ACM Press, 1989.
11. D. Sands. Total correctness by local improvement in the transformation of functional programs. *ACM Trans. on Prog. Lang. and Systems*, 18:175–234, 1996.
12. J. Voigtländer and A. Kühnemann. Composition of functions with accumulating parameters. Technical Report TUD-FI01-08, Dresden University of Technology, 2001. *http://wwwtcs.inf.tu-dresden.de/~voigt/TUD-FI01-08.ps.gz.*
13. P. Wadler. Deforestation: Transforming programs to eliminate trees. *Theoret. Comput. Sci.*, 73:231–248, 1990.
14. C.P. Wadsworth. *Semantics and Pragmatics of the Lambda Calculus.* PhD thesis, Oxford University, 1971.

Rewriting Strategies for Instruction Selection

Martin Bravenboer and Eelco Visser

Institute of Information and Computing Sciences, Universiteit Utrecht, P.O. Box 80089, 3508 TB Utrecht, The Netherlands. http://www.cs.uu.nl/~visser, mbravenb@cs.uu.nl, visser@acm.org

Abstract. Instruction selection (mapping IR trees to machine instructions) can be expressed by means of rewrite rules. Typically, such sets of rewrite rules are highly ambiguous. Therefore, standard rewriting engines based on fixed, exhaustive strategies are not appropriate for the execution of instruction selection. Code generator generators use special purpose implementations employing dynamic programming. In this paper we show how rewriting strategies for instruction selection can be encoded concisely in Stratego, a language for program transformation based on the paradigm of programmable rewriting strategies. This embedding obviates the need for a language dedicated to code generation, and makes it easy to combine code generation with other optimizations.

1 Introduction

Code generation is the phase in the compilation of high-level programs in which an intermediate representation (IR) of a program is translated to a list of machine instructions. Code generation is usually divided into instruction selection and register allocation. During instruction selection the nodes of IR expression trees are associated with machine instructions.

The process of instruction selection can be formulated as a rewriting problem. First the IR tree is rewritten to an instruction tree, in which parts (tiles) of the original tree are replaced by instruction identifiers. This instruction tree is then flattened into a list of instructions (code emission) during which (temporary) registers are introduced to hold intermediate values. Typically, the set of rewrite rules for rewriting the IR tree is highly ambiguous, i.e., there are many instruction sequences that implement the computation specified by the tree, some of which are more efficient than others. The instruction sequence obtained *depends on the rewriting strategy* used to apply the rules.

Standard rewriting engines such as ASF+SDF [4] provide a fixed strategy for exhaustively applying rewrite rules, e.g., innermost normalization. This is appropriate for many applications, e.g., algebraic simplifications on expression trees, but not for obtaining the most efficient code sequence for an instruction selection problem.

For the construction of compiler back-ends, code generator generators such as TWIG [1], BEG [5], and BURG [8] use tree pattern matching [6] and dynamic programming [2] to compute all possible matches in parallel and then choose the

S. Tison (Ed.): RTA 2002, LNCS 2378, pp. 237–251, 2002.

rewrite sequences with the lowest cost. These systems use the same basic strategy, although there are some differences between them, mainly in the choice of pattern match algorithm and the amount of precomputation done at generator generation time. Although some other tree manipulation operations can be formulated in these systems, they are essentially specialized for code generation. Other compilation tasks need to be defined in a different language.

In this paper we show how rewriting strategies for instruction selection can be encoded concisely in Stratego, a language for program transformation based on the paradigm of programmable rewriting strategies [10, 11]. The explicit specification of the rewriting strategy obviates the need for a special purpose language for the specification of instruction selection, allows the programmer to switch to a different strategy, and makes it easier to combine code generation with other optimizations such as algebraic simplification, peephole optimization, and inlining.

In Section 2 we illustrate the instruction selection problem by means of a small intermediate representation, and a RISC-like instruction set, define instruction selection rules, and discuss covering a tree with rules. In Section 3 we explore several strategies for applying the RISC selection rules, including global backtracking to generate all possibilities, innermost rewriting, and greedy top-down (maximal munch). It turns out that no dynamic programming is needed in the case of simple instruction sets. In Section 4 we turn to the problem of code generation for complex instruction set (CISC) machines, which is illustrated with a small instruction set and corresponding rewrite rules. In Section 5 we introduce a specification of a rewriting strategy based on dynamic programming in the style of [1] to compute the cheapest rewrite sequence given costs associated with rewrite rules. The generic specification of dynamic programming is very concise, it fits in half a page, and provides a nice illustration of the use of scoped dynamic rewrite rules [9].

2 Instruction Selection

In this section we describe instruction selection as a rewriting problem. First we describe a simple intermediate representation and a subset of an instruction set for a RISC-like machine. Then we describe the problem of code generation for this RISC machine as a set of rewrite rules. The result of rewriting an intermediate representation tree with these rules is a tile tree which is turned into a list of instructions by *code emission*, a transformation which flattens a tree. In Section 4 we will consider code generation for a CISC-like instruction set.

2.1 Intermediate Representation

An intermediate representation (IR) is a low level, but target independent representation for programs. Constructors of IR represent basic computational operations such as adding two values, fetching a value from memory, moving a

```
module IR
imports Operators
signature
  sorts Exp Stm
  constructors
    BINOP : BinOp * Exp * Exp -> Exp // arithmetic operation
    CONST : Int -> Exp               // integer constant
    REG   : String -> Exp            // machine register
    TEMP  : String -> Exp            // symbolic (temporary) register
    MEM   : Exp -> Exp               // dereference memory address
    MOVE  : Exp * Exp -> Stm         // move between memory a/o registers
```

Fig. 1. Signature of an intermediate representation

value from one place to another, or jumping to another instruction. These operations are often more atomic than actual machine instructions, i.e., a machine instruction involves several IR constructs. This makes it possible to translate intermediate representations to different architectures. Figure 1 gives the signature of a simple intermediate representation format. We have omitted constructors for forming whole programs; for the purpose of instruction selection we are only interested in expressions and simple statements. Figure 2 shows an example IR expression tree in textual and corresponding graphical form. The expression denotes moving a value from memory at the address at offset 8 from the sum of the B and C registers to memory at offset 4 from the FP register.

2.2 Instruction Set

Code generation consists of mapping IR trees to sequences of machine instructions. The complexity of this task depends on the complexity of the instruction set of the target architecture. Reduced Instruction Set (RISC) architectures provide simple instructions that perform a single action, while Complex Instruction Set (CISC) architectures provide complex instructions that perform many actions. For example loading and storing values from and to memory is expressed

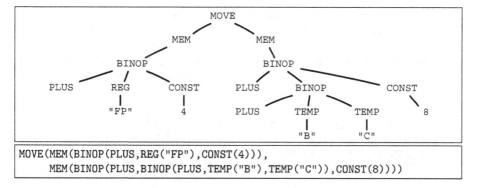

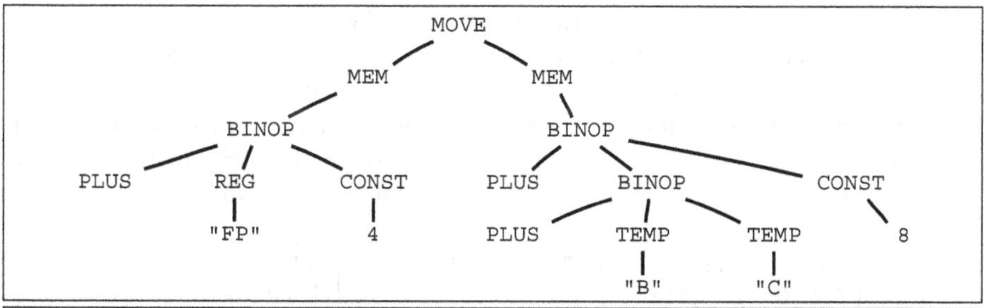


```
MOVE(MEM(BINOP(PLUS,REG("FP"),CONST(4))),
     MEM(BINOP(PLUS,BINOP(PLUS,TEMP("B"),TEMP("C")),CONST(8))))
```

Fig. 2. Example IR expression tree

```
module RISC
signature
  sorts Instr Reg
  constructors
    TEMP    : String -> Reg                 // temporary register
    REG     : String -> Reg                 // machine register
    R       : Reg -> Reg                    // result register
    ADD     : Reg * Reg * Reg -> Instr      // add registers
    ADDI    : Reg * Reg * Int   -> Instr    // add register and immediate
    LOADC   : Reg * Int -> Instr            // load constant
    LOAD    : Reg * Reg -> Instr            // load memory into register
    LOADI   : Reg * Reg * Int -> Instr      // load memory into register
    MOVER   : Reg * Reg -> Instr            // move register to register
    STORE   : Reg * Reg -> Instr            // store register in memory
    STOREI  : Reg * Reg * Int -> Instr      // store register in memory
```

Fig. 3. Constructors for a subset of the instructions of RISC machine.x

by separate instructions on RISC machines, while CISC instruction sets allow many instructions to fetch their operands from memory and store their results back to memory.

We will first consider code generation for the small RISC-like instruction set of Figure 3. The instructions consist of arithmetic operations working on registers and storing their result in a register (only addition for the example), load and store operations for moving values from and to memory, and an operation for loading constants and one for moving values between registers. Several instructions have an 'immediate' variant in which one of the operands is a constant. In load and store operations these immediate values indicate an offset from the address register. For example, `LOADI(REG("a"), REG("b"), 8)` means loading the value at offset 8 from the address in register `b`. The sequence of instructions in Figure 4 implements the example IR tree in Figure 2.

2.3 Selecting Instructions with Rewrite Rules

The process of instruction selection can be divided into two phases. First, determine for each tree node which instruction to use. Then linearize the tree into a sequence of instructions (code emission). The connection between machine instructions and intermediate representation can be defined in terms of rewrite rules. Figure 5 defines rules mapping IR tree patterns to RISC instructions. To translate a complete IR tree to machine instructions we now have to find an

```
[ADD(TEMP("b"),TEMP("B"),TEMP("C")),
 LOADI(TEMP("a"),TEMP("b"),8),
 STOREI(TEMP("a"),REG("FP"),4)]
```

```
ADD b B C
LOADI a b 8
STOREI a FP 4
```

Fig. 4. An instruction sequence in abstract and concrete syntax.

```
module IR-to-RISC-Rules
imports IR RISC
rules
  Select-ADD :
    BINOP(PLUS, e1, e2) -> ADD(R(TEMP(<new>)), e1, e2)
  Select-ADDI :
    BINOP(PLUS, e, CONST(i)) -> ADDI(R(TEMP(<new>)), e, i)
  Select-LOADC :
    CONST(i) -> LOADC(R(TEMP(<new>)), i)
  Select-MOVER :
    MOVE(r@<REG(id) + TEMP(id)>, e) -> MOVER(r, e)
  Select-STORE :
    MOVE(MEM(e1), e2) -> STORE(e2, e1)
  Select-STOREI :
    MOVE(MEM(BINOP(PLUS, e1, CONST(i))), e2) -> STOREI(e2, e1, i)
  Select-LOAD :
    MEM(e) -> LOAD(R(TEMP(<new>)), e)
  Select-LOADI :
    MEM(BINOP(PLUS, e, CONST(i))) -> LOADI(R(TEMP(<new>)), e, i)
```

Fig. 5. Instruction selection rules mapping IR trees to instructions.

application of rules to nodes of the tree such that each node of the tree is *covered* by some pattern; except for 'leaf' nodes such as `REG`, `TEMP` and constants. Figure 6 shows a possible covering for the example IR tree of Figure 2 together with the resulting instruction tree.

In an expression tree, results from subexpressions are passed implicitly up in the tree. When breaking a tree into a sequence of instructions this is no longer the case; intermediate values should be stored into registers or in memory. We will assume here that they can be stored in registers. In the rewrite rules in Figure 5 passing of intermediate values is achieved by generating a new temporary register at the position of the destination register to hold the result of the instruction. For example, rule `Select-ADD` generates the term `ADD(R(TEMP(<new>)), e1, e2)`. The `R(_)` constructor indicates that this argument contains the result of the instruction. The `new` primitive generates a new unique name.

2.4 Code Emission

After covering the IR tree with instructions, the program is still in tree shape. *Code emission* linearizes the tree into a list of instructions. For example, for the tiling in Figure 6 we get the code sequence of Figure 4. A specification of a code emission strategy based on the generic tree flattening strategy `postorder-collect` is given in the technical report version of this paper.

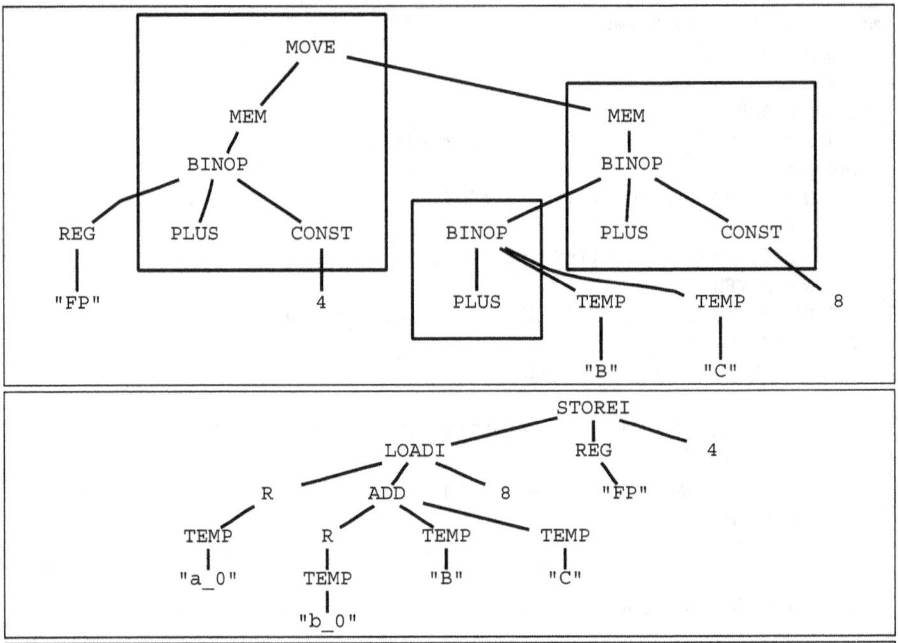


```
STOREI(LOADI(R(TEMP("x_0")), ADD(R(TEMP("y_0")),TEMP("B"),TEMP("C")),8),
       REG("FP"),5)
```

Fig. 6. Optimum tiling of the example tree and the corresponding instruction tree with temporary intermediate result registers.

```
STORE(LOAD(a,ADD(b,ADD(c,B,C),LOADC(d,8))),ADD(e,FP,LOADC(f,4)))
STORE(LOAD(a,ADD(b,ADD(c,B,C),LOADC(d,8))),ADDI(g,FP,4))
STORE(LOAD(a,ADDI(h,ADD(i,B,C),8)),ADD(j,FP,LOADC(k,4)))
STORE(LOAD(a,ADDI(h,ADD(i,B,C),8)),ADDI(l,FP,4))
STORE(LOADI(m,ADD(n,B,C),8),ADD(o,FP,LOADC(p,4)))
STORE(LOADI(m,ADD(n,B,C),8),ADDI(q,FP,4))
STOREI(LOAD(r,ADD(s,ADD(t,B,C),LOADC(u,8))),FP,4)
STOREI(LOAD(r,ADDI(v,ADD(w,B,C),8)),FP,4)
STOREI(LOADI(x,ADD(y,B,C),8),FP,4)
```

```
ADD    c B C
LOADC  d 8
ADD    b c d
LOAD   a b
LOADC  f 4
ADD    e FP f
STORE  a e
```

```
ADD    n B C
LOADI  m n 8
LOADC  p 4
ADD    o FP p
STORE  m o
```

```
ADD    y B C
LOADI  x y 8
STOREI x FP 4
```

Fig. 7. All possible instruction selections for the example IR tree and three pretty-printed code sequences

3 Strategies for Instruction Selection

In the previous section we have seen that rewrite rules can be used for expressing the mapping from intermediate representation trees to lists of machine instructions. However, the set of rewrite rules in Figure 5 is highly ambiguous, i.e., there are many rewritings possible. For example, Figure 7 shows all 9 possible tilings for the example expression in Figure 2 together with three of the code sequences pretty-printed. This kind of ambiguity is typical for instruction selection rules. Therefore, the instruction sequence finally obtained *depends on the rewriting strategy* used to order the application of selection rules. In this section we will examine several strategies and their applicability to instruction selection.

3.1 Intermezzo: Rewriting Strategies

Thusfar we have considered algebraic signatures and standard rewrite rules. In a normal rewriting system a term over a signature is normalized with respect to the set of rewrite rules. In Stratego the rewriting strategy is not fixed, but can be specified in a language of strategy combinators. A strategy is a term transformation that may fail. Rewrite rules are basic strategies. Examples of basic combinators are sequential composition, non-deterministic choice, deterministic choice, and a number of generic term traversal operators. Using these combinators a wide variety of rewriting strategies can be defined. The Stratego library provides a large number of generic strategies built using these combinators. In the rest of this paper we will define several compound strategies. The strategy elements in these definitions will be explained when needed. For a full account we refer to the literature; good starting points are [10, 9, 11].

3.2 Exhaustive Application

The traditional way to interpret a set of rewrite rules is by applying them exhaustively to a subject term. The following strategy takes this approach:

```
innermost-tilings =
  innermost(Select-ADD + Select-ADDI + Select-STORE + Select-LOAD +
            Select-STOREI + Select-LOADC + Select-LOADI + Select-MOVER)
```

The *local non-deterministic choice operator* `(s1 + s2)` combines two strategies `s1` and `s2` by trying either of them. If that fails the other is tried. If one of the branches has succeeded the choice is committed, i.e., no backtracking to this choice is done if the continuation fails.

The innermost strategy takes a transformation, e.g., a choice of rules, and applies them exhaustively to a subject term starting with the inner terms. This is expressed generically as

```
innermost(s) = all(innermost(s)); try(s; innermost(s))
```

The generic traversal combinator all(s) applies the transformation s to each direct subterm of a constructor application. Thus, <all(s)>C(t1,...,tn) denotes C(<s>t1,...,<s>tn). The first part of the strategy normalizes all direct subterms of a node. After that (; is sequential composition), the strategy tries to apply the transformation s. If that succeeds the result, e.g., the right-hand side of the term, is again normalized. The combinator try(s) is defined as try(s) = (s <+ id). That is, try to apply s, but if that fails do nothing, i.e., use the identity strategy id. Thus, if s fails the subject term is in normal form.

The innermost strategy is not adequate for instruction selection, however. It produces the worst result, e.g., the first result in Figure 7. This is caused by the fact that reducible subexpressions of a pattern are reduced before the (more complex) pattern itself is applied. Furthermore, the exhaustive application is overkill since instruction selection rules do not need exhaustive application; each node needs to be rewritten at most once.

3.3 All Rewritings

Another approach is to generate all possible results. This is achieved by the following strategy:

```
all-tilings =
  bagof(topdown(try(Select-ADD ++ Select-ADDI ++ Select-STORE
                    ++ Select-STOREI ++ Select-LOADC
                    ++ Select-LOAD ++ Select-LOADI ++ Select-MOVER)))
```

In constrast to the + operator used above, the *global non-deterministic choice* operator (s1 ++ s2) is a non-committing choice operator. This means that after one of the branches has succesfully terminated, but the continuation fails, the strategy backtracks to the other branch. Formally, we have that (s1 ++ s2); s3 is equal to (s1; s3) ++ (s2; s3), which does not hold for +. This property is used by the bagof(s) operator to produce the list of all possible results by forcing a backtrack to the last choice point. The topdown strategy, defined as

```
topdown(s) = s; all(topdown(s))
```

traverses a tree in pre-order, applying a transformation s to a node before transforming its subterms using the generic traversal operator all. The list of all results in Figure 7 was produced using this strategy.

After generating the list of all possible results we could define a filter that selects the best solution. As a specification this is interesting, as an implementation it is not feasible due to the combinatorial explosion; for each node all possible rewrites for its subnodes need to be considered.

3.4 Maximal Munch

If each rule corresponds to a machine instruction, the best solution is usually the shortest sequence of instructions. In other words, we want to maximize the tree tilings selected. The following strategy takes this approach:

```
maximal-munch =
  topdown(try((Select-ADDI <+ Select-ADD) + Select-LOADC
             + (Select-STOREI <+ Select-STORE)
             + (Select-LOADI <+ Select-LOAD) + Select-MOVER))
```

Starting at the root of the expression tree, the strategy tries to apply one of the selection rules. Instead of combining all rules with the non-deterministic choice operator `+`, some pairs of rules are combined using the *deterministic choice* operator `(s1 <+ s2)`, which always tries its left argument first. In this way the `maximal-munch` strategy gives priority to patterns that take the largest bite, e.g., preferring `Select-ADDI` over `Select-ADD`. In fact, this simple strategy, combining pre-order top-down traversal with rule priority, is adequate for simple RISC-like instruction sets.

3.5 Pattern Matching

As can be seen from the examples above, programmable strategies allow the combination of rewrite rules into different strategies. This flexibility in combining rules does not restrict the efficient implementation of pattern matching. When combining rules in a choice, as in the examples above, the rules are not tried one by one. Instead a pattern match optimization compiles the rule selection into a matching automaton which folds left-hand sides with the same root constructor into a single condition (and recursively for subpatterns).

4 Complex Instruction Sets

In RISC instruction sets complex memory addressing modes are restricted to load and store operations. In complex instruction set (CISC) machines, many instructions can use the full range of addressing modes to read their operands directly from memory using base and index registers and constant offsets from these. Some machines also have heterogeneous register sets such that not all instructions can work with all registers, requiring values to be moved between registers.

For such machines maximal munch is not adequate. Code generators in production compilers use a dynamic programming approach to compute the code

```
module CISC
signature
  sorts Reg Addr Instr
  constructors
    TEMP : String -> Reg          IMM   : Int -> Addr
    REG  : String -> Reg          REGA  : Reg -> Addr
    NONE : Reg                    ADDR  : Reg * Reg * Int * Int -> Addr
    R    : Reg -> Reg             ADD   : Addr * Addr * Addr -> Instr
                                  MOVEM : Addr * Addr -> Instr
```

Fig. 8. Constructors for a subset of the instructions of a CISC machine

```
MOVEM 4[FP] 8[B,C]  # C1=1 C2=1 C3=0 C4=0 C5=0 C6=0 C7=0 cost=1

ADD r B C           # C1=1 C2=1 C3=0 C4=0 C5=0 C6=0 C7=2 cost=2
MOVEM 4[FP] 8[r]

ADD   n FP $4       # C1=1 C2=1 C3=1 C4=1 C5=3 C6=3 C7=4 cost=6
ADD   e B C
ADD   l e $8
MOVEM [n] [l]
```

Fig. 9. Lowest cost selections based on different cost assignments

sequence with the least cost. To illustrate rewriting with dynamic programming we define the small CISC-like instruction set in Figure 8. In contrast to the RISC instruction set there are no special load and store operations. Instead each instruction can use all addressing modes and thus load and store values to memory. The `ADDR(r1,r2,scale,off)` addressing mode indicates a value in memory at address `r1 + (r2*scale) + off`, where `r1` is a base register, `r2` an index register, and `scale` and `off` constants.

Figure 10 defines instruction selection rules for this instruction set from the intermediate representation of Figure 1. Since each instruction can use all addressing modes, it is not desirable to define rules for each instruction parsing each addressing mode. Instead addressing modes are selected separately. To ensure that tiles connect, each rule indicates the *mode* of its tile and the expected modes of the leaves of the tile using the constructor `l(mode,tile)`.

Again, this rule set is ambiguous. To indicate which rewrite should be used, costs are assigned to rewrite rules corresponding to the operational cost of selected instruction. Figure 11 assigns symbolic costs `C1` to `C7` to several of the patterns of Figure 10. The goal now becomes to select the tree with lowest overall cost. Figure 9 shows several lowest cost selections for different cost assignments to these symbolic costs. This illustrates that instruction selection can no longer be achieved by a fixed strategy.

5 Dynamic Programming

Maximal munch does not work since chain rules such as `Select-LOAD` can be applied indefinitly. A straightforward solution would be to generate all solutions, compute their cost, and take the cheapest solution. However, in Section 3 we already saw that this leads to a combinatorial explosion in which subresults are computed many times. In the case of the rules of Figure 10, the situation is worse, since rules `Select-REGA` and `Select-LOAD` together lead to infinitly many possibile rewrites.

The dynamic programming approach of [2, 1] is similar to the `all-tilings` strategy of Section 2, but instead of computing all possible rewrites in sequence, we first compute all possible matches in one bottom-up pass over the tree, tabulating the cheapest match for each node. This information can then be used to

```
module IR-to-CISC-Rules
imports CISC IR Dynamic-Programming
signature
  constructors
    a : Mode  reg : Mode  imm : Mode  s : Mode
rules
  Select-MOVEM :
    MOVE(e1, e2) -> l(s, MOVEM(l(a, e1), l(a, e2)))
  Select-ADDM :
    MOVE(e1, BINOP(PLUS, e2, e3)) -> l(s,ADD(l(a,e1), l(a,e2), l(a,e3)))
  Select-ADD :
    BINOP(PLUS,e1,e2) -> l(reg,ADD(R(REGA(TEMP(<new>))),l(a,e1),l(a,e2)))
  Select-imm :
    CONST(i) -> l(imm, i)
  Select-reg :
    REG(r) -> l(reg, REG(r))
  Select-reg' :
    TEMP(r) -> l(reg, TEMP(r))
  Select-ABS :
    MEM(i) -> l(a, ADDR(NONE,NONE,1,l(imm, i)))
  Select-BASE :
    MEM(r) -> l(a, ADDR(l(reg, r),NONE,1,0))
  Select-BASEIMM :
    MEM(BINOP(PLUS, e1, e2)) -> l(a, ADDR(l(reg,e1), NONE, 1, l(imm,e2)))
  Select-BASEINDEX :
    MEM(BINOP(PLUS, e1, e2)) -> l(a, ADDR(l(reg,e1), l(reg,e2), 1, 0))
  Select-BASEINDEXIMM :
    MEM(BINOP(PLUS, BINOP(PLUS, e1, e2), i)) ->
    l(a, ADDR(l(reg, e1), l(reg, e2), 1, l(imm, i)))
  Select-IMM :
    r -> l(a, IMM(l(imm, r)))
  Select-REGA :
    r -> l(a, REGA(l(reg, r)))
  Select-LOAD :
    x -> l(reg, MOVEM(R(REGA(TEMP(<new>))), l(a, x)))
```

Fig. 10. Instruction selection rules mapping IR trees to CISC instructions.

```
module CISC-Costs
imports CISC
rules
  CiscCost : ADD(x, y, z)        -> C1
  CiscCost : MOVEM(x, y)         -> C2
  CiscCost : ADDR(NONE, y, 1, 0) -> C3 where <not(NONE)> y
  CiscCost : ADDR(x, NONE, 1, 0) -> C4 where <not(NONE)> x
  CiscCost : ADDR(NONE, y, 1, i) -> C5 where <not(NONE)> y; <not(0)> i
  CiscCost : ADDR(x, NONE, 1, i) -> C6 where <not(NONE)> x; <not(0)> i
  CiscCost : ADDR(x, y, 1, i)    -> C7 where <not(NONE)> x; <not(NONE)> y
```

Fig. 11. Instruction costs

perform the optimal rewrite sequence. Figure 13 presents the specification of a generic dynamic programming strategy. Figure 12 shows an instantiation of this strategy with the rules and costs of Figures 10 and 11, and a default cost of 0.

5.1 Optimum Tiling

The dynamic programming strategy is implemented by `optimum-tiling`, which is parameterized with a cost computation `cost`, a set of normal rules `rs`, and a set of chain rules `is`. The strategy consists of a bottomup traversal followed by a topdown traversal. During the bottomup traversal, the strategy `match-tile` finds out which rules are applicable and which one is the cheapest. This is expressed by defining a dynamic rewrite rule `BestTile` for each subtree `t`, which rewrites `t` tagged with a mode `m` as `l(m,t)` to the corresponding instruction tree. These dynamic rules are then applied in a topdown traversal over the tree. In other words, this corresponds to `maximal-munch` where the rule to be used is determined dynamically instead of statically and can be different at each node.

5.2 Match Tile

Strategy `match-tile` first creates a pair of the subject term and the application of the rules to the subject term. If one of the rules succeeds, it is registered by `register-match`. Then `fail` forces backtracking to another rule, until no more rules match. After applying the normal rules, the closure of the chain rules is computed by repeating the same procedure until no more chain rules apply. The repetition is necessary since one chain rule application can enable another one. This process terminates since `register-match` only succeeds if the cost of the match is lower than the cost of all previous matches.

5.3 Register Match

`Register-match` is the core of the dynamic programming strategy. It gets a pair of a subject `tree` and the result of applying a rule to it, which should be a `tile` tagged with a `mode`. `ComputeCost` returns the cumulative cost of the tile and the costs of the trees at the leafs of the tile. If this cost is less than the `BestCost` so far for the `tree` in this `mode`, the match is registered by generating new *dynamic rules* for `BestCost` and `BestTile` rewriting the `l(mode,tree)` term to the computed cost and the given tile. This overrides any previously generated rules for `l(mode,tree)`. Thus, the best result per mode is computed for each node in a bottomup fashion; suboptimal results are discarded.

Dynamic rewrite rules [9] are just like ordinary rules, except that they inherit the values of any of their variables that are bound in their definition context. Thus the definitions of the rule `BestCost` inherits from its context the values of the variables `mode`, `tree`, and `cost`. A rule specific for these values is generated, which may exist next to other `BestCost` rules, except if the values for `mode` and `tree` are the same. In that case the old rule is overridden.

```
module IR-to-CISC
imports lib IR-to-CISC-Rules CISC-Costs Dynamic-Programming
strategies
  cisc-select =
    optimum-tiling(CiscCost <+ !0, cisc-tiles, cisc-injections, !s)

  cisc-tiles =
    Select-ADDM ++ Select-MOVEM ++ Select-ADD ++ Select-imm
    ++ Select-reg ++ Select-reg' ++ Select-ABS ++ Select-BASE
    ++ Select-BASEINDEX ++ Select-BASEIMM ++ Select-BASEINDEXIMM

  cisc-injections =
    Select-LOAD + Select-IMM + Select-REGA
```

Fig. 12. Instantiation of the dynamic programming algorithm

```
module Dynamic-Programming
imports lib dynamic-rules
signature
  constructors
    l : Mode * a -> a
strategies
  optimum-tiling(cost, rs, is, rootmode) =
    {| BestTile, BestCost :
       bottomup(match-tile(cost, rs, is));
       !l(<rootmode>, <id>);
       topdown(try(BestTile))
    |}

  match-tile(cost, rs, is) =
    try(   test(!(<id>, <rs>); register-match(cost); fail));
    repeat(test(!(<id>, <is>); register-match(cost)))

  register-match(cost) =
    ?(tree, l(mode, tile));
    where(
      <ComputeCost(cost)> tile => cost
    ; <lt-inf> (cost, <BestCost <+ !Infinite> l(mode, tree))
    ; rules(
        BestCost : l(mode, tree) -> cost
        BestTile : l(mode, tree) -> tile
      )
    )
  ComputeCost(cost) = where(cost => tile-cost)
    ; collect(?l(_,_)); foldr(!tile-cost, add, BestCost)
```

Fig. 13. Dynamic programming algorithm for computing an optimum tiling.

5.4 Compute Cost

The cost of a match is determined by adding the cost of the tile itself as provided by the parameter strategy `cost` and the `BestCosts` of the leafs of the tile. The leafs are obtained by collecting all outermost subterms matching `l(_,_)` from the tile. For example, the tile `ADDR(l(reg,e1), NONE, 1, l(imm,e2))` has cost `C6` and leafs `[l(reg,e1),l(imm,e2)]`. Thus the fold over this list produces `<add>(<BestCost>l(reg,e1), <add>(<BestCost>l(imm,e2), C6))`. If `BestCost` fails for some leaf of a tile, this indicates that no tile was found for the corresponding tree with the mode needed by this tile.

6 Discussion

The instruction selection techniques in this paper were developed as part of a project building a complete Tiger compiler [3] using rewriting techniques. The generic strategies are part of the Stratego Standard Library, which contains many more generic strategies. The examples in this paper are written in Stratego and have been tested with release 0.7 of the Stratego Compiler[1].

The maximal munch algorithm was directly inspired by [3]. However, our separation of selection rules and code emission makes the rules independent of the strategy to be used. The dynamic programming algorithm is based on the ideas of [2, 1]. These ideas have been used in several systems including TWIG [1], BEG [5], and BURG [8]. The main differences between these systems is the pattern matching algorithm used and the amount of computation done at generator generation time. BURG [8] is based on bottom-up rewriting (BURS) theory and performs all cost computations at generator generation time, whereas our implementation does all cost computations at code generation time. Furthermore, BURG uses bottom-up pattern matching [6], an efficient matching algorithm that does not reexamine subterms. This is achieved by normalizing patterns to shallow patterns and introducing new modes.

We have not yet performed comparative benchmarks, but our implementation will undoubtely be much slower than the highly optimized BURG code generators. For a list of IR trees of 9200 nodes, the `maximal-munch` strategy took 0.03 seconds to produce a (RISC) instruction tree of 11,100 nodes (2100 instructions), while the `optimum-tiling` dynamic programming strategy took 1.96 seconds to produce a (CISC) instruction tree of 15,000 nodes (1000 instructions). We expect to improve the performance considerably by applying specialization and fusion in the style of [7], where an optimization for specialization of the generic `innermost` strategy is presented. A possible optimization is the normalization of rule left-hand sides in order to achieve better pattern matching.

7 Conclusion

In this paper we have shown the complete code of *several* instruction selection algorithms. Creating a complete code generator only requires more rules, not more

[1] http://www.stratego-language.org

implementation. This is clear evidence of the expressivity of the specification of program transformations with rewrite rules and rewriting strategies. Instead of hardwiring the strategy in the rewrite engine, the programmer can define the most appropriate strategy for the task at hand. For example, one could start with a simple maximal-munch strategy for a code generator and only switch to dynamic programming when it is really needed. Furthermore, programmable strategies enable the combination of several transformations each using different strategies. Much of the expressivity of strategies is due to the ability of capturing generic schemas with generic traversal operators. Finally, the incorporation of dynamic programming techniques provides new opportunities for the specification of other kinds of transformations, and poses new opportunities for the optimization of strategies. In particular, it would be interesting to generalize tabulation techniques from code generators to more general transformation strategies.

References

1. A. V. Aho, M. Ganapathi, and S. W. K. Tjiang. Code generation using tree pattern matching and dynamic programming. *ACM Transactions on Programming Languages and Systems*, 11(4):491–516, October 1989.
2. A. V. Aho and S. C. Johnson. Optimimal code generation for expression trees. *Journal of the ACM*, 23(3):488–501, 1976.
3. A. W. Appel. *Modern Compiler Implementation in ML.* Cambridge University Press, 1998.
4. A. van Deursen, J. Heering, and P. Klint, editors. *Language Prototyping. An Algebraic Specification Approach*, volume 5 of *AMAST Series in Computing.* World Scientific, Singapore, September 1996.
5. H. Emmelmann, F.-W. Schroer, and R. Landwehr. BEG - a generator for efficient back ends. In *ACM SIGPLAN 1989 Conference on Programming Language Design and Implementation (PLDI'89)*, pages 227–237. ACM, July 1989.
6. C. H. Hoffmann and M. J. O'Donnell. Pattern matching in trees. *Journal of the ACM*, 29(1):68–95, January 1982.
7. P. Johann and E. Visser. Fusing logic and control with local transformations: An example optimization. In B. Gramlich and S. Lucas, editors, *Workshop on Reduction Strategies in Rewriting and Programming (WRS'01)*, volume 57 of *Electronic Notes in Theoretical Computer Science*, Utrecht, The Netherlands, May 2001. Elsevier Science Publishers.
8. T. A. Proebsting. BURS automata generation. *ACM Transactions on Programming Languages and Systems*, 17(3):461–486, May 1995.
9. E. Visser. Scoped dynamic rewrite rules. In M. van den Brand and R. Verma, editors, *Rule Based Programming (RULE'01)*, volume 59/4 of *Electronic Notes in Theoretical Computer Science.* Elsevier Science Publishers, September 2001.
10. E. Visser. Stratego: A language for program transformation based on rewriting strategies. System description of Stratego 0.5. In A. Middeldorp, editor, *Rewriting Techniques and Applications (RTA'01)*, volume 2051 of *Lecture Notes in Computer Science*, pages 357–361. Springer-Verlag, May 2001.
11. E. Visser, Z.-e.-A. Benaissa, and A. Tolmach. Building program optimizers with rewriting strategies. In *Proceedings of the third ACM SIGPLAN International Conference on Functional Programming (ICFP'98)*, pages 13–26. ACM Press, September 1998.

Probabilistic Rewrite Strategies. Applications to ELAN

Olivier Bournez and Claude Kirchner

LORIA & INRIA,
615 rue du Jardin Botanique, BP 101,
54602 Villers-lès-Nancy Cedex, Nancy, France.
{Olivier.Bournez,Claude.Kirchner}@loria.fr

Abstract. Recently rule based languages focussed on the use of rewriting as a modeling tool which results in making specifications executable. To extend the modeling capabilities of rule based languages, we explore the possibility of making the rule applications subject to probabilistic choices.
We propose an extension of the ELAN strategy language to deal with randomized systems. We argue through several examples that we propose indeed a natural setting to model systems with randomized choices. This leads us to interesting new problems, and we address the generalization of the usual concepts in abstract reduction systems to randomized systems.

1 Introduction

Term rewriting has been developed since the last thirty years, leading to a deep and solid corpus of knowledge about the rewrite relation induced by a set of rewrite rules. More recently, rule based languages focussed on the use of rewriting as a modeling tool, which results in making the out-coming specification executable in a very efficient way [11]. Such languages enlighten the fundamental role of rewrite strategies, either for computation or for deduction. In the ELAN language, the notion of rule and strategy are both first class and this is backed-up by the concept of rewriting calculus [5]. In this framework, that generalizes the lambda-calculus, the basic notions are those of rewrite rule and of rule application. To extend the modeling capabilities of the calculus, we explore in this work the possibility of making the rule application subject to probabilistic choices.

Since the probabilistic firing of a rewrite rule can be seen as a specific kind of strategy, we introduce in the first part of this work a new notion of strategy. From a practical point of view, we added to ELAN the notion of *probabilistic choice* strategy, permitting us to fire under some probability a given set of rules. It is in particular a fundamental design choice to manage probabilities at the level of strategies instead of at the level of the rules themself where the flexibility and the semantics would have been then less clear.

This leads us to interesting new capabilities and problems that we are addressing in this paper.

S. Tison (Ed.): RTA 2002, LNCS 2378, pp. 252–266, 2002.

The combination of the concept of explicit rule application and of probabilistic choice has many applications. In particular for probabilistic data bases [21], probabilistic agents [6], genetic algorithms [9], randomized algorithms [15], randomized proof procedures [14], ... etc. Therefore, the notion of probabilistic strategy is quite useful as a modeling tool. What we address in the second part of this paper is a first step in the understanding of such strategies in the rewriting context.

Indeed we focus on the study of abstract reduction systems extensions under probabilistic choice. For example, we can define a notion of almost-sure termination as the property that the set of infinite derivations of an abstract reduction system is of null measure. A typical situation is the rewrite system $a \to a, a \to b$ where the rewrite strategy consists in applying each rule with probability 0.5. This system is terminating with probability one but clearly not terminating.

Similarly, probabilistic notions for confluence or other usual notions of classical abstract reduction systems can be defined. The design of tools to check these properties at the level of a rewrite *relation* is a challenge and will strengthen the use of the concept as a specification tool. This paper aims at putting a first stone towards these directions.

We presented first ideas about such an approach to control rule firing together with a first prototype in [1]. A different point of view has been developed recently to extend the Constraint Handling Rule process with probabilistic capabilities put on the rewrite rules themselves [8,16,18,17,19]. This is, to the best of our knowledge, the only attempts to formalize probabilistic transitions using rule based languages.

In Section 2 we introduce through some simple examples the strategy operator that we added to ELAN strategy language to deal with probabilistic systems. In Section 3, we argue through several examples that this is indeed a natural setting for specifying systems with probabilistic choices. In Section 4, we discuss on the operational semantic of the proposed strategy operator through a discussion on the way this operator is implemented in the ELAN prototype. This leads us to important questions about the generalization of usual rewriting concepts such as confluence and termination that we discuss for probabilistic abstract reduction systems in Section 5.

2 A General Probabilistic Rewrite Strategy Operator

Let us first consider the simple example of euro coin flipping[1]. It could be modeled using the following rules:

```
[h] x => head
[t] x => tail
```

[1] See `http://www.inference.phy.cam.ac.uk/mackay/abstracts/euro.html` for a summary of the recent attention devoted to the potential bias of the euro coin (thanks to one of the referees).

where `h` and `t` are the names, also call labels of the rules, `head, tail` are just constant function symbols and `x` is a variable.

In order to express the equiprobability of the two sides of a euro coin, we write

```
equi => PC(h:0.5, t:0.5)
```

which means that the two rules fire with the same probability and that the application of the strategy `equi` to `flip` (denoted `(equi flip)`) reduces either to `tail` or to `head` with the same probability 0.5.

If euro coins were biased, we would write something like:

```
bias => PC(h:0.4, t:0.6)
```

Then, we would have with probability 0.4 the derivation

```
(bias flip) => (PC(h:0.4, t:0.6) flip)
            => (h flip)
            => head
```

and, with probability 0.6 the derivation

```
(bias flip) => (PC(h:0.4, t:0.6) flip)
            => (t flip)
            => tail
```

But actually a strategy operator that makes choices with fixed probabilities is not sufficient to model all games, since probabilities often also depend on parameters.

Suppose for example, that in addition to bias problems there were no harmonization between the countries of the euro zone. Then, the bias would depend on the country and we would need somehow a function `probaH`, from countries to numbers in the interval $[0, 1]$ to represent the bias.

It could be natural to model this latter function as a strategy itself, mapping (normalizing) terms representing countries to real numbers in the interval $[0, 1]$.

This could be done in ELAN using something like:

```
[probaH] france  => 0.49
[probaH] belgium => 0.50
...
```

The `bias` operator would then be simply described by:

```
bias    => PC(h:probaH, t:1 - probaH)
```

and we would write `(bias luxembourg)` to simulate Luxembourg euro flipping.

Hence, we actually introduce the following strategy operator: for strategies $p_1, \ldots, p_n$ mapping terms (which sort is un-important here) to terms of the sort of real numbers so that on any term t, $\Sigma_i(p_i\ t) = 1$, and when $s_1, \ldots, s_n$ are other strategies which application is subject to a certain probability law, $PC(s_1 : p_1, \ldots, s_n : p_n)$ is the strategy that consists, applied on some term t, in choosing an $i \in \{1, \ldots, n\}$ with probability $(p_i\ t)$, and then returning $(s_i\ t)$, the application of strategy s_i on the term t.

Before formally describing this strategy combinator, let us play with more examples.

3 Examples

We now develop several examples to argue that the PC operator is indeed a natural setting for modeling probabilistic systems. We choose first an example that requires probabilistic choices with fixed probabilities. The second example requires probabilistic choices with run-time varying probabilities. We choose as third example to take a simple algorithm of the book [15] to exemplify how easy it is to transform any randomized algorithm into a ELAN program using the previous operator.

Example 1. Suppose we want to model the following game between two players. Initially, two players have M euros ($M \geq 2$). An unbiased coin is flipped. If it falls on head, player one wins one euro from player 2, otherwise player 2 wins two euros from player 1. The game stops when one of the two players is ruined.

This can be modeled as follows.

```
[h] game(M1,M2) => game(M1+1,M2-1)
                   if M2>0
[h] game(M1,0)  => player-1-wins
[t] game(M1,M2) => game(M1-2,M1+2)
                   if M1>1
[t] game(0,M2)  => player-2-wins
[t] game(1,M2)  => player-2-wins
```

The strategy modeling the repetition of coin flipping is:

```
play => repeat(PC(h:0.5, t:0.5))
```

Considering the derivations of (play game(M,M)), we get a modeling of our game.

This is clearly not a terminating system, since there are infinite derivations starting from (play game(M,M)). However, according to probability theory, the probability of such an infinite derivation is null: with probability 1, a derivation terminates. So we can say, as we will develop in Section 5, that this system is *almost-surely terminating* (but not terminating).

Of course we must also have a picking balls example:

Example 2. Consider a urn with N green balls and two yellow balls. We pick a ball until we get a yellow one. We loose if we pick all the green balls.

```
[pickgreen]  urn(N) => urn(N-1)
                       if N>1
[pickgreen]  urn(1) => loose
[pickyellow] urn(N) => win
```

Transition probabilities now depend on time, i.e. on terms. If balls are picked equiprobably, then the probability of picking a green ball should be $N/(N+2)$.

So we write:

```
play => repeat(PC(pickgreen: p, pickyellow: 1-p))
```

where p is itself a strategy.

```
[p] urn(N) => N / (N+2)
```

We could add many other examples to argue that this setting is quite natural for prototyping probabilistic games, random walks, or randomized algorithms.

To argue that randomized algorithms can easily been turned into a ELAN program, we take the first example of the book [15]. This is a sorting algorithm in expected time $O(2nH_n) = O(n \log n)$, where H_n is the nth Harmonic number [15].

Example 3. We need first to choose an element uniformly in a list L. This is obtained as the application of the probabilistic strategy any to L with:

```
[thisone]  y.L => y
[otherone] y.L => L

[pthisone]  L => 1/length(L)
[potherone] L => (length(L)-1)/length(L)

any => repeat*(PC(thisone: pthisone, otherone: potherone))
```

And now, we sort as follows:

```
[] sort(L) => sort(subset-lower(L,y)).y.sort(subset-greater(L,y))
           where y:= (any L)
[] sort(nil) => nil

[] subset-lower(z.L,y) => z.subset-lower(L,y)       if z < y
[] subset-lower(z.L,y) => subset-lower(L,y)         if z >= y
[] subset-lower(nil,y) => nil

[] subset-greater(z.L,y) => z.subset-greater(L,y)   if z > y
[] subset-greater(z.L,y) => subset-greater(L,y)     if z <= y
[] subset-greater(nil,y) => nil
```

4 Operational Semantics

The previous PC operator is available in the ELAN system and the previous examples can be check in the current prototype.

We now present the operational semantic of the probabilistic choice strategy. This is easy to get from following discussion on its implementation, using the background semantics of the ELAN language [3,5].

Actually, the PC operators are implemented in ELAN itself, using the classical trick of imperative programming to simulate any probabilistic distribution using a uniform one [13]. Concretely, we just added to ELAN a built-in operator `uniform-random-generator()` that returns a real number in interval [0, 1] with uniform distribution.

Now, when the p_i are strategies (possibly constant) that, on any term t, evaluate to a positive (or null) real number such that $\sum_{i=1}^{n}(p_i\ t) = 1$, then the application of the strategy $PC(s_1 : p_1, \ldots, s_n : p_n)$ on a term t is defined as follows:

$$[\,]\ (PC(s_1 : p_1, \ldots, s_n : p_n)\ t) \Rightarrow (\texttt{choose}(y, s_1 : (p_1\ t), \ldots, s_n : (p_n\ t))\ t)$$
$$\texttt{where}\ y := \texttt{uniform-random-generator}()$$

With, for reals $\xi_i \in [0, 1]$ (that result from the evaluation of p_i on t) such that $\sum_{i=1}^{n} \xi_i = 1$:

$$[\,]\ \texttt{choose}(y, \tau_1 : \xi_1, \ldots, \tau_n : \xi_n) \Rightarrow \tau_1 \quad \texttt{if}\ y \leq \xi_1$$
$$[\,]\ \texttt{choose}(y, \tau_1 : \xi_1, \ldots, \tau_n : \xi_n) \Rightarrow \tau_2 \quad \texttt{if}\ \xi_1 < y \leq \xi_1 + \xi_2$$
$$\ldots$$
$$[\,]\ \texttt{choose}(y, \tau_1 : \xi_1, \ldots, \tau_n : \xi_n) \Rightarrow \tau_n \quad \texttt{if}\ \xi_1 + \xi_2 + \ldots + \xi_{n-1} < y$$

This ELAN code simulates a non-uniform distribution using a uniform one.
For example, we have the derivations

```
(bias france) => (PC(h : probaH, t:1 - probaH) france)
              => (choose(0.7, h: (probaH france),
                               t: ((1- probaH) france)) france)
              => (choose(0.7, h: 0.49, t: 0,51) france)
              => (t france)
              => tail
```

and

```
(bias france) => (PC(h: probaH, t:1 - probaH) france)
              => (choose(0.25, h: (probaH france),
                                t: ((1- probaH) france)) france)
              => (choose(0.25, h: 0.49, t: 0,51) france)
              => (h france)
              => head
```

5 Probabilistic Abstract Reduction Systems

Dealing with probabilistic systems gives rise to interesting new problems. For example, what are the generalizations of the classical notions of rewriting such as confluence, termination? Which results still hold, and which can be extended to that setting? These are the kind of problems that we address in this section by studying the generalization of the classical results about abstract reduction systems to probabilistic ones.

In a first step, we are not dealing with rewriting (ad posteriori nor with conditional rewriting, nor rewriting controlled by strategies) but with reduction systems. However, we claim that this study can give some clues to understand what probabilistic rewriting could be.

Observe also that we took the approach of starting from results from classical abstract reduction systems and studying whether they are still valid in that setting. This is a priori different from considering probabilistic abstract reduction systems by themselves.

We first come back to the classical setting (see for example [2,12]). An *abstract reduction system (ARS)* is $\mathcal{A} = (A, \rightarrow)$ consisting of a set A and a binary relation $\rightarrow \subset A \times A$ on A . A *derivation* is a finite, or infinite sequence $\pi = \pi_0 \rightarrow \pi_1 \cdots \rightarrow \pi_n$ with $(\pi_i, \pi_{i+1}) \in \rightarrow$ for all i.

$\rightarrow^n$ denotes the n-step composition of $\rightarrow$: $\rightarrow^0$ is the identity, and $\rightarrow^{n+1} = \rightarrow^n \circ \rightarrow$. $\rightarrow^{\leq n}$ denotes less-than-n-step composition: $\rightarrow^{\leq n} = \bigcup_{0 \leq i \leq n} \rightarrow^i$. The reflexive transitive closure of a relation $\rightarrow$ is denoted by $\rightarrow^*$ and its symmetric closure is denoted by $\leftrightarrow$.

The reduction relation $\rightarrow$ is said *locally confluent (LC)* if $b \leftarrow a \rightarrow c \Rightarrow \exists d \; b \rightarrow^* d \wedge c \rightarrow^* d$. It is said *confluent (C)* if $b \leftarrow^* a \rightarrow^* c \Rightarrow \exists d \; b \rightarrow^* d \wedge c \rightarrow^* d$.

The following results are well-known.

Proposition 1 (Classical Setting [2,12]). *The following are equivalent:*

1. $\rightarrow$ *is confluent (C)*
2. $\rightarrow^*$ *is locally confluent (LC)*
3. $\rightarrow^*$ *is confluent (C)*
4. *the relation is semi-confluent:* $b \leftarrow a \rightarrow^* c \Rightarrow \exists d \; b \rightarrow^* d \leftarrow^* c$.
5. *the relation is Church Rosser:* $a \leftrightarrow^* b \Rightarrow \exists c \; a \rightarrow^* c \leftarrow^* b$.

Let $\mathcal{A}$ be an ARS. $a \in A$ is said to be in *normal form* if there is no b with $a \rightarrow b$. We say that a *has a normal form (hnf)* if there is a b in normal form with $a \rightarrow^* b$. $a, b \in A$ are said to be *convertible* if $a \leftrightarrow^* b$.

The reduction relation $\rightarrow$, or $\mathcal{A}$, is *normalizing (N)* if every $a \in A$ has a normal form. It is *terminating (T)* if there is no infinite chains $\pi_0 \rightarrow \pi_1 \cdots \rightarrow \pi_n \cdots$.

It has the *unique normal form property (UN)* if for all $a, b \in A$, if a and b are convertible and in normal form, then a and b are identical. It has the *normal form property (NF)* if for all $a, b \in A$, if a is in normal form and convertible with b, then $b \rightarrow^* a$.

It is *inductive (Ind)* if every reduction sequence $\pi_0 \rightarrow \pi_1 \cdots \rightarrow \pi_n$ (possibly infinite) there is an a with $\pi_i \rightarrow^* a$ for all i. It is *increasing (Inc)* if there is a map f from A to $\mathbb{N}$ such that $a \rightarrow b$ implies $f(a) < f(b)$. It is *finitely branching (FB)* if for all a, the set of b with $a \rightarrow b$ is finite.

The following relations are well-known.

Proposition 2 (Classical Setting [2,12]).

1. *confluent (C) $\Rightarrow$ normal form property (NF) $\Rightarrow$ unique normal form property (UN)*
2. *terminating (T) and locally confluent (LC) $\Rightarrow$ confluent (C) (Newman's Lemma)*
3. *unique normal form property (UN) and normalizing (N) $\Rightarrow$ confluent (C)*
4. *unique normal form property (UN) and normalizing (N) $\Rightarrow$ inductive (Ind)*
5. *inductive (Ind) and increasing (Inc) $\Rightarrow$ terminating (T)*
6. *locally confluent (LC) and normalizing (N) and increasing (Inc) $\Rightarrow$ terminating (T)*

We can now switch to the probabilistic case. The idea behind the probabilistic setting is to consider reductions with probabilities.

Let us first come back to school [10,7,20]: a *σ-algebra* on a set Ω is a set of subsets of Ω which contains the empty-set, and is stable by countable union and complementation. In particular, the set of subsets is a natural σ-algebra for any countable set. A *measurable space* (Ω, σ) is a set with a σ-algebra on it. If (Ω, σ) and (Ω', σ') are measurable spaces, a function $f : \Omega \to \Omega'$ is *measurable* if for all W in σ', $f^{-1}(W) \in \sigma$.

A *probability* is a function P from a σ-algebra to $[0, 1]$, which is countably additive, and such that $P(\Omega) = 1$. A triplet (Ω, σ, P) is called *a probability space.* A *random variable* is a measurable function on some probability space.

Given two probabilities P and P' on σ and σ' respectively, one can consider $P \oplus P'$ which is the product measure defined on the σ-algebra $\sigma \times \sigma'$, and is characterized by $P \oplus P'(A \times A') = P(A)P(A')$. Given $A, B \in \sigma$, when $P(B) > 0$, *the conditional probability of A given B* is by definition $P(A|B) = P(A \cap B)/P(B)$.

A *stochastic sequence on a set S* is a family $(X_i)_{i \in \mathbb{N}}$, of random variables defined on some fixed probability space (Ω, σ, P) with values on S. It is said to be *Markovian* if its conditional distribution function satisfies the so-called Markov property, that is for all n

$$P(X_n = s | X_0 = \pi_0, X_1 = \pi_1, \ldots, X_{n-1} = \pi_{n-1}) = P(X_n = s | X_{n-1} = \pi_{n-1}),$$

and *homogeneous* if furthermore this probability is independent of n.

We are ready to define probabilistic abstract reduction systems (PARS). The idea is that a PARS is given by some set A, and a function $[s \leadsto t]$ that gives, for all s,t in A, the probability of going from s to t. Formally:

Definition 1 (PARS). *A probabilistic abstract reduction system (PARS) is a pair $\mathcal{A} = (A, [\_ \leadsto \_])$ consisting of a countable set A and a function $[\_ \leadsto \_]$ from $A \times A$ to $[0, 1]$, such that for all s, $\sum_{t \in A} [s \leadsto t]$ is 0 or 1.*

Note that we allow $\sum_{t \in A} [s \leadsto t]$ to be 0. This happens exactly for *terminal* states: a state $a \in A$ is *terminal* if there is no b with $[a \leadsto b] > 0$. For non-terminal states, $\sum_{t \in A} [s \leadsto t]$ is necessarily 1.

A *derivation of* $\mathcal{A}$ is then a finite or infinite homogeneous Markovian stochastic sequence whose transition probabilities are given by $[\_ \leadsto \_]$. Formally:

Definition 2 (Derivations). *A derivation* π *of* $\mathcal{A}$ *is an homogeneous Markovian stochastic sequence* $\pi = (\pi_i)_{i \in \mathbb{N}}$ *on*

$$S \cup \{\bot\}$$

such that for all i,

$$\begin{aligned} P(\pi_{i+1} = t | \pi_i = s) &= [s \leadsto t] \textit{ if } s \neq \bot \textit{ non-terminal} \\ &= 1 \textit{ if } t = \bot \textit{ and } s \neq \bot \textit{ terminal} \\ &= 1 \textit{ if } t = \bot \textit{ and } s = \bot \\ &= 0 \textit{ otherwise.} \end{aligned}$$

PARS correspond to the extension of ARS with probabilities on derivations. Indeed, to a finitely branching[2] ARS $\mathcal{A} = (A, \rightarrow)$, we can associate a PARS $\mathcal{A}' = (A, [\_ \leadsto \_])$ where, for all $x, y \in A$, $[x \leadsto y]$ is 0 if $(x, y) \not\in \rightarrow$, $[x \leadsto y] = 1/cardinal(\{t | (s,t) \in \rightarrow\})$ otherwise. Derivations of the ARS A and of the PARS $\mathcal{A}'$ are in correspondence: a non-terminating chain $\pi_0 \rightarrow \pi_1 \cdots \rightarrow \pi_n \ldots$ of $\mathcal{A}$ corresponds to derivation $\pi_0 \pi_1 \ldots \pi_n \ldots$ of $\mathcal{A}'$ (observe that we have $\pi_i \in A\ \forall i$). A terminating chain $\pi_0 \rightarrow \pi_1 \cdots \rightarrow \pi_n$ of $\mathcal{A}$ corresponds to derivation $\pi_0 \pi_1 \ldots \pi_n \bot \bot \ldots$ of $\mathcal{A}'$.

Clearly, to an ARS can correspond several PARS. Conversely, to a PARS $\mathcal{A}' = (A, [\_ \leadsto \_])$ corresponds a unique ARS $\mathcal{A} = (A, \rightarrow)$ which is obtained by forgetting probabilities: the relation $\rightarrow \subset A \times A$ of $\mathcal{A}$ is defined by $s \rightarrow t$ iff $[s \leadsto t] > 0$. This unique ARS, or the corresponding relation $\rightarrow$, will be called the *projection of* $\mathcal{A}'$.

The matrix $(p_{t,s}) = (P(\pi_i = t | \pi_{i-1} = s))$ on $S \cup \{\bot\}$ is what is called a stochastic matrix (even when S is an infinite set) [4]. It has the nice property that columns sum to 1. Actually our setting is the one of Homogeneous Markov Chain (HMC) theory [4]. The main difference is that we will focus on the generalization of classical rewriting notions and not on the usual Markov chain problematics: see [4] for a presentation of HMC theory.

Definition 3 (Relations $\leadsto$, $\leadsto^n$, $\leadsto^*$). *Let* $s \in A$ *be a state, and* $\pi = \pi_0 \pi_1 \ldots \pi_n$ *a derivation with* $\pi_0 = s$. *We write*

1. $s \leadsto t$ *iff* $\pi_1 = t$,
2. $s \leadsto^n t$ *iff* $\pi_n = t$,
3. $s \leadsto^{\leq n} t$ *iff there is some* $i \leq n$ *with* $\pi_i = t$,
4. $s \leadsto^* t$ *iff there is some* $i \in \mathbb{N}$ *with* $\pi_i = t$.

[2] To an unrestricted (non-necessarily finitely-branching) ARS associate similarly a PARS $\mathcal{A}'$ by putting probabilities $1/2, 1/4, 1/8, \ldots$ on the outgoing transitions of a non-terminal s when the set $\{t | (s,t) \in \rightarrow\}$ is not finite.

For $s, t \in A$, we write $[s \rightsquigarrow t]$ (respectively: $[s \rightsquigarrow^n t]$,$[s \rightsquigarrow^{\leq n} t]$,$[s \rightsquigarrow^* t]$) for the probability that $s \rightsquigarrow t$ (resp. $s \rightsquigarrow^n t$, $s \rightsquigarrow^{\leq n} t$, $s \rightsquigarrow^* t$) holds on a derivation $\pi = \pi_0\pi_1 \dots \pi_n \dots$ given that $\pi_0 = s$.

Derivations have the nice property of being preserved by shifting: a *stopping time* with respect to a stochastic sequence $(X_n)_{n\in\mathbb{N}}$ is a random variable τ taking its values in $\mathbb{N} \cup \{\infty\}$ such that for all integer $i \geq 0$, the event $\tau = m$ can be expressed in terms of $X_0, X_1, \dots, X_m$ [4]. A typical example is *the hitting time* of some state $s \in S$ defined as $T = \inf\{i | X_i = s\}$. An other typical example is a constant time which is a particular stopping time [4].

Observation 1 (Strong Markov Property) *Let τ be a stopping time with respect to a derivation $\pi = \pi_0\pi_1 \dots \pi_n \dots$ of $\mathcal{A}$. The derivation after τ, (also called the τ-shift of π) is by definition the derivation $(\pi'_i)_{i\in\mathbb{N}}$ defined by $\pi'_i = \pi_{\tau+i}$.*

Then for any state $s \in S$, given that $X_\tau = s$ (hence we suppose $\tau \neq \infty$), the derivation after τ is a derivation of $\mathcal{A}$.

Proof. This is a restatement of the so-called "Strong Markov Property" for HMC: see [4] for example.

Clearly, the function $[\_ \rightsquigarrow \_]$ corresponds to the function $[\_ \rightsquigarrow \_]$ of $\mathcal{A}$. From the (strong) homogeneous Markov properties, we can also easily prove:

Proposition 3. *for all $s, t \in A$, $n \geq 1$*

1. $[s \rightsquigarrow^1 t] = [s \rightsquigarrow t]$
2. $[s \rightsquigarrow^n t] = \sum_{u\in A}[s \rightsquigarrow^{n-1} u][u \rightsquigarrow t]$
3. $[s \rightsquigarrow^* t] = \lim_{n\to\infty}[s \rightsquigarrow^{\leq n} t]$

Example 4. Let $\mathcal{A}$ be the PARS defined by $A = \{a, b\}$ and

1. $[a \rightsquigarrow a] = 1/2$
2. $[a \rightsquigarrow b] = 1/2$

Then.

1. $[a \rightsquigarrow^n a] = 1/2^n$
2. $[a \rightsquigarrow^n b] = 1/2^n$
3. $[a \rightsquigarrow^{\leq n} a] = 1$
4. $[a \rightsquigarrow^{\leq n} b] = 1 - 1/2^n$
5. $[a \rightsquigarrow^* a] = 1$
6. $[a \rightsquigarrow^* b] = 1$

Recall that in the classical setting that an ARS $\mathcal{A}$ is *locally confluent (LC)* if $b \leftarrow a \rightarrow c \Rightarrow \exists d\ b \rightarrow^* d \wedge c \rightarrow^* d$. We will then say that a PARS $\mathcal{A}$ is *probabilistically locally confluent (pLC)* if $P(\exists d\ b \rightsquigarrow^* d \wedge c \rightsquigarrow^* d) > 0$ whenever $a \rightsquigarrow b$ and $a \rightsquigarrow c$ by two independently chosen derivations.

Of course, such a definition must be understood as follows: a PARS $\mathcal{A}$ is *probabilistically locally confluent*, if for all $a, b, c \in A$, if $a \rightsquigarrow b$ by some derivation

π, and $a \rightsquigarrow c$ by some independently chosen derivation π', then the probability that there is a $d \in A$ such that simultaneously $b \rightsquigarrow^* d$ by derivation π after stopping time $\tau = 1$ and $c \rightsquigarrow^* d$ by derivation π' after stopping time $\tau = 1$ is positive. The probability measure considered here is the product measure $P \oplus P$, since we consider two independently randomly chosen derivations π and π'.

Observe that from Strong Markov Property, this is equivalent to say that, for all $b, c \in A$, the probability that for two independently randomly chosen derivations π and π' starting from b and c respectively, there is a $d \in A$ such that simultaneously $b \rightsquigarrow^* d$ by derivation π and $c \rightsquigarrow^* d$ by derivation π', is positive, whenever $[a \rightsquigarrow b] > 0$ and $[a \rightsquigarrow c] > 0$ for some $a \in A$.

In the same vein, we will say that $\rightsquigarrow$ is *almost-surely locally confluent (PLC)* if $P(\exists d\ b \rightsquigarrow^* d \wedge c \rightsquigarrow^* d) = 1$ whenever $a \rightsquigarrow b$ and $a \rightsquigarrow c$ by two independently chosen derivations.

We let the reader guess what *probabilistically confluent (pC)* , and *almost-surely confluent (PC)* are.

Proposition 1 has the following equivalent:

Proposition 4. *The following are equivalent:*

1. *$\rightsquigarrow$ is almost-surely confluent (PC) (resp. probabilistically confluent (pC))*
2. *$\rightsquigarrow^*$ is almost-surely locally confluent (PLC) (resp. probabilistically locally confluent (pLC))*
3. *$\rightsquigarrow^*$ is almost-surely confluent (PC) (resp. probabilistically confluent (pC))*
4. *$a \rightsquigarrow b$, $a \rightsquigarrow^* c$ implies there is a d with $b \rightsquigarrow^* d, c \rightsquigarrow^* d$ with probability one (resp. positive probability).*

Recall that an $a \in A$ is in normal form if there is no b with $[a \rightsquigarrow b] > 0$.

We say that $a \in A$ *probabilistically has a normal form (phnf)* if $P(\exists b\ nf\ a \rightsquigarrow^* b) > 0$. We say that $a \in A$ *almost-surely has a normal form (Phnf)* if $P(\exists b\ nf\ a \rightsquigarrow^* b) = 1$.

A relation is *probabilistically normalizing (pN)* if every $a \in A$ probabilistically has a normal form (phnf) . A relation is *almost-surely normalizing (PN)* if every $a \in A$ almost-surely has a normal form (Phnf) .

A relation is *almost-surely terminating (PT)* if every derivation $\pi_0, \pi_1, \ldots, \pi_n$ is terminal (ultimately equal to $\perp$) with probability 1. It is *probabilistically terminating (pT)* if this latter probability is positive.

Two states $a, b \in A$ are said to be *convertible* if a and b are convertible in the classical sense for the projection of $\mathcal{A}'$, that is for the classical binary relation $[\_ \rightsquigarrow \_] > 0$.

A relation has *unique normal form property (UN)* if for all $a, b \in A$, if a and b are convertible and in normal form, then a and b are identical.

A relation has the *probabilistic normal form property (pNF)* if $P(b \rightsquigarrow^* a) > 0$ whenever a is in normal form and convertible with b. A relation has the *almost-sure normal form property (PNF)* if $P(b \rightsquigarrow^* a) = 1$ whenever a is in normal form and convertible with b.

We are now ready to study the properties of the above notions.

We will then say that a PARS is *locally confluent (LC)* (respectively confluent (C) , normalizing (N) , terminating (T) , unique normal form property (UN) , inductive (Ind) , increasing (Inc) , finitely branching (FB)) if its projection is.

It is noticeable that the projection operator commutes with finite iteration and (reflexive) transitive iteration. In other words, probabilistic joinability and joinability corresponds. Formally,

Proposition 5. *Let $\mathcal{A}' = (A, [\_ \rightsquigarrow \_])$ be a PARS and $\mathcal{A} = (A, \rightarrow)$ its projection: $s \rightarrow t$ iff $[s \rightsquigarrow t] > 0$.*

Then, for all $s, t \in A$,

1. $[s \rightsquigarrow^n t] > 0$ *iff* $s \rightarrow^n t$
2. $[s \rightsquigarrow^* t] > 0$ *iff* $s \rightarrow^* t$

Proof. The two assertions are proved by induction over n. They are clear for $n = 0$ and $n = 1$. For $n > 1$, we have $s \rightarrow^n t$ iff there is a u with $s \rightarrow^{n-1} u$ and $u \rightarrow t$ iff there is a u with $[s \rightsquigarrow^{n-1} u] > 0$ and $[u \rightsquigarrow t]$ iff $[s \rightsquigarrow^n t] = \sum_{u \in A} [s \rightsquigarrow^{n-1} u][u \rightsquigarrow t] > 0$. For the second assertion, we have $s \rightarrow^* t$ iff there is a n with $s \rightarrow^n t$ iff there is a n with $[s \rightsquigarrow^n t] > 0$ iff $[s \rightsquigarrow^* t] > 0$. Indeed, if there is such a n, $[s \rightsquigarrow^* t]$ is at least $[s \rightsquigarrow^n t]$. For the other direction, we clearly have by sigma-additivity of probabilities $[s \rightsquigarrow^* t] \leq \sum_{n \in \mathbb{N}} [s \rightsquigarrow^n t]$.

We then claim.

Proposition 6. *1. almost-surely locally confluent (PLC) $\Rightarrow$ probabilistically locally confluent (pLC) $\Leftrightarrow$ locally confluent (LC) .*

2. The reverse implication is false in general.

Proof. Only the equivalence between probabilistically locally confluent (pLC) and locally confluent (LC) is not trivial. Clearly $P(\exists d\ a \rightsquigarrow^* d \wedge b \rightsquigarrow^* d) > 0$ iff $\exists d\ P(a \rightsquigarrow^* d \wedge b \rightsquigarrow^* d) > 0$ iff $\exists d\ P(a \rightsquigarrow^* d) > 0 \wedge P(b \rightsquigarrow^* d) > 0$. From previous lemma, that happens iff $[\_ \rightsquigarrow \_] > 0$ is locally confluent (LC) .

In the same vein, we can prove:

Proposition 7. *1. almost-surely confluent (PC) $\Rightarrow$ probabilistically confluent (pC) $\Leftrightarrow$ confluent (C)*

2. almost-surely has a normal form (Phnf) $\Rightarrow$ probabilistically has a normal form (phnf) $\Leftrightarrow$ has a normal form (hnf)

3. almost-surely normalizing (PN) $\Rightarrow$ probabilistically normalizing (pN) $\Leftrightarrow$ normalizing (N)

4. almost-sure normal form property (PNF) $\Rightarrow$ probabilistic normal form property (pNF) $\Leftrightarrow$ normal form property (NF)

5. terminating (T) $\Rightarrow$ almost-surely terminating (PT) $\Leftrightarrow$ almost-surely normalizing (PN) $\Rightarrow$ probabilistically terminating (pT)

6. The reverse implications are false in general.

We then get the following picture:

Proposition 8. *1.*

$$\begin{array}{ccccc} C & \longrightarrow & NF & & \\ \updownarrow & & \updownarrow & & \\ pC & \longrightarrow & pNF & \longrightarrow & UN \\ \uparrow & & \uparrow & & \\ PC & \longrightarrow & PNF & & \end{array}$$

2. Newman's Lemma.

$$\begin{array}{ccc} T\&LC & & \\ \updownarrow & & \\ T\&pLC & \longrightarrow & C \\ \uparrow & & \uparrow \\ T\&PLC & \longrightarrow & PC \\ \downarrow & & \updownarrow \\ PT\&PLC & \longrightarrow & PC \end{array}$$

3.

$$\begin{array}{ccc} UN\&N & & \\ \updownarrow & & \\ UN\&pN & \longrightarrow & C \\ \uparrow & & \uparrow \\ UN\&PN & \longrightarrow & PC \end{array}$$

4.

$$\begin{array}{ccc} UN\&N & & \\ \updownarrow & & \\ UN\&pN & \longrightarrow & Ind \\ \uparrow & & \\ UN\&PN & & \end{array}$$

5.

$$\begin{array}{ccc} LC\&N\&Inc & & \\ \updownarrow & & \\ pLC\&pN\&Inc & \longrightarrow & T \\ \uparrow & & \\ PLC\&PN\&Inc & & \end{array}$$

The previous implications can all be derived by adapting the proofs from the classical abstract reduction systems towards probabilistic ones.

Observe that they were obtained by considering the classical rewriting settings and testing whether it generalizes to that context. That means first that we are not at all exhaustive: some reverse implications, or other implications may hold. Second, the classical theory of HMC can a priori also be used to derived other facts, or relations between our notions.

6 Conclusion

In this paper we presented an extension of the ELAN system to deal with the prototyping and modeling of systems with probabilistic evolutions. We proposed to extend the system by adding a new "probabilistic choose" operator to the strategy language of the system.

We argued through simple classical examples that this approach that puts probabilities at the level of strategies, i.e. at the level of the control of rewriting is very natural. This is in contrast with similar works where probabilities are in some sense hard-coded in rules [8]. Furthermore, this approach has the advantage that the problem of probabilities normalization (probabilities must be normalized to sum to 1) in the above mentioned work is avoided in a nice manner: this is put at the level of the definition of the operators which are coded in the language itself.

Dealing with probabilistic systems gives rise to many very interesting questions about the generalization of the classical rewriting notions. We put the first stones in that direction by defining several notions and stating generalizations of classical results for abstract derivation systems. Our results mainly say that the notions defined by considering joinability with positive probability corresponds to classical notions by putting away probabilities, and that the notions defined by considering almost-sure joinability are stronger.

These are of course only the first steps in the direction of fully understanding probabilistic rewrite systems. Actually, when dealing with rewrite systems, the objects of interest are; (1) abstract rewrite systems, (2) the rewrite relation that should be stable by context and substitution, (3) the rewrite logic, (4) the rewrite calculus. The generalization of all these concepts and their relations dealing with probabilistic systems/theories give rise to many potentially interesting problems. We touched only (1), and on this topics we think that some other results could also been derived from Homogeneous Markov Chain Theory.

References

1. D. Abdemouche. Stratégies probabilistes en elan et applications. Mémoire de dea en informatique, Université Henri Poincaré – Nancy 1, July 2001.
2. F. Baader and T. Nipkow. *Term Rewriting and all That*. Cambridge University Press, 1998.

3. P. Borovanský, C. Kirchner, H. Kirchner, and C. Ringeissen. Rewriting with strategies in ELAN: a functional semantics. *International Journal of Foundations of Computer Science*, 12(1):69–98, February 2001.
4. P. Brémaud. *Markov Chains.* Springer, 1991.
5. H. Cirstea and C. Kirchner. The rewriting calculus — Part I *and* II. *Logic Journal of the Interest Group in Pure and Applied Logics*, 9(3):427–498, May 2001.
6. A. Dutech, O. Buffet, and F. Charpillet. Multi-Agent Systems by Incremental Gradient Reinforcement Learning. In *17th International Joint Conference on Artificial Intelligence, Seattle, WA, USA*, volume 2, pages 833–838, August 2001.
7. W. Feller. *An introduction to probability theory and its applications, volume 1 and 2.* Wiley, 1968.
8. T. Frühwirth, A. Di Pierro, and H. Wiklicky. Toward probabilistic constraint handling rules. In S. Abdennadher and T. Frühwirth, editors, *Proceedings of the third Workshop on Rule-Based Constraint Reasoning and Programming (RCoRP'01)*, Paphos, Cyprus, December 2001. Under the hospice of the International Conferences in Constraint Programming and Logic Programming.
9. D. E. Goldberg. *Genetic Algorithms in Search, Optimization, and Machine Learning.* Addison-Wesley, Reading, Massachusetts, 1989.
10. G. Grimmett. *Probability Theory.* Cambridge University Press, 1993.
11. H. Kirchner and P.-E. Moreau. Promoting rewriting to a programming language: A compiler for non-deterministic rewrite programs in associative-commutative theories. *Journal of Functional Programming*, 11(2):207–251, 2001.
12. J. W. Klop. Term rewriting systems. In S. Abramsky, D. M. Gabbay, and T. S. E. Maibaum, editors, *Handbook of Logic in Computer Science*, volume 2, chapter 1, pages 1–117. Oxford University Press, Oxford, 1992.
13. D. E. Knuth. *The art of Computer Programming*, volume 2. Addison-Wesley Publishing Company, second edition, 1981.
14. P. Lincoln, J. Mitchell, and A. Scedrov. Stochastic interaction and linear logic. In J.-Y. Girard, Y. Lafont, and L. Regnier, editors, *Advances in Linear Logic*, pages 147–166. Volume 222, London Mathematical Society Lecture Notes, Cambridge University Press, 1995.
15. R. Motwani and P. Raghavan. *Randomized Algorithms.* Cambridge University Press, 1995.
16. A. D. Pierro and H. Wiklicky. An operational semantics for probabilistic concurrent constraint programming. In *Proceedings of the 1998 International Conference on Computer Languages*, pages 174–183. IEEE Computer Society Press, 1998.
17. D. Pierro and Wiklicky. A markov model for probabilistic concurrent constraint programming. In *APPIA-GULP-PRODE'98, Joint Conference on Declarative Programming*, 1998.
18. D. Pierro and Wiklicky. Probabilistic concurrent constraint programming: Towards a fully abstract model. In *MFCS: Symposium on Mathematical Foundations of Computer Science*, 1998.
19. D. Pierro and Wiklicky. Concurrent constraint programming: Towards probabilistic abstract interpretation. In *2nd International ACM SIGPLAN Conference on Principles and Practice of Declarative Programming (PPDP'00)*, 2000.
20. W. Rudin. *Real and Complex Analysis, 3rd edition.* McGraw Hills, USA, March 1987.
21. V. Subrahmanian. Probabilistic databases and logic programming. In *International Conference on Logic Programming*, volume 2237 of *Lecture Notes in Computer Science*, page 10. Springer-Verlag, 2001.

Loops of Superexponential Lengths in One-Rule String Rewriting*

Alfons Geser

ICASE, Mail Stop 132C, NASA Langley Research Center, Hampton, VA 23681.
geser@icase.edu

Abstract. Loops are the most frequent cause of non-termination in string rewriting. In the general case, non-terminating, non-looping string rewriting systems exist, and the uniform termination problem is undecidable. For rewriting with only one string rewriting rule, it is unknown whether non-terminating, non-looping systems exist and whether uniform termination is decidable. If in the one-rule case, non-termination is equivalent to the existence of loops, as McNaughton conjectures, then a decision procedure for the existence of loops also solves the uniform termination problem. As the existence of loops of bounded lengths is decidable, the question is raised how long shortest loops may be. We show that string rewriting rules exist whose shortest loops have superexponential lengths in the size of the rule.

Keywords: string rewriting, semi-Thue system, uniform termination, termination, loop, one-rule, single-rule
Submission category: Regular research paper.

1 Introduction

Uniform termination, i.e., the non-existence of an infinite reduction sequence, is an undecidable property of string rewriting systems (SRSs) [8], even if they comprise only three rules [13]. It is open whether uniform termination is decidable for SRSs with less than three rules.

An SRS admits a *loop* if there is a reduction of the form $u \to^+ sut$. Every looping SRS is non-terminating. The converse does not hold, even for two-rule SRSs [6]. McNaughton [15] conjectures that every one-rule non-terminating SRS admits a loop.

If McNaughton's conjecture holds, and if the existence of loops is decidable for one-rule SRSs, then the uniform termination of one-rule SRSs is decidable. Existence of loops of *bounded length* is decidable [6]. This immediately raises the question whether there is an algorithm that outputs upper bounds of lengths of shortest loops. On this account it is most interesting how long shortest loops can be.

* This work was supported by the National Aeronautics and Space Administration under NASA Contract No. NAS1-97046 while the author was in residence at ICASE.

S. Tison (Ed.): RTA 2002, LNCS 2378, pp. 267–280, 2002.

The purpose of this note is to prove that there are one-rule SRSs that admit loops of superexponential lengths in the size of the rule, but no shorter loops. This is in harsh contrast to the common belief that loops are simple. Specifically, we prove the following result.

Theorem 1. *For all* $p \geq 2q$, $q \geq 1$, $r \geq 2$, *the string rewriting rule*

$$R = \{10^p \to 0^p 1^r 0^q\}$$

admits loops of length $1 + \sum_{i=0}^{\ell-1} r^i$ *where* $\ell = \lceil \frac{p}{q} \rceil$ *but no shorter loops.*

The fact that these SRSs loop is known [19, Case 6.2.2] [10, Theorem 10.1, Case 2.2]; the contribution of this paper is a sharp lower bound for the length of the loop.

Theorem 1 follows immediately from Lemmas 5 and 16, which we will prove below.

By choosing $q = 1$ and keeping $p \geq 2$ fixed, we get a family of rules where the shortest length of loops is polynomial in r with degree $p - 1$. By choosing $q = 1$ and keeping $r \geq 2$ fixed, we get shortest loop lengths exponential in p with base r. By choosing $q = 1$ and $r = p$ the minimal loop length is greater than p^{p-1}. This shows the claimed superexponential growth.

The paper is organized as follows. We will show: in Section 3 that each R has a loop; in Section 4 that the length of the loop is as claimed; in Section 5 that the start string of a shortest loop has a special shape; and in Section 6 that these strings initiate no shorter loops.

2 Preliminaries

We assume that the reader is familiar with termination of string rewriting. SRSs are also called *semi-Thue systems.*

For an introduction to string rewriting see Book and Otto [1] or Jantzen [9]. The study of termination in one-rule string rewriting has been initiated by Kurth in his thesis [11]. Further work includes McNaughton [14,15,16], Senizergues [19], Kobayashi et al. [20,10], and Zantema and Geser [6,21,4,5]. Since SRSs can be encoded as term rewriting systems where letters are unary function symbols, the results of termination of term rewriting [3] apply.

An SRS R is a set of *string rewriting rules*, i.e., pairs of strings denoted as $u \to v$. The reduction step relation, also denoted by $\to$, is defined by $sut \to svt$ for all strings s, t and string rewriting rules $u \to v$. Here st denotes the *concatenation* of strings s and t.

A *loop* is a reduction of the form $t \to^+ utv$ where u, v are strings. An SRS R is said to *admit a loop* if a loop $t \to^+ utv$ exists.

The string t is also called a *prefix*, u a *suffix* of tu. Any string utv is said to contain t as a *factor*. The set of *overlaps* of a string u with a string v is defined by

$$\mathrm{OVL}(u, v) = \{w \in \Sigma^+ \mid u = u'w, v = wv', u'v' \neq \varepsilon, u', v' \in \Sigma^*\} .$$

3 The Rule Admits a Loop

Throughout this paper we assume strings over the two-letter alphabet $\{0, 1\}$, and we speak about one-rule SRSs $R = \{10^p \to 0^p1^r0^q\}$, for some $p \geq 2q$, $q \geq 1$, $r \geq 2$. In this section we show that each R has a loop.

Definition 1. *Let strings t_i, $i \geq 0$ be defined recursively by*

$$t_0 = 1,$$
$$t_{i+1} = 0^q t_i^r \ .$$

The following lemma is crucial for the proof that R admits loops.

Lemma 1. *$t_k^m 0^p \to^* 0^{p-q} t_{k+1}^m 0^q$ holds for all $k, m \geq 0$.*

Proof. Proof by induction on (k, m) ordered lexicographically. The case $m = 0$ is trivial, so assume $m > 0$. Case 1: $k = 0$. Then

$$t_0^m 0^p = t_0^{m-1} 10^p \to t_0^{m-1} 0^p 1^r 0^q \to^* 0^{p-q} t_1^{m-1} 0^q 1^r 0^q = 0^{p-q} t_1^m 0^q,$$

by definition of t_0, inductive hypothesis for $(k, m-1)$, and definition of t_1, respectively. Case 2: $k > 0$. Then

$$t_k^m 0^p = t_k^{m-1} 0^q t_{k-1}^r 0^p \to^* t_k^{m-1} 0^q 0^{p-q} t_k^r 0^q \to^* 0^{p-q} t_{k+1}^{m-1} 0^q t_k^r 0^q = 0^{p-q} t_{k+1}^m 0^q,$$

by definition of t_k, inductive hypothesis for $(k-1, r)$, inductive hypothesis for $(k, m-1)$, and definition of t_{k+1}, respectively.

By definition t_k is a factor of t_{k+1}. Now if $t_k 0^p$ is a factor of $t_{k+1} 0^q$ then we have a loop. To this end k has to be great enough.

Example 1. Let $p = 2$, $q = 1$, $r = 2$. Then $R = \{100 \to 00110\}$. We have $t_0 = 1$, $t_1 = 0t_0t_0 = 011$, $t_2 = 0t_1t_1 = 0011011$, $t_3 = 0t_2t_2 = 000110110011011$, and so forth. The string $t_0 0^p = 100$ is not a factor of $t_1 0^q = 0110$. Neither is $t_1 0^p = 01100$ a factor of $t_2 0^q = 00110110$. However $t_3 0^q = 0\underline{001101100}110110$ contains $t_2 0^p$ as a factor at the underlined occurrence.

The problem is traced back to finding a factor 0^p within t_k. The following property of t_i is the key to the solution.

Lemma 2. *For all $k \geq 0$, the following hold:*

1. *0^{qk} is a prefix of t_k.*
2. *0^{qk+1} is not a factor of t_k.*

Proof. Straightforward induction on k.

If k is chosen great enough then 0^p fits into 0^{qk}. Let $\lceil x \rceil$ denote the least integer i such that $i \geq x$.

Lemma 3. *Let $\ell = \lceil \frac{p}{q} \rceil$. Then $t_\ell 0^p \to^* 0^{p-q} t_{\ell+1} 0^q$ is a loop.*

Proof. By Lemma 1 for $m = 1$ we get a reduction

$$t_\ell 0^p \to^* 0^{p-q} t_{\ell+1} 0^q \ . \tag{1}$$

Now suppose that $\ell = \lceil \frac{p}{q} \rceil$, whence $q\ell \geq p$. The following analysis shows that in this case Reduction (1) indeed forms a loop, i.e., that its left hand side $t_\ell 0^p$ is a factor of its right hand side, $0^{p-q} t_{\ell+1} 0^q$. For some string w we get

$$0^{p-q} t_{\ell+1} 0^q = 0^{p-q} 0^q t_\ell^r 0^q = 0^{p-q} 0^q t_\ell^{r-2} t_\ell t_\ell 0^q = \\ 0^{p-q} 0^q t_\ell^{r-2} t_\ell 0^{q\ell} w 0^q = 0^{p-q} 0^q t_\ell^{r-2} \underline{t_\ell 0^p} 0^{q\ell - p} w 0^q$$

by definition of $t_{\ell+1}$, the premise $r \geq 2$, Lemma 2, and the property $q\ell \geq p$, respectively. The occurrence of $t_\ell 0^p$ is underlined.

4 The Loop Has Superexponential Length

Let us start by calculating the lengths of reductions of Lemma 1.

Lemma 4. *For all $k, m \geq 0$, the reduction $t_k^m 0^p \to^* 0^{p-q} t_{k+1}^m 0^q$ constructed in Lemma 1 has length mr^k.*

Proof. Check that $0r^k = 0$, $(m+1)r^0 = 1 + mr^0$, and $(m+1)r^{k+1} = rr^k + mr^{k+1}$.

From Lemma 4 we get immediately:

Proposition 1. *The length of the reduction in Lemma 3 is r^ℓ.*

The length of the loop in Lemma 3 is not yet minimal. A refinement leads to the following shorter loop. We will prove its minimality in the subsequent sections.

Lemma 5. *Let $\ell = \lceil \frac{p}{q} \rceil$. Then there is a loop*

$$10^{(p-q)(\ell+1)+q} \to^n 0^p 1^{r-1} 0^{(p-q)\ell} t_\ell 0^q$$

of length $n = 1 + \sum_{i=0}^{\ell-1} r^i$.

Proof. By Lemmas 1 and 4 we have the reduction of length n,

$$\begin{aligned} 10^{(p-q)(\ell+1)+q} &\to \\ 0^p 1^{r-1} t_0 0^{(p-q)\ell+q} &\to^{r^0} \\ 0^p 1^{r-1} 0^{p-q} t_1 0^{(p-q)(\ell-1)+q} &\to^{r^1} \\ &\vdots \\ 0^p 1^{r-1} 0^{(p-q)(\ell-1)} t_{\ell-1} 0^{(p-q)+q} &\to^{r^{\ell-1}} 0^p 1^{r-2} \underline{10^{(p-q)\ell}} t_\ell 0^q \ . \end{aligned}$$

Now t_ℓ has a prefix $0^{q\ell}$ by Lemma 2, and so a prefix 0^p by definition of ℓ. Together with the underlined string this forms a reoccurrence of the initial string, $10^{(p-q)(\ell+1)+q} = 10^{(p-q)\ell} 0^p$, as a factor in the final string. So the presented reduction is indeed a loop.

5 How Shortest Loops Start

To prove that there are no loops shorter than those of Lemma 5, we first restrict the set of strings that may initiate shortest loops. To this end we employ the fact that the existence of loops is characterized by the existence of looping forward closures. Forward closures [12,2] are restricted reductions. The following characterization of forward closures by Hermann is convenient.

Definition 2 (Forward Closure [12,2,7]). *The set of* forward closures *of an SRS R is the least set $\mathrm{FC}(R)$ of R-reductions such that*

fc1. *if $(l \to r) \in R$ then $(l \to r) \in \mathrm{FC}(R)$,*
fc2. *if $(s_1 \to^+ t_1'x) \in \mathrm{FC}(R)$ and $(xl_2' \to r_2) \in R$ such that $x \neq \varepsilon$ then $(s_1 l_2' \to^+ t_1'xl_2' \to^+ t_1'r_2) \in \mathrm{FC}(R)$,*
fc3. *if $(s_1 \to^+ t_1'l_2t_1'') \in \mathrm{FC}(R)$ and $(l_2 \to r_2) \in R$ then $(s_1 \to^+ t_1'l_2t_1'' \to^+ t_1'r_2t_1'') \in \mathrm{FC}(R)$.*

We call a forward closure of the form $s \to^+ usv$ a *looping* forward closure.

Theorem 2 ([6]). *An SRS admits a loop if and only if it has a looping forward closure. Moreover if there is a loop of length n then there is a looping forward closure of length at most n.*

Lemma 6. *Every forward closure of R has the form $10^{(p-q)k+q} \to^* w1^r0^q$ for some $k \geq 1$ and some string w.*

Proof. By induction on the definition of forward closure. Case 2 is trivial. For Case 2, let $s_1 = 10^{(p-q)k+q}$, $t_1'x = w1^r0^q$, $xl_2' = 10^p$, $r_2 = 0^p1^r0^q$. Observe that x must be $x = 10^q$. This implies $t_1' = w1^{r-1}$, $l_2' = 0^{p-q}$ and we get

$$s_1l_2' = 10^{(p-q)k+q}0^{p-q} = 10^{(p-q)(k+1)+q} \to^* w1^{r-1}0^p1^r0^q = t_1'r_2$$

as the composed forward closure. It has the claimed form.

Case 2: Let $s_1 = 10^{(p-q)k+q}$, $t_1'l_2t_1'' = w1^r0^q$, $l_2 = 10^p$, $r_2 = 0^p1^r0^q$. By $p > q$, l_2 cannot be a factor of 1^r0^q. Nor can it left overlap with it: $\mathrm{OVL}(l_2, 1^r0^q) = \emptyset$. Therefore t_1'' is longer than 1^r0^q. In other words a string w' exists such that $t_1'' = w'1^r0^q$. Hence $w = t_1'l_2w'$ and the composed forward closure is

$$s_1 = 10^{(p-q)k+q} \to^* t_1'r_2w'1^r0^q = t_1'r_2t_1''$$

which has the claimed form.

Next we show that a forward closure can only issue an infinite reduction, and so a loop, if its left hand side is large enough.

Lemma 7. *$0^{(p-q)k+q}t_k0^q$ is irreducible for all $k \leq \lceil \frac{p}{q} \rceil$.*

Proof. Suppose that $0^{(p-q)k+q}t_k0^q$ is reducible. Then t_k is reducible; but then t_{k-1} contains a factor 0^p; by Lemma 2 then $q(k-1) \geq p$; so $k \geq 1 + \lceil \frac{p}{q} \rceil$.

Theorem 3 ([18]). *Let R be non-overlapping and let s be an arbitrary string. Then s has an infinite reduction if and only if all reductions starting from s can be prolonged infinitely.*

Lemma 8. *If $10^{(p-q)k+q}$ issues an infinite reduction then $k \geq 1 + \lceil \frac{p}{q} \rceil$.*

Proof. First we observe that R is non-overlapping, i.e., its left hand side, 10^p, has no overlap with itself: $\mathrm{OVL}(10^p, 10^p) = \emptyset$. Now there is a reduction $s = 10^{(p-q)k+q} \rightarrow^* 0^{(p-q)k} t_k 0^q = s'$ by Lemma 1 applied k times for $m = 1$. If $k \leq \lceil \frac{p}{q} \rceil$ then this reduction cannot be prolonged as its final string, s', is irreducible by Lemma 7. By Theorem 3 therefore s issues no infinite reduction for $k \leq \lceil \frac{p}{q} \rceil$.

6 Shorter Loops Do Not Exist

We still have to prove that strings $10^{k(p-q)+q}$, $k \geq 1 + \ell = 1 + \lceil \frac{p}{q} \rceil$, initiate no loops shorter than $1 + \sum_{i=0}^{\ell-1} r^i$.

First let us switch from strings $s \in \{0,1\}^*$ to their tuple representation $T(s) \in \mathbb{N}^*$. Here $\mathbb{N}$ denotes the set of non-negative integers. Please note that our notion of tuple representation differs from the literature [17,11].

Definition 3. *Let $k \in \mathbb{N}$, and let $x_i, y_i \in \mathbb{N}$ for all $0 \leq i \leq k$. A string $s \in \{0,1\}^*$ of the form*

$$s = 0^{x_0(p-q)+y_0 q} 10^{x_1(p-q)+y_1 q} \ldots 10^{x_k(p-q)+y_k q}$$

is said to have a tuple representation

$$T(s) = (x_0, \ldots, x_k; y_0, \ldots, y_k) \ .$$

if for all $1 \leq i \leq k$, $y_i \leq \ell - 1$, and $x_i > 0$ implies $y_i > 0$.

The guard $y_i \leq \ell - 1$, which is equivalent to $y_i q < p$, ensures that the x_i and y_i are uniquely given by $x_i(p-q) + y_i q$. Some strings over the alphabet $\{0,1\}$ may have no tuple representation, e.g. 0 has no tuple representation if $p = 4$, $q = 2$. For our purposes, however, it is reassuring to know that $10^{k(p-q)+q}$ has a tuple representation for any k and that certain rewrite steps preserve the existence of tuple representation.

We will conveniently speak about rewriting steps at position m:

Definition 4. *Let s have a tuple representation, $T(s) = (x_0, \ldots, x_k; y_0, \ldots, y_k)$, and let $0 \leq m \leq k-1$. Then $s \rightarrow_m s'$ if $x_{m+1} > 0$, (and so $y_{m+1} > 0$,) and*

$$s' = \ldots 10^{x_m(p-q)+y_m q} \underline{0^p 1^r 0^q} 0^{(x_{m+1}-1)(p-q)+(y_{m+1}-1)q} \ldots \ .$$

Proposition 2. *Let s have a tuple representation. Then $s \rightarrow s'$ if and only if $s \rightarrow_m s'$ for some $0 \leq m \leq k-1$.*

Definition 5. *Let* $T(s) = (x_0, \ldots, x_k; y_0, \ldots, y_k)$ *and let* $0 \le m \le k-1$. *Then a rewrite step* $s \to_m s'$ *is called* ordinary *if* $y_m < \ell - 1$. *Else the step is called* extraordinary. *A reduction is called ordinary if every step is ordinary. An extraordinary reduction has at least one extraordinary step.*

Ordinary rewrite steps preserve the existence of tuple representation:

Proposition 3. *Let* $T(s) = (x_0, \ldots, x_k; y_0, \ldots, y_k)$, *let* $0 \le m \le k-1$, *and let* $s \to_m s'$ *be ordinary. Then* s' *has the tuple representation*

$$T(s') = (x_0, \ldots, x_{m-1}, x_m + 1, \underbrace{0, \ldots, 0}_{r-1}, x_{m+1} - 1, x_{m+2}, \ldots, x_k;$$

$$y_0, \ldots, y_{m-1}, y_m + 1, \underbrace{0, \ldots, 0}_{r-1}, y_{m+1}, y_{m+2}, \ldots, y_k) \ .$$

In contrast, extraordinary steps may create strings that have no tuple representation.

Example 2. Let s have the tuple representation $T(s) = (2, 1; \ell - 1, 1)$, and let $m = 0$. Then we have

$$s = 0^{2(p-q)+(\ell-1)q}10^p \to_m 0^{3(p-q)+\ell q}10^q = 0^{4(p-q)+(\ell+1)q-p}10^q \ .$$

For $p = 5, q = 2$ we get $\ell = 3$ and $(\ell + 1)q - p = 3$ which has no representation as an integer multiple of q. So the string $0^{4(p-q)+(\ell+1)q-p}10^q$ has no tuple representation.

Our goal is to demonstrate that, in any reduction starting from $10^{k(p-q)+q}$, the first extraordinary step takes place only as late as the completion of the loop.

We are now going to construct two functions h, h' that estimate the length of the shortest ordinary reduction to the next string that has the factor $10^{(p-q)\ell+q}$, and the length of the shortest extraordinary reduction, respectively. These two functions will be based on the following auxiliary functions g_k.

Definition 6. *The functions* $g_k : \mathbb{N}^{k+1} \to \mathbb{N}$, $k \in \mathbb{N}$ *are defined by*

$$g_k(x_0, \ldots, x_k) = \frac{1}{r-1}(r^{x_1+\cdots+x_k} + r^{x_2+\cdots+x_k} + \cdots + r^{x_k} - k) \ .$$

g_k does not depend on its first argument. This is intentional.

The following derived properties will be useful below.

Proposition 4. *For all* $k \ge 1, x_0, \ldots, x_k$ *the following hold:*

1. $g_k(x_0, \ldots, x_k) = g_{k-1}(x_0, \ldots, x_{k-1})$ *if* $x_k = 0$;
2. $g_k(x_0, \ldots, x_k) \ge g_{k-1}(x_1, \ldots, x_k)$;
3. $g_k(x_0, \ldots, x_k) = g_{k+r-1}(x_0, \ldots, x_i + 1, \underbrace{0, \ldots, 0}_{r-1}, x_{i+1} - 1, \ldots, x_k) + 1$ *for all* $0 \le i \le k-1$ *such that* $x_{i+1} \ge 1$;

4. $g_k(x_0,\ldots,x_k) \geq g_k(x_0+1,x_1,\ldots,x_{i-1},x_i-1,x_{i+1},\ldots,x_k)$ *for all* $1 \leq i \leq k$ *such that* $x_i \geq 1$*;*

5. g_k *is monotone in each argument.*

The next lemma is the workhorse of this section. It states that a reduction step decreases by at most one, the minimal value of all those g_k terms which have the same sum of arguments.

Lemma 9. *Let* $s \to s'$ *be an ordinary step where* $T(s) = (x_0,\ldots,x_k;y_0,\ldots,y_k)$ *and* $T(s') = (x'_0,\ldots,x'_{k'};y'_0,\ldots,y'_{k'})$*. Then for every* $1 \leq i' \leq j' \leq k'$, $z'_{j'} \leq x'_{j'}$ *there exist* $1 \leq i \leq j \leq k$, $z_j \leq x_j$ *such that*

$$g_{j'-i'}(x'_{i'},\ldots,x'_{j'-1},z'_{j'}) \geq g_{j-i}(x_i,\ldots,x_{j-1},z_j) - 1,$$
$$x'_{i'} + \cdots + x'_{j'-1} + z'_{j'} = x_i + \cdots + x_{j-1} + z_j \ .$$

Proof. Let $0 \leq m \leq k-1$ be determined by the rewrite step $s \to_m s'$. Then by Proposition 3 we have $k' = k+r-1$, $x'_0 = x_0$, $\ldots$, $x'_{m-1} = x_{m-1}$, $x'_m = x_m + 1$, $x'_{m+1} = \cdots = x'_{m+r-1} = 0$, $x'_{m+r} = x_{m+1} - 1$, $x'_{m+r+1} = x_{m+2}$, $\ldots$, $x'_{k'} = x_k$. And we have $y'_0 = y_0$, $\ldots$, $y'_{m-1} = y_{m-1}$, $y'_m = y_m+1$, $y'_{m+1} = \cdots = y'_{m+r-1} = 0$, $y'_{m+r} = y_{m+1}$, $y'_{m+r+1} = y_{m+2}$, $\ldots$, $y'_{k'} = y_k$.

Let $1 \leq i' \leq j' \leq k'$ and $z'_{j'} \leq x'_{j'}$. The proof is done by case analysis on i' and j'.

Case 1: $1 \leq i' \leq j' \leq m$. If $j' \neq m$ or $z'_{j'} \neq x_m + 1$ then choose $i = i', j = j', z_j = z'_{j'}$. In this case we get

$$g_{j'-i'}(x'_{i'},\ldots,x'_{j'-1},z'_{j'}) = g_{j-i}(x_i,\ldots,x_{j-1},z_j) \ . \tag{2}$$

Else choose $i = i', j = m+1, z_j = 1$. Here we use $x_{m+1} > 0$ to establish $z_j \leq x_j$. We get

$$\begin{aligned} g_{j'-i'}(x'_{i'},\ldots,x'_{j'-1},z'_{j'}) &= g_{m-i}(x_i,\ldots,x_{m-1},x_m+1) \\ &= g_{m+r-i}(x_i,\ldots,x_{m-1},x_m+1,\underbrace{0,\ldots,0}_{r}) \\ &= g_{m+1-i}(x_i,\ldots,x_m,1) - 1 \\ &= g_{j-i}(x_i,\ldots,x_{j-1},z_j) - 1, \end{aligned}$$

by Items 1 and 3 of Proposition 4.

Case 2: $m+1 \leq i' \leq j' \leq m+r-1$. Then $x_{i'} + \cdots + x_{j'-1} + z_{j'} = 0$. Choose any $1 \leq i \leq k$, and let $j = i$ and $z_j = 0$. Then

$$g_{j'-i'}(x'_{i'},\ldots,x'_{j'-1},z'_{j'}) = 0 = g_0(0) \ .$$

Case 3: $m+r \leq i' \leq j' \leq k'$. If $i' \neq m+r$ or $j' = i'$ then choose $i = i'-r+1, j = j'-r+1, z_j = z'_{j'}$. We get (2). Else we have $i' = m+r$ and $j' > i'$, and so $x'_{i'} = x_{m+1} - 1$. In this case let $i = m+1$. If $z'_{j'} > 0$ then let j and z_j be defined by $j = j'-r+1$ and $z_j = z'_{j'} - 1$; else let $i \leq j < j'-r+1$ be the greatest number such that $x_j > 0$ and let $z_j = x_j - 1$. By $x_i > 0$, j and z_j are well-defined. Thus we get

$$\begin{aligned} g_{j'-i'}(x'_{i'}, \ldots, x'_{j'-1}, z'_{j'}) &= g_{j'-r+1-i}(x_i - 1, x_{i+1}, \ldots, x_{j'-r}, z'_{j'}) \\ &= g_{j-i}(x_i - 1, x_{i+1}, \ldots, x_{j-1}, z_j + 1) \\ &\geq g_{j-i}(x_i, x_{i+1}, \ldots, x_{j-1}, z_j) \end{aligned}$$

by Items 1 and 4 of Proposition 4.

Case 4: $1 \leq i' \leq m$ and $m+r \leq j' \leq k'$. Choose $i = i', j = j' - r + 1, z_j = z'_{j'}$. We get

$$\begin{aligned} g_{j'-i'}(x'_{i'}, \ldots, x'_{j'-1}, z'_{j'}) &= g_{j-i+r-1}(x_i, \ldots, x_m + 1, \underbrace{0, \ldots, 0}_{r-1}, x_{m+1} - 1, \ldots, x_{j-1}, z_j) \\ &= g_{j-i}(x_i, \ldots, x_m, x_{m+1}, \ldots, x_{j-1}, z_j) - 1 \end{aligned}$$

by Item 3 of Proposition 4. Note that if $j' = m+r$ then $x'_{j'} = x_{m+1} - 1 = x_j - 1$. So one can always choose $z_j = z'_{j'}$ yielding $z_j = z'_{j'} \leq x'_{j'} \leq x_j$ as required.

Case 5: $1 \leq i' \leq m < j' \leq m + r - 1$. This case reduces to Case 1 by the identity

$$g_{j'-i'}(x'_{i'}, \ldots, x'_{j'-1}, z'_{j'}) = g_{m-i'}(x'_{i'}, \ldots, x'_m)$$

due to Item 1 of Proposition 4.

Case 6: $m + 1 \leq i' < m + r \leq j' \leq k'$. This case reduces to Case 3 by the inequality

$$g_{j'-i'}(x'_{i'}, \ldots, x'_{j'-1}, z'_{j'}) \geq g_{j'-m-r}(x'_{m+r}, \ldots, x'_{j'-1}, z'_{j'})$$

due to Item 2 of Proposition 4.

These are all cases. In each case it is easy to show that $x'_{i'} + \cdots + x'_{k'} = x_i + \cdots + x_k$. This finishes the proof.

Definition 7. *Let* $T(s) = (x_0, \ldots, x_k; y_0, \ldots, y_k)$ *and let* $x_1 + \cdots + x_k \geq \ell$. *Then* $h(s) \in \mathbb{N}$ *is defined by*

$$\begin{aligned} h(s) = \min\{ & g_{j-i}(x_i, \ldots, x_{j-1}, z_j) \mid \\ & 1 \leq i \leq j \leq k, z_j \leq x_j, x_i + \cdots + x_{j-1} + z_j = \ell\} \ . \end{aligned}$$

Well-definedness of $h(s)$ follows immediately from the fact that the minimum is taken from a finite, non-empty set.

Lemma 10. *Let* $s \to s'$ *be an ordinary step where* $T(s) = (x_0, \ldots, x_k; y_0, \ldots, y_k)$ *and* $T(s') = (x'_0, \ldots, x'_{k'}; y'_0, \ldots, y'_{k'})$. *If* $x'_1 + \cdots + x'_{k'} \geq \ell$ *then* $x_1 + \cdots + x_k \geq \ell$ *and* $h(s) \leq h(s') + 1$.

Proof. The condition $x'_1 + \cdots + x'_{k'} \geq \ell$ ensures that $h(s')$ is defined. If $s \to_m s'$ for $1 \leq m \leq k-1$ then $x_1 + \cdots + x_k = x'_1 + \cdots + x'_{k'} \geq \ell$ by Proposition 3. Else $s \to_m s'$ for $m = 0$ and then $(x_1 - 1) + \cdots + x_k = x'_1 + \cdots + x'_{k'} \geq \ell$. So $x_1 + \cdots + x_k \geq \ell$ whence $h(s)$ is defined.

By definition of $h(s')$, there is $1 \leq i' \leq j' \leq k', z'_{j'} \leq x'_{j'}$ such that both $h(s') = g_{j'-i'}(x'_{i'}, \ldots, x'_{j'-1}, z'_{j'})$ and $x'_{i'} + \cdots + x'_{j'-1} + z'_{j'} = \ell$. Hence by Lemma 9, there is $1 \leq i \leq j \leq k$, $z_j \leq x_j$ such that

$$h(s') \geq g_{j-i}(x_i, \ldots, x_{j-1}, z_j) - 1 \quad \text{and} \quad x_i + \cdots + x_{j-1} + z_j = \ell \ .$$

So $h(s) \leq g_{j-i}(x_i, \ldots, x_{j-1}, z_j) \leq h(s') + 1$.

Lemma 11. *Let $T(s) = (x_0, \ldots, x_k; y_0, \ldots, y_k)$. If $s \to^n u10^{(p-q)\ell+q}v$ for some strings u, v is an ordinary reduction then $x_1 + \cdots + x_k \geq \ell$ and $h(s) \leq n$.*

Proof. By induction on n. The base case $n = 0$ is proven by $h(s) \leq g_0(\ell) = 0$. For the inductive step let $s \to s' \to^{n-1} u10^{(p-q)\ell+q}v$, let $x'_1 + \cdots + x'_{k'} \geq \ell$, and let $h(s') \leq n-1$. Thus $x_1 + \cdots + x_k \geq \ell$ and $h(s) \leq n$ by Lemma 10.

With Lemma 11 we have a criterion for ordinary reductions. For extraordinary reductions we pursue a similar line of reasoning. We start with a lemma akin to Lemma 9. If $i = j$ then for convenience let $g_{j-i}(y_i, x_{i+1}, \ldots, x_{j-1}, z_j) = g_0(z_j)$ and let $y_i + x_{i+1} + \cdots + x_{j-1} + z_j = z_j$. Note that we require $z_j \leq y_j$ if $i = j$ and $z_j \leq x_j$ else.

Lemma 12. *Let $s \to s'$ be an ordinary step where $T(s) = (x_0, \ldots, x_k; y_0, \ldots, y_k)$ and $T(s') = (x'_0, \ldots, x'_{k'}; y'_0, \ldots, y'_{k'})$. Then for every $1 \leq i' < j' \leq k'$, $z'_{j'} \leq x'_{j'}$ and for every $1 \leq i' = j' \leq k'$, $z'_{j'} \leq y'_{j'}$ there exist $1 \leq i < j \leq k$, $z_j \leq x_j$ or $1 \leq i = j \leq k$, $z_j \leq y_j$ such that*

$$\begin{aligned} g_{j'-i'}(y'_{i'}, x'_{i'+1}, \ldots, x'_{j'-1}, z'_{j'}) &\geq g_{j-i}(y_i, x_{i+1}, \ldots, x_{j-1}, z_j) - 1, \\ y'_{i'} + x'_{i'+1} + \cdots + x'_{j'-1} + z'_{j'} &= y_i + x_{i+1} + \cdots + x_{j-1} + z_j \ . \end{aligned}$$

Proof. Let $0 \leq m \leq k-1$ be determined by the rewrite step $s \to_m s'$. Case 1: $1 \leq i' = j' \leq k'$, $z'_{j'} \leq y'_{j'}$. If $j' \neq m$ or $z'_{j'} \neq z_m + 1$ then $g_0(z'_{j'}) = 0 = g_0(z_j)$. Else choose $j = m+1, z_j = 1$. Here we use $y_{m+1} > 0$ to establish $z_j \leq y_j$. We get

$$g_0(z'_{j'}) = g_r(y_m + 1, \underbrace{0, \ldots, 0}_{r}) = g_1(y_m, 1) - 1 = g_{j-i}(y_i, z_j) - 1$$

by Items 1 and 3 of Proposition 4.

Case 2: $1 \leq i' < j' \leq k'$, $z'_{j'} \leq x'_{j'}$.

Case 2.1: $m+1 \leq i' < j' \leq m+r-1$. Then $y_{i'} + x_{i'+1} + \cdots + x_{j'-1} + z_{j'} = 0$. Choose any $1 \leq i \leq k$, and let $j = i$ and $z_j = 0$. Then

$$g_{j'-i'}(y'_{i'}, x'_{i'+1}, \ldots, x'_{j'-1}, z'_{j'}) = 0 = g_0(0) \ .$$

Case 2.2: $i' = m + r$. Choose $= i' - r + 1, j = j' - r + 1, z_j = z'_{j'}$. We get

$$g_{j'-i'}(y'_{i'}, x'_{i'}, \ldots, x'_{j'-1}, z'_{j'}) = g_{j-i}(y_i, x_{i+1}, \ldots, x_{j-1}, z_j),$$
$$y'_{i'} + x'_{i'} + \cdots + x'_{j'-1} + z'_{j'} = y_i + x_{i+1} + \cdots + x_{j-1} + z_j \ .$$

Case 2.3: $1 \leq i' \leq m$ or $i' \neq m + r$ and $m + r - 1 \leq j' \leq k'$. We carry over the proof of Lemma 9, observing the facts $i' \neq j'$, $i \neq j$, and $y'_{i'} - x'_{i'} = y_i - x_i$. Then we may conclude from the proof of Lemma 9 that

$$\begin{aligned} g_{j'-i'}(y'_{i'}, x'_{i'+1}, \ldots, x'_{j'-1}, z'_{j'}) &= g_{j'-i'}(x'_{i'}, \ldots, x'_{j'-1}, z'_{j'}) \\ &\geq g_{j-i}(x_i, \ldots, x_{j-1}, z_j) - 1 \\ &= g_{j-i}(y_i, x_{i+1}, \ldots, x_{j-1}, z_j) - 1 \end{aligned}$$

and

$$\begin{aligned} y'_{i'} + x'_{i'+1} + \cdots + x'_{j'-1} + z'_{j'} &= (y'_{i'} - x'_{i'}) + x'_{i'} + \cdots + x'_{j'-1} + z'_{j'} \\ &= (y'_{i'} - x'_{i'}) + x_i + \cdots + x_{j-1} + z_j \\ &= (y_i - x_i) + x_i + \cdots + x_{j-1} + z_j \\ &= y_i + x_{i+1} + \cdots + x_{j-1} + z_j \ . \end{aligned}$$

This finishes the proof.

Lemma 13. *Let $s \to s'$ be an ordinary step where $T(s) = (x_0, \ldots, x_k; y_0, \ldots, y_k)$ and $T(s') = (x'_0, \ldots, x'_{k'}; y'_0, \ldots, y'_{k'})$. If $y'_i + x'_{i+1} + \cdots + x'_{k'} \geq \ell$ for some $1 \leq i' \leq k'$ then $y_i + x_{i+1} + \cdots + x_k \geq \ell$ for some $1 \leq i \leq k$.*

Proof. The claim immediately follows from the following claim. For every $1 \leq i' \leq k'$ there is $1 \leq i \leq k$ such that

$$\Delta = y_i + x_{i+1} + \cdots + x_k - (y'_{i'} + x'_{i'+1} + \cdots + x'_{k'}) \geq 0 \ .$$

The proof is done by case analysis on i'.

Case 1: $1 \leq i' \leq m - 1$. Choose $i = i'$. Then

$$\begin{aligned} \Delta &= x_m + x_{m+1} - (x'_m + \underbrace{0 + \cdots + 0}_{r-1} + x'_{m+r}) \\ &= x_m + x_{m+1} - (x_m + 1 + x_{m+1} - 1) = 0 \ . \end{aligned}$$

Case 2: $i' = m$. Again choose $i = i'$. Then

$$\begin{aligned} \Delta &= y_m + x_{m+1} - (y'_m + \underbrace{0 + \cdots + 0}_{r-1} + x'_{m+r}) \\ &= y_m + x_{m+1} - (y_m + 1 + x_{m+1} - 1) = 0 \ . \end{aligned}$$

Case 3: $m+1 \leq i' \leq m+r-1$. Choose $i = m+1$. Then $\Delta = x_{m+1} - x'_{m+r} = 1 \geq 0$.

Case 4: $m + r \leq i' \leq k'$. Choose $i = i' - r + 1$. Then obviously $\Delta = 0$. This finishes the proof.

Definition 8. *Let* $T(s) = (x_0, \ldots, x_k; y_0, \ldots, y_k)$ *and let* $y_i + x_{i+1} + \cdots + x_k \geq \ell$ *for some* $1 \leq i \leq k$*. Then* $h'(s) \in \mathbb{N}$ *is defined by*

$$h'(s) = \min\{g_{j-i}(y_i, x_{i+1}, \ldots, x_{j-1}, z_j) \mid \\ 1 \leq i < j \leq k, z_j \leq x_j, y_i + x_{i+1} + \cdots + x_{j-1} + z_j = \ell\} \ .$$

Because the minimum is taken from a finite, non-empty set, $h'(s)$ is well-defined.

Using Lemmas 12 and 13, one can prove the following in the same way as Lemma 10:

Lemma 14. *Let* $s \to s'$ *be an ordinary step where* $T(s) = (x_0, \ldots, x_k; y_0, \ldots, y_k)$ *and* $T(s') = (x'_0, \ldots, x'_{k'}; y'_0, \ldots, y'_{k'})$*. If* $y'_i + x'_{i+1} + \cdots + x'_{k'} \geq \ell$ *for some* $1 \leq i' \leq k'$ *then* $y_i + x_{i+1} + \cdots + x_k \geq \ell$ *for some* $1 \leq i \leq k$*, and* $h'(s) \leq h'(s') + 1$*.*

Thus we get a lemma like Lemma 11:

Lemma 15. *Let* $T(s) = (x_0, \ldots, x_k; y_0, \ldots, y_k)$*. If there is an extraordinary reduction* $s \to^n t$ *then* $y_i + x_{i+1} + \cdots + x_k \geq \ell$ *for some* $1 \leq i \leq k$*, and moreover* $h'(s) \leq n$*.*

Proof. Let $s \to^n s'$ be shortest, i.e., $s \to^{n-1} s'$ is an ordinary reduction and only the last step $s' \to t$ is extraordinary. The inductive base $n = 1$ is proven by $h'(s) \leq g_1(\ell - 1, 1) = 1$. For the inductive step $s \to s'' \to^{n-2} s'$ let $y'_i + x'_{i+1} + \cdots + x'_{k'} \geq \ell$ for some $1 \leq i' \leq k'$ and let $h'(s'') \leq n - 1$. Thus $y_i + x_{i+1} + \cdots + x_k \geq \ell$ for some $1 \leq i \leq k$ and $h'(s) \leq n$ by Lemma 14.

Now let us prove that the length of the loop in Lemma 5 is minimal.

Lemma 16. *R admits no loops of length less than* $1 + \sum_{i=0}^{\ell-1} r^i$ *for* $\ell = \lceil \frac{p}{q} \rceil$*.*

Proof. Suppose that $s \to^+ usv$ is a loop of minimal length. By Theorem 2 we may assume that $s \to^+ usv$ is a forward closure. By Lemma 6, $s = 10^{(p-q)k+q}$ for some $k \geq 0$. By Lemma 8, $k \geq 1 + \ell$. Since the loop is a forward closure, all zeroes in s must be consumed during the reduction. By $OVL(\ell, \ell) = \emptyset$ we may assume that the steps are rearranged to $s \to^k (0^p 1^{r-1})^k 10^q = s' \to^n usv$. Case 1: $s' \to^n usv$ is ordinary. Then we get $h(s') \leq n$ by Lemma 11 and by $10^{(p-q)\ell+q}$ prefix of s. We compute $h(s')$ as follows:

$$h(s') = g_{(r-1)(\ell-1)}(\underbrace{1, \underbrace{0, \ldots, 0}_{r-2}, 1, \underbrace{0, \ldots, 0}_{r-2}, 1, \ldots, \underbrace{0, \ldots, 0}_{r-2}}_{\ell-1}, 1)$$

$$= \frac{1}{r-1}((r-1)r^{\ell-1} + (r-1)r^{\ell-2} + \cdots + (r-1)r^1 - (r-1)(\ell-1))$$

$$= \sum_{i=0}^{\ell-1} r^i - \ell \ .$$

So the total length of the reduction is $k + n \geq k + h(s') = k + \sum_{i=0}^{\ell-1} r^i - \ell \geq 1 + \sum_{i=0}^{\ell-1} r^i$. Case 2: $s' \to^n usv$ is extraordinary. Then we get $h'(s') \leq n$ by Lemma 15. It turns out that $h'(s') = h(s')$. So, no matter whether the reduction $s \to^n usv$ is ordinary or not, we get $k+n \geq 1 + \sum_{i=0}^{\ell-1} r^i$. This finishes the proof.

Acknowledgements. I am grateful to Robert McNaughton, David Musser, and Paliath Narendran for their encouragement to pursue this research and to Hans Zantema and Hanne Gottliebsen for reading the manuscript.

References

1. Ronald Book and Friedrich Otto. *String-rewriting systems.* Texts and Monographs in Computer Science. Springer, New York, 1993.
2. Nachum Dershowitz. Termination of linear rewriting systems. In *Proc. 8th Int. Coll. Automata, Languages and Programming*, LNCS 115, pages 448–458. Springer, 1981.
3. Nachum Dershowitz. Termination of rewriting. *J. Symb. Comput.*, 3(1&2):69–115, Feb./April 1987. Corrigendum: 4, 3, Dec. 1987, 409-410.
4. Alfons Geser. Note on normalizing, non-terminating one-rule string rewriting systems. *Theoret. Comput. Sci.*, 243:489–498, 2000.
5. Alfons Geser. Decidability of termination of grid string rewriting rules. *SIAM J. Comput.*, 2002. In print.
6. Alfons Geser and Hans Zantema. Non-looping string rewriting. *Theoret. Informatics Appl.*, 33(3):279–301, 1999.
7. Miki Hermann. *Divergence des systèmes de réécriture et schématisation des ensembles infinis de termes.* Habilitation, Université de Nancy, France, March 1994.
8. Gérard Huet and Dallas S. Lankford. On the uniform halting problem for term rewriting systems. Technical Report 283, INRIA, Rocquencourt, FR, March 1978.
9. Matthias Jantzen. *Confluent string rewriting*, volume 14 of *EATCS Monographs on Theoretical Computer Science.* Springer, Berlin, 1988.
10. Yuji Kobayashi, Masashi Katsura, and Kayoko Shikishima-Tsuji. Termination and derivational complexity of confluent one-rule string rewriting systems. *Theoret. Comput. Sci.*, 262(1/2):583–632, 2001.
11. Winfried Kurth. *Termination und Konfluenz von Semi-Thue-Systemen mit nur einer Regel.* Dissertation, Technische Universität Clausthal, Germany, 1990.
12. Dallas S. Lankford and D. R. Musser. A finite termination criterion. Technical report, Information Sciences Institute, Univ. of Southern California, Marina-del-Rey, CA, 1978.
13. Yuri Matiyasevitch and Geraud Sénizergues. Decision problems for semi-Thue systems with a few rules. In *Proc. 11th IEEE Symp. Logic in Computer Science*, pages 523–531, New Brunswick, NJ, July 1996. IEEE Computer Society Press.
14. Robert McNaughton. The uniform halting problem for one-rule Semi-Thue Systems. Technical Report 94-18, Dept. of Computer Science, Rensselaer Polytechnic Institute, Troy, NY, August 1994. See also "Correction to 'The Uniform Halting Problem for One-rule Semi-Thue Systems'", unpublished paper, August, 1996.
15. Robert McNaughton. Well-behaved derivations in one-rule Semi-Thue Systems. Technical Report 95-15, Dept. of Computer Science, Rensselaer Polytechnic Institute, Troy, NY, November 1995. See also "Correction by the author to 'Well-behaved derivations in one-rule Semi-Thue Systems'", unpublished paper, July, 1996.

16. Robert McNaughton. Semi-Thue Systems with an Inhibitor. *J. Automated Reasoning*, 26:409–431, 1997.
17. Paliath Narendran and Friedrich Otto. The problems of cyclic equality and conjugacy for finite complete rewriting systems. *Theoret. Comput. Sci.*, 47:27–38, 1986.
18. Michael J. O'Donnell. *Computing in systems described by equations.* LNCS 58. Springer, 1977.
19. Geraud Sénizergues. On the termination problem for one-rule Semi-Thue Systems. In *RTA-7*, LNCS 1103, pages 302–316. Springer, 1996.
20. Kayoko Shikishima-Tsuji, Masashi Katsura, and Yuji Kobayashi. On termination of confluent one-rule string rewriting systems. *Inform. Process. Lett.*, 61(2):91–96, 1997.
21. Hans Zantema and Alfons Geser. A complete characterization of termination of $0^p1^q \to 1^r0^s$. *Applicable Algebra in Engineering, Communication, and Computing*, 11(1):1–25, 2000.

Recursive Derivational Length Bounds for Confluent Term Rewrite Systems

Research Paper

Elias Tahhan-Bittar

Dpto. de Matemáticas Puras y Aplicadas,
Univ. Simón Bolívar,
P.O. Box 89000, Caracas 1080-A, Venezuela.
etahhan@usb.ve

Abstract. Let F be a signature and $\mathcal{R}$ a term rewrite system on ground terms of F. We define the concepts of a context-free potential redex in a term and of bounded confluent terms. We bound recursively the lengths of derivations of a bounded confluent term t by a function of the length of derivations of context-free potential redexes of this term. We define the concept of inner redex and we apply the recursive bounds that we obtained to prove that, whenever $\mathcal{R}$ is a confluent overlay term rewrite system, the derivational length bound for arbitrary terms is an iteration of the derivational length bound for inner redexes.

1 Introduction

Given a finite term rewrite system $\mathcal{R}$, a natural question beyond termination is how to bound the length of derivations of a term t. This question has been addressed for term rewrite systems whose termination is proved by recursive path orderings, see for example [2], [5] and [11]. Another way to address the bounding problem is to bound the lengths of derivations of a term recursively. For instance, given a confluent terminating term rewrite system, given terms $u_0, \ldots, u_m$ and $\downarrow(u_0), \ldots, \downarrow(u_m)$ their respective normal forms how to bound the lengths of the derivations of a term $f(u_0, \ldots, u_m)$ by the lengths of the derivations of $u_0, \ldots, u_m$ and of $f(\downarrow(u_0), \ldots, \downarrow(u_m))$. In [3] we studied this approach for strict constructor orthogonal rewrite systems and applied it to study the lengths of derivations of primitive recursive terms.

In the present paper we continue the work developed in [3]. We define the notion of *context-free potential redexes* and elaborate a new proof using techniques of [7] to extend the scope of the recursive bounds given in [3] to confluent term rewrite systems.

O'Donnell proved in [10] that innermost terminating orthogonal TRS are terminating. Later Gramlich extended this result to confluent overlay TRS in [8]. We explore in this work a related question: how to bound the length of derivations of terms in overlay confluent TRS as a function of the lengths of derivations of inner redexes.

S. Tison (Ed.): RTA 2002, LNCS 2378, pp. 281–295, 2002.

2 Preliminaries

Here we review definitions and notations required through the article, wherever possible, we adopt the definitions and notations proposed in [4] or [1].

Signature, term, ground term. A *signature* F is a set of function symbols. The set of those elements of F whose arity is k is denoted F_k. The elements of F_0 are also referred to as *constants*. Ω is a symbol which is not in F and which we call a *hole*. The set $T(F,\{\Omega\})$ of *terms* is the least set (with respect to the inclusion order) satisfying the following rules: i) $\Omega \in T(F,\{\Omega\})$, ii) $F_0 \subseteq T(F,\{\Omega\})$ and iii) if $u_0,\ldots,u_k$ is a sequence of terms in $T(F,\{\Omega\})$ and $f \in F_{k+1}$ then $f(u_0,\ldots,u_k) \in T(F,\{\Omega\})$. The subset $T(F)$ of $T(F,\{\Omega\})$ of *ground terms* is the least set satisfying rules ii) and iii). Note that $T(F)$ is non-empty if and only if F_0 is non empty. Throughout this article, the equality relation on terms is the syntactic identity.

Position, parallel positions. $\mathbb{N}$ denotes the natural numbers and $\mathbb{N}^*$ is the set of *strings* of natural numbers. The empty string is denoted by ϵ . If $p,q \in \mathbb{N}^*$, $p \cdot q$ is the *concatenation* of p and q. If $A \in \mathbb{N}^*$, and $i \in \mathbb{N}$, the set $\{i \cdot a : a \in A\}$ is denoted by $i \cdot A$. Elements of $\mathbb{N}^*$ are ordered by the prefix ordering, $\leq$, that is $q \leq p$ if, and only if, there exists $\omega \in \mathbb{N}^*$ such that $p = q \cdot w$; in this case we define $p - q := w$.

Subsets of $\mathbb{N}^*$ enable us to define the notion of *position* of a node in a term. Given a term t, the subset $Pos(t)$ of $\mathbb{N}^*$ is defined by

$$Pos(t) = \{\epsilon\} \text{ if } t \in F_0 \cup \{\Omega\}, \quad Pos(f(t_0,\ldots,t_{k-1})) = \{\epsilon\} \cup \bigcup_{0 \leq i \leq k-1} i \cdot Pos(t_i).$$

If $P \subset Pos(t)$ then P is a set of *parallel positions* if the elements of P are pairwise incomparable with respect to $\leq$. Given a position p and Q a set of parallel positions, we write $p < Q$, $p \leq Q$, $Q < p$ and $Q \leq p$, if there is $q \in Q$ such that, respectively, $p < q$, $p \leq q$, $q < p$ and $q \leq p$.

Subterm at a position, Context of a set of positions. For $p \in Pos(t)$, the subterm of t whose root is at position p in t, denoted by $t|_p$, is defined by :

$$t|_\epsilon = t, \qquad f(t_0,\ldots,t_{k-1})|_{i \cdot p} = t_i|_p, \quad \text{for } 0 \leq i \leq k-1.$$

If $f \in F_k$, and $u_1,\ldots,u_k$ are terms in $T(F,\{\Omega\})$, we call $u_1,\ldots,u_k$ *immediate subterms* of the term $f(u_1,\ldots,u_k)$. Given a term t and a sequence, $P = p_1,\ldots,p_m$, of parallel positions of t, we denote by $t|_P$ the sequence $t|_{p_1},\ldots,t|_{p_m}$.

For $p \in Pos(t)$, the *context* of p in t, denoted by $t|^p$, is defined by :

$$t|^\epsilon = \Omega, \quad f(t_0,\ldots,t_{k-1})|^{i \cdot p} = f(t_0,\ldots,t_{i-1},t_i|^p,t_{i+1},\ldots,t_{k-1}), \quad \text{for } 0 \leq i \leq k-1.$$

This notion of context of a position can be extended to that of a set of parallel positions. Suppose that $t \in T(F, \{\Omega\})$ and $\{p_1, \ldots, p_m\}$ is a set of parallel positions of t, then $t|^{p_1,\ldots,p_m}$ is defined as follows :

$$t|^{\emptyset} := t \qquad \text{and} \qquad t|^{p_1,\ldots,p_m} := (t|^{p_1})|^{p_2,\ldots,p_m} .$$

Subterm replacement. Suppose that $t, s \in T(F, \{\Omega\})$ and $p \in Pos(t)$, then $t[s]_p$ is defined by :

$$t[s]_\epsilon = s, \quad f(t_0, \ldots, t_{k-1})[s]_{i\cdot p} = f(t_0, \ldots, t_{i-1}, t_i[s]_p, t_{i+1}, \ldots, t_{k-1}), \quad \text{where } 0 \leq i \leq k-1.$$

$t[s]_p$ is the result of replacing $t|_p$ by s in t.

Parallel replacement is defined as follows : if $t, s_1, \ldots, s_m \in T(F, \{\Omega\})$ and $p_1, \ldots, p_m$ are parallel positions in $Pos(t)$, then :

$$t[s_1, \ldots, s_m]_{p_1,\ldots,p_m} = (t[s_1]_{p_1})[s_2, \ldots, s_m]_{p_2,\ldots,p_m} .$$

Given a set of parallel positions $P = \{p_1, \ldots, p_m\}$ in $Pos(t)$, and a term s, the parallel replacement $t[s, \ldots, s]_{p_1,\ldots,p_m}$ is denoted by $t[s]_P$. If $t, s_1, \ldots, s_m \in T(F, \{\Omega\})$ and if $P_1, \ldots, P_m$ is a partition of a set of parallel positions in $Pos(t)$, then we use the notation :

$$t[s_1, \ldots, s_m]_{P_1,\ldots,P_m} := (t[s_1]_{P_1})[s_2, \ldots, s_m]_{P_2,\ldots,P_m} .$$

The above definitions give rise to the following identities :

$$t|^{p_1,\ldots,p_m} = t[\Omega]_{p_1,\ldots,p_m} \quad \text{and} \quad t|^{p_1,\ldots,p_m}[t|_{p_1}, \ldots, t|_{p_m}] = t .$$

Rewrite rule, rewrite system. Given a signature F and a set of variable symbols with arity zero, V, such that $F \cap V = \emptyset$, we consider terms of $T(F \cup V)$. Given $t \in T(F \cup V)$, we denote by $VPos\,(t)$ the set of positions $p \in Pos(t)$ such that $t|_p \in V$. The set of *variables occurring in* t, denoted by $var(t)$, is defined by $var(t) := \{t|_p \ : \ p \in VPos(t)\}$. A *rewrite rule* on $T(F \cup V)$, denoted by $l \to r$, is a pair $(l, r) \in T(F \cup V)$ such that $l \notin V$ and $var(r) \subseteq var(l)$.

Given $t \in T(F \cup V)$ and $x \in var(t)$, we denote by $Pos(t, x)$ the set of positions $p \in VPos(t)$ such that $t|_p = x$. Two rewrite rules $l \to r$ and $l' \to r'$ are *equivalent* if they are equals up to renaming variables, i.e. $l[\Omega]_{VPos(l)} = l'[\Omega]_{VPos(l')}$, $r[\Omega]_{VPos(r)} = r'[\Omega]_{VPos(r')}$ and there is a one to one function, ϕ, from $var(l)$ onto $var(l')$, such that for each $x \in var(l)$, $Pos(l, x) = Pos(l', \phi(x))$ and $Pos(r, x) = Pos(r', \phi(x))$. Throughout this article we identify equivalent rewrite rules.

A set of rewrite rules on $T(F \cup V)$ is called a *term rewrite system*, (TRS). A rewrite rule $l \to r$ is *left linear* if for each variable $x \in var(l)$, the set $Pos(t, x)$ is a singleton. A TRS $\mathcal{R}$ on $T(F \cup V)$ is *left-linear* if each rewrite rule in $\mathcal{R}$ is left-linear.

The rewrite relation on $T(F)$. Given a rewrite rule $\rho = (l \to r)$ on $T(F \cup V)$, and $var(l) := \{x_1, \ldots, x_m\}$. Let $t_1, t_2 \in T(F)$ and $p \in Pos(t_1)$, the relation $t_1 \to^{\rho,p} t_2$ holds if, and only if, there are terms $s_{x_1}, \ldots, s_{x_m} \in T(F)$ such that the following conditions hold :

(i) $t_1|^p = t_2|^p$,
(ii) $t_1|_p = l[s_{x_1}, \ldots, s_{x_m}]_{Pos(l,x_1),\ldots,Pos(l,x_m)}$ and
(iii) $t_2|_p = r[s_{x_1}, \ldots, s_{x_m}]_{Pos(r,x_1),\ldots,Pos(r,x_m)}$.

Given a rewrite system $\mathcal{R}$, a term $t_1 \in T(F)$ rewrites, or reduces, to a term $t_2 \in T(F)$, $t_1 \to_{\mathcal{R}} t_2$, if there is a rewrite rule $\rho \in R$ and a position $p \in Pos(t_1)$ such that $t_1 \to^{\rho,p} t_2$. The transitive closure of the rewrite relation $\to_{\mathcal{R}}$ is denoted by $\to^+_{\mathcal{R}}$, the reflexive closure of $\to^+_{\mathcal{R}}$ is denoted by $\to^*_{\mathcal{R}}$, the symmetric relation of $\to^*_{\mathcal{R}}$ is denoted by ${}^*_{\mathcal{R}}\!\leftarrow$. When there is no ambiguity we omit the mention of $\mathcal{R}$, for instance we write $\to^*$ instead of $\to^*_{\mathcal{R}}$. We notice that we consider only reductions of ground terms.

Let $\mathcal{R}$ be a rewrite system. A *rewrite sequence*, or *derivation*, is given by a sequence $t_0, \ldots, t_j$ of terms, a sequence of rules $\rho_1, \ldots, \rho_j$ and a sequence of positions $p_1, \ldots, p_j$ such that

$$t_0 \to^{\rho_1,p_1} t_1 \to^{\rho_2,p_2} \ldots \to^{\rho_{j-1},p_{j-1}} t_{j-1} \to^{\rho_j,p_j} t_j \ .$$

We abbreviate by (t_0, σ) or by $t_0 \to^\sigma t_j$, where σ is the sequence $(\rho_1, p_1), \ldots, (\rho_j, p_j)$.

Further, when we don't give any detail about the signature and the rewrite system; we suppose given a signature F and a rewrite system, $\mathcal{R}$, on $T(F)$.

Termination, confluence. Let $\mathcal{R}$ be a rewrite system. An *infinite rewrite sequence* or *infinite derivation* starting from a term t_0, is an infinite sequence $t_0, \ldots, t_j, \ldots$ such that $t_0 \to t_1 \to \cdots \to t_{j-1} \to t_j \to t_{j+1} \cdots$. A term t_0 is *strongly normalizing*, abbreviated by (SN), if there is no infinite $\mathcal{R}$-derivation starting from t_0, A rewrite system $\mathcal{R}$ *terminates* or is *strongly normalizing*, also abbreviated by (SN), if there is no infinite $\mathcal{R}$-derivation.

A rewrite system $\mathcal{R}$ is *confluent* if ${}^*\!\leftarrow \circ \to^* \subset \to^* \circ {}^*\!\leftarrow$, (*i.e.* if $u \ {}^*\!\leftarrow t \to^* v$, then there is w such that $u \to^* w \ {}^*\!\leftarrow v$). A rewrite system $\mathcal{R}$ is *locally confluent* if $\leftarrow \circ \to \subset \to^* \circ {}^*\!\leftarrow$.

Let $\mathcal{R}$ be a rewrite system. A term t is a *redex* if it is redeucible at the root, i.e. there is a rule ρ and a term t' such that $t \to^{\rho,\epsilon} t'$. A term t is *irreducible* if there is no u such that $t \to u$. An irreducible term u is a normal form of a term t if $t \to^* u$. If $\mathcal{R}$ is a confluent rewrite system, any term which admits a normal form has a unique normal form, this normal form is denoted by $\downarrow(t)$.

Orthogonal, Overlay Rewrite Systems. A TRS $\mathcal{R}$ on $T(F)$ is *overlay* if for each redex $t \in T(F)$, for each rewrite rule $l \to r$ such that $t \to^{l \to r,\epsilon} t'$ and for each position p such that $\epsilon < p < VPos(l)$ the subterm $t|_p$ is not a redex. A TRS on $T(F)$, $\mathcal{R}$, is *orthogonal* if it is overlay, left-linear, and for each redex $t \in T(F)$ there is only one rule $l \to r$ such that $t \to^{l \to r,\epsilon} t'$. We notice that these definitions are equivalent with the usual definitions given for example in [8], and we consider that they are more suitable for this article.

A TRS on $T(F)$, $\mathcal{R}$, is said to be a *constructor rewrite system* if its signature F can be partitioned into two sets C and D called, respectively, the set of *constructors* and the set of *defined functions symbols*, such that, for each rule $l \to r$ in $\mathcal{R}$, the term l is of the form $d(u_0, \ldots, u_k)$ where d is a defined function symbol and $u_0, \ldots, u_k$ belong to $T(C \cup V)$. Remark that any constructor TRS is an overlay term rewrite system. A constructor TRS is *strict* if every irreducible ground term belongs to $T(C)$.

3 Potential Redexes, Context-Free Potential Redexes

3.1 Ancestors and Potential Redexes

We define an ancestor relation which generalizes the notion of residuals given in [7]:

Definition 1. *Given a rewrite rule $\rho = (l \to r)$ on $T(F \cup V)$, with $var(l) := \{x_1, \ldots, x_m\}$, given a reduction $t_1 \to^{\rho,p} t_2$ and positions $q \in Pos(t_2)$ and $p' \in Pos(t_1)$; we say that* p' is an ancestor of q by the reduction $t_1 \to^{\rho,p} t_2$, *written p' is a $(t_1, (\rho, p))$-ancestor of q, if*

$$p' = \begin{cases} q & \text{if } q \in Pos(t_2|^p)\backslash\{p\} \\ p & \text{if } q \in p \cdot (Pos(r)\backslash VPos(r)) \\ p \cdot p_j \cdot w & \text{if } q = p \cdot q_j \cdot w \text{ where } q_j \in Pos(r, x_j) \text{ and } p_j \in Pos(l, x_j) \ . \end{cases}$$

The *ancestor relation with respect to a derivation* is the reflexive transitive closure of the ancestor relation by a reduction defined above. When $t \to^\sigma t'$, $p \in Pos(t)$ and $q \in Pos(t')$ we write that p is a (t,σ)-ancestor of q, or $p \preceq_{(t,\sigma)} q$, to indicate that p is an ancestor of q by the derivation (t, σ). The set of (t, σ)-ancestors of a position q is denoted by $anc_{(t,\sigma)}(q)$

We notice that if $\mathcal{R}$ is a left-linear TRS, then the (t, σ)-ancestor relation is actually a function, i.e. given $t \to^\sigma t'$, and $q \in Pos(t')$ then $anc_{(t,\sigma)}(q)$ has a unique element which we denote also by $anc_{(t,\sigma)}(q)$. This is not the case for non left-linear TRS; moreover $anc_{(t,\sigma)}(q)$ may have comparable elements, for instance:

Example 1. Given the TRS $R := \{c(a, x) \to d(b, x), d(x, x) \to e(x)\}$. Given the derivation $t_0 = c(a, b) \to^{\delta_1} t_1 = d(b, b) \to^{\delta_2} t_2 = e(b)$, then $anc_{(t_1,\delta_2)}(0) = \{0, 1\}$ and $anc_{(t_0,\delta_1\cdot\delta_2)}(0) = \{\epsilon, 1\}$.

The ancestor relation satisfies the following comparability property :

Lemma 1. *Given a derivation $t \to^\sigma t'$ and $q, q' \in Pos(t')$ with $q \leq q'$, then*

$$p \preceq_{(t,\sigma)} q \Rightarrow p \leq anc_{(t,\sigma)}(q') \quad \text{and} \quad p' \preceq_{(t,\sigma)} q' \Rightarrow anc_{(t,\sigma)}(q) \leq p' \ .$$

Proof. By induction on the length of the derivation σ. Since the ancestor relation is the reflexive transitive closure of the ancestor relation by a reduction it is sufficient to prove the lemma for one reduction. Suppose that $\rho = (l \to r)$ and that $t \to^{\rho,\bar{p}} t'$. We proceed by cases:

– if $\bar{p} \not< anc_{(t,\sigma)}(q)$ then there is a position p such that $anc_{(t,\sigma)}(q) = \{p\}$ and $p \leq p'$ for each position $p' \in anc_{(t,\sigma)}(q')$.
– if $\bar{p} < anc_{(t,\sigma)}(q)$. Then $q = \bar{p} \cdot q_j \cdot v$ and $q' = \bar{p} \cdot q_j \cdot v'$ for some $q_j \in VPos(r)$ and $v < v'$. Therefore, $\bar{p} \cdot p_j \cdot v \in anc_{(t,\sigma)}(q)$ if and only if $\bar{p} \cdot p_j \cdot v' \in anc_{(t,\sigma)}(q')$.

A term t is a *potential redex* if there is a redex t' such that $t \rightarrow^* t'$. Given a term t and $p \in Pos(t)$, we say that $t|_p$ *is a potential redex of* t, (PR), or that p *is a potential redex position of* t, (PRP), if $t|_p$ is a potential redex. The ancestor relation backward preserves potential redex positions; indeed:

Lemma 2. *Given a derivation $t \rightarrow^\sigma t'$, if $t'|_q$ is a potential redex and p is a (t,σ)-ancestor of q, then $t|_p$ is a potential redex.*

Proof. As in the proof of lemma 1 it is sufficient to prove the lemma for one reduction. Suppose that $\rho = (l \rightarrow r)$ and that $t \rightarrow^{\rho,p'} t'$. Given p a $(t,(\rho,p'))$-ancestor of q, we proceed by case:

– if p and p' are parallel then $q = p$ and $t|_q = t'|_q$,
– if $p = p'$, we have already that $t|_p$ is a redex,
– if $p < p'$ then $p = q$ and $t|_q \rightarrow^{\sigma,(p'-q)} t'|_q$,
– if $p' < p$ then $t|_p = t'|_q$.

Given a derivation $t \rightarrow^\sigma t'$ and positions $p \in Pos(t)$, $q \in Pos(t')$. We say that the subterm $t'|_q$ *σ-stems* from the subterm $t|_p$, if there is a position p' such that p' is a (t,σ)-ancestor of q and $p \leq p'$; we say also that the position q (t,σ)-stems from the position q and write $p \vdash_{(t,\sigma)} q$.

The stemming relation satisfies right-monotonicity:

Lemma 3. *Given a derivation $t \rightarrow^\sigma t'$ and positions $p \in Pos(t)$, $q, q' \in Pos(t')$. If $p \vdash_{(t,\sigma)} q$ and $q < q'$ then $p \vdash_{(t,\sigma)} q'$.*

Proof. Since $p \vdash_{(t,\sigma)} q$ there is $p' \in Pos(t)$ such that $p \leq p' \preceq_{(t,\sigma)} q$. By lemma 1, and since $q < q'$, there is $p'' \in Pos(t)$ such that $p' \leq p''$ and $p'' \preceq_{(t,\sigma)} q'$.

Definition 2. *Given a derivation $t \rightarrow^\sigma t'$, $p \in Pos(t)$ and $q \in Pos(t')$. We say that $t'|_q$ is* an outermost potential redex (OPR) of t' (t,σ)-stemming from $t|_p$ *if q is a minimal potential redex position (t,σ)-stemming from p.*

Lemma 4. *Given a derivation $t \rightarrow^\sigma t'$ and $p \in Pos(t)$; if $t'|_q$ is an OPR (t,σ)-stemming from $t|_p$ and $r < q$ such that $t'|_r$ is a potential redex then there is $s < p$ such that s is a (t,σ)-ancestor of r.*

Proof. Since $t'|_q$ (t,σ)-stems from $t|_p$ there is a position p' such that $p \leq p'$ and such that p' is a (t,σ)-ancestor of q. By lemma 1 there is s such that $s \leq p'$ and such that s is a (t,σ)-ancestor of r. By hypothesis $t'|_r$ is a potential redex and $t'|_q$ is an OPR (t,σ)-stemming from $t|_p$; therefore, $p \not\leq s$ and hence $s < p'$.

Lemma 5. *Given a left-linear TRS $\mathcal{R}$. Given a derivation $t \rightarrow^\sigma t' \rightarrow^\tau t''$ and $p \in Pos(t)$. If $t'|_q$ is an OPR (t,σ)-stemming from $t|_p$ and $t''|_r$ is an OPR. (t',τ)-stemming from $t'|_q$ then $t''|_r$ is an OPR $(t,\sigma\cdot\tau)$-stemming from $t|_p$.*

Proof. Let s be a potential redex position of t'' such that $s < r$. By the previous lemma and left-linearity, $q' := anc_{(t',\tau)}(s) < q$. By lemma (2) q' is a potential redex position and by the previous lemma and left-linearity again $p' := anc_{(t',\sigma)}(q') < p$. Hence, by left-linearity, $p' := anc_{(t,\sigma\cdot\tau)}(s)$.

The following example shows that the left-linearity is a necessary hypothesis in the previous lemma.

Example 2. Given the (overlay) TRS

$$R := \{a(x) \to b(s(x), x), d \to e, d \to s(e), b(x, x) \to c(x), e \to f, s(x) \to x\}.$$

Given the derivation $t \to^{\sigma} t' \to^{\tau} t''$ where

$$t = a(d) \to b(s(d), d) \to b(s(e), d) \to t' = b(s(e), s(e)) \to t'' = c(s(e)) \ .$$

Then $t'|_{00}$ is an OPR (t, σ)-stemming from $t|_0$ and $t''|_{00}$ is an OPR (t', τ)-stemming from $t'|_{00}$ but $t''|_0$ is the OPR $(t, \sigma\cdot\tau)$-stemming from $t|_0$.

Given a derivation $t \to^{\sigma} t'$ and a set of parallel positions $P \subseteq Pos(t)$. A position, $q \in Pos(t')$, (t, σ)*-stems from* P if there is some $p \in P$ such that the position q (t, σ)-stems from p. We abbreviate the sentence outermost potential redex positions by OPRP. General TRS satisfy a weaker version of lemma (5):

Lemma 6. *Given a TRS $\mathcal{R}$. Given a derivation $t \to^{\sigma} t' \to^{\tau} t''$ and $P \subseteq Pos(t)$ a set of parallel positions. Given $Q \subseteq Pos(t')$ the set of OPRP (t, σ)-stemming from P and R the set of OPRP (t', τ)-stemming from Q then R is the set of OPRP $(t, \sigma \cdot \tau)$-stemming from P.*

Proof. By lemma (3), it is clear that each element of R $(t, \sigma\cdot\tau)$-stems from P. Given $r \in Pos(t'')$ a potential redex position such that r is $(t, \sigma\cdot\tau)$-stemming from P. There are positions p and q such that $P \leq p \preceq_{t,\sigma} q \preceq_{t',\tau} r$. By lemma (2), the positions p and q are potential redex positions. Since Q is the OPRP (t, σ)-stemming from P necessarily $Q \leq q$ and since R is the OPRP (t', τ)-stemming from Q necessarily $R \leq r$. Therefore, R is the set of OPRP $(t, \sigma\cdot\tau)$-stemming from P.

3.2 Context-Free Potential Redexes

A term t is *independent with respect to a context* $C[\Omega]_p$ if for any rewrite rule $\rho = (l \to r)$ and position $q \leq p$ such that $C[t]_p \to^{\rho,q} t'$, there is $q_1 \in VPos(l)$ such that $q\cdot q_1 \leq p$; roughly speaking each rewrite of $C[t]_p$ occurs in the context $C[\Omega]_p$ or in t. Given a term t and a position $p \in Pos(t)$, such that $t|_p$ is independent with respect to $t|^p$, then $t|_p$ is called an *independent subterm of* t.

Definition 3. *A potential redex t is* context-free with respect to a context $C[\Omega]_p$ *– or a $C[\Omega]_p$*-free potential redex *– if for any derivation $C[t]_p \to^{\sigma} t'$, the outermost potential redexes of t' σ-stemming from t are independent subterms of t'. A potential redex t is* context-free *if it is context-free with respect to any context. Given a term t and $p \in Pos(t)$, we say that the subterm $t|_p$ is* context-free in t *if it is context-free with respect to $t|^p$.*

Given a term t and a set of parallel positions $P \subseteq Pos(t)$. The set of subterms $t|_P$ is context free in t if for any derivation $t \to^{\sigma} t'$, and for any q outermost potential redex position (t, σ)-stemming from P, then $t'|_q$ is an independent subterm of t'.

Outermost potential redexes which are subterms of a term t are context-free subterms of t. In fact, if P is the set of outermost potential redex positions in t then the context $t|^P$ is "invariant", i.e. if $t \to^{\sigma} t' \to^{(\rho,p)}$, is a derivation then $P \leq p$.

Example 3. Given the overlay TRS defined in example (2) and the term $t = a(d)$. The subterm $t|_0$ is a context-free redex in t.

In fact we have:

Proposition 1. *For any term rewrite system $\mathcal{R}$ the following statements are equivalent:*

1. *$\mathcal{R}$ is overlay.*
2. *Every potential redex is independent w.r.t. any context.*
3. *Every potential redex is context-free.*
4. *Every redex is independent w.r.t. any context.*

Proof. **1 implies 2** Suppose that $\mathcal{R}$ is overlay. Given a term t and a redex u such that t is a non independent subterm of u. Let $l \to r$ be a rule such that $u \to^{l \to r, \epsilon} v$ and such that there is a position p satisfying $\epsilon < p < VPos(l)$ and $t = u|_p$. If $(t, \langle(\rho_i, p_i)\rangle_i)$ is a derivation then $(u, \langle(\rho_i, p \cdot p_i)\rangle_i)$ is a derivation and, since $\mathcal{R}$ is overlay, for each i, $VPos(l) \leq p \cdot p_i$, therefore $p_i \neq \epsilon$, and so t is not a potential redex.

2 implies 3 and 3 implies 4 By definitions.

4 implies 1 Suppose that every redex is independent. Given a redex t and a rule $l \to r$ such that $t \to^{l \to r, \epsilon} u$; given $p \in Pos(t) \backslash \{\epsilon\}$ such that $t|_p$ is a redex, necessarily $VPos(l) \leq p$; hence $\mathcal{R}$ is overlay.

Corollary 1. *Given a strict constructor overlay TRS $\mathcal{R}$, the context-free potential redexes are the terms $d(u_0, \ldots, u_n)$ where d is a defined function symbol.*

The following is a key proposition which allows to give recursive bounds on the lengths of derivations:

Proposition 2. *Given a term t and a set of parallel potential redex positions $P \subseteq Pos(t)$ such that the set of subterms $t|_P$ is context-free in t. Let $t \to^{\sigma} t'$ be a derivation. Let $Q \subseteq Pos(t')$ be the set of OPRP (t, σ)-stemming from P, then $t'|_Q$ is context-free in t'.*

Proof. Given a derivation $t \to^{\sigma} t' \to^{\tau} t''$. Let $R \subseteq Pos(t'')$ be the set of OPRP (t', τ)-stemming from Q. By lemma (6), R is the set of OPRP $(t, \sigma \cdot \tau)$-stemming from P and hence for each $r \in R$, the subterm $t''|_r$ is independent in t''. Therefore, the set of subterms $t'|_Q$ is context-free in t'.

4 Recursive Bounded Confluence Theorem

The derivational height of a term t, denoted by $\lambda(t)$, is the supremum of the lengths of derivations of t. Given a finite set X, we denote by $|X|$ the cardinality of X. *The multiplicity* of a rewrite rule, $\rho = (l \to r)$, denoted by $m(\rho)$, is defined by

$$m(\rho) = max\{\frac{|Pos(r,x)|}{|Pos(l,x)|} \; : \; x \in var(l)\} \, .$$

The multiplicity of a term t is given by:

$$m(t) = \sup\{m(\rho) \in \mathbb{Q} \; : \; \exists u, u' \in T(F), \; \exists \rho \in R, \; \exists p \in Pos(u), \; t \to^* u \to^{\rho,p} u'\} \, .$$

A term t is *Bounded Confluent*, (BC), if $\lambda(t)$ and $m(t)$ are finite and if t has a unique normal form. Given a term t and a set of parallel positions P, if the terms of $t|_P$ are BC, we say that $t|_P$ is BC.

Given $P := p_1, \ldots, p_m$ a sequence of parallel positions of a term t. If $t|_P$ is BC we use the following notations: $\downarrow (t|_P)$ is the sequence $\downarrow (t|_{p_1}), \ldots, \downarrow (t|_{p_m})$, $m(t|_P) := max\{m(t_{p_1}), \ldots, m(t_{p_m}\}$, $\lambda(t|_P) := \sum_{j=1}^{j=m} \lambda(t_{p_j})$ and $\downarrow_P (t) := t|^P[\downarrow (t|_P)]_P$.

Remark 1. We notice that if u is a BC potential redex and $u \to^* u'$, if Q' is the set of OPRP of u' then $u'|_{Q'}$ is BC, $\lambda(u'|_{Q'}) = \lambda(u')$ and $\downarrow (u) = \downarrow_{Q'} (u')$.

Given a set of parallel positions P, a derivation of t, $\sigma = \langle(\rho_j, q_j)\rangle_{0 \leq j \leq k}$, is said to be a derivation of $t|_P$ if for each j, $P \leq q_j$. We notice that in this case, for each $p \in P$, the subsequence $\{(\rho_j, q_j) \; : \; 0 \leq j \leq k \text{ and } p \leq q_j\}$ is a derivation of $t|p$.

The techniques used in the next lemma are of the same vein as those developed in [7] and [9].

Lemma 7. *Given a term t and a sequence P of parallel potential redex positions such that $t|_P$ is BC and context-free in t and $\downarrow_P (t)$ is BC. Let $t \to^{(\rho,p)} t'$ be a reduction and Q be the OPRP $(t, (\rho, p))$-stemming from P. Then, $t'|_Q$ and $\downarrow_Q (t')$ are BC.*

Moreover, one of the following cases occurs:

- *If $P \leq p$ then $\lambda(t'|_Q) < \lambda(t|_P)$ and $\lambda(\downarrow_Q (t')) = \lambda(\downarrow_P (t))$.*
- *If $P \not\leq p$ then $\lambda(t'|_Q) \leq m(\rho) \cdot \lambda(t|_P)$ and $\lambda(\downarrow_Q (t')) < \lambda(\downarrow_P (t))$.*

Proof. Since $t|_P$ is independent in t the reduction occurs either in $t|_P$ or in $t|^P$. We study each case.

Reduction of a context-free potential redex: $P \leq p$. There is $\bar{p} \in P$ such that $\bar{p} \leq p$. Let Q' be the set of OPR of $t'|_{\bar{p}}$. By remark (1), and since $t|_{\bar{p}}$ is BC, we know that $t'|_{\bar{p} \cdot Q'}$ is BC and $\downarrow (t|_{\bar{p}}) = t'|_{\bar{p}}[\downarrow (t'|_{Q'})]_{Q'}$. Therefore, since $Q = P \backslash \{\bar{p}\} \cup \bar{p} \cdot Q'$, we conclude that $t'|_Q$ is BC and $\downarrow_Q (t') = \downarrow_P (t)$ is BC. Hence, $\lambda(\downarrow_Q (t')) = \lambda(\downarrow_P (t))$. Since for each derivation σ of $t'|_Q$, $(\rho, p) \cdot \sigma$ is a derivation of $t|_P$, we have $\lambda(t'|_Q) < \lambda(t|_P)$.

Reduction of a context: $P \not\leq p$. Since $t|_P$ is independent in t we have $t|^P \rightarrow^{(\rho,p)} t'|^Q$. Therefore, for each $q \in Q$, if $\bar{p} \preceq_{(t,(\rho,p))} q$ then $\bar{p} \in P$ and $t|_{\bar{p}} = t'|_q$. Hence, $t'|_Q$ is BC and $\lambda(t'|_Q) \leq m(\rho) \cdot \lambda(t|_P)$. Moreover, since $\downarrow_P t \rightarrow^{(\rho,p)} \downarrow_Q t'$, $\downarrow_Q (t') = \downarrow_P (t)$ is BC and $\lambda(\downarrow_Q t') < \lambda(\downarrow_P t)$.

Given a set of derivations Δ, a derivation $\sigma \in \Delta$ is called a *worst* derivation in Δ if the length of σ is greater or equal than the length of any derivation belonging to Δ. For instance, given a term t, i) if t is BC then the length of a worst derivation of t is $\lambda(t)$ and ii) if P is a set of parallel positions such that $t|_P$ is BC, then the length of a a worst derivation of $t|_P$ is $\lambda(t|_P)$.

If $t|_P$ and $\downarrow_P (t)$ are BC the notation

$$t \xrightarrow[\lambda(t|_P)]{\text{w.d. of } t|_P} \downarrow_P (t) \xrightarrow[\lambda(\downarrow_P(t))]{\text{w.d.}} \downarrow (\downarrow_P (t))$$

means that first we perform a worst derivation of $t|_P$, whose length is $\lambda(t|_P)$ and then we perform a worst derivation of $\downarrow_P (t)$, whose length is $\lambda(\downarrow_P (t))$. Using this notation, we can rephrase partially the previous lemma as follows:

Lemma 8. *Under the same hypothesis of lemma (7), one of the following cases occurs:*

Reduction of a context-free potential redex: $P \leq p$. *Then:*

$$\begin{array}{ccccc}
t & \xrightarrow[\lambda(t|_P)]{w.d.\ of\ t|_P} & \downarrow_P (t) & \xrightarrow[\lambda(\downarrow_P(t))]{w.d.} & \downarrow (\downarrow_P (t)) \\
{\scriptstyle (\rho,p)}\downarrow & & \downarrow{\scriptstyle \emptyset} & & \downarrow{\scriptstyle \emptyset} \\
t' & \xrightarrow[\lambda(t'|_Q)]{w.d.\ of\ t'|_Q} & \downarrow_Q (t') & \xrightarrow[\lambda(\downarrow_Q(t'))]{w.d.} & \downarrow (\downarrow_Q (t')
\end{array}$$

with $\lambda(t'|_Q) < \lambda(t|_P)$ *and* $\lambda(\downarrow_Q (t')) = \lambda(\downarrow_P (t))$.

Reduction of a context: $P \not\leq p$. *Then:*

$$\begin{array}{ccccc}
t & \xrightarrow[\lambda(t|_P)]{w.d.\ of\ t|_P} & \downarrow_P (t) & \xrightarrow[\lambda(\downarrow_P(t))]{w.d.} & \downarrow (\downarrow_P (t)) \\
{\scriptstyle (\rho,p)}\downarrow & & {\scriptstyle (\rho,p)}\downarrow & & \downarrow{\scriptstyle \emptyset} \\
t' & \xrightarrow[\lambda(t'|_Q)]{w.d.\ of\ t'|_Q} & \downarrow_Q (t') & \xrightarrow[\lambda(\downarrow_Q(t'))]{w.d.} & \downarrow (\downarrow_Q (t')
\end{array}$$

with $\lambda(t'|_Q) \leq m(\rho) \cdot \lambda(t|_P)$ *and* $\lambda(\downarrow_Q (t')) < \lambda(\downarrow_P (t))$.

Proposition 3. *Given a term t and a set of parallel potential redex positions $P \subseteq Pos(t)$ such that the set of subterms $t|_P$ is BC and context-free in t. Let $t \rightarrow^\sigma t'$ be a derivation. Let $Q \subseteq Pos(t')$ be the set of OPRP (t,σ)-stemming from P. Then $t'|_Q$ is BC and context-free in t', and $\downarrow_Q (t')$ is BC. Also, the following statements are satisfied.*

1. *There is a subsequence σ_1 of σ such that $\downarrow_P t \rightarrow^{\sigma_1} \downarrow_Q t'$.*
2. *For each $q \in Q$ there is a position $p \in P$ and a derivation $(t|_p, \sigma_q)$ such that $t|_p \rightarrow^{\sigma_q} u$ and $t'|_q$ is an outermost potential redex of u.*

Proof. Let t be a term and P a set of positions satisfying the hypothesis of the proposition. Given a derivation $t \rightarrow^{\sigma} t' \rightarrow^{(\rho,\bar{q})} t''$. Let $Q \subseteq Pos(t')$, be the set of OPRP (t, σ)-stemming from P. Let $R \subseteq Pos(t'')$, be the set of OPRP $(t, \sigma \cdot (\rho, \bar{q}))$-stemming from P. Suppose that Q satisfies the conclusions of the proposition and lets prove that R satisfies them also.

By lemma (6), R is the set of OPRP $(t', (\rho, \bar{q}))$-stemming from Q. Hence, by proposition (2) and lemma (7) we know that $t''|_R$ is BC and context-free in t'', and $\downarrow_R (t'')$ is BC.

We prove the statements of the proposition by case.

If $Q \leq \bar{q}$. Then, as seen in the proof of lemma (7), we have $\downarrow_Q (t') = \downarrow_R (t'')$ and hence $\downarrow_P t \rightarrow^{\sigma_1} \downarrow_Q t' = \downarrow_R (t'')$. So, statement 1 is satisfied in this case. Let be $q \in Q$ such that $q \leq \bar{q}$. For any $r \in R$ such that r and q are incomparable we have $t'|_r = t''|_r$; hence the statement 2 is satisfied for $t''|_r$. If $r \in R$ and $q \leq r$. Then, by hypothesis, there is a position $p \in P$ and a derivation $(t|_p, \sigma_q)$ such that $t|_p \rightarrow^{\sigma_q} u$ and $t'|_q$ is an OPR of u. But, as seen in the proof of lemma (7), we have $t'|_q \rightarrow^{(\rho, \bar{q}-q)} t''|_q$ and $t''|_r$ is an OPR of $t''|_q$. Therefore, if $t'|_q = u|_s$ then $t|_p \rightarrow^{\sigma_q} u \rightarrow^{(\rho, s \cdot \bar{q}-q)} v$ and $t''|_r$ is an OPR of v.

If $Q \not\leq \bar{q}$. By hypothesis there is a subsequence σ_1 of σ such that $\downarrow_P t \rightarrow^{\sigma_1} \downarrow_Q t'$. But, as seen in the proof of lemma (7), we have $\downarrow_Q t' \rightarrow^{(\rho,\bar{q})} \downarrow_R t''$. So, statement 1 is satisfied in this case. As seen in the proof of lemma (7), for each $r \in R$ there is $q \in Q$ such that $t'|_q = t''|_r$ and hence statement 2 is satisfied in this case.

Remark 2. Given a sequence of natural numbers $\langle (m_i, n_i) \rangle_{0 \leq i \leq k}$. If there is a constant c such that for each i either $m_{i+1} < m_i$ and $n_{i+1} = n_i$ or either $m_{i+1} \leq c \cdot m_i$ and $n_{i+1} < n_i$, then the inequality $k \leq n_0 + c^{n_0} \cdot m_0$ holds.

Theorem 1 (Recursive bounded confluence). *Given a term t and a sequence P of parallel positions of t such that $t|_P$ is BC, context-free in t, and $\downarrow_P (t)$ is BC, then t is BC, and the following inequalities hold:*

$$\lambda(t) \leq \lambda(\downarrow_P (t)) + m(\downarrow_P (t))^{\lambda(\downarrow_P (t))} \times \lambda(t|_P)$$

and

$$m(t) \leq max\{m(t|_P), m(\downarrow_P (t))\} .$$

Proof. Let t be a term and P a set of positions satisfying the hypothesis of the proposition. Given a derivation $t \rightarrow^{\sigma} t' \rightarrow^{(\rho,\bar{q})} t''$. Let $Q \subseteq Pos(t')$, be the set of OPRP (t, σ)-stemming from P. Let $R \subseteq Pos(t'')$, be the set of OPRP $(t, \sigma \cdot (\rho, \bar{q}))$-stemming from P. By lemmas (6) and (7) and by proposition (3) we conclude that:

- If $q \leq \bar{q}$ for some $q \in Q$ then $\lambda(t''|_R) < \lambda(t'|_Q)$ and $\lambda(\downarrow_Q (t')) = \lambda(\downarrow_R (t''))$. Moreover, there is a position $p \in P$ and a derivation $(t|_p, \sigma_q)$ such that $t|_p \rightarrow^{\sigma_q} u \rightarrow^{(\rho, s \cdot \bar{q}-q)} v$, hence $m(\rho) \leq m(t|_P)$.

- If $q \not\leq \bar{q}$ for every $q \in Q$, then $\lambda(t''|_R) \leq m(\rho) \cdot \lambda(t'|_Q)$. Moreover, $\downarrow_{Q'} t' \rightarrow^{(\rho,p')} \downarrow_{Q''} t''$, hence $m(\rho) \leq m(\downarrow_P t)$ and $\lambda(\downarrow_R t'') < \lambda(\downarrow_Q t')$.

Using remark (2) we conclude that the inequalities of the theorem hold.

Example 4. Given the TRS

$$\mathcal{R} = \{a \rightarrow b, b \rightarrow e, e \rightarrow f, c(x,0) \rightarrow d(x,x), c(x,s(y)) \rightarrow c(d(x,x),y)\} .$$

We write n instead of $s(s(....(s(0))))$, where s is repeated n times. Given the term $t := (c(a,n))$ then the subterm $t|_0 = a$ is BC and context-free potential redex in t. We have $\lambda(a) = 3$, $\downarrow (a) = f$, $\lambda(c(f,n)) = n+1$, $m(c(f,n)) = 2$ and $\lambda(t) = (n+1) + 2^{n+1} \cdot 3$. Hence the theorem gives the sharpest possible bound.

5 Complexity of Bounded Confluent Overlay TRS

The recursive BC theorem allows us to conclude:

Corollary 2. *Given an overlay TRS. Let* $t = f(u_0, \ldots, u_k)$ *be a term such that* $u_0, \ldots u_k$ *are BC and* $\downarrow_i (t) := f(\downarrow (u_0), \ldots, \downarrow (u_k))$ *is BC then* t *is BC and the following inequalities hold:*

$$\lambda(t) \leq \lambda(\downarrow_i (t)) + m(\downarrow_i (t))^{\lambda(\downarrow_i(t))} \times (\lambda(u_0) + \cdots + \lambda(u_k))$$

and

$$m(t) \leq max\{m(\downarrow_i (t)), m(u_o), \ldots m(u_k)\} .$$

Proof. By proposition (1) every potential redex is context free. Given

$$P := \{k \cdot q \;:\; 0 \leq k \leq m \text{ and } q \text{ is a position of an OPR of } u_k\} ,$$

the equality $\downarrow_i (t) = \downarrow_P (t)$ holds and we can apply theorem (1).

This corollary has been proved by other means and applied in [3] to obtain bounds for the lengths of derivations of primitive recursive terms by primitive recursive rewriting schemes. We study further in this section bounds for the derivational length of bounded confluent overlay TRS.

A term is an *inner redex* if every proper subterm is irreducible. An *innermost derivation* is a derivation where only inner redex subterms are reduced at each step. A TRS is *innermost terminating* if every innermost derivation has a finite length. A TRS is *finitely branching* if for every redex t there is a finite number of rules $l \rightarrow r$ such that $t \rightarrow^{(l \rightarrow r, \epsilon)} t'$. Given a finitely branching innermost terminating TRS, we denote by $\lambda_i(t)$ the length of a longest innermost derivation of t. We know by a result of [8] which extends a result of [10] that any locally confluent overlay innermost-terminating TRS is terminating. The previous corollary allows us to prove a weaker version of Gramlich's theorem:

Proposition 4. *Every locally confluent overlay finitely branching innermost terminating TRS is bounded confluent.*

Proof. Let $\mathcal{R}$ be a finitely branching and locally confluent TRS. The *size* of a term t, denoted by $|t|$, is the cardinality of $Pos(t)$. We prove by induction on $(\lambda_i(t), |t|)$, lexicographically ordered, that each term is BC. Suppose that for every term u such that $(\lambda_i(u), |u|) < (\lambda_i(t), |t|)$, the term u is BC. If t is an inner redex then for every rule $l \to r$ if $t \to^{(l \to r, \epsilon)} t'$, we have $\lambda_i(t') < \lambda_i(t)$, so t' is BC; therefore, since $\mathcal{R}$ is a finitely innermost branching, $\lambda(t)$ and $m(t)$ are finite, and since $\mathcal{R}$ is locally confluent then t has a unique normal form (by Newmann's lemma). If t is not an inner redex, then $\lambda_i(\downarrow_i(t)) < \lambda_i(t)$ and, for each subterm u_l, we have $\lambda_i(u_l) \leq \lambda_i(t)$; therefore, $\downarrow_i(t)$ is BC and each subterm u_k is BC, so by corollary (2) the term t is BC.

A *uniformly bounded confluent*, (UBC), TRS is a bounded confluent TRS such that there are functions $\mu \in \mathbb{N}^{\mathbb{N}}$ and $\varphi \in \mathbb{N}^{\mathbb{N}}$ such that for every term t we have $m(t) \leq \mu(|t|)$ and $\lambda(t) \leq \varphi(|t|)$. The functions μ and φ are called respectively the *multiplicity bound function* and the *derivational-length bound function.* The *inner redexes derivational-length bound function* is a function $\varphi_i \in \mathbb{N}^{\mathbb{N}}$ such that for every inner redex t we have $\lambda(t) \leq \varphi_i(|t|)$. A function, ϕ, from $\mathbb{N}$ to $\mathbb{N}$, is a *size function,* if for each reduction $t \to^{(\rho,p)} t'$, $|t'| \leq \phi(t)$. We notice that if $\mathcal{R}$ is UBC then $\phi(n) = n \cdot \mu(n)$ is a size function.

Given f a function from $\mathbb{N}$ to $\mathbb{N}$, we denote: (i) $f^{(0)} = id$, where id is the identity function of $\mathbb{N}$, (ii) $f^{(n+1)} = f \circ f^{(n)}$, where $\circ$ is the composition of functions and (iii) The *iteration operator It* is defined by: $It(f)(x) = f^{(x)}(x)$. We stress the fact that f^n is the exponential function and is not the function $f^{(n)}$.

A function, $f \in \mathbb{N}^{\mathbb{N}}$, is *expansive* if $f(n) + f(m) \leq f(n+m)$ for each $n, m \in \mathbb{N}$. We notice that if f and g are expansive, then $f + g$, $f \cdot g$, $f \circ g$ and $It(f)$ are expansive; moreover, for each $n \in \mathbb{N}$, $n \leq f(n)$. The symbol S will denote the successor function.

Proposition 5. *Let R be a uniformly bounded confluent overlay TRS. Let μ be an expansive multiplicity function, let φ_i be an expansive inner redexes derivational-length bound function, and let $\Gamma \in \mathbb{N}^{\mathbb{N}}$ be the function defined by:*

$$\Gamma(x) := \varphi_i(x) + x \cdot (\mu(x))^{\varphi_i(x)} .$$

Let ϕ be an expansive size function. Then the function:

$$\varphi := It\,(\Gamma \circ S \circ It(\phi)) ,$$

is an expansive derivational-length bound function.

Proof. The function φ is expansive, since it is built with expansive functions using operations which preserve expansivity. We prove by lexicographic induction on $(\lambda(t), |t|)$ that for each term t the inequality $\lambda(t) \leq \varphi(|t|)$ holds.

- If t is an irreducible term then $\lambda(t) = 0 \leq \varphi(|t|)$.
- If t is an inner redex then $\lambda(t) \leq \varphi_i(|t|) \leq \varphi(|t|)$.

– We use the notation of corollary (2). If $t = f(u_0, \dots, u_k)$ is a reducible term but is not an inner redex, then $\lambda(\downarrow_i (t)) < \lambda(t)$ and for each subterm u_l, we have $\lambda(u_l) \le \lambda(t)$; therefore we can apply the inductive hypothesis. By corollary (2), and since $\downarrow_i (t)$ is an inner redex; we have:

$$\begin{aligned}\lambda(t) &\le \lambda(\downarrow_i (t)) + (m(\downarrow_i (t)))^{\lambda(\downarrow_i(t))} \cdot (\lambda(u_0) + \dots + \lambda(u_k)) \\ &\le \varphi_i(|\downarrow_i (t)|) + \mu(|\downarrow_i (t))|)^{\varphi_i(|\downarrow_i(t)|)} \cdot (\varphi(|u_0|) + \dots + \varphi(|u_k|)) \ .\end{aligned}$$

For each immediate subterm u_j of t, we have by induction hypothesis that:

$$|\downarrow (u_j)| \le \phi^{(\lambda(u_j))}(|u_j|) \le It(\phi)\,(\varphi\,(|u_j|)) \ .$$

Applying the expansivity properties

$$\begin{aligned}|\downarrow_i (t)| &\le 1 + \sum_{j=0}^{j=k} (It(\phi) \circ \varphi)\,(|u_j|) \\ &\le 1 + (It(\phi) \circ \varphi)\left(\sum_{j=0}^{j=k} |u_j|\right) \\ &= (S \circ It(\phi) \circ \varphi)\left(\sum_{j=0}^{j=k} |u_j|\right)\end{aligned}$$

On the other hand:

$$\begin{aligned}\sum_{j=0}^{j=k} \lambda(u_j) &\le \sum_{j=0}^{j=k} \varphi\,(|u_j|) \\ &\le \varphi\left(\sum_{j=0}^{j=k} |u_j|\right) \\ &\le (S \circ It(\phi) \circ \varphi)\left(\sum_{j=0}^{j=k} |u_j|\right)\end{aligned}$$

Thus :

$$\begin{aligned}\lambda(t) &\le (\Gamma \circ S \circ It(\phi) \circ \varphi)\left(\sum_{j=0}^{j=k} |u_j|\right) \\ &\overset{IH}{=} (\Gamma \circ S \circ It(\phi))\left((\Gamma \circ S \circ It(\phi))^{\left(\sum_{j=0}^{j=k} |u_j|\right)}\left(\sum_{j=0}^{j=k} |u_j|\right)\right) \\ &= (\Gamma \circ S \circ It(\phi))^{\left(1+\sum_{j=0}^{j=k} |u_j|\right)}\left(\sum_{j=0}^{j=k} |u_j|\right)\end{aligned}$$

$$\leq (\Gamma \circ S \circ It(\phi))^{\left(1+\sum_{j=0}^{j=k}|u_j|\right)} \left(1 + \sum_{j=0}^{j=k} |u_j|\right)$$
$$= \varphi(|t|) \ .$$

6 Conclusion

We studied how to bound recursively the lengths of derivations of bounded confluent terms for any TRS. We saw how this result implies a known theorem relating innermost termination and strong termination for finitely branching locally confluent overlay TRS. We think that the analysis given in this paper could bring some insight to the study of strategy defined derivations, for instance perpetual derivations developed by Nederpelt, Khasidashvili and others more recently.

References

1. F. Baader and T. Nipkow *Term Rewriting and All That.* Cambridge University Press (1998).
2. G. Bonfante and A. Cichon and J-Y Marion and H. Touzet *Complexity classes and rewrite systems with polynomial interpretation,* Lecture Notes in Computer Science, **1584** (1999) 372–384
3. E.A. Cichon, E. Tahhan. *Strictly Orthogonal Left-Linear Rewrite Systems and Primitive Recursion.* Annals of Pure and Applied Logic. **108** (2001) 79–102.
4. N. Dershowitz, J.P. Jouannaud. *Rewrite Systems.* Handbook of Theoretical Computer Science, Volume B, Elsevier, Amsterdam (1990) 243–320.
5. Dieter Hofbauer. *Termination proofs by multiset path orderings imply primitive recursive derivation lengths* Theoretical Computer Science **105** (1) (1992) 129–140.
6. G. Huet. *Confluent reductions: Abstract properties and applications to term rewriting systems.* J. Assoc. Comput. Mach. **27** (4) (1980) 797–821.
7. G. Huet, J.-J. Lévy. *Computations in orthogonal rewriting systems, I and II.* Computational Logic: Essays in Honor of Alan Robinson, Ed. J.-L. Lassez and G. Plotkin, MIT Press (1991) 395–443.
8. Bernhard Gramlich *Abstract Relations between Restricted Termination and Confluence Properties of Rewrite Systems.* Fundamenta Informaticae **24** (1995) 3–23.
9. Zurab Khasidashvili. *Perpetuality and Strong Normalization in Orthogonal Term Rewriting Systems.* Lecture Notes in Computer Science **775** (1994) 163-174.
10. M.J. O'Donnell. *Computing in Systems Described by Equations.* Lecture Notes in Computer Science **58** (1977).
11. A. Weiermann, *Termination proofs for term rewriting systems with lexicographic path orderings imply multiply recursive derivation lengths,* Theoretical Computer Science **139** (1995) 355-362.

Termination of (Canonical) Context-Sensitive Rewriting*

Salvador Lucas

DSIC, Universidad Politécnica de Valencia
Camino de Vera s/n, E-46022 Valencia, Spain
slucas@dsic.upv.es

Abstract. Context-sensitive rewriting (*CSR*) is a restriction of rewriting which forbids reductions on selected arguments of functions. A *replacement map* discriminates, for each symbol of the signature, the argument positions on which replacements are allowed. If the replacement restrictions are less restrictive than those expressed by the so-called *canonical* replacement map, then *CSR* can be used for computing (infinite) normal forms of terms. Termination of such *canonical CSR* is desirable when using *CSR* for these purposes. Existing transformations for proving termination of *CSR* fulfill a number of new properties when used for proving termination of canonical *CSR*.

Keywords: (infinitary) normalization, term rewriting, termination.

1 Introduction

A *replacement map* is a mapping $\mu : \mathcal{F} \to \mathcal{P}(\mathbb{N})$ satisfying $\mu(f) \subseteq \{1, \dots, k\}$, for each k-ary symbol f of the signature $\mathcal{F}$ [Luc98]. We use them to discriminate the argument positions on which replacements are allowed. In this way, we obtain a restriction of rewriting which we call *context-sensitive rewriting* (*CSR* [Luc98]). Terminating TRSs are μ-*terminating* (i.e., no term initiates an infinite sequence of *CSR* under μ). However, *CSR* can *achieve* termination, by pruning (all) infinite rewrite sequences. Several methods have been developed to formally prove μ-termination of TRSs [BLR02,FR99,GM99,GM02,Luc96,SX98,Zan97].

Example 1. Consider the following TRS $\mathcal{R}$:

```
sel(0,x:y)      → x               first(0,x)       → []
sel(s(x),y:z)   → sel(x,z)        first(s(x),y:z)  → y:first(x,z)
from(x)         → x:from(s(x))
```

Let $\mu(\mathtt{s}) = \mu(\mathtt{:}) = \mu(\mathtt{from}) = \{1\}$ and $\mu(\mathtt{sel}) = \mu(\mathtt{first}) = \{1,2\}$. The μ-termination of $\mathcal{R}$ can be proved by using Zantema's techniques [Zan97].

Up to now, termination of *CSR* has been studied disregarding other computational issues such as achieving some kind of completeness in computations. This

* Work partially supported by CICYT TIC2001-2705-C03-01, Acciones Integradas HI 2000-0161, HA 2001-0059, HU 2001-0019, and Generalitat Valenciana GV01-424.

S. Tison (Ed.): RTA 2002, LNCS 2378, pp. 296–310, 2002.

problem has been studied in [Luc98]. Since computing (infinite) normal forms is the essential task of most rewriting-based engines (interpreters of rewriting-based (lazy) programming languages, theorem provers, etc.), the application of *CSR* in practice requires undertaking these issues. In Section 3, we define *canonical CSR* as an specialization of *CSR* that only uses replacement maps which are less restrictive than the *canonical* replacement map $\mu_{\mathcal{R}}^{can}$ of the TRS $\mathcal{R}$. This is the least replacement map ensuring that all non-variable subterms of every left-hand side of the rules of $\mathcal{R}$ are replacing [Luc98]. Section 4 discusses the use of canonical *CSR* for computing normal forms and infinite normal forms of terms (with left-linear, possibly non-terminating TRSs). Termination of canonical *CSR* becomes essential for these tasks. Thus, for the purpose of computing (infinite) normal forms, proving termination of canonical *CSR* can be priorized over proofs of termination: a proof of termination of canonical *CSR* can be used to achieve normalizations (with possibly non-terminating, left-linear TRSs) and also to arbitrarily approximate infinite normal forms. In the remainder of the paper we try to analyze whether proving termination of canonical *CSR* for a *terminating* TRS is much more *difficult* than proving its termination.

In Section 5, we review the main (transformational) techniques for proving termination of *CSR* and prove a couple of new interesting properties. Most of them appear when these techniques apply for proving termination of canonical *CSR*. We prove that all transformations introduced by Giesl and Middeldorp [GM99,GM02] are *complete* for proving termination of canonical *CSR*. We also prove that the contractive transformation of [Luc96] is complete for proving termination of canonical *CSR* if the replacing variables of the right-hand sides of each rule of the TRS are also replacing in the left-hand side (conservativeness [Luc96]). We show that, in both cases, Zantema's transformation [Zan97] and Ferreira and Ribeiro's transformation [FR99] remain incomplete. In Section 6, we also analyze the (relative) behavior of the transformations when *simple* termination, rather than termination, is considered. Simple termination covers the use of most usual automatizable orderings for proving termination of rewriting [Der87]. We obtain a new hierarchy of the transformations which is helpful for guiding their practical use. We prove that the transformations for proving termination of *CSR* preserve (simple) termination. Thus, proving termination of canonical *CSR* does not seem to be essentially more difficult than proving termination. Section 7 provides a first experimental evidence supporting this claim.

2 Preliminaries

Throughout the paper, $\mathcal{X}$ denotes a countable set of variables and $\mathcal{F}$ denotes a signature, i.e., a set of function symbols $\{\mathtt{f}, \mathtt{g}, \ldots\}$, each having a fixed arity given by a mapping $ar : \mathcal{F} \to \mathbb{N}$. A TRS is a pair $\mathcal{R} = (\mathcal{F}, R)$ where R is a set of rewrite rules. Given TRSs $\mathcal{R} = (\mathcal{F}, R)$ and $\mathcal{R}' = (\mathcal{F}', R')$, we let $\mathcal{R} \cup \mathcal{R}'$ be the TRS $(\mathcal{F} \cup \mathcal{F}', R \cup R')$. $L(\mathcal{R})$ denotes the set of *lhs*'s of $\mathcal{R}$. Given a signature $\mathcal{F}$, we consider the left-linear TRS $\mathcal{E}mb(\mathcal{F}) = (\mathcal{F}, \{f(x_1, \ldots, x_k) \to x_i \mid f \in \mathcal{F}, i \in \{1, \ldots, ar(f)\}\})$. A TRS $\mathcal{R}$ is simply terminating if $\mathcal{R} \cup \mathcal{E}mb(\mathcal{F})$ is terminating.

3 (Canonical) Context-Sensitive Rewriting

A mapping $\mu : \mathcal{F} \to \mathcal{P}(\mathbb{N})$ is a *replacement map* (or $\mathcal{F}$-map) if $\forall f \in \mathcal{F},\ \mu(f) \subseteq \{1,\dots,ar(f)\}$ [Luc98]. The ordering $\sqsubseteq$ on $M_{\mathcal{F}}$, the set of all $\mathcal{F}$-maps, is: $\mu \sqsubseteq \mu'$ if for all $f \in \mathcal{F}$, $\mu(f) \subseteq \mu'(f)$. Thus, $\mu \sqsubseteq \mu'$ means that μ considers less positions than μ' (for reduction), i.e., μ is more restrictive than μ'. According to $\sqsubseteq$, $\mu_\bot$ (resp. $\mu_\top$) given by $\mu_\bot(f) = \varnothing$ (resp. $\mu_\top(f) = \{1,\dots,ar(f)\}$) for all $f \in \mathcal{F}$, is the minimum (maximum) of $M_{\mathcal{F}}$. Given a TRS $\mathcal{R} = (\mathcal{F}, R)$, we write $M_{\mathcal{R}}$ rather than $M_{\mathcal{F}}$. The set of *μ-replacing positions* $\mathcal{P}os^\mu(t)$ of $t \in \mathcal{T}(\mathcal{F},\mathcal{X})$ is: $\mathcal{P}os^\mu(t) = \{\Lambda\}$, if $t \in \mathcal{X}$ and $\mathcal{P}os^\mu(t) = \{\Lambda\} \cup \bigcup_{i \in \mu(root(t))} i.\mathcal{P}os^\mu(t|_i)$, if $t \notin \mathcal{X}$. The set of *replacing* variables $\mathcal{V}ar^\mu(t)$ of t is $\mathcal{V}ar^\mu(t) = \{x \in \mathcal{V}ar(t) \mid \mathcal{P}os_x(t) \cap \mathcal{P}os^\mu(t) \neq \varnothing\}$. In *context-sensitive rewriting* (*CSR* [Luc98]), we (only) rewrite *replacing* redexes: t μ-rewrites to s, written $t \hookrightarrow_\mu s$, if $t \xrightarrow{p}_{\mathcal{R}} s$ and $p \in \mathcal{P}os^\mu(t)$. The $\hookrightarrow_\mu$-normal forms are called μ-normal forms. Note that, except for the trivial case $\mu = \mu_\top$, the μ-normal forms strictly include all normal forms of $\mathcal{R}$. A TRS $\mathcal{R}$ is μ-terminating if $\hookrightarrow_\mu$ is terminating. The canonical replacement map $\mu^{can}_{\mathcal{R}}$ is *the most restrictive replacement map ensuring that the non-variable subterms of the left-hand sides of the rules of $\mathcal{R}$ are replacing.* Note that $\mu^{can}_{\mathcal{R}}$ is easily obtained from $\mathcal{R}$: $\forall f \in \mathcal{F}$, $i \in \{1,\dots,ar(f)\}$,

$$i \in \mu^{can}_{\mathcal{R}}(f) \quad \text{iff} \quad \exists l \in L(\mathcal{R}), p \in \mathcal{P}os_{\mathcal{F}}(l), (root(l|_p) = f \wedge p.i \in \mathcal{P}os_{\mathcal{F}}(l))$$

where $\mathcal{P}os_{\mathcal{F}}(l)$ is the set of positions of non-variable subterms of l. Let $CM_{\mathcal{R}} = \{\mu \in M_{\mathcal{R}} \mid \mu^{can}_{\mathcal{R}} \sqsubseteq \mu\}$ be the set of replacement maps which are less or equally restrictive than $\mu^{can}_{\mathcal{R}}$. We say that $\hookrightarrow_\mu$ is *a canonical* context-sensitive rewrite relation if $\mu \in CM_{\mathcal{R}}$.

4 Termination and Normalization Properties

The most important semantic property of canonical *CSR* is that, for left-linear TRSs, μ-normal forms are head-normal forms[1] [Luc98]. We define a normalization procedure $norm_\mu$ which is based on performing a *layered* μ-normalization of terms (see Figure 1). In Figure 1, the *maximal replacing context* $MRC^\mu(t)$ of t is the maximal prefix of t whose positions are μ-replacing in t. Also, μ-*norm* is a mapping from sets of terms to μ-normal forms of them. The normalization of a term proceeds by first μ-normalizing terms and next recursively normalizing (in this way) the maximal non-replacing subterms of the obtained μ-normal forms.

Theorem 1 (Correctness). *Let $\mathcal{R} = (\mathcal{F}, R)$ be a left-linear TRS, $\mu \in CM_{\mathcal{R}}$, and $t \in \mathcal{T}(\mathcal{F},\mathcal{X})$. If $s \in norm_\mu(\{t\})$, then s is a normal form of t.*

This procedure is not ensured to be correct or terminating if $\mu \notin CM_{\mathcal{R}}$.

Example 2. Consider $\mathcal{R}$ and μ as in Example 1. Let t = `first(5,from(0))` (numbers n are a shorthand for $\texttt{s}^n\texttt{(0)}$) and $\mu'(f) = \mu(f)$ for all symbol f

[1] A head-normal form is a term which does not rewrite to a redex.

```
Procedure norm_μ(T)
   T := μ-norm(T)
   for each t ∈ T
     let t = C[t_1,...,t_n], where C[ ] = MRC^μ(t)
     for i := 1,...,n do S_i := norm_μ({t_i})
     T_t := C[S_1,...,S_n]
   return ⋃_{t∈T} T_t
end procedure norm_μ
```

Given a context $C[\,]$ and sets of terms $T_1, \ldots, T_n$, notation $C[T_1, \ldots, T_n]$ denotes the set $\{C[t_1, \ldots, t_n] \mid t_1 \in T_1, \ldots, t_n \in T_n\}$.

Fig. 1. Algorithm for *normalization via μ-normalization*

except $\mu'(\texttt{first}) = \{1\}$. Note that $\mu' \notin CM_{\mathcal{R}}$, but $\mathcal{R}$ is still μ'-terminating. We have

$$\begin{aligned} norm_{\mu'}(\{t\}) &= \texttt{first(5,}norm_{\mu'}(\{\texttt{from(0)}\})) \\ &= \texttt{first(5,0:}norm_{\mu'}(\{\texttt{from(1)}\})) \\ &= \texttt{first(5,0:1:}norm_{\mu'}(\{\texttt{from(2)}\})) \\ &\vdots \end{aligned}$$

Thus, $norm_{\mu'}(t)$ does not terminate and there is no output. Even if $norm_{\mu'}$ terminates, the outcome can be incorrect. Let $s = \texttt{first(1,first(1,[0,1,2]))}$ (`[0,1,2]` is a shorthand for `0:1:2:[]`). Now:

$$\begin{aligned} norm_{\mu'}(\{s\}) &= \texttt{first(1,}norm_{\mu'}(\{\texttt{first(1,[0,1,2])}\})) \\ &= \texttt{first(1,0:}norm_{\mu'}(\{\texttt{first(0,[1,2])}\})) \\ &= \{\texttt{first(1,[0])}\} \end{aligned}$$

The outcome is *not* a normal form: $\texttt{first(1,[0])} \to^* \texttt{[0]}$.

In the following theorem we say that *μ-norm* is *complete for μ-normalization* if for all $T \subseteq \mathcal{T}(\mathcal{F}, \mathcal{X})$ and $t \in T$, the μ-normal forms of t are in *μ-norm*(T).

Theorem 2 (Completeness). *Let $\mathcal{R} = (\mathcal{F}, R)$ be a left-linear TRS, $\mu \in CM_{\mathcal{R}}$, and $t \in \mathcal{T}(\mathcal{F}, \mathcal{X})$. Assume that μ-norm is complete for μ-normalization. If s is a normal form of t, then $s \in norm_\mu(\{t\})$.*

Correctness and completeness of $norm_\mu$ can easily be achieved if canonical μ-termination of the TRS $\mathcal{R}$ can be ensured: in this case, we let *μ-norm*(T) to be the set of all μ-normal forms of each $t \in T$.

Example 3. Consider $\mathcal{R}$ and μ as in Example 1. Now we can successfully apply the 'normalization via μ-normalization' procedure to t in Example 2:

$$\begin{aligned} norm_\mu(\{t\}) &= \texttt{0:}norm_\mu(\{\texttt{first(4,from(1))}\}) \\ &= \texttt{0:1:}norm_\mu(\{\texttt{first(3,from(2))}\}) \\ &\vdots \\ &= \texttt{0:1:2:3:4:}norm_\mu(\{\texttt{first(0,from(5))}\}) \\ &= \{\texttt{[0,1,2,3,4]}\} \end{aligned}$$

Lazy programming languages admit giving *infinite values* as the meaning of expressions. Infinite values are limits of converging infinite sequences of *partially defined* values which are more and more defined. In this setting, the following 'termination-like' property has been envisaged:

Top-termination: A TRS is *top-terminating* if no infinitary reduction sequence performs infinitely many rewrites at topmost position Λ [DKP91].

Top-termination is, in general, an undecidable property. We can prove top-termination of a left-linear TRS $\mathcal{R}$ by proving its *canonical μ-termination.*

Theorem 3. *Let $\mathcal{R}$ be a left-linear TRS and $\mu \in CM_{\mathcal{R}}$. If $\mathcal{R}$ is μ-terminating, then $\mathcal{R}$ is top-terminating.*

A TRS is *infinitary normalizing* if every (finite) term t admits a *strongly convergent sequence* (i.e., a sequence that, ultimately[2], reduces deeper and deeper redexes) starting from t and ending into a (possibly infinite) normal form (i.e., a term without redexes). For left-linear TRSs, top-termination is a sufficient condition for infinitary normalization. This fact easily follows from[3] [DKP91]. Still, there are infinitary normalizing TRSs which are not top-terminating.

Example 4. Consider the TRS $\mathcal{R}$:

$$\begin{array}{l} \mathtt{f(a)} \to \mathtt{f(f(a))} \\ \mathtt{f(a)} \to \mathtt{a} \end{array}$$

This TRS is infinitary normalizing but not top-terminating:

$$\underline{\mathtt{f(a)}} \to \mathtt{f(}\underline{\mathtt{f(a)}}\mathtt{)} \to \underline{\mathtt{f(a)}} \to \cdots$$

Hence, from Theorem 3, we conclude that μ-termination criteria (see Section 5 below) can also be used for proving infinitary normalization. Infinitary normalizing TRSs can be thought of as playing the role of normalizing TRSs in the infinitary setting, i.e., ensuring that every term has a meaning (its, possibly infinite, normal form). Our normalization via μ-normalization technique easily generalizes to infinitary normalization (see [Luc02] for details).

5 Termination of *CSR* by Transformation

The μ-termination of a TRS $\mathcal{R} = (\mathcal{F}, R)$ is usually proved by demonstrating *termination* of a transformed TRS $\mathcal{R}^{\mu}_{\Theta} = (\mathcal{F}^{\mu}_{\Theta}, R^{\mu}_{\Theta})$ obtained from $\mathcal{R}$ and $\mu \in M_{\mathcal{F}}$ by using a transformation Θ [FR99,GM99,GM02,Luc96,SX98,Zan97].

Definition 1. *A transformation Θ is*

1. correct *(regarding μ-termination) w.r.t. $\mathcal{M} \subseteq M_{\mathcal{F}}$ if, for all $\mu \in \mathcal{M}$, termination of $\mathcal{R}^{\mu}_{\Theta}$ implies μ-termination of $\mathcal{R}$.*

[2] We only consider sequences of length at most ω.

[3] But note that the notion of infinitary normal form used in [DKP91] (a term t such that $t = t'$ whenever $t \to t'$) differs from the usual one.

2. complete *(regarding μ-termination) w.r.t.* $\mathcal{M} \subseteq M_{\mathcal{F}}$ *if, for all* $\mu \in \mathcal{M}$, *μ-termination of* $\mathcal{R}$ *implies the termination of* $\mathcal{R}^{\mu}_{\Theta}$.

If $\mathcal{M} = M_{\mathcal{F}}$ in Definition 1, we just say that Θ is correct (or complete). Obviously, complete transformations Θ preserve termination of $\mathcal{R}$, i.e., if $\mathcal{R}$ is terminating, then $\mathcal{R}^{\mu}_{\Theta}$ is terminating. The simplest (and trivial) correct transformation for proving μ-termination is the identity: $\mathcal{R}^{\mu}_{ID} = \mathcal{R}$ (terminating TRSs are μ-terminating for every replacement map μ). We review the main (nontrivial) correct transformations for proving termination of *CSR* and prove a number of new properties (e.g., completeness) which appear when they are used for proving termination of canonical *CSR*.

5.1 The Contractive Transformation

With the contractive transformation [Luc96], the non-μ-replacing arguments of all symbols in $\mathcal{F}$ are *removed* and a new, *μ-contracted* signature $\mathcal{F}^{\mu}_{L}$ is obtained (possibly reducing the *arity* of symbols). The function $\tau_\mu : \mathcal{T}(\mathcal{F},\mathcal{X}) \to \mathcal{T}(\mathcal{F}^{\mu}_{L},\mathcal{X})$ *drops* the non-replacing immediate subterms of a term $t \in \mathcal{T}(\mathcal{F},\mathcal{X})$ and constructs a 'μ-contracted' term by joining the (also transformed) replacing arguments below the corresponding operator of $\mathcal{F}^{\mu}_{L}$. A TRS $\mathcal{R} = (\mathcal{F}, R)$ is μ-contracted into $\mathcal{R}^{\mu}_{L} = (\mathcal{F}^{\mu}_{L}, \{\tau_\mu(l) \to \tau_\mu(r) \mid l \to r \in R\})$.

Example 5. Consider the TRS $\mathcal{R}$ (borrowed from [GM02]):

```
nats  → adx(zeros)      adx(x:y)  → incr(x:adx(y))
zeros → 0:zeros         incr(x:y) → s(x):incr(y)
```

and $\mu(\texttt{:}) = \mu(\texttt{incr}) = \mu(\texttt{adx}) = \mu(\texttt{s}) = \{1\}$. Then $\mathcal{R}^{\mu}_{L}$ (we use the same symbols with possibly decreased arities):

```
nats  → adx(zeros)      adx(:(x))  → incr(:(x))
zeros → :(0)            incr(:(x)) → :(s(x))
```

is terminating (use an *rpo* with precedence $\texttt{nats} > \texttt{adx}, \texttt{zeros} > \texttt{incr} > \texttt{:}, \texttt{0}, \texttt{s}$). Hence, $\mathcal{R}$ is μ-terminating.

The contractive transformation only works well with *μ-conservative* TRSs, i.e., satisfying that $\mathcal{V}ar^{\mu}(r) \subseteq \mathcal{V}ar^{\mu}(l)$ for all rule $l \to r$ of $\mathcal{R}$ [Luc96]; otherwise, extra variables will appear in a rule of $\mathcal{R}^{\mu}_{L}$ thus becoming non-terminating. Let

$$CoCM_{\mathcal{R}} = \{\mu \in CM_{\mathcal{R}} \mid \text{for all rules } l \to r \text{ in } \mathcal{R}, \mathcal{V}ar^{\mu}(r) \subseteq \mathcal{V}ar^{\mu}(l)\}$$

That is: $CoCM_{\mathcal{R}}$ contains the replacement maps $\mu \in CM_{\mathcal{R}}$ that make $\mathcal{R}$ μ-conservative. Since $\mu_\top \in CoCM_{\mathcal{R}}$ for all TRS $\mathcal{R}$, $CoCM_{\mathcal{R}}$ is not empty. We prove that the contractive transformation becomes complete w.r.t. $CoCM_{\mathcal{R}}$.

Theorem 4 (Completeness). *Let* $\mathcal{R}$ *be a left-linear TRS and* $\mu \in CoCM_{\mathcal{R}}$. *If* $\mathcal{R}$ *is μ-terminating, then* $\mathcal{R}^{\mu}_{L}$ *is terminating.*

Even though μ-conservativeness can be difficult to achieve, there still are interesting examples fulfilling this requirement (see [BLR02,Luc02]). Theorem 4 does not hold if $\mu \notin CM_{\mathcal{R}}$ (even for μ-conservative TRSs).

Example 6. Consider the terminating (hence μ-terminating) TRS $\mathcal{R}$ [GM02]:

`f(b,x)` $\rightarrow$ `f(c,x)`

together with $\mu(\mathtt{f}) = \{2\}$. Note that $\mathcal{R}$ is μ-conservative but $\mu \notin CM_{\mathcal{R}}$; $\mathcal{R}^{\mu}_{L}$:

`f(x)` $\rightarrow$ `f(x)`

is not terminating.

Left-linearity cannot be dropped in Theorem 4.

Example 7. Consider the TRS $\mathcal{R}$:

`f(x,x)` $\rightarrow$ `f(a,b)`

with $\mu(\mathtt{f}) = \{1\}$. Since $\mathcal{R}$ is terminating, $\mathcal{R}$ is μ-terminating. However, $\mathcal{R}^{\mu}_{L}$:

`f(x)` $\rightarrow$ `f(a)`

is not terminating. Note that $\mu \in CoCM_{\mathcal{R}}$.

An obvious corollary of Theorem 4 is that the μ-contractive transformation preserves termination of left-linear TRSs $\mathcal{R}$ whenever $\mu \in CoCM_{\mathcal{R}}$. Examples 6 and 7 show the need for these restrictions.

5.2 Zantema's Transformation

Zantema's transformation *marks* the *non-replacing arguments* of function symbols (disregarding their positions within the term) [Zan97]. Given $\mathcal{R} = (\mathcal{F}, R)$ and $\mu \in M_{\mathcal{F}}$, $\mathcal{R}^{\mu}_{Z} = (\mathcal{F} \cup \mathcal{F}' \cup \{\mathtt{activate}\}, R^{\mu}_{Z})$ where R^{μ}_{Z} consists of two parts The first part results from R by replacing every function symbol f occurring in a left or right-hand side with f' (a fresh function symbol of the same arity as f which, then, is included in $\mathcal{F}'$) if it occurs in a non-replacing *argument* of the function symbol directly above it. These new function symbols are used to block further reductions at this position. In addition, if a variable x occurs in a non-replacing position in the *lhs* l of a rewrite rule $l \rightarrow r$, then all occurrences of x in r are replaced by $\mathtt{activate}(x)$. Here, `activate` is a new unary function symbol which is used to activate blocked function symbols again.

The second part of R^{μ}_{Z} consists of rewrite rules that are needed for blocking and unblocking function symbols:

$$f(x_1, \ldots, x_k) \rightarrow f'(x_1, \ldots, x_k)$$
$$\mathtt{activate}(f'(x_1, \ldots, x_k)) \rightarrow f(x_1, \ldots, x_k)$$

for every[4] $f' \in \mathcal{F}'$, together with the rule $\mathtt{activate}(x) \rightarrow x$ Zantema's transformation can succeed when the contractive transformation fails.

Example 8. Let $\mathcal{R}$ and μ be as in Example 1. Since $\mathcal{R}$ is not μ-conservative, $\mathcal{R}^{\mu}_{L}$ cannot be used for proving μ-termination of $\mathcal{R}$. However, $\mathcal{R}^{\mu}_{Z}$:

[4] Actually, the rules $\mathtt{activate}(f'(x_1, \ldots, x_k)) \rightarrow f(x_1, \ldots, x_k)$ can be avoided if no symbol `activate` appears in the rules of the first part of the transformation.

```
first(0,x)         → []
first(s(x),y:z)    → y:first'(x,activate(z))
sel(0,x:y)         → x
sel(s(x),y:z)      → sel(x,activate(z))
from(x)            → x:from'(s(x))
from(x)                     → from'(x)
activate(from'(x))          → from(x)
first(x,y)                  → first'(x,y)
activate(first'(x,y))       → first(x,y)
activate(x)                 → x
```

is terminating (use *rpo* based on `sel` > `activate` ≈ `first` > `from`, :, `first'`, `[]` and `from` > :, `from'`, `s`, and giving `sel` the usual lexicographic status).

Zantema's transformation remains incomplete for proving canonical μ-termination even with left-linear, μ-conservative TRSs.

Example 9. Consider $\mathcal{R}$ and μ as in Example 5. Note that $\mu \in CoCM_{\mathcal{R}}$. Remember that $\mathcal{R}^{\mu}_{L}$ is (simply) terminating. However, $\mathcal{R}^{\mu}_{Z}$:

```
nats      → adx(zeros)
zeros     → 0:zeros'
incr(x)   → incr'(x)
adx(x)    → adx'(x)
activate(zeros') → zeros
activate(x)      → x
incr(x:y)  → s(x):incr'(activate(y))
adx(x:y)   → incr(x:adx'(activate(y)))
zeros      → zeros'
activate(incr'(x))  → incr(x)
activate(adx'(x))   → adx(x)
```

is not terminating:

```
adx(zeros) → adx(0:zeros') → incr(0:adx'(activate(zeros')))
           → incr(0:adx'(zeros)) → s(0):incr'(activate(adx'(zeros)))
           → s(0):incr'(adx(zeros)) → ···
```

Zantema's transformation preserves termination of TRSs.

Theorem 5. *Let $\mathcal{R}$ be a TRS and $\mu \in M_{\mathcal{R}}$. If $\mathcal{R}$ is terminating, then $\mathcal{R}^{\mu}_{Z}$ is terminating.*

In [FR99], Ferreira and Ribeiro propose a variant of Zantema's transformation which has been proved strictly more powerful than Zantema's one (see [GM02]). Again, $\mathcal{R}^{\mu}_{FR}$ has two parts. The first part results from the first part of $\mathcal{R}^{\mu}_{Z}$ by marking all function symbols (except `activate`) which occur below an already marked symbol. Therefore, all function symbols of non-replacing subterms are marked. The second part consists of the rule `activate`$(x) \to x$ plus the rules:

$$f(x_1,\dots,x_k) \to f'(x_1,\dots,x_k)$$
$$\texttt{activate}(f'(x_1,\dots,x_k)) \to f([x_1]^f_1,\dots,[x_k]^f_k)$$

for every $f \in \mathcal{F}$ for which f' appears in the first part of $\mathcal{R}^{\mu}_{FR}$, where $[t]^f_i =$ `activate`(t) if $i \in \mu(f)$ and $[t]^f_i = t$ otherwise. They also include rules

$$\texttt{activate}(f(x_1,\dots,x_k)) \to f([x_1]^f_1,\dots,[x_k]^f_k)$$

for k-ary symbols f where f' does *not* appear in the first part of $\mathcal{R}^{\mu}_{FR}$. However, Giesl and Middeldorp have recently shown that these rules are not necessary for obtaining a correct transformation [GM02].

Note that $\mathcal{R}$ and μ of Example 5 also show incompleteness of Ferreira and Ribeiro's transformation $\mathcal{R}^{\mu}_{FR}$.

5.3 Giesl and Middeldorp's Transformations

Giesl and Middeldorp introduced a transformation that explicitly *marks* the replacing positions of a term (by using a new symbol `active`). Given a TRS $\mathcal{R} = (\mathcal{F}, R)$ and $\mu \in M_{\mathcal{F}}$, the TRS $\mathcal{R}^{\mu}_{GM} = (\mathcal{F}^{\mu}_{GM}, R^{\mu}_{GM})$ consists of $\mathcal{F}^{\mu}_{GM} = \mathcal{F} \cup \{\texttt{active}, \texttt{mark}\}$ and the rules R^{μ}_{GM} given by (for all $l \to r \in R$ and $f \in \mathcal{F}$):

$$\begin{aligned} \texttt{active}(l) &\to \texttt{mark}(r) \\ \texttt{mark}(f(x_1, \ldots, x_k)) &\to \texttt{active}(f([x_1]_f, \ldots, [x_k]_f)) \\ \texttt{active}(x) &\to x \end{aligned}$$

where $[x_i]_f = \texttt{mark}(x_i)$ if $i \in \mu(f)$ and $[x_i]_f = x_i$ otherwise [GM99]. Let $\mathcal{M} = (\mathcal{F}^{\mu}_{GM}, \{\texttt{mark}(f(x_1, \ldots, x_k)) \to \texttt{active}(f([x_1]_f, \ldots, [x_k]_f)) \mid f \in \mathcal{F}\})$; note that $\mathcal{M}$ is a confluent and terminating TRS [GM99]. Giesl and Middeldorp showed that, in general, this transformation is not complete. We prove the *completeness* of this transformation for proving termination of canonical *CSR* when dealing with left-linear TRSs.

Theorem 6 (Completeness). *Let $\mathcal{R}$ be a left-linear TRS and $\mu \in CM_{\mathcal{R}}$. If $\mathcal{R}$ is μ-terminating, then $\mathcal{R}^{\mu}_{GM}$ is terminating.*

Example 1 of [GM99] shows that this result does not hold if $\mu \notin CM_{\mathcal{R}}$. Giesl and Middeldorp also proposed two refinements of this transformation that are intended to provide better results in practice (being still incomplete).

1. The first modification ($\mathcal{R}^{\mu}_{mGM}$), see [GM02] Definition 3, consists in replacing the single symbol `active` by fresh symbols f_a for every $f \in \mathcal{F}$. Thus, $\mathcal{F}^{\mu}_{mGM} = \mathcal{F} \cup \{f_a \mid f \in \mathcal{F}\} \cup \{\texttt{mark}\}$. The pattern $\texttt{active}(f(\cdots))$ is also replaced everywhere by $f_a(\cdots)$. The rule $\texttt{active}(x) \to x$ is expanded into all rules of the form $f_a(x_1, \ldots, x_k) \to f(x_1, \ldots, x_k)$ and, hence, removed. Now, let $\mathcal{M}' = (\mathcal{F}^{\mu}_{mGM}, \{\texttt{mark}(f(x_1, \ldots, x_k)) \to f_a([x_1]_f, \ldots, [x_k]_f) \mid f \in \mathcal{F}\})$.
2. The second modification ($\mathcal{R}^{\mu}_{nGM}$) consists of normalizing the *rhs*'s of rules $f_a(l_1, \ldots, l_k) \to \texttt{mark}(r)$ in $\mathcal{R}^{\mu}_{mGM}$ (where $f(l_1, \ldots, l_k) \to r \in R$) w.r.t. $\mathcal{M}'$, see Section 7 of [GM02].

Since $\mathcal{R}^{\mu}_{GM}$, $\mathcal{R}^{\mu}_{mGM}$ and $\mathcal{R}^{\mu}_{nGM}$ are equivalent regarding termination [GM02], they are also complete under the conditions of Theorem 6. Also, $\mathcal{R}^{\mu}_{GM}$ and $\mathcal{R}^{\mu}_{mGM}$ are equivalent regarding simple termination.

Theorem 7. *Let $\mathcal{R}$ be a TRS and $\mu \in M_{\mathcal{R}}$. Then, $\mathcal{R}^{\mu}_{GM}$ is simply terminating if and only if $\mathcal{R}^{\mu}_{mGM}$ is simply terminating.*

Theorem 7 does not hold for $\mathcal{R}^{\mu}_{nGM}$.

Example 10. Consider the following non-simply terminating TRS $\mathcal{R}$:

```
f(0,1,x) → c(f(x,x,x))
```

Let $\mu(\texttt{f}) = \{1, 2\}$ and $\mu(\texttt{c}) = \varnothing$. Then, with $\mathcal{R}^{\mu}_{mGM}$:

$$\begin{array}{lll} \mathtt{f}_a\mathtt{(0,1,x)} & \to \mathtt{mark(c(f(x,x,x)))} & \mathtt{f}_a\mathtt{(x,y,z)} \to \mathtt{f(x,y,z)} \\ \mathtt{mark(f(x,y,z))} & \to \mathtt{f}_a\mathtt{(mark(x),mark(y),z)} & \mathtt{0}_a \to \mathtt{0} \\ \mathtt{mark(0)} & \to \mathtt{0}_a & \mathtt{1}_a \to \mathtt{1} \\ \mathtt{mark(1)} & \to \mathtt{1}_a & \mathtt{c}_a\mathtt{(x)} \to \mathtt{c(x)} \\ \mathtt{mark(c(x))} & \to \mathtt{c}_a\mathtt{(x)} & \end{array}$$

we have the following infinite derivation (in $\mathcal{R}^\mu_{mGM} \cup \mathcal{E}mb(\mathcal{F}^\mu_{mGM})$):

$$\begin{array}{l} \underline{\mathtt{f}_a\mathtt{(0,1,f(0,1,x))}} \to_{\mathcal{R}^\mu_{mGM}} \mathtt{mark(c(f(f(0,1,x),f(0,1,x),f(0,1,x))))} \\ \quad \to^+_{\mathcal{E}mb(\mathcal{F}^\mu_{mGM})} \underline{\mathtt{mark(f(0,1,f(0,1,x)))}} \to_{\mathcal{R}^\mu_{mGM}} \mathtt{f}_a\mathtt{(mark(0),mark(1),f(0,1,x))} \\ \quad \to^+_{\mathcal{E}mb(\mathcal{F}^\mu_{mGM})} \mathtt{f}_a\mathtt{(0,1,f(0,1,x))} \to \cdots \end{array}$$

Thus, $\mathcal{R}^\mu_{mGM}$ is not simply terminating. However, $\mathcal{R}^\mu_{nGM}$:

$$\begin{array}{lll} \mathtt{f}_a\mathtt{(0,1,x)} & \to \mathtt{c}_a\mathtt{(f(x,x,x))} & \mathtt{f}_a\mathtt{(x,y,z)} \to \mathtt{f(x,y,z)} \\ \mathtt{mark(f(x,y,z))} & \to \mathtt{f}_a\mathtt{(mark(x),mark(y),z)} & \mathtt{0}_a \to \mathtt{0} \\ \mathtt{mark(0)} & \to \mathtt{0}_a & \mathtt{1}_a \to \mathtt{1} \\ \mathtt{mark(1)} & \to \mathtt{1}_a & \mathtt{c}_a\mathtt{(x)} \to \mathtt{c(x)} \\ \mathtt{mark(c(x))} & \to \mathtt{c}_a\mathtt{(x)} & \end{array}$$

is *rpo*-terminating (use the precedence $\mathtt{mark} > \mathtt{f}_a, \mathtt{0}_a, \mathtt{1}_a > \mathtt{c}_a > \mathtt{f}, \mathtt{0}, \mathtt{1} > \mathtt{c}$).

Thus, $\mathcal{R}^\mu_{mGM}$ (or $\mathcal{R}^\mu_{GM}$) and $\mathcal{R}^\mu_{nGM}$ are *not* equivalent regarding simple termination. However, we have the following.

Theorem 8. *Let $\mathcal{R}$ be a TRS and $\mu \in M_\mathcal{R}$. If $\mathcal{R}^\mu_{GM}$ is simply terminating, then $\mathcal{R}^\mu_{nGM}$ is simply terminating.*

Unfortunately, the transformations also have some (practical) drawbacks.

Theorem 9. *Let $\mathcal{R}$ be a left-linear TRS and $\mu \in CM_\mathcal{R}$. If $\mathcal{R}$ is not simply terminating, then $\mathcal{R}^\mu_{GM}$ and $\mathcal{R}^\mu_{mGM}$ are not simply terminating.*

The theorem does not hold if $\mu \notin CM_\mathcal{R}$.

Example 11. Consider the following TRS:

$$\begin{array}{ll} \mathtt{f(f(x))} & \to \mathtt{f(g(h(x)))} \\ \mathtt{h(x)} & \to \mathtt{f(x)} \end{array}$$

This TRS is not simply terminating:

$$\underline{\mathtt{f(f(x))}} \to \mathtt{f(\underline{g(h(x))})} \to_{\mathcal{E}mb(\mathcal{F})} \mathtt{f(\underline{h(x)})} \to \mathtt{f(f(x))} \to \cdots$$

Let $\mu(\mathtt{f}) = \mu(\mathtt{g}) = \mu(\mathtt{h}) = \varnothing$. Note that $\mu \notin CM_\mathcal{R}$. Now, $\mathcal{R}^\mu_{GM}$:

$$\begin{array}{lll} \mathtt{active(f(f(x)))} & \to \mathtt{mark(f(g(h(x))))} & \mathtt{mark(g(x))} \to \mathtt{active(g(x))} \\ \mathtt{active(h(x))} & \to \mathtt{mark(f(x))} & \mathtt{mark(h(x))} \to \mathtt{active(h(x))} \\ \mathtt{mark(f(x))} & \to \mathtt{active(f(x))} & \mathtt{active(x)} \to \mathtt{x} \end{array}$$

is simply terminating; e.g., the reduction sequence that corresponds to the previous 'dangerous' sequence is not longer possible in $\mathcal{R}^\mu_{GM} \cup \mathcal{E}mb(\mathcal{F}^\mu_{GM})$.

Theorem 9 does not hold for $\mathcal{R}^\mu_{nGM}$ (see Example 10; note that $\mu \in CM_\mathcal{R}$). Still, $\mathcal{R}^\mu_{nGM}$ can also be non-simply terminating if $\mathcal{R}$ is not simply terminating.

Example 12. Consider the non-simply terminating TRS $\mathcal{R}$:

$\mathtt{f(f(x))} \to \mathtt{f(g(f(x)))}$

together with $\mu(\mathtt{f}) = \{1\}$ and $\mu(\mathtt{g}) = \varnothing$. Note that $\mu \in CM_{\mathcal{R}}$. Then, $\mathcal{R}^{\mu}_{nGM}$:

$$
\begin{array}{llll}
\mathtt{f}_a\mathtt{(f(x))} \to \mathtt{f}_a\mathtt{(g}_a\mathtt{(f(x)))} & & \mathtt{f}_a\mathtt{(x)} \to \mathtt{f(x)} \\
\mathtt{mark(f(x))} \to \mathtt{f}_a\mathtt{(mark(x))} & & \mathtt{g}_a\mathtt{(x)} \to \mathtt{g(x)} \\
\mathtt{mark(g(x))} \to \mathtt{g}_a\mathtt{(x)} & & &
\end{array}
$$

is not simply terminating:

$$\underline{\mathtt{f}_a\mathtt{(f(x))}} \to_{\mathcal{R}^{\mu}_{nGM}} \mathtt{f}_a\mathtt{(}\underline{\mathtt{g}_a\mathtt{(f(x))}}\mathtt{)} \to_{\mathcal{E}mb(\mathcal{F}^{\mu}_{nGM})} \mathtt{f}_a\mathtt{(f(x))} \to \cdots$$

As one can notice by re-examining the examples of Giesl and Middeldorp's papers [GM99,GM02], the success of their transformations usually depends on using Arts and Giesl's dependency pairs method [AG00]. Our results in this section clarify this situation by showing that, in fact, simplification orderings (alone) should *not* be used to orientate the rules of $\mathcal{R}^{\mu}_{GM}$ or $\mathcal{R}^{\mu}_{mGM}$ in proofs of canonical μ-termination. Rather, some kind of auxiliary technique *must* be additionally used to make them suitable. This is critically true when a proof of canonical μ-termination of a nonterminating (hence non-simply terminating) TRS is attempted. Automatic tools for proving termination that do not provide for adequate preprocessings (based on, e.g., applying argument filtering transformations, exploiting modularity, using dependency pairs, etc.) should not be used together with $\mathcal{R}^{\mu}_{GM}$ or $\mathcal{R}^{\mu}_{mGM}$. For instance, when using C*i*ME 2.0 system[5] for proving termination of $\mathcal{R}^{\mu}_{GM}$ or $\mathcal{R}^{\mu}_{mGM}$, the use of dependency pairs should be activated before.

6 Comparing the Transformations

Figure 2 shows how different transformations for proving termination of *CSR* compare. The leftmost diagram of Figure 2 is obtained from that of [GM02] plus Theorem 5. An arrow from Θ to Θ' means that whenever $\mathcal{R}^{\mu}_{\Theta}$ is terminating, then $\mathcal{R}^{\mu}_{\Theta'}$ is also terminating (for μ selected from the subset of replacement maps written below the diagram). Dashed arrows (as well as minimal paths containing them) require left-linearity of $\mathcal{R}$. **GM* stands for the three transformations introduced in the previous section (that are equivalent regarding termination of the transformed TRS); *C* stands for the *complete* Giesl and Middeldorp's transformation (not described here, see [GM99,GM02]).

Comparisons regarding simple termination are also interesting from a more practical point of view. Apart from Theorems 7 and 8, we have the following.

Theorem 10. *Let $\mathcal{R}$ be a TRS and $\mu \in M_{\mathcal{R}}$. If $\mathcal{R}$ is simply terminating, then $\mathcal{R}^{\mu}_{Z}$, $\mathcal{R}^{\mu}_{FR}$, and $\mathcal{R}^{\mu}_{GM}$ are simply terminating.*

We conjecture that simple termination of $\mathcal{R}^{\mu}_{Z}$ implies that of $\mathcal{R}^{\mu}_{FR}$ (the opposite is not true). Simple termination can be achieved after applying some transformations.

[5] Available at `http://cime.lri.fr`

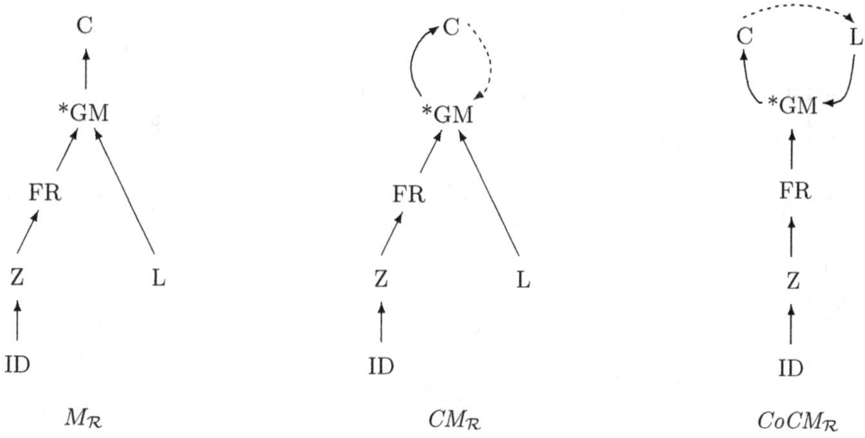

Fig. 2. Comparing transformations regarding termination

Example 13. Consider $\mathcal{R}$ and μ as in Example 12. Recall that $\mathcal{R}$ (and $\mathcal{R}^\mu_{GM}$, $\mathcal{R}^\mu_{mGM}$, $\mathcal{R}^\mu_{nGM}$) is not simply terminating. However, $\mathcal{R}^\mu_L$:

$\mathtt{f(f(x))} \to \mathtt{f(g)}$

is *rpo*-terminating (use precedence $\mathtt{f} > \mathtt{g}$). Also, $\mathcal{R}^\mu_Z$:

$\mathtt{f(f(x))}$	$\to$	$\mathtt{f(g(f'(x)))}$	$\mathtt{activate(f'(x))}$	$\to$	$\mathtt{f(x)}$
$\mathtt{f(x)}$	$\to$	$\mathtt{f'(x)}$	$\mathtt{activate(x)}$	$\to$	$\mathtt{x}$

is *rpo*-terminating (use the precedence $\mathtt{activate} > \mathtt{f} > \mathtt{g}, \mathtt{f'}$). $\mathcal{R}^\mu_{FR}$ also is *rpo*-terminating.

Simple termination of $\mathcal{R}^\mu_{nGM}$ does *not* imply simple termination of $\mathcal{R}^\mu_Z$ or $\mathcal{R}^\mu_{FR}$.

Example 14. Consider $\mathcal{R}$ and μ as in Example 5. Note that $\mathcal{R}^\mu_Z$ (and $\mathcal{R}^\mu_{FR}$) is not terminating (see Example 9). However, $\mathcal{R}^\mu_{nGM}$:

$\mathtt{nats}_a$	$\to$	$\mathtt{adx}_a\mathtt{(zeros}_a\mathtt{)}$	$\mathtt{mark(0)}$	$\to$	$\mathtt{0}_a$
$\mathtt{adx}_a\mathtt{(x:y)}$	$\to$	$\mathtt{incr}_a\mathtt{(mark(x):}_a\mathtt{adx(y))}$	$\mathtt{mark(s(x))}$	$\to$	$\mathtt{s}_a\mathtt{(mark(x))}$
$\mathtt{zeros}_a$	$\to$	$\mathtt{0}_a\mathtt{:}_a\mathtt{zeros}$	$\mathtt{nats}_a$	$\to$	$\mathtt{nats}$
$\mathtt{incr}_a\mathtt{(x:y)}$	$\to$	$\mathtt{s}_a\mathtt{(mark(x)):}_a\mathtt{incr(y)}$	$\mathtt{adx}_a\mathtt{(x)}$	$\to$	$\mathtt{adx(x)}$
$\mathtt{mark(nats)}$	$\to$	$\mathtt{nats}_a$	$\mathtt{zeros}_a$	$\to$	$\mathtt{zeros}$
$\mathtt{mark(adx(x))}$	$\to$	$\mathtt{adx}_a\mathtt{(mark(x))}$	$\mathtt{x:}_a\mathtt{y}$	$\to$	$\mathtt{x:y}$
$\mathtt{mark(zeros)}$	$\to$	$\mathtt{zeros}_a$	$\mathtt{incr}_a\mathtt{(x)}$	$\to$	$\mathtt{incr(x)}$
$\mathtt{mark(x:y)}$	$\to$	$\mathtt{mark(x):}_a\mathtt{y}$	$\mathtt{0}_a$	$\to$	$\mathtt{0}$
$\mathtt{mark(incr(x))}$	$\to$	$\mathtt{incr}_a\mathtt{(mark(x))}$	$\mathtt{s}_a\mathtt{(x)}$	$\to$	$\mathtt{s(x)}$

is simply terminating.

For replacement maps $\mu \in CM_{\mathcal{R}}$, an immediate consequence of Theorems 9 and 10 is the equivalence of simple termination of $\mathcal{R}$ and $\mathcal{R}^\mu_{GM}$ for left-linear TRSs $\mathcal{R}$. With conservativeness, we have:

Theorem 11. *Let $\mathcal{R}$ be a left-linear TRS and $\mu \in CoCM_{\mathcal{R}}$. If $\mathcal{R}^\mu_Z$, $\mathcal{R}^\mu_{FR}$, or $\mathcal{R}^\mu_{nGM}$ are simply terminating, then $\mathcal{R}^\mu_L$ is simply terminating.*

Figure 3 shows the hierarchy of transformations regarding *simple* termination.

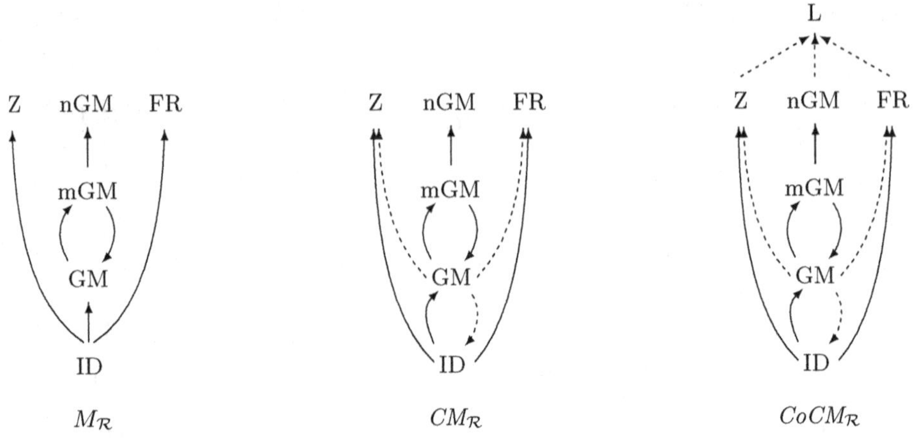

Fig. 3. Comparing transformations regarding simple termination

7 Experiments on Proving Termination of Canonical *CSR*

Consider a left-linear TRS $\mathcal{R}$ for which we want to establish some interesting termination property. We claim that, for the purpose of computing (infinite) normal forms, proving termination of canonical *CSR* could be priorized over proofs of termination: a proof of termination of canonical *CSR* can be used to achieve normalizations (even with non-terminating TRSs) and also to arbitrarily approximate infinite normal forms (see Section 4). Of course, if (it is clear that) $\mathcal{R}$ is not terminating, then only the proof of termination of canonical *CSR* should be attempted. Our results suggest that proving μ-termination of a *terminating* TRS $\mathcal{R}$ is not much harder than directly proving termination of $\mathcal{R}$ (e.g., simple termination is preserved by the transformations, see Theorems 5 and 10). Moreover, proving termination of canonical *CSR* can be easier than attempting a direct proof of termination (see Examples 12 and 13). Table 1 shows the result of a preliminary experience aimed at providing some experimental evidence supporting our claim. We have selected two rather different collections of termination examples: Dershowitz's *33 examples of termination* [Der95] (first group of four examples) and Arts and Giesl's *collection of examples for termination using dependency pairs* [AG01] (most of them non-simply terminating; see the second group of four examples). In our experiments, given a TRS $\mathcal{R}$, we systematically try to prove termination of $\mathcal{R}_{\Theta}^{\mu_{\mathcal{R}}^{can}}$ for $\Theta \in \{ID, L, Z, FR, GM, mGM, nGM\}$ (note that proving termination of $\mathcal{R}_{ID}^{\mu_{\mathcal{R}}^{can}}$ is just proving termination of $\mathcal{R}$). We have used MU-TERM 1.0 for obtaining the transformed[6] TRSs. We have attempted the proofs of termination using C*i*ME 2.0 (MU-TERM 1.0 has also been used for giving the appropriate format) in both its standard version (Std,

[6] MU-TERM 1.0 is available from `http://www.dsic.upv.es/users/elp/slucas/muterm`.

Table 1. Experiments on termination vs. $\mu_{\mathcal{R}}^{can}$-termination with C*i*ME 2.0

	ID		L		Z		FR		GM		mGM		nGM	
Ref. Example	Std	DP	Std	DP	Std	DP	Std	DP	Std	DP	Std	DP	Std	DP
5. Non Simp.	N	0.06	0.03	0.00	0.04	0.00	0.08	0.05	N	4153	N	0.14	N	0.16
7. Dutch Flag	0.09	0.06	0.05	0.02	0.22	0.12	?	0.20	?	0.33	?	0.2	?	0.17
8. Diff.	N	N	0.02	0.00	3.14	1.43	?	?	?	0.70	?	0.59	?	0.47
33. Hydra	N	N	NC	NC	N	?	?	?	N	N	N	N	N	N
3.1. Division v.1	N	2152	NC	NC	?	1.11	?	?	?	?	?	?	?	?
3.5. Remainder	N	2.65	NC	NC	?	9.08	=Z	=Z	?	?	?	?	?	?
3.7. Logarithm	105.0	17.9	=ID	=ID	6469	117.6	=Z	=Z	?	?	?	?	?	?
3.10. Min. sort	N	?	NC	NC	N	?	=Z	=Z	N	N	N	?	N	?

where polynomial orderings are used for proving simple termination of the TRS), and activating the use of dependency pairs (DP, thus dealing with non-simply terminating TRSs). In this way, we proceeded as a (typical) user of automatic sofware tools having no expertise in proofs of termination.

A positive answer for a given experiment is reported in our figures by indicating the time (in seconds) used by C*i*ME to find out the proof. C*i*ME's negative answers are reported with 'N'. Finally, '?' means that the search for a proof was interrupted without getting any answer (usually after several hours, often days, of execution). Other marks such as 'NC' or '=ID' (in column L), or '=Z' (in column FR) mean that the TRS $\mathcal{R}$ is not $\mu_{\mathcal{R}}^{can}$-conservative ('NC'), or $\mathcal{R}_L^{\mu_{\mathcal{R}}^{can}} = \mathcal{R}$ ('=ID'), or that $\mathcal{R}_{FR}^{\mu_{\mathcal{R}}^{can}} = \mathcal{R}_Z^{\mu_{\mathcal{R}}^{can}}$ ('=Z'). In Table 1, we have selected a representative sample of the (still growing) complete experiment[7]. We can see that, whenever a TRS $\mathcal{R}$ is proved terminating, $\mu_{\mathcal{R}}^{can}$-termination of $\mathcal{R}$ can also be proved using some (non-trivial) transformation. This usually amounts to have some overhead in the proof, but we have also obtained improvements (e.g., with Dutch Flag and $\mathcal{R}_L^{\mu_{\mathcal{R}}^{can}}$, or Division v.1 and $\mathcal{R}_Z^{\mu_{\mathcal{R}}^{can}}$). In several cases we can prove (with C*i*ME) $\mu_{\mathcal{R}}^{can}$-termination even when we *cannot* prove termination of the (terminating) TRS (e.g., with Non Simp. or Diff.). On the other hand, hard problems like Hydra were not managed in any way (with the current conditions for the experiment). We believe that these results are quite encouraging regarding our hypothesis.

8 Conclusions

We have proven that, for left-linear TRSs $\mathcal{R}$, Giesl and Middeldorp's transformations are complete for proving termination of canonical *CSR*; if conservativeness is additionally fulfilled, then the contractive transformation is also complete. We have compared the different transformations regarding *simple* termination. Although undecidable, simple termination copes with the use of most usual orderings for proving termination of rewriting automatically (e.g., *rpo*, *kbo*, and *poly* [Der87]). Giesl and Middeldorp noticed that, even though their transformations subsume the others (see the leftmost diagram of Figure 2), the latter

[7] See `http://www.dsic.upv.es/users/elp/slucas/experiments` for more details.

'may still be useful for the purpose of automation' [GM99]. This is true, indeed, and Figure 3 provides a new (quite different, compared to Figure 2) hierarchy of the transformations which is helpful for guiding their practical use in combination with simplification orderings. Their interaction with other techniques for automatically proving termination (e.g., DP-simple termination [AG00,AG01]) should be further investigated.

We argued that termination of canonical *CSR* is a computational property which can be more interesting to analyze than standard termination. We have partially supported this claim both with theoretical and experimental results.

Acknowledgements. I thank Cristina Borralleras and Albert Rubio for many interesting discussions about the topics of this paper. I also thank Jürgen Giesl and the anonymous referees for their useful remarks.

References

[AG00] T. Arts and J. Giesl. Termination of Term Rewriting Using Dependency Pairs *Theoretical Computer Science*, 236:133-178, 2000.

[AG01] T. Arts and J. Giesl. A collection of examples for termination of term rewriting using dependency pairs. TR AIB-2001-09, RWTH Aachen, 2001.

[BLR02] C. Borralleras, S. Lucas, and A. Rubio. Recursive Path Orderings can be Context-Sensitive. *Proc. of CADE'02*, Springer LNAI to appear, 2002.

[Der87] N. Dershowitz. Termination of rewriting. *JSC*, 3:69-115, 1987.

[Der95] N. Dershowitz. 33 Examples of Termination. LNCS 909:16-26, Springer-Verlag, Berlin, 1995.

[DKP91] N. Dershowitz, S. Kaplan, and D. Plaisted. Rewrite, rewrite, rewrite, rewrite, rewrite. *Theoretical Computer Science* 83:71-96, 1991.

[FR99] M.C.F. Ferreira and A.L. Ribeiro. Context-Sensitive AC-Rewriting. *Proc. of RTA'99*, LNCS 1631:286-300, Springer-Verlag, Berlin, 1999.

[GM99] J. Giesl and A. Middeldorp. Transforming Context-Sensitive Rewrite Systems. *Proc. of RTA'99*, LNCS 1631:271-285, Springer-Verlag, Berlin, 1999.

[GM02] J. Giesl and A. Middeldorp. Transformation Techniques for Context-Sensitive Rewrite Systems. Technical Report AIB-2002-02, Aachen, 2002.

[Luc96] S. Lucas. Termination of context-sensitive rewriting by rewriting. *Proc. of ICALP'96*, LNCS 1099:122-133, Springer-Verlag, Berlin, 1996.

[Luc98] S. Lucas. Context-sensitive computations in functional and functional logic programs. *Journal of Functional and Logic Programming*, 1998(1):1-61, January 1998.

[Luc02] S. Lucas. Context-sensitive rewriting strategies. *Information and Computation*, to appear.

[SX98] J. Steinbach and H. Xi. Freezing – Termination Proofs for Classical, Context-Sensitive and Innermost Rewriting. Institut für Informatik, T.U. München, January 1998.

[Zan97] H. Zantema. Termination of Context-Sensitive Rewriting. *Proc. of RTA'97*, LNCS 1232:172-186, Springer-Verlag, Berlin, 1997.

Atomic Set Constraints with Projection

Witold Charatonik[1,2] and Jean-Marc Talbot[3]

[1] Max-Planck Institut für Informatik, Saarbrücken, Germany
[2] University of Wrocław, Poland
[3] Laboratoire d'Informatique Fondamentale de Lille, Lille, France

Abstract. We investigate a class of set constraints defined as atomic set constraints augmented with projection. This class subsumes some already studied classes such as atomic set constraints with left-hand side projection and INES constraints. All these classes enjoy the nice property that satisfiability can be tested in cubic time. This is in contrast to several other classes of set constraints, such as definite set constraints and positive set constraints, for which satisfiability ranges from DEXPTIME-complete to NEXPTIME-complete. However, these latter classes allow set operators such as intersection or union which is not the case for the class studied here. In the case of atomic set constraints with projection one might expect that satisfiability remains polynomial. Unfortunately, we show that that the satisfiability problem for this class is no longer polynomial, but CoNP-hard. Furthermore, we devise a PSPACE algorithm to solve this satisfiability problem.

1 Introduction

Set constraints are first-order formulas interpreted over sets of trees. They are defined as inclusions between set expressions. These set expressions are built up from variables, a ranked signature (constants and function symbols) and various set operators. According to the set operators used, different classes of set constraints are defined.

The main domain of application of set constraints is program analysis, often called set-based analysis [15,13,2]. Set constraints play also a role in type checking [24] (See [1,16,20] for surveys).

The satisfiability problem for different classes of set constraints has been extensively studied and its complexity turns out to be often quite high; in most cases (definite set constraints [15], set constraints with intersection [8], positive and/or negative set constraints [12,4,6], positive and negative set constraints with projection [7]) satisfiability is DEXPTIME-hard [8] or even NEXPTIME-hard [5] for some of them.

Nonetheless for some other classes, it turned out that the satisfiability problem is tractable, more precisely cubic time. Atomic constraints are such a class: for these simple constraints, no set operator is allowed. A cubic time algorithm for satisfiability for these constraints is given in [13]. Motivated by program analysis, Müller, Niehren and Podelski introduced in [19] the class of INES constraints (Inclusions over Non-Empty Sets of trees). They correspond syntactically to atomic set constraints but the empty set is no longer an admissible value to instantiate variables. They showed that satisfiability for INES constraints can be tested in cubic time. In the same paper, the authors also investigate atomic set constraints extended with the possibility to force the denotation of

S. Tison (Ed.): RTA 2002, LNCS 2378, pp. 311–325, 2002.

some variables (but maybe not all of them) to be non-empty and show that satisfiability remains cubic.

In [13], another extension of atomic set constraints have been studied. There the syntax of constraints was extended to allow the use of projection on the left-hand side of inclusions. It has been shown that satisfiability remains cubic also in this case.

A natural question is then "Can theses two extensions (atomic set constraints with non-emptiness constraints and projection in the left-hand side of inclusions) be combined to obtain a new class for which satisfiability could be tested in cubic time ?". We depict in the table below known results from previous works about complexity of the satisfiability problem for extensions of the class of atomic set constraints. At the first view the answer seems to be "yes"; in the conclusion of an extended version of [19] available from Podelski's web page the authors explicitly say that it should be straightforward. Surprisingly, in Section 3 we show that even in the "simple" case of projections on the left-hand side of inclusions, in presence of non-emptiness constraints, the satisfiability problem becomes CoNP-hard.

	no left-hand side projection	with left-hand side projection
no emptiness constraints	$O(n^3)$	$O(n^3)$ [13]
with emptiness constraints	$O(n^3)$ [19]	?

On one hand, projection is a quite intricate operator which can be usually treated easily in the left-hand set of inclusions ([8,21]) by a simple elimination method.[1] However, it leads to more difficult processing in the right-hand side: several methods have been proposed for classes of set constraints that do not admit projection at all [3,4,6,12, 19,23] or admit it only on the left-hand side of inclusions [5,8,13,15,21] but only few concerning projections in the right-hand side [7,9,11]. On the other hand, for the latter classes adding projection operation does not increase the complexity of satisfiability problem. So, one may wonder whether projections could be freely added to atomic set constraints without modifying the cubic time complexity for satisfiability. Again our result from Section 3 implies the negative answer: there exists a simple encoding of non-emptiness constraints using projections in the right-hand side of inclusions provided that the signature is rich enough.

Our second contribution is an algorithm for solving atomic set constraints with projections. It covers both extensions with non-emptiness constraints and projection on both sides of inclusions. We show that it works in PSPACE, while the only previously known algorithm covering this class of constraints (devised for the general set constraints with projections [7]) worked in NEXPTIME. There is still a gap between the CoNP lower bound and PSPACE upper bound; we leave open the question about the exact complexity of the problem.

This work is a continuation of the research project initialized by Aiken, Kozen, Vardi, and Wimmers in [3] to characterize the complexity of the satisfiability problem for set constraints depending on the syntax used. Given the NEXPTIME lower bound in the general case it is natural to ask under which assumptions faster algorithms may be found. In [3] the authors give a detailed analysis of the complexity of subclasses of positive set

[1] Known methods for elimination of left-hand side projections either require boolean operations on set expressions or are correct only wrt. least solutions. Therefore they do not apply to the class in question (cf. Examples 1 and 2).

constrains which are obtained by restricting the ranked alphabet of constructor symbols (for example, positive set constraints over unary trees have a DEXPTIME-complete satisfiability problem). This classification, however, does not say much about applications in program analysis where we usually have to deal with symbols of high arity. The approach based on the choice of used set operators seems to be more appropriate here. For example, the complement operation is almost never used in set-based analysis. Contrary to this, projection (on the left-hand side of inclusion) is used in most applications, it is present already in the early paper by Reynolds [21]. The restriction to nonempty sets is often used as an optimization to speed up the solving process.

Due to lack of space, some proofs are omitted; they can be found in [10].

2 Atomic Set Constraints with Projection

2.1 Definitions

We consider a fixed signature Σ, that is a set of function symbols equipped with arities. Constants are symbols from Σ with arity 0. We use a, b, c to denote constants and f, g, h to denote symbols of arity at least one. We assume moreover that Σ contains at least one constant and one symbol of arity greater than or equal to 2.[2] We denote by $\Sigma^{>0}$ the set of all symbols from Σ that are not constants. We also use a countable set $V = \{x, y, z, \dots\}$ of variables ranging over sets of trees over the signature Σ.

Trees: Let Π be the free monoid generated by the empty path ϵ and concatenation over $\mathbb{N}^+$, the set of strictly positive natural numbers. Words from Π are called *paths*. Given two paths π, π' we write $\pi.\pi'$ for the concatenation of π and π'.

Let τ be a partial mapping from Π to Σ. The mapping τ is said to be (i) *non-empty* if τ is defined for some π, (ii) *prefix-closed* if for any π and i such that τ is defined for $\pi.i$, then τ is defined for π as well and, (iii) *arity consistent* if for all π, if τ is defined for π and the arity of $\tau(\pi)$ is n, then τ is defined for $\pi.1, \dots, \pi.n$ and undefined for all $\pi.j$ with $n < j$.

A tree τ is a partial mapping from Π to Σ, which is non-empty, prefix closed and arity consistent. A tree τ is *finite* if τ is defined for only finitely many paths. We denote T_Σ^ω (resp. T_Σ) the set of all trees (resp. of all finite trees) generated over Σ. For a tree τ and a path π, we write $\tau[\pi]$ for the subtree of τ rooted at position π. If τ is not defined for π then also $\tau[\pi]$ is undefined.

Set Constraints: let Σ^{-1} be the set of unary function symbols, called *projections* and defined as $\{f_i^{-1} \mid f \in \Sigma^{>0}$ and i ranges over $1 \dots arity(f)\}$. We call *set expressions*, terms that are built from V,Σ and Σ^{-1}. We use the letters s, t, u, v (possibly with subscripts) to denote set expressions.

[2] This assumption will be used only for the lower-bound proof; the algorithm from Section 4 works also in the unary case. On the other hand, e.g. in the unary case without emptiness constraints all projections can be easily eliminated and the problem becomes polynomial.

Definition 1. *An* atomic set constraint with projection φ *is given by the following grammar (where s, t are some set expressions and x is a variable)*

$$\varphi ::= s \subseteq t \mid x \not\subseteq \varnothing \mid \varphi \wedge \varphi$$

In the definition above, $\wedge$ stands for logical conjunction. However, we will often identify such conjunction with the set of all its conjuncts.

The semantics of our constraints is given by the structure of sets of trees $\mathsf{P}(\mathcal{T}^{\omega}_{\Sigma})$ or by the structure of sets of finite trees $\mathsf{P}(\mathcal{T}_{\Sigma})$: the domain of $\mathsf{P}(\mathcal{T}^{\omega}_{\Sigma})$ (resp. of $\mathsf{P}(\mathcal{T}_{\Sigma})$) is the set of all subsets of $\mathcal{T}^{\omega}_{\Sigma}$ (resp. of $\mathcal{T}_{\Sigma}$) and function and projection symbols are interpreted as follows: for sets $S, S_1, \ldots, S_n$ of trees (or of finite trees),

$$f(S_1, \ldots, S_n) = \{f(\tau_1, \ldots, \tau_n) \mid \tau_1 \in S_1, \ldots, \tau_n \in S_n\}$$
$$f_i^{-1}(S) = \{\tau_i \mid f(\tau_1, \ldots, \tau_n) \in S\}$$

Finally, both $\mathsf{P}(\mathcal{T}^{\omega}_{\Sigma})$ and $\mathsf{P}(\mathcal{T}_{\Sigma})$ interpret $\subseteq$ as set inclusion and $\not\subseteq \varnothing$ as non-emptiness.

A valuation is a mapping that associates a set of trees with each variable from V. It can be then canonically extended to all set expressions.

We say that a valuation σ satisfies an inclusion $s \subseteq t$ iff $\sigma(s)$ is a subset of $\sigma(t)$ and satisfies a non-emptiness constraint $x \not\subseteq \varnothing$ iff $\sigma(x)$ is a non-empty set. We say that σ satisfies a constraint φ if it satisfies each conjunct from φ.

The satisfiability problem is to determine for a given constraint φ whether there exists a valuation σ that satisfies φ. In this case, we say that σ is a solution of φ. When φ admits a solution, we say that φ is satisfiable, otherwise it is said to be unsatisfiable.

2.2 Expressiveness

Non-emptiness constraints $x \not\subseteq \varnothing$ in Definition 1 may be seen as redundant as we can encode them using projection provided that the signature Σ contains a binary function symbol and a constant as follows:

Example 1. The constraint $a \subseteq f_1^{-1}(f(a, x))$ implies that the denotation of the variable x must be non-empty in any solution of this constraint.

Atomic set constraints with projection generalize several classes of set constraints that have already been studied and for which a cubic time satisfiability algorithm has been devised.

□ The atomic set constraints and atomic set constraints with left-hand side projections correspond to restrictions of the syntax from Definition 1: in both of them, non-emptiness constraints $x \not\subseteq \varnothing$ are not allowed. Moreover, in the first one, set expressions s and t from Definition 1 do not contain projection whereas for the second one only s may contain projections.

Satisfiable constraints from these two classes always have a least solution. This property does not hold when projections are added, which proves that atomic set constraints with projections are strictly more expressive.

Example 2. Looking again at the constraint from Example 1, it is easy to find two minimal solutions σ,σ' such that respectively $\sigma(x) = \{a\}$ and $\sigma'(x) = \{f(a, a)\}$ that are not comparable. Hence, there is no least solution for this constraint.

□ INES Constraints have been introduced in [19]: syntactically defined as atomic set constraints, their main feature is that they are interpreted over non-empty set of trees. As for atomic set constraints, a cubic time algorithm to solve satisfiability has been proposed in [19].

We can encode an INES constraint φ in the constraint from Definition 1 by simply adding to φ the non-emptiness constraint $x \not\subseteq \varnothing$ for all variables x occurring in φ.

INES constraints always admit a greatest solution. However, this is not true for atomic constraints with projection as exemplified below. This proves that our class is strictly more expressive than the class of INES constraints.

Example 3. The constraint $f(x, x') \subseteq a$ implies that the denotation of either x or x' is empty. Thus, one can find two incomparable maximal solutions σ,σ' satisfying respectively $\sigma(x) = \varnothing$, $\sigma(x') = \mathcal{T}_\Sigma$ and $\sigma'(x) = \mathcal{T}_\Sigma$, $\sigma'(x') = \varnothing$. It should be noticed that this constraint is simply not satisfiable over INES.

□ Beyond INES, in [19] the authors consider also atomic set constraints enhanced with non-emptiness constraints $x \not\subseteq \varnothing$ and give a cubic time satisfiability algorithm.

It is proved in [19] that every satisfiable constraint φ in this class has a solution where each nonempty variable x contains either only trees starting with one particular function symbol f (if φ contains an inclusion $x \subseteq f(\dots)$) or it contains trees starting with all function symbols from Σ (if φ does not contain such an inclusion). This is no longer the case for atomic set constraints with projection. For example the constraint $f_i^{-1}(x) \subseteq y \ \wedge y \subseteq a \ \wedge y \subseteq b$ implies that the variable x may contain trees starting with any function symbol from Σ except f. This proves that atomic set constraints with projection are strictly more expressive than atomic set constraints with non-emptiness constraints.

Our final example in this section shows some of the ideas behind our lower-bound proof (cf. the use of markers in the next section). Note that the projection here occurs only on the left-hand side of inclusions.

Example 4. Consider the constraint

$$\begin{array}{lllll} x \not\subseteq \varnothing & \wedge & x \subseteq f(a, z) & \wedge & z \subseteq a \\ & \wedge & f_2^{-1}(x) \subseteq v & \wedge & v \subseteq f(a, a). \end{array}$$

It is not satisfiable because the first two columns imply that intersection of z and v must be nonempty while the third one implies that this intersection is empty.

3 Complexity Lower Bound

We show in this section that the satisfiability problem for atomic set constraints with projection is CoNP-hard. We prove this statement by presenting a polynomial reduction from the unsatisfiability problem for propositional formulas in conjunctive normal form to the satisfiability problem of these set constraints. We recall that CoNP is the class of problems whose complements are in NP and that a problem is CoNP-hard (respectively CoNP-complete) if its complement is NP-hard (respectively NP-complete).

We consider propositional variables ranged over by u. A literal l is either a propositional variable u or its negation $\neg u$. A clause is a disjunctive formula $l_1 \vee \ldots \vee l_n$ where

l_i is a literal. We say that a clause C is non-trivial if the disjunction contains at least one literal and each variable u appears at most once in C either positively u or negatively $\neg u$. A CNF formula is a finite conjunction of non-trivial clauses.

The Problem CNF-UNSAT is as follows :

- INPUT :
  - $u_1, \ldots, u_n$: a non-empty sequence of propositional variables.
  - $C_1 \wedge \ldots \wedge C_p$: a CNF formula written over $u_1, \ldots, u_n$.
- ANSWER : Yes, if $C_1 \wedge \ldots \wedge C_p$ is not satisfiable. No, otherwise.

It is well-known that satisfiability of CNF formulas is NP-complete, hence CNF-UNSAT is CoNP-complete.

Before presenting formally how CNF-UNSAT can be polynomially reduced to the satisfiability problem of atomic set constraints with projection, let us sketch the main ideas of this reduction. Some of these ideas have been used before in [17,18] to prove hardness for satisfiability or entailment problems for constraints overs trees or sets of trees. We describe first these ideas in terms of trees and then turn when needed to sets of trees.

The first point is that Boolean valuations for a sequence of propositional variables $u_1, \ldots, u_n$ can be viewed as a path of length n in a complete binary tree assuming that 1 stands for *true* and 2 for *false*. For instance, for the sequence u_1, u_2, u_3, the path 1.2.1 encodes the valuation associating *true* with u_1 and u_3 and *false* with u_2.

The second idea is to notice that valuations satisfying a clause C do not need to be depicted as a full binary tree, but that a dag representation is sufficient. For instance, the clauses $C_1 = u_1 \vee u_3$ and $C_2 = \neg u_1 \vee \neg u_2$ can be represented as follows

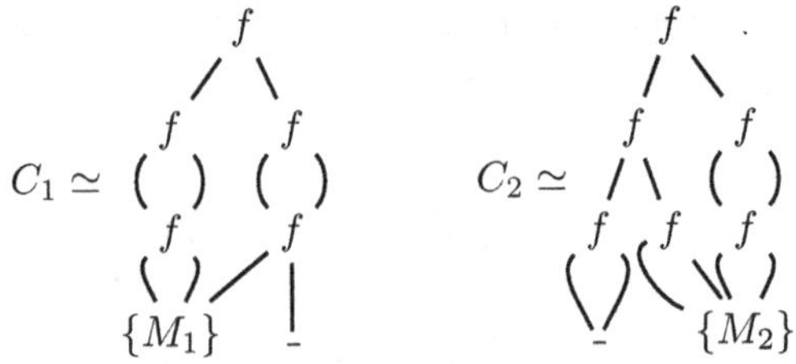

We label leaves of the dags with some markers distinguishing the valuations (paths) satisfying the clause from the valuations that do not. For instance, in the representation of C_1 and of C_2, we used $\{M_1\}$ and $\{M_2\}$ respectively to point out valuations satisfying this clause and _ for the others.

Now, if we unfold these two dags to full binary trees and overlap one with the other we obtain the tree

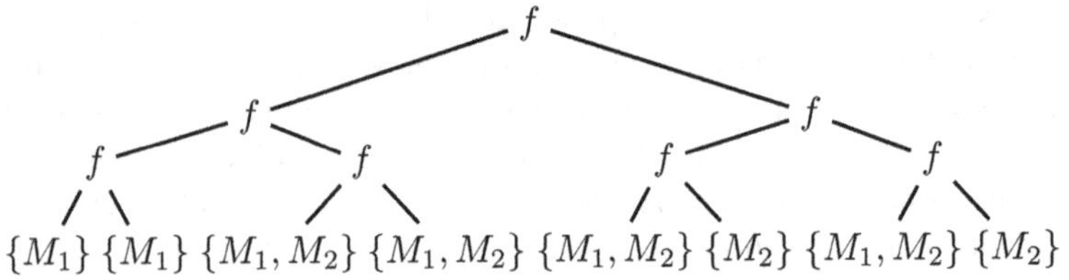

where each leaf corresponds to a valuation (described by the path leading to this leaf) and is labeled with the set of markers identifying the clauses satisfied by this valuation. For example, the node identified by the path 1.2.1 is labeled with $\{M_1, M_2\}$ which indicates that both clauses C_1 and C_2 are satisfied by this valuation. In the general case we will have that the formula in question is satisfiable if and only if this corresponding tree contains a node labeled with all markers.

Now, let us come to a "set" point of view. Encoding of markers and their use to produce an unsatisfiable set constraint in case of a satisfiable formula is the novelty of this reduction. Let us give some intuitions about that. We will use $p+1$ markers, one for each clause and one "tester". Each marker represents a set of trees. These sets are constructed in such a way that the intersection of any p of them is nonempty while the intersection of all $p+1$ is empty. A node in the overlapped tree above labeled with several markers represents the intersection of the sets represented by all these markers. Now if the formula $C_1 \wedge \ldots \wedge C_p$ is satisfiable then the overlapped tree contains a node marked with $\{M_1, \ldots, M_p\}$, hence the set represented by this node intersected with the set represented by the tester is empty, which together with a non-emptiness constraint make the corresponding set constraint unsatisfiable.

The crucial part is the definition of the markers. It heavily depends on the use of projections and is a generalization of Example 4, where the two lines define respectively the marker and the tester in the case of $p = 1$. The rough idea is that a constraint of the form $f_2^{-1}(x) \subseteq x'$ is very similar to a constraint of the form $x \subseteq f(\_, x')$. The important difference is that in the second case the variable x may contain only trees whose root is labelled with the symbol f while in the first case it may contain terms starting with a different symbol. On the picture below we use dashed lines for the constraints of the first form and solid lines for the constraints of the second form.

For $p = 2$, the markers $\{M_1\}$, $\{M_2\}$ can be depicted as the following two trees on the left:

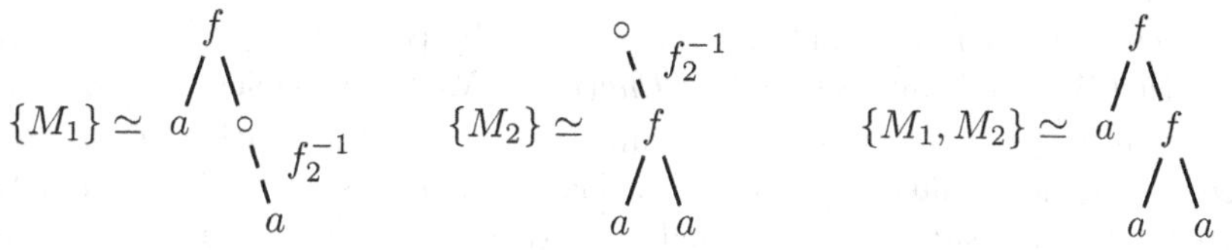

Sets of trees represented by the markers are partially described by means of projections. The main part in the markers is the right-most path, that is paths of the form 2^*. For the makers $\{M_i\}$ we force the node 2^{i-1} to be an f: for $\{M_1\}$, this is ϵ and for $\{M_2\}$ the node at path 2. The rest of this path is defined as projection f_2^{-1}.

Roughly speaking, the meaning of $\{M_1\}$ is a non-empty set of trees satisfying the following requirement: the root of the tree is labeled with f and the position 1 with a. Additionally, if the position 2 is labeled with f then the position 2.2 is labeled with a. Now, concerning $\{M_2\}$, its meaning is a non-empty set of trees satisfying the following requirement: if the root is labeled with f then the positions 2, 2.1 and 2.2 are labeled respectively with f, a and a.

Now, the marker $\{M_1, M_2\}$ produced by overlapping trees from the clauses is described above as the last tree on the right. Note that this tree is totally defined, there is no more projection f_2^{-1} over it. This property implies that the intersection with the set represented by the tester is empty. The meaning of $\{M_1, M_2\}$ must satisfy both requirements of $\{M1\}$ and $\{M_2\}$; the only set satisfying them is the singleton $\{f(a(f(a,a)))\}$.

The last point of the encoding is the tester. The left-hand part of the picture below is used for overlapping with the whole binary tree, the right-hand part is the actual tester.

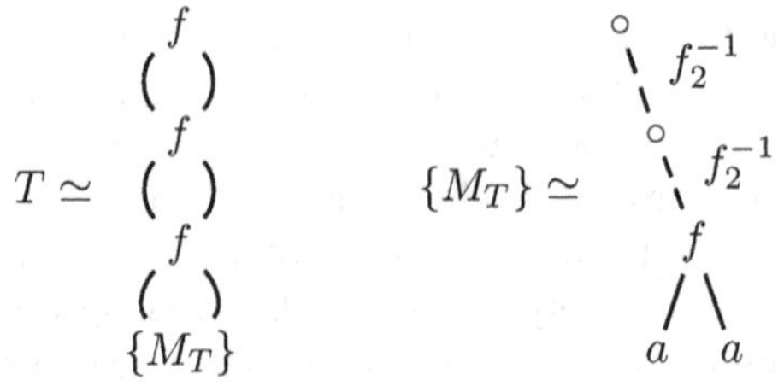

The tester $\{M_T\}$ represents any non-empty set of trees that satisfy the requirement: if the positions ϵ and 2 of the tree are labeled with f then positions 2.2, 2.2.1 and 2.2.2 are labeled respectively by f, a and a.

Note that the intersection of the sets represented by $\{M_T\}$ with any of $\{M_1\}$, $\{M_2\}$ is nonempty but the intersection of $\{M_T\}$ with $\{M_1, M_2\}$ is empty. Indeed, the set $\{f(a,a)\}$ satisfies the requirements of both $\{M_1\}$ and $\{M_T\}$ whereas the set $\{a\}$ satisfies the requirements of both $\{M_2\}$ and $\{M_T\}$. But no non-empty set can satisfy the requirements of $\{M_1\}$, $\{M_2\}$ (that is, of $\{M_1, M_2\}$) and $\{M_T\}$ at the same time.

We show now formally that we can built in polynomial time from any instance of the CNF-UNSAT problem a set constraint with projection that is satisfiable iff the answer to the instance of CNF-UNSAT is *Yes*.

We consider a signature Σ containing a symbol f which is at least binary and another function symbol. For simplicity, we will assume f to be binary and that the other function symbol is a constant a.

We consider an instance of CNF-UNSAT given by $\{u_1, \ldots, u_n, C_1 \wedge \ldots \wedge C_p\}$ (for short, $\{\bar{u}, \bar{C}\}$) and define its encoding $\mathit{Enc}(\bar{u}, \bar{C})$. We use variables x, x_j^i, y_j^i, w_j with $1 \leq i \leq p, 1 \leq j \leq n+1$ and z_k^i, v^i with $1 \leq i, k \leq p$. The variables x_j^i, y_j^i are used to encode Boolean valuations (for the ith clause and the jth variable), the variables w_j to encode the top part of the tester and the variables z_k^i, v^i to encode respectively the markers for clauses and the marker of the tester. We let

$$\mathit{Enc}(\bar{u}, \bar{C}) = \begin{cases} x \not\subseteq \varnothing \wedge x \subseteq w_1 \wedge \mathit{Enc}_{\mathit{Test}}(\bar{u}) \wedge \mathit{Enc}_{\mathit{MTest}}(\bar{C}) \\ \bigwedge_{1 \leq i \leq p} x \subseteq x_1^i \wedge \bigwedge_{1 \leq i \leq p} \mathit{Enc}_{\mathit{Cl}}(i, \bar{u}, \bar{C}) \wedge \bigwedge_{1 \leq i \leq p} \mathit{Enc}_{\mathit{Mark}}(i, \bar{C}) \end{cases}$$

$$\mathit{Enc}_{\mathit{Test}}(\bar{u}) = \bigwedge_{1 \leq j \leq n} w_j \subseteq f(w_{j+1}, w_{j+1}) \wedge w_{n+1} \subseteq v^1$$

$$\mathit{Enc}_{\mathit{MTest}}(\bar{C}) = \bigwedge_{1 \leq i \leq p} f_2^{-1}(v^i) \subseteq v^{i+1} \wedge v^{p+1} \subseteq f(a,a)$$

$$Enc_{Mark}(i, \bar{C}) = \bigwedge_{1 \leq k \leq p, k \neq i} f_2^{-1}(z_k^i) \subseteq z_{k+1}^i \wedge z_i^i \subseteq f(a, z_{i+1}^i) \wedge z_{p+1}^i \subseteq a$$

$$Enc_{Cl}(i, \bar{u}, \bar{C}) = \bigwedge_{1 \leq j \leq n} y_j^i \subseteq f(y_{j+1}^i, y_{j+1}^i) \wedge y_{n+1}^i \subseteq z_1^i \wedge \bigwedge_{1 \leq j \leq n} Enc_{LC}(i, j, \bar{u}, \bar{C})$$

$$Enc_{LC}(i, j, \bar{u}, \bar{C}) = \begin{cases} x_j^i \subseteq f(y_{j+1}^i, x_{j+1}^i) & \text{if } u_j \text{ occurs in } C_i \\ x_j^i \subseteq f(x_{j+1}^i, y_{j+1}^i) & \text{if } \neg u_j \text{ occurs in } C_i \\ x_j^i \subseteq f(x_{j+1}^i, x_{j+1}^i) & \text{if } u_j \text{ and } \neg u_j \text{ don't occur in } C_i \end{cases}$$

Proposition 1. *A CNF formula $C_1 \wedge \ldots \wedge C_p$ written over variables $u_1, \ldots, u_n$ is satisfiable if and only if the corresponding set constraint $Enc(\{u_1, \ldots, u_n, C_1, \ldots, C_p\})$ is not satisfiable.*

Theorem 1. *Let Σ be a signature containing at least two function symbols, one of these symbols having an arity greater than or equal to two. Then satisfiability problem for atomic set constraints with projection is Co-NP-hard both over sets of trees $\mathsf{P}(\mathcal{T}_\Sigma^\omega)$ and sets of finite trees $\mathsf{P}(\mathcal{T}_\Sigma)$.*

Proof. *Obvious from Proposition 1 noticing that the size of the constraint $Enc(\{\bar{u}, \bar{C}\})$ is polynomial in the size of $\{\bar{u}, \bar{C}\}$, the input of CNF-UNSAT.*

Remark 1. Consider an introduction of projections of constant symbols to the syntax of set constraints with the semantics $a^{-1}(S) = \mathcal{T}_\Sigma$ if $a \in S$ and $a^{-1}(S) = \varnothing$ otherwise. Then the satisfiability problem becomes DEXPTIME-hard by a simple reduction from the emptiness of intersection of deterministic top-down tree automata.

4 A Satisfiability Algorithm

4.1 Preliminaries

To handle constraints more easily we will assume them to be *flat*.

Definition 2. *A constraint φ is said to be flat if it is generated by the following grammar ($x, y, x_1, \ldots, x_n$ being variables)*

$$\varphi ::= x \subseteq y \mid x = f(x_1, \ldots, x_n) \mid x = f_i^{-1}(y) \mid x \nsubseteq \varnothing \mid \varphi \wedge \varphi \mid \bot$$

where equality $x = t$ is interpreted as $x \subseteq t \wedge t \subseteq x$ and $\bot$ represents the value False. As soon as $\bot$ occurs in a constraint φ, φ is not satisfiable.

Considering flat constraints is not a loss of generality: for any constraint φ one can compute in linear time in the size of φ a flat constraint φ' equivalent to φ.

Given a constraint φ, we write Σ_φ for the set of all function symbols f from Σ occurring in φ either as a function symbol f or as a projection f_i^{-1}. The set $\Sigma_\varphi^{>0}$ is the set Σ_φ restricted to symbols that are not constants.

4.2 Annotated Paths – Path Constraints

We introduce a new kind of constraints called *path constraints* defined over annotated paths.

Annotated Paths: An *annotated path* κ is an element of the free monoid generated by the empty path ϵ and concatenation over $\{\langle f, i\rangle \mid f \in \Sigma_{\varphi}^{>0}$ and $1 \leq i \leq arity(f)\}$.

Given a tree τ and an annotated path κ, we define the subtree $\tau[\kappa]$ of τ rooted at κ as (i) τ if $\kappa = \epsilon$, as (ii) $\tau_i[\kappa']$ if $\kappa = \langle f, i\rangle.\kappa'$ and $\tau = f(\tau_1, \ldots, \tau_n)$, and (iv) undefined otherwise.

It should be noticed that whereas paths express only positions in a tree, annotated paths express also conditions over the labeling of the tree; for the trees $\tau = f(a, b)$ and $\tau' = g(a, b)$, both $\tau[1]$ and $\tau'[1]$ are defined and equal to a. But for the annotated path $\langle f, 1\rangle$, the tree $\tau[\langle f, 1\rangle]$ is defined and equal to a, whereas $\tau'[\langle f, 1\rangle]$ is undefined as the root of τ' is a symbol different from f.

Given a set of trees S, we write $S[\kappa]$ for the set $\{\tau[\kappa] \mid \tau \in S\}$. It is always defined although it can be empty. For an annotated path κ, we define its *restriction over* $\mathbb{N}^+$ (denoted $\mathsf{res}(\kappa)$) as the path π obtained from κ by removing annotating function symbols.

Path Constraints: a path constraint ψ is written as $x[\kappa] \subseteq y$. Its semantics is as follows: a valuation σ satisfies $x[\kappa] \subseteq y$ iff $\sigma(x)[\kappa] \subseteq \sigma(y)$. Note that if $\kappa = \langle f, 1\rangle$ is an annotated path of length 1 then $x[\kappa] \subseteq y$ is just a different notation for $f_i^{-1}(x) \subseteq y$.

We define a binary relation $\vdash$ between flat constraints and path constraints inductively as follows:

- $\varphi \vdash x[\epsilon] \subseteq y$ if $x \subseteq y \in \varphi$.
- $\varphi \vdash x[\langle f, i\rangle] \subseteq x_i$ if $x = f(x_1, \ldots, x_n) \in \varphi$ or $f_i^{-1}(x) = x_i \in \varphi$.
- $\varphi \vdash x[\kappa.\kappa'] \subseteq y$ if there exists z such that $\varphi \vdash x[\kappa] \subseteq z$ and $\varphi \vdash z[\kappa'] \subseteq y$.

Proposition 2. *If $\varphi \vdash x[\kappa] \subseteq y$ then any solution σ of φ, $\sigma(x)[\kappa] \subseteq \sigma(y)$.*

Proof. By induction over the length of κ.

Let us consider an annotated path κ such that for some x and y, $\varphi \vdash x[\kappa] \subseteq y$. There may exist a solution σ for which $\sigma(x)$ is non-empty but nonetheless $\sigma(x)[\kappa]$ is empty. For instance, for the constraint $\varphi_1 = x \not\subseteq \varnothing \wedge f_1^{-1}(x) = y \wedge y = g(z)$ there exists a solution σ such that $\sigma(x) = \{a\}, \sigma(y) = \sigma(z) = \varnothing$, so $\sigma(x)[\langle f, 1\rangle]$ is empty. Now considering the constraint $\varphi_2 = \varphi_1 \wedge x = f(x_1, x_2)$, one can deduce that for any solution σ' of φ_2, both $\sigma'(x)[\langle f, 1\rangle]$ and $\sigma'(x)[\langle f, 1\rangle.\langle g, 1\rangle]$ is not empty. The annotated paths $\langle f, 1\rangle$ and $\langle f, 1\rangle.\langle g, 1\rangle$ become unavoidable for x in the constraint φ_2.

We define $\mathcal{UAP}(x, \varphi)$ the set of *unavoidable annotated paths* for the variable x and the constraint φ as the least set satisfying :

- $\epsilon \in \mathcal{UAP}(x, \varphi)$
- if $\kappa \in \mathcal{UAP}(x, \varphi)$, $\varphi \vdash x[\kappa] \subseteq y$ and $y = f(\bar{y})$ then $\kappa.\langle f, i\rangle \in \mathcal{UAP}(x, \varphi)$ for all $1 \leq i \leq arity(f)$.

Proposition 3. *For all solutions σ for φ, for all variables x, for all annotated paths κ from $\mathcal{UAP}(x, \varphi)$, if $\sigma(x)$ is non-empty then for all trees τ in $\sigma(x)$, $\tau[\kappa]$ is defined.*

Proposition 4. *For all variables x in φ, if there exists an unavoidable annotated path κ in $\mathcal{UAP}(x, \varphi)$ such that $\varphi \vdash x[\kappa] \subseteq y$, $\varphi \vdash x[\kappa] \subseteq z$ and for two different function symbols f, g, the two constraints $y = f(\bar{y})$, $z = g(\bar{z})$ belong to φ then for any solution σ for φ, $\sigma(x)$ is empty.*

(I1) $x \subseteq x; \quad x \subseteq y, y \subseteq z \to x \subseteq z$
(I2) $x \not\subseteq \varnothing, x = f(\bar{x}), y = f(\bar{y}), x \subseteq y \to \bar{x} \subseteq \bar{y}$
(I3) $x \not\subseteq \varnothing, x = f(\bar{x}), y = f_i^{-1}(z), x \subseteq z \to x_i \subseteq y$
(I4) $x = f_i^{-1}(y), y \subseteq z, z = f(\bar{z}) \to x \subseteq z_i$
(I5) $x = f_i^{-1}(x'), y = f_i^{-1}(y'), x' \subseteq y' \to x \subseteq y$

(N1) $\bar{x} \not\subseteq \varnothing, x = f(\bar{x}) \to x \not\subseteq \varnothing$
(N2) $x \not\subseteq \varnothing, x = f_i^{-1}(x') \to x' \not\subseteq \varnothing$
(N3) $x \not\subseteq \varnothing \to y \not\subseteq \varnothing$ if exists κ in $\mathcal{UAP}(x, \varphi)$ such that $\varphi \vdash x[\kappa] \subseteq y$

(C1) $x \not\subseteq \varnothing, y = f(\bar{y}), z = g(\bar{z}) \to \bot$
if exists κ in $\mathcal{UAP}(x, \varphi)$ such that $\varphi \vdash x[\kappa] \subseteq y$, $\varphi \vdash x[\kappa] \subseteq z$
(C2) $x \subseteq y, y = g(\bar{y}), f_i^{-1}(x) = x', x' \not\subseteq \varnothing \to \bot$

Fig. 1. The axioms Ax

4.3 A Satisfiability Test

The Axioms Ax*:* we give in Figure 1 a set of axioms Ax on which our satisfiability test is based. To shorten the presentation, we write $\bar{x}$ a sequence $x_1, \ldots, x_k$ of variables; the length k is most of the time fixed by the context. For instance, writing $f(\bar{x})$ fixes k to be the arity of f. Similarly, we write $\bar{x} \subseteq \bar{y}$ to denote the constraints $x_1 \subseteq y_1, \ldots, x_k \subseteq y_k$, $\bar{x} \not\subseteq \varnothing$ to denote the constraints $x_1 \not\subseteq \varnothing, \ldots, x_k \not\subseteq \varnothing$.

Let us discuss the axioms from Ax: the axioms (I1 – 5) as well as (N1 – 2) and (C2) are immediate. The axiom (N3) is less obvious and subsumes other axioms that we may think about; it is more general than $x \not\subseteq \varnothing, x \subseteq y \to y \not\subseteq \varnothing$ as ϵ is always unavoidable and $x \subseteq y$ implies $\varphi \vdash x[\epsilon] \subseteq y$. It makes also redundant the axiom $x \not\subseteq \varnothing, x = f(\bar{x}) \to \bar{x} \not\subseteq \varnothing$, as $x = f(\bar{x})$ implies $\varphi \vdash x[i] \subseteq x_i$ for all $1 \leq i \leq arity(f)$. Furthermore, from the following constraint, using the axiom (N3),

$$x \not\subseteq \varnothing \wedge x \subseteq y \wedge y = f(y_1, y_2) \wedge g_2^{-1}(y_1) = y' \\ \wedge x \subseteq z \wedge f_1^{-1}(z) = z' \wedge z' = g(z_1, z_2)$$

one can deduce $y' \not\subseteq \varnothing, z_1 \not\subseteq \varnothing, z_2 \not\subseteq \varnothing$ as the paths $\langle f, 1\rangle.\langle g, 1\rangle, \langle f, 1\rangle.\langle g, 2\rangle$ are unavoidable for x.

Additionally, when the structure of sets of finite trees $\mathsf{P}(\mathcal{T}_\Sigma)$ is considered, a new axiom (OC) addressing the occur-check problem is required.

(OC) $x \not\subseteq \varnothing \to \bot$ if $\mathcal{UAP}(x, \varphi)$ is infinite.

This axiom is pretty unusual: traditionally [19,8], occur-check is stated as the existence of a non-trivial path from a non-empty variable to itself. So, one may expect an axiom such as: (OC$_1$) $x \not\subseteq \varnothing \to \bot$ if $\varphi \vdash x[\kappa] \subseteq x$ for an annotated path $\kappa \neq \epsilon$. Unfortunately, this axiom is too strong as $\bot$ could be deduced from the constraint $x \not\subseteq \varnothing \wedge f_1^{-1}(x) = y \wedge y \subseteq x$ and obviously, this constraint is satisfiable over sets of finite trees $\mathsf{P}(\mathcal{T}_\Sigma)$—for example the valuation assigning $\{a, f(a)\}$ to x and $\{a\}$ to y is a solution. One may try to make the axiom (OC$_1$) weaker by reasoning not about annotated paths, but only about unavoidable annotated paths. We would have an axiom such as: $(\mathsf{OC}_2)\ x \not\subseteq \varnothing \rightarrow \bot \quad$ if $\varphi \vdash x[\kappa] \subseteq x$ for $\kappa \in \mathcal{UAP}(x, \varphi)$ and $\kappa \neq \epsilon$. Unfortunately, the axiom (OC_2) is too weak; consider the constraint

$$x \not\subseteq \varnothing \wedge x \subseteq y_1 \wedge h_1^{-1}(y_1) = y_2 \wedge \ y_2 = h(y_1)$$
$$x \subseteq z_1 \wedge \ z_1 = h(z_2) \ \wedge h_1^{-1}(z_2) = z_1$$

One can see that this constraint admits no solution over sets of finite trees; nevertheless, $\bot$ can not be deduced using the axiom (OC_2) as there is no non-trivial unavoidable annotated path κ leading from a variable to itself (intuitively, there is an unavoidable path from the intersection $y_1 \cap z_1$ to itself).

So, we require simply for the occur-check the set $\mathcal{UAP}(x, \varphi)$ to be infinite and we will prove below that this statement is correct.

Theorem 2. *The axioms* Ax *are valid for* $\mathsf{P}(\mathcal{T}_\Sigma^\omega)$. *The axioms* $\mathsf{Ax} + (\mathsf{OC})$ *are valid for* $\mathsf{P}(\mathcal{T}_\Sigma)$.

Proof. By routine check for the axioms $(\mathsf{I1}-5)$, $(\mathsf{N1}-2)$ *and* $(\mathsf{C2})$. *Using Proposition 3 for the axiom* $(\mathsf{N3})$. *Using Proposition 4 for the axiom* $(\mathsf{C1})$. *Finally, for the axiom* (OC), *in case of sets of finite trees* $\mathsf{P}(\mathcal{T}_\Sigma)$, *let us show that for any variable* x, *if* $x \not\subseteq \varnothing$ *in* φ *and* $\mathcal{UAP}(x, \varphi)$ *is infinite, then* φ *admits no solution. Function symbols from* Σ *occurring in* φ *have fixed arities, and thus, these arities must be bounded. As* $\mathcal{UAP}(x, \varphi)$ *is infinite, by Koenig' Lemma, we can extract from it an infinite sequence of annotated paths* $\kappa_0, \kappa_1, \kappa_2, \ldots$ *such that* κ_i *is a prefix of* κ_{i+1}. *So, by Proposition 3, for any tree* τ *in* $\sigma(x)$ *and for any* κ_i, $\tau[\kappa_i]$ *is defined. Obviously, this requirement can not be satisfied by some finite tree. So,* φ *is not satisfiable over* $\mathsf{P}(\mathcal{T}_\Sigma)$.

A Satisfiability Test: we describe here how the axioms Ax (or $\mathsf{Ax} + (\mathsf{OC})$) can be used for defining a test for satisfiability for a constraint φ.

We first consider the axioms from Ax and (OC) as inference rules: premises are left-hand sides of implications and conclusions the right-hand sides. Note that some rules like $(\mathsf{N3})$, $(\mathsf{C1})$, (OC) have side conditions concerning unavoidable paths.

We denote $Sat(\varphi)$ as the least set of constraints containing φ and closed under those inference rules, that is if premises of an inference rule belong to $Sat(\varphi)$ and if the side condition is satisfied, then the conclusion belongs to $Sat(\varphi)$. We will moreover require that the variables from $Sat(\varphi)$ are exactly those from φ; this is guaranteed by an application of the axiom $x \subseteq x$ from Ax only for variables from φ.

Actually, the axioms $\mathsf{Ax} + (\mathsf{OC})$ can be viewed as an operator Sat over constraints that, given a constraint φ computes all consequences of φ under these axioms. Moreover, we can notice that for φ viewed as a collection of basic constraints, the set $\mathcal{UAP}(x, \varphi)$ is monotonic in the sense that $\varphi \subseteq \varphi'$ implies $\mathcal{UAP}(x, \varphi) \subseteq \mathcal{UAP}(x, \varphi')$. Thus, Sat is a monotonic operator over constraints: for two constraints φ, φ', if $\varphi \subseteq \varphi'$ then $Sat(\varphi) \subseteq Sat(\varphi')$. So, over the lattice with φ as least element, the operator Sat admits a least fix-point that we denote $Sat^\omega(\varphi)$.

Theorem 3. *if* $\bot \in Sat^\omega(\varphi)$ *then* φ *is unsatisfiable.*

Proof. Immediate from Theorem 2

We prove now the converse of Theorem 3: if $\perp$ does not belong to $Sat^{\omega}(\varphi)$ then φ admits a solution. We build from $Sat^{\omega}(\varphi)$ a valuation σ_S defined for any variable x as follows (i) $\sigma_S(x) = \varnothing$ if $x \not\subseteq \varnothing \notin Sat^{\omega}(\varphi)$ and (ii) otherwise, $\sigma_S(x)$ is the set of all trees τ satisfying

for all f, π, κ, y s.t. $\pi = \mathsf{res}(\kappa), \tau[\kappa]$ is defined and $Sat^{\omega}(\varphi) \vdash x[\kappa] \subseteq y$

if $y = f(\bar{y}) \in Sat^{\omega}(\varphi)$ then $\tau(\pi) = f$ and

if $z = f_i^{-1}(y), y \not\subseteq \varnothing \in Sat^{\omega}(\varphi)$ and $z \not\subseteq \varnothing \notin Sat^{\omega}(\varphi)$ then $\tau(\pi) \neq f$

Theorem 4. *if* $\perp \notin Sat^{\omega}(\varphi)$ *then* σ_S *is a solution for* $Sat^{\omega}(\varphi)$ *and thus for* φ.

4.4 A PSPACE Algorithm

We devise in this section a PSPACE algorithm deciding whether a constraint φ is unsatisfiable, that is according to Theorems 3 and 4 deciding whether $\perp$ belongs to $Sat^{\omega}(\varphi)$. To present and prove correctness of our algorithm requires some auxiliary propositions.

Proposition 5. *For any constraint* φ *of size* $|\varphi|$ *with* n *variables,*

- *for any* i*, the size of* $Sat^{i}(\varphi)$ *is upper-bounded by* $|\varphi|^2 + 2|\varphi| + 1$*;*
- *there exists an integer* p *such that* $p \leq n^2 + n + 1$ *and* $Sat^{p}(\varphi) = Sat^{\omega}(\varphi)$.

Testing that an annotated path κ is unavoidable for a variable x and a constraint φ is polynomial:

Proposition 6. *Let* φ *be a constraint and* κ *an annotated path. Deciding whether* $\kappa \in \mathcal{UAP}(x, \varphi)$ *can be done in polynomial time in* $|\varphi|$ *and* $|\kappa|$ *where* $|\varphi|$ *is the size of* φ *and* $|\kappa|$ *the length of* κ.

Moreover, we can show that if premises and side-conditions for the axioms (N3) (C1) hold for some annotated paths then they hold for annotated paths whose length is at most exponential in the number of variables in φ,

Proposition 7. *Let* φ *be a constraint and* n *the number of variables occurring in* φ.

- *for all* κ *in* $\mathcal{UAP}(x, \varphi)$ *such that* $\varphi \vdash x[\kappa] \subseteq y$*, there exists* κ' *in* $\mathcal{UAP}(x, \varphi)$ *such that* $\varphi \vdash x[\kappa'] \subseteq y$ *and* $|\kappa'| \leq 2^n$.
- *for all* κ *in* $\mathcal{UAP}(x, \varphi)$ *such that* $\varphi \vdash x[\kappa] \subseteq y$ *and* $\varphi \vdash x[\kappa] \subseteq z$ *with* $y = f(\bar{y})$ *and* $z = g(\bar{z})$ *then there exists* κ' *in* $\mathcal{UAP}(x, \varphi)$*,* y', z' *two variables and* f', g' *two different function symbols such that* $\varphi \vdash x[\kappa'] \subseteq y'$ *and* $\varphi \vdash x[\kappa'] \subseteq z'$ *with* $y' = f'(\bar{y'})$ *and* $z' = g'(\bar{z'})$ *and* $|\kappa'| \leq 2^{n^2}$.

Finally, we show that $\mathcal{UAP}(x, \varphi)$ is infinite if it contains a path sufficiently long and this length on which infiniteness can be tested is simply exponential in the number of variables occurring in φ,

Proposition 8. *Let* φ *be a constraint and* n *the number of variables occurring in* φ*, if* $\mathcal{UAP}(x, \varphi)$ *contains a path* κ *of length equal* $2^n + 1$ *then* $\mathcal{UAP}(x, \varphi)$ *is infinite.*

Proposition 9. *Testing side-conditions for the inference rules* (N3)*,* (C1) *and* (OC) *is in PSPACE.*

Proof. *We present a non-deterministic polynomial-space algorithm to test the side-condition for the rule* (N3)*; algorithms for* (C1) *and* (OC) *are similar.*

This algorithm is based on the result of Proposition 7, that is if there exists an unavoidable path κ *for* x *in* φ *leading to* y *then there is an unavoidable path* κ' *of length at most* 2^n *for* x *in* φ *leading to* y.

$V := \{x\}$; $V_n := \varnothing$; $c := 1$
While $y \notin V$ *and* $c \leq 2^n$ *do*
  Guess $f \in \Sigma$ *and* $i \leq arity(f)$ *s.t.* $u = f(\bar{v}) \in \varphi$ *for some* $u \in V$
  $V_n := \{w \mid \varphi \vdash z[\langle f, i\rangle] \subseteq w \text{ and } z \in V\}$
  $c := c + 1$; $V := V_n$
Endwhile
Return ($y \in V$)

Obviously, this algorithm uses only polynomial space in n*: the cardinality of the sets* V *and* V_n *is bounded by* n *and the counter* c *can be coded on* $\log(2^n) \leq n + 1$ *bits. This non-deterministic algorithm can be turned into a deterministic one using the Savitch's theorem [22] (NPSPACE(*n*) = PSPACE(*n^2*)).*

Proposition 10. *For any constraint* φ, $Sat^{i+1}(\varphi)$ *can be computed from* $Sat^i(\varphi)$ *with an algorithm running in polynomial-space in* $|\varphi|^2 + 2|\varphi| + 1$.

Proof. *The computation is done as follows: we start from a constraint* S *equal to* $Sat^i(\varphi)$ *and we iterate until* S *is unchanged the following steps:*

1. *we saturate* S *by the axioms* $(I1-5)$, $(N1-2)$ *and* $(C2)$ *in polynomial time in the size of* $Sat^i(\varphi)$ *(upper-bounded by* $|\varphi|^2 + 2|\varphi| + 1$*; Proposition 5);*
2. *for any variable* x *and any constraint* $y \not\subseteq \varnothing$, *we test if the side-condition of* $(N3)$ *holds for* $\mathcal{UAP}(x, Sat^i(\varphi))$ *in polynomial-space in the size of* $Sat^i(\varphi)$ *(Proposition 9); if it holds, we add* $y \not\subseteq \varnothing$ *to* S*;*
3. *for any variable* x *such that* $x \not\subseteq \varnothing \in S$, *any pair of constraints* $y = f(\bar{y}), z = g(\bar{y})$ *from* φ, *we test if the side-condition of* $(C1)$ *holds for* $\mathcal{UAP}(x, Sat^i(\varphi))$ *in polynomial-space in the size of* $Sat^i(\varphi)$ *(Proposition 9); if it holds, we add* $\perp$ *to* S*;*
4. *for an interpretation over sets of finite trees: for any variable* x *such that* $x \not\subseteq \varnothing \in S$, *we test if the side-condition of* (OC) *holds for* $\mathcal{UAP}(x, Sat^i(\varphi))$ *in polynomial-space in the size of* $Sat^i(\varphi)$ *(Proposition 9); if it holds, we add* $\perp$ *to* S*;*

The constraint $Sat^{i+1}(\varphi)$ *is the constraint* S *obtained at the end of the iteration process. As the size of* $Sat^{i+1}(\varphi)$, *is bounded by* $|\varphi|^2 + 2|\varphi| + 1$, *this latter provides also an upper-bound on the number of possible iterations.*

Theorem 5. *Satisfiability for atomic set constraints with projection is in PSPACE.*

Proof. *Immediate, combining results from Propositions 5 and 10.*

References

1. A. Aiken. Set Constraints: Results, Applications, and Future Directions. In *Proc. of the* 2rd *International Workshop on Principles and Practice of Constraint Programming*, LNCS 874, pages 326–335. 1994.

2. A. Aiken. Introduction to Set Constraint-Based Program Analysis. In *Science of Computer Programming*, 35(2): 79-111, 1999.
3. A. Aiken, D. Kozen, M. Vardi and E. L. Wimmers. The complexity of set constraints. In *1993 Conference on Computer Science Logic*, LNCS 832, pages 1–17, 1993.
4. A. Aiken, D. Kozen and E. Wimmers. Decidability of Systems of Set Constraints with Negative Constraints. *Information and Computation*, 122(1):30–44, oct 1995.
5. L. Bachmair, H. Ganzinger and U. Waldmann. Set constraints are the monadic class. In *Proc. of the* 8th *IEEE Symp. on Logic in Computer Science*, pages 75–83, 1993.
6. W. Charatonik and L. Pacholski. Negative Set Constraints with Equality. In *Proc. of the* 9th *IEEE Symp. on Logic in Computer Science*, pages 128–136, 1994.
7. W. Charatonik and L. Pacholski. Set Constraints with Projections are in NEXPTIME. In *Proc. of the* 35th *Symp. on Foundations of Computer Science*, pages 642–653, 1994.
8. W. Charatonik and A. Podelski. Set Constraints with Intersection. In *Proc. of the* 12th *IEEE Symp. on Logic in Computer Science*, pages 362–372, 1997.
9. W. Charatonik and A. Podelski. Co-definite Set Constraints. In *Proc. of the* 9th *International Conference on Rewriting Techniques and Applications*, LNCS 1379, pages 211–225, 1998.
10. W. Charatonik and J.-M. Talbot. Atomic Set Constraints with Projection. *Research Report Max-Planck-Institut für Informatik.* MPI-I-2002-2-008, 2002.
11. P. Devienne, J.-M. Talbot and S. Tison. Solving classes of set constraints with tree automata. In *Proc. of the* 3rd *International Conference on Principles and Practice of Constraint Programming*, LNCS 1330, pages 62–76, 1997.
12. R. Gilleron, S. Tison and M. Tommasi. Solving Systems of Set Constraints with Negated Subset Relationships. In *Proc. of the* 34th *Symp. on Foundations of Computer Science*, pages 372–380, 1993.
13. N. Heintze. *Set Based Program Analysis.* PhD thesis, Carnegie Mellon University, 1992.
14. N. Heintze and J. Jaffar. A Finite Presentation Theorem for Approximating Logic Programs. In *Proc. of the* 17th *Symp. on Principles of Programming Languages*, pages 197–209, 1990.
15. N. Heintze and J. Jaffar. A Decision Procedure for a Class of Herbrand Set Constraints. In *Proc. of the* 5th *IEEE Symp. on Logic in Computer Science*, pages 42–51, 1990.
16. N. Heintze and J. Jaffar. Set constraints and set-based analysis. In *Proc. of the Workshop on Principles and Practice of Constraint Programming*, LNCS 874, pages 281–298, 1994.
17. F. Henglein and J. Rehof. The complexity of subtype entailment for simple types. In *Proc. of the* 12th *IEEE Symp. on Logic in Computer Science*, pages 362–372, 1997.
18. J. Niehren, M. Müller and J.-M. Talbot. Entailment of Atomic Set Constraints is PSPACE-Complete. In *Proc. of the* 14th *IEEE Symp. on Logic in Computer Science*, pages 285–294, 1999.
19. M. Müller, J. Niehren and A. Podelski. Inclusion Constraints over Non-Empty Sets of Trees. In *Proc. of* 7th *International Joint Conference CAAP/FASE - (TAPSOFT'97)*, LNCS 1214, pages 345–356, 1997.
20. L. Pacholski and A. Podelski. Set Constraints: A Pearl in Research on Constraints. In *Proc. of the* 3rd *International Conference on Principles and Practice of Constraint Programming*, LNCS 1330, pages 549–561, 1997. Tutorial.
21. J. C. Reynolds. Automatic computation of data set definitions. *Information Processing*, 68:456–461, 1969.
22. Savitch, W. Relationships between nondeterministic and deterministic tape complexities. *Journal of Computer and System Sciences* 4(2), 177–192.
23. K. Stefansson. Systems of Set Constraints with Negative Constraints are NEXPTIME-Complete. In *Proc. of the* 9th *IEEE Symp. on Logic in Computer Science*, pages 137–141, 1994.
24. T. E. Uribe. Sorted Unification using Set Constraints. In *Proc. of the* 11th *International Conference on Automated Deduction*, LNAI 607, pages 163–177, 1992.

Currying Second-Order Unification Problems*

Jordi Levy[1] and Mateu Villaret[2]

[1] IIIA, CSIC, Campus de la UAB, Barcelona, Spain.
http://www.iiia.csic.es/~levy
[2] IMA, UdG, Campus de Montilivi, Girona, Spain.
http://www.ima.udg.es/~villaret

Abstract. The Curry form of a term, like $f(a,b)$, allows us to write it, using just a single binary function symbol, as $@(@(f,a),b)$. Using this technique we prove that the signature is not relevant in second-order unification, and conclude that one binary symbol is enough.
By currying variable applications, like $X(a)$, as $@(X,a)$, we can transform second-order terms into first-order terms, but we have to add beta-reduction as a theory. This is roughly what it is done in explicit unification. We prove that by currying only constant applications we can reduce second-order unification to second-order unification with just one binary function symbol. Both problems are already known to be undecidable, but applying the same idea to context unification, for which decidability is still unknown, we reduce the problem to context unification with just one binary function symbol.
We also discuss about the difficulties of applying the same ideas to third or higher order unification.

1 Introduction

The Curry form of a term, like $f(a,b)$, allows us to write it, using just a single binary symbol, as $@(@(f,a),b)$, where $@$ denotes the explicit application. This helps to solve unification problems. In first-order logic, this transformation reduces a unification problem to a new unification problem containing a single binary symbol. The size of the new problem [and of the unifier] is similar to the size of the original problem [and of the original unifier]. So, from the point of view of complexity there is not a significant difference, but in practical implementations this allows representing terms as binary trees, and contexts as subterms, and has been used in term indexing data structures [GNN01].

In second-order logic the transformation is not so obvious. We can currify constant symbol applications and second-order variable applications, obtaining a first-order term. For instance, for $f(X(a),Y)$, where X is a second-order variable, we obtain $@(@(f,@(X,a)),Y)$, where both X and Y are now first-order variables. However, solvability of unification problems is not preserved by such transformation, unless we consider some form of first-order unification modulo

* This research has been partially supported by the CICYT Research Projects DENOC (BFM2000-1054-C02), LOGFAC, and TIC2001-2392-C03-01

S. Tison (Ed.): RTA 2002, LNCS 2378, pp. 326–339, 2002.

β-reduction for solving the new problem. For instance, the second-order unification problem $F(G(a),b) \stackrel{?}{=} g(a)$ is solvable, whereas its first-order Curry form $@(@(F,@(G,a)),b) \stackrel{?}{=} @(g,a)$ is unsolvable. Moreover, the right-hand side of the β-equation $(\lambda x \,.\, t_1)t_2 = t_1[t_2/x]$ is a meta-term, unless we make substitution explicit [ACCL98]. Roughly speaking this is what is done in the so called *explicit unification* [DHK00,BM00].

Here, we propose to currify function symbol applications, but not variable applications. Therefore, the new problem we get is also a second-order unification problem. For instance, for $F(G(a),b) \stackrel{?}{=} g(a)$, we get $F(G(a),b) \stackrel{?}{=} @(g,a)$, that is also solvable. In this case, we do not reduce the order of the unification problem, but we reduce the number of function symbols to just one: the application symbol @. It can be argued that this reduction is useless, since second-order unification [Gol81] was already known to be undecidable for just one binary function symbol [Far91], although applying the reduction to the results of [LV00] proves that second-order unification is undecidable for one binary function symbol and one second-order variable occurring four times. Moreover, the same reduction is applicable to context unification [Com98], for which decidability is still unknown [Com98,LV01,SS96,Lev96,SS98,SSS98,SSS99], and it allows concentrating the efforts in a very simple signature. We also think that currying could help to simplify the signature used in higher-order matching, and this could help to prove its decidability (or undecidability).

If we currify function applications in a second-order [or context] unification problem, it is easy to prove that, if the original problem is solvable, then its Curry form is also solvable: we can currify the unifier of the original problem to obtain a unifier of its Curry form. However, the converse is not true and, in general, solvability is not preserved by currying, as the following examples prove.

Example 1. The following context unification problem

$$\begin{array}{l} g(\ F(G(a)),\ F(a),\quad G(a)\quad) \stackrel{?}{=} \\ \stackrel{?}{=} g(\ f(a,b),\quad H(a,b),\ H(X,a)\ ) \end{array}$$

is unsolvable. However, its Curry form

$$\begin{array}{l} @(@(@(g,\ F(G(a))\qquad),\ F(a)\qquad),\ G(a)\qquad) \stackrel{?}{=} \\ \stackrel{?}{=} @(@(@(g,\ @(@(f,a),b)\ ),\ H(a,b)\ ),\ H(X,a)\ ) \end{array}$$

is solvable and has the following unifier

$$\begin{array}{l} \sigma(F) = \lambda x \,.\, @(x,b) \\ \sigma(G) = \lambda x \,.\, @(f,x) \\ \sigma(H) = \lambda x \,.\, \lambda y \,.\, @(x,y) \\ \sigma(X) = f \end{array}$$

Similarly, the following second-order unification problem

$$\begin{array}{l} g(\ F(G(a)),\ F(G(a')),\ F(a),\quad F(a'),\quad G(a),\quad G(a')\quad) \stackrel{?}{=} \\ \stackrel{?}{=} g(\ f(a,b),\quad f(a',b),\quad H(a,b),\ H(a',b),\ H(X,a),\ H(X,a')\ ) \end{array}$$

is also unsolvable, whereas its Curry form is solvable.

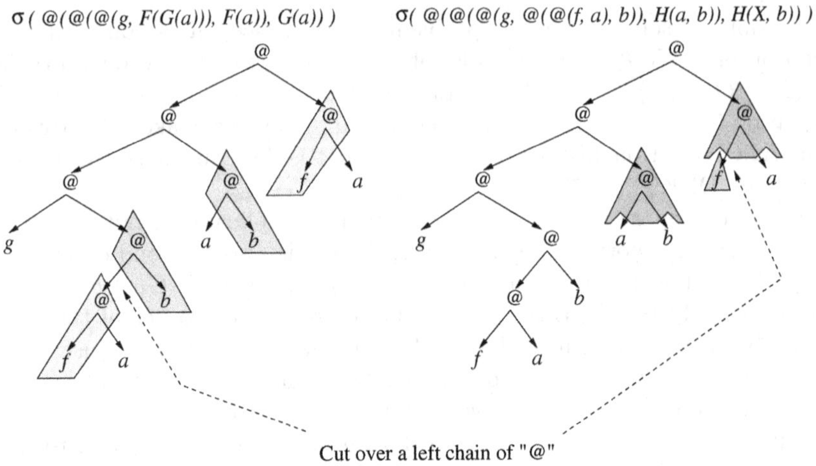

Fig. 1. Graphic representation of the unifier of curried context unification problem of Example 1.

In the previous example, $\sigma(F)$, $\sigma(G)$, $\sigma(H)$ and $\sigma(X)$ are not "well-typed", i.e. they are not the Curry form of any well-typed term. For instance, $\sigma(F) = \lambda x\,.\,@(x, b)$ is the Curry form of $\lambda x\,.\,x(b)$, but this term is third-order typed (and F is a second-order typed variable), and $\sigma(G) = \lambda x\,.\,@(f, x)$ is the Curry form of $\lambda x\,.\,f(x)$, but f has two arguments. This disallows us to reconstruct a unifier for the original unification problem from the unifier we get for its Curry form.

We can also see that the original unification problems contain variables that "touch". For instance, F touches G in $F(G(a))$, and H touches X in $H(X, a)$. We will prove, for second-order and for context unification, that, if no variable touches any other variable, then solvability of the problems is preserved in both directions by our partial currying. It is easy to reduce second-order and context unification problems to problems accomplishing such property. Therefore, we conclude that second-order and context unification can be both reduced to the partial Curry form, where only a binary function symbol @ is used.

It is well known that word unification [Mak77] is an special case of context unification. Plandowski [Pla99] proves that if σ is a most general unifier of a word unification problem $t \stackrel{?}{=} u$, then any substring of $\sigma(t)$ "is over a cut", i.e. there exists an occurrence of the substring in $\sigma(t)$ that is not completely inside the instance of a variable. Something similar can be proved for second-order and for context unification. The pathology of Example 1 is due to the existence of a cut over a left chain of @ ended by a constant. For instance, in the example, the left chain $@(@(f, \ldots), \ldots)$ is "cut" by $F(G(\ldots))$, i.e. one piece is inside $\sigma(F)$ and another inside $\sigma(G)$ (see Figure 1). If variables "do not touch" this situation is avoided, and satisfiability is preserved. Our main result could be proved using a version of Plandowski's theorem for second-order unification, but the proof would be longer than the one we present in this paper.

This paper proceeds as follows. In Section 2 we introduce some standard definitions and results about second-order and context unification. Most of our results hold for second-order and for context unification, and sometimes we do not make the distinction explicit. In Section 3 we define the partial Curry forms where only function symbols applications are made explicit. In Section 4 we define a labeling on Curry forms that is used to characterize "well-typed" terms, i.e. terms that are the Curry form of some well-built term. In Section 5 we prove our main result: second-order and context unification can be reduced to a simplified form where only a single binary function symbol and constants are used. We conclude in Section 6 with a discussion about the difficulties to extended these results to higher order.

2 Preliminary Definitions

A *second-order signature* Σ is a finite disjoint union of finite sets of symbols $\Sigma = \bigcup_{n\geq 0} \Sigma_n$, where symbols $f \in \Sigma_n$ are said to be n-ary, noted $\text{arity}(f) = n$. We distinguish between *constant symbols*, when $\text{arity}(a) = 0$, and *function symbols*, when $\text{arity}(f) > 0$. Similarly, we define the set of variables $\mathcal{X} = \bigcup_{n\geq 0} \mathcal{X}_n$, and distinguish between first-order variables (the set $\mathcal{X}_0$) and second-order variables (the rest of variables $\bigcup_{n\geq 1} \mathcal{X}_n$). We use lambda bindings and the usual notion of bound and free variables. For simplicity, we assume that bound and free variables have distinct names, and use lower case letters for the bound variables and upper case letters for free variables.

The set of terms $\mathcal{T}(\Sigma, \mathcal{X})$ is defined as in [Gol81]. A first-order term is either a constant symbol $a \in \Sigma_0$, a first-order variable $X \in \mathcal{X}_0$, or has the form $f(t_1, \ldots, t_n)$ or $X(t_1, \ldots, t_n)$, where $f \in \Sigma_n$, $X \in \mathcal{X}_n$, and t_i's are first-order terms. A second-order term has the form $\lambda x_1 \ldots \lambda x_n . t$, where t is a first-order term, $n \geq 1$, and x_i's are (bound) first-order variables. In other words, we assume that any term is written in $\beta\eta$-long normal form, and we do not consider constants of third or higher order. Therefore, λ-abstractions always appear in the head of the term. The arity of a term is the number of λ-abstractions that it has in the head. Therefore, first-order terms are the terms of arity zero.

A *second-order unification problem* is a pair of first-order terms, noted $t \stackrel{?}{=} u$. Notice that all variables of $t \stackrel{?}{=} u$ occur free.

A *second-order substitution* σ is a set of (variable,term) pairs, like $[X_1 \mapsto t_1] \ldots [X_n \mapsto t_n]$, where X_i and t_i have the same arity. Therefore, instances of second-order variables contain λ-abstractions. The application of a substitution σ to a first-order term t is defined recursively as follows

$$\sigma(f(t_1, \ldots, t_n)) = f(\sigma(t_1), \ldots, \sigma(t_n))$$

$$\begin{aligned} \sigma(X) &= t && \text{if the pair } X \mapsto t \text{ is in } \sigma \\ \sigma(X) &= X && \text{otherwise} \end{aligned}$$

$$\begin{aligned} \sigma(F(t_1, \ldots, t_n)) &= \rho(u) && \text{if the pair } F \mapsto \lambda x_1 \cdots \lambda x_n . u \text{ is in } \sigma \text{, and} \\ & && \text{where } \rho = [x_1 \mapsto \sigma(t_1)] \ldots [x_n \mapsto \sigma(t_n)] \\ \sigma(F(t_1, \ldots, t_n)) &= F(\sigma(t_1), \ldots, \sigma(t_n)) && \text{otherwise} \end{aligned}$$

Notice that in the previous definition we avoid the definition of instances of a second-order terms, and therefore all problems related with the variable capture.

A substitution σ is said to be a *unifier* (or solution) of a second-order unification problem $t \stackrel{?}{=} u$ if $\sigma(t) = \sigma(u)$. The definition of *most general unifier* is standard.

A *context unification problem* is also a pair of first-order terms $t \stackrel{?}{=} u$ over a second-order signature. In the ambit of context unification, second-order variables are called *context variables*. A *context substitution* is a second-order substitution $[X_1 \mapsto t_1] \ldots [X_n \mapsto t_n]$ where, for all second-order term $t_i = \lambda x_1 \cdots \lambda x_n \,.\, u$, every x_j occurs exactly once in u. Then, a *context unifier* of a context unification problem $t \stackrel{?}{=} u$ is a context substitution σ satisfying $\sigma(t) = \sigma(u)$. Notice that second-order and context unification problems have the same presentation, and any solvable context unification problem is also solvable viewed as a second-order unification problem.

Sometimes, context unification is defined restricting context variables to be unary. Here we consider n-ary variables, and use bound variables to denote the "holes" of the context (instead of the box $\square$ used by other authors).

If nothing is said, the signature of a problem is given by the set of constants that it contains and a denumerable infinite set of variables, for every arity. For technical reasons we also assume that the signature contains, at least, a binary function symbol and a constant (that can be added if the problem does not contain any). The following is a basic property of most general second-order [and context] unifiers that will be required in some proofs. It ensures that the signature does not play an important role w.r.t. the decidability of the problem.

Property 1. Let $t \stackrel{?}{=} u$ be a second-order or a context unification problem, and σ be a most general unifier. Then, for any variable X, $\sigma(X)$ does not contain constants not occurring in the problem $t \stackrel{?}{=} u$.

Proof: Suppose that a most general unifier σ introduces a constant f not occurring in the problem. Then we can replace everywhere this constant by a fresh variable F of the same arity and get another unifier that is more general than σ (we can instantiate F by $\lambda x_1 \ldots \lambda x_n \,.\, f(x_1, \ldots, x_n)$, but not vice versa). This contradicts the fact that σ is most general. ■

3 Currying Terms

Definition 1. *Given a signature* $\Sigma = \bigcup_{n \geq 0} \Sigma_n$, *the* curried signature $\Sigma^c = \bigcup_{n \geq 0} \Sigma^c_n$ *is defined by*

$$\begin{array}{ll} \Sigma^c_0 = \bigcup_{n \geq 0} \Sigma_n & \\ \Sigma^c_2 = \{@\} & \\ \Sigma^c_n = \emptyset & \textit{for } n \neq 0, 2 \end{array}$$

The currying function $\mathcal{C} : \mathcal{T}(\Sigma, \mathcal{X}) \rightarrow \mathcal{T}(\Sigma^c, \mathcal{X})$ *is defined recursively as follows:*

$$\mathcal{C}(a) = a$$
$$\mathcal{C}(x) = x$$
$$\mathcal{C}(f(t_1, \ldots, t_n)) = @(\overset{n}{\cdots} @(f, \mathcal{C}(t_1)) \overset{n}{\cdots}, \mathcal{C}(t_n))$$
$$\mathcal{C}(F(t_1, \ldots, t_n)) = F(\mathcal{C}(t_1), \ldots, \mathcal{C}(t_n))$$
$$\mathcal{C}(\lambda x . t) = \lambda x . \mathcal{C}(t)$$

for any constant $a \in \Sigma_0$, *bound variable* x, *function symbol* $f \in \Sigma_n$, *and variable* $F \in \mathcal{X}_n$.

The currying function is injective, but it is not onto, as the following definition suggests.

Definition 2. *Given a term* $t \in \mathcal{T}(\Sigma^c, \mathcal{X})$, *we say that it is* well-typed *(w.r.t.* Σ), *if* $\mathcal{C}^{-1}(t)$ *is defined, i.e. if there exists a term* $u \in \mathcal{T}(\Sigma, \mathcal{X})$ *such that* $\mathcal{C}(u) = t$.

Lemma 1. *If the second-order [context] unification problem* $t \overset{?}{=} u$ *over* Σ *is solvable, then the second-order [context] unification problem* $\mathcal{C}(t) \overset{?}{=} \mathcal{C}(u)$ *over* Σ^c *is also solvable.*

Proof: Let σ be a unifier of $t \overset{?}{=} u$, then it is easy to prove that the substitution $\sigma_{\mathcal{C}}$ defined as $\sigma_{\mathcal{C}}(F) = \mathcal{C}(\sigma(F))$ is a unifier of $\mathcal{C}(t) \overset{?}{=} \mathcal{C}(u)$.   ∎

In fact, we have proved a stronger result: given a unifier σ of $t \overset{?}{=} u$, we can find a unifier $\sigma_{\mathcal{C}}$ of $\mathcal{C}(t) \overset{?}{=} \mathcal{C}(u)$ that satisfies the commutativity property $\mathcal{C}(\sigma(t)) = \sigma_{\mathcal{C}}(\mathcal{C}(t))$. This commutativity property is represented by the following diagram:

$$
\begin{array}{ccc}
t \overset{?}{=} u & \xrightarrow{\ \mathcal{C}\ } & \mathcal{C}(t) \overset{?}{=} \mathcal{C}(u) \\
\downarrow{\scriptstyle\sigma} & \overset{\mathcal{C}}{\Longrightarrow} {\scriptstyle\sigma_{\mathcal{C}}} & \downarrow \\
\sigma(t) & \xrightarrow{\ \mathcal{C}\ } & \sigma_{\mathcal{C}}(\mathcal{C}(t))
\end{array}
$$

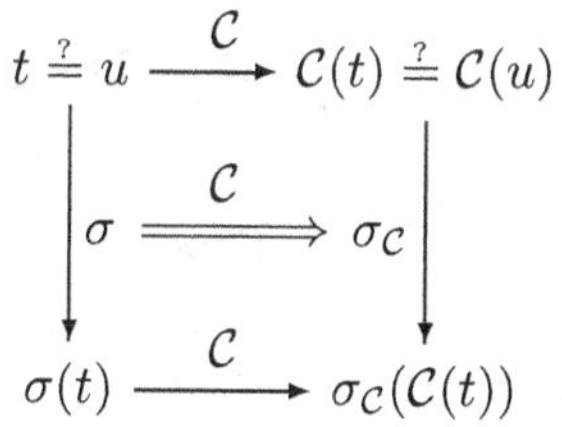

Unfortunately, as it is shown in Example 1, the converse is not true. Given a unifier of $\mathcal{C}(t) \overset{?}{=} \mathcal{C}(u)$ it is not always possible to obtain a unifier of $t \overset{?}{=} u$. In the next Section, we describe sufficient conditions to ensure that the inverse construction is possible.

4 Labeling Terms

The first step to find a sufficient condition ensuring that the currying function preserves satisfiability is to characterize well-typed curried terms. This is done by labeling application symbols @ with the "arity" of their left argument, and using a "hat" to mark the roots of right arguments. If left arguments always have positive arity, and right arguments always have arity zero, then the term is well-typed.

Definition 3. *Given a signature* $\Sigma = \bigcup_{n\geq 0} \Sigma_n$, *the* labeled signature $\Sigma^L = \bigcup_{n\geq 0} \Sigma^L_n$ *is defined by:*

$$\begin{array}{ll} \Sigma^L_0 = \bigcup_{n\geq 0} \Sigma_n & \\ \Sigma^L_2 = \{@^l, \widehat{@^l} \mid l \in \{\dots,-1,0,1,\dots\}\} & \\ \Sigma^L_n = \emptyset & \text{for } n \neq 0,2 \end{array}$$

The labeling functions $\mathcal{L}, \widehat{\mathcal{L}} : \mathcal{T}(\Sigma^c, \mathcal{X}) \to \mathcal{T}(\Sigma^L, \mathcal{X})$ *are defined by the following rules:*

1. *If the left child of an @ is an n-ary symbol* $f \in \Sigma_n$, *then it has label* $l = \text{arity}(f) - 1 = n - 1$.
2. *If the left child of an @ is a variable* $X \in \mathcal{X}$, *or a bound variable, then it has label* -1, *regardless the arity of the variable is.*
3. *If the left child of an @ is another @ with label n, then it has label* $n - 1$.

In the case of $\widehat{\mathcal{L}}$ *we also use the following rule:*

4. *If an @ is the right child of another @, or it is the child of a variable, or it is the root of the term, then, apart from the label, it also has a hat.*

Example 2. The $\widehat{\mathcal{L}}$-labeling of the term $\sigma(C(t))$, used in Example 1 and shown in Figure 1, is as follows.

$$\widehat{@^0}(@^1(@^2(g, \widehat{@^0}(@^1(f,a),b)), \widehat{@^{-1}}(a,b)), \widehat{@^1}(f,a))$$

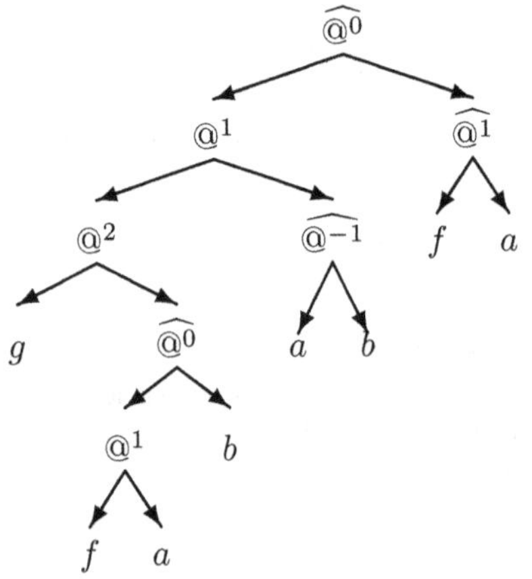

Notice that labels can be negative numbers. These negative labels do not appear in labellings of "well-typed" terms.

Based on these labels, it is easy to characterize well-typed terms.

Lemma 2. *A term* $t \in \mathcal{T}(\Sigma^c, \mathcal{X})$ *is well-typed if, and only if,* $\widehat{\mathcal{L}}(t)$ *does not contain application symbols with negative labels* ($@^{-n}$, *for* $n > 0$) *or with hat and non-zero labels* ($\widehat{@^n}$ *with* $n \neq 0$).

Proof: The only if implication is obvious. For the if implication, assume that the labeling $\widehat{\mathcal{L}}(t)$ does not contain $@^{-n}$, with $n > 0$, or $\widehat{@^n}$, with $n \neq 0$. Then, any @ symbol is in a sequence of the form:

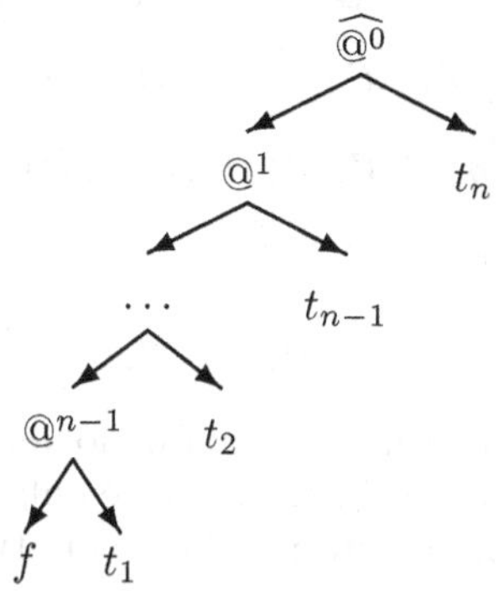

where the node $\widehat{@^0}$ is a right child of another @, or the child of a variable F, or the root of the term. We can prove that this is the currying of $f(\mathcal{C}^{-1}(t_1), \ldots, \mathcal{C}^{-1}(t_n))$ that is a well constructed term, because f has n arguments and arity n. ■

5 When Variables Do Not Touch

In this section, we try to find sufficient conditions ensuring that, when we have a unifier for $\mathcal{C}(t) \stackrel{?}{=} \mathcal{C}(u)$, we can find a unifier for $t \stackrel{?}{=} u$. The strategy to prove this result is summarized in the following diagram:

$$\begin{array}{ccccc}
t \stackrel{?}{=} u & \xrightarrow{\mathcal{C}} & \mathcal{C}(t) \stackrel{?}{=} \mathcal{C}(u) & \xrightarrow{\widehat{\mathcal{L}}} & \widehat{\mathcal{L}}(\mathcal{C}(t)) \stackrel{?}{=} \widehat{\mathcal{L}}(\mathcal{C}(u)) \\
\downarrow \sigma & \xLeftarrow{\mathcal{C}^{-1}} & \downarrow \sigma_{\mathcal{C}} & \xRightarrow{\widehat{\mathcal{L}}} & \downarrow \sigma_{\widehat{\mathcal{L}}} \\
\sigma(t) & \xleftarrow{\mathcal{C}^{-1}} & \sigma_{\mathcal{C}}(\mathcal{C}(t)) & \xrightarrow{\widehat{\mathcal{L}}} & \sigma_{\widehat{\mathcal{L}}}(\widehat{\mathcal{L}}(\mathcal{C}(t)))
\end{array}$$

We will find a condition that makes the right square commute (Lemma 5). Then we will prove that when the right square commutes, then the left one also commutes (Lemma 6). This second commutativity property ensures that the currying transformation preserves satisfiability.

The sufficient condition we have found is based on the following definition.

Definition 4. *Given a term $t \in \mathcal{T}(\Sigma, \mathcal{X})$, we say that two variables $X, Y \in \mathcal{X}$* touch, *if t contains a subterm of the form $X(t_1, \ldots, Y(u_1, \ldots, u_m), \ldots, t_n)$.*

In the context unification problem of Example 1, the variable F touches G, and the variable H touches X.

For technical reasons, before proving that the first square commutes (Lemma 5) we prove the same result using a variant of the labeling function where hats are not considered (notice that in Lemma 3 $\mathcal{L}$'s have no hats).

Lemma 3. *If the variables of $t \stackrel{?}{=} u$ do not touch, and $\sigma_{\mathcal{C}}$ is a most general unifier of $\mathcal{C}(t) \stackrel{?}{=} \mathcal{C}(u)$, then the substitution $\sigma_{\mathcal{L}}$ defined by*

$$\sigma_{\mathcal{L}}(F) = \mathcal{L}(\sigma_{\mathcal{C}}(F))$$

is a most general unifier of $\mathcal{L}(\mathcal{C}(t)) \stackrel{?}{=} \mathcal{L}(\mathcal{C}(u))$, and satisfies

$$\sigma_{\mathcal{L}}(\mathcal{L}(\mathcal{C}(t))) = \mathcal{L}(\sigma_{\mathcal{C}}(\mathcal{C}(t)))$$

Proof: First, we prove that

$$\sigma_{\mathcal{L}}(\mathcal{L}(\mathcal{C}(t))) = \mathcal{L}(\sigma_{\mathcal{C}}(\mathcal{C}(t)))$$

As far as $\sigma_{\mathcal{C}}$ and $\sigma_{\mathcal{L}}$ only differ in the introduction of labels, both terms have the same form, except for the labels. Therefore, we only have to compare the labels of the corresponding @'s in both terms. There are two cases:

- If the occurrence of the @ is outside the instance of any variable, then this @ already occurs in $\mathcal{C}(t)$, and it is in a sequence of the form:

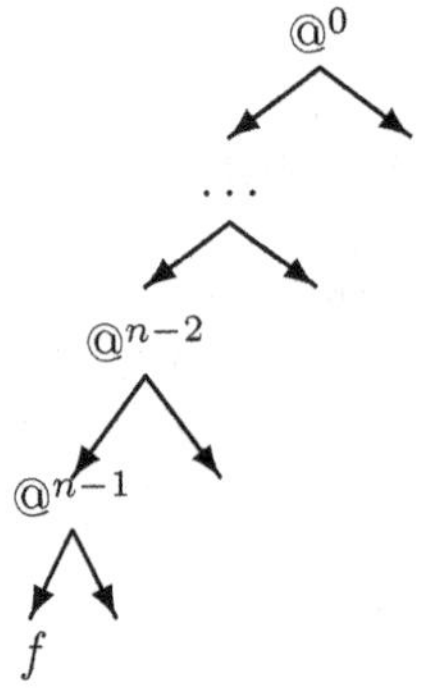

  where the f and all the @'s in between already occur in $\mathcal{C}(t)$ (they have not been introduced by an instantiation either). Thus, the @ gets the same label in $\sigma_{\mathcal{L}}(\mathcal{L}(\mathcal{C}(t))$ as in $\mathcal{L}(\sigma_{\mathcal{C}}(\mathcal{C}(t)))$, because this label only depends on the left descendants, and they have not been introduced by $\sigma_{\mathcal{C}}$ or $\sigma_{\mathcal{L}}$.
- If the @ is inside the instance of a variable F, we have to prove that it gets the same label in $\sigma_{\mathcal{L}}(\mathcal{L}(F(t_1,\ldots,t_n))) = \sigma_{\mathcal{L}}(F)(\sigma_{\mathcal{L}}(\mathcal{L}(t_1)),\ldots,\sigma_{\mathcal{L}}(\mathcal{L}(t_1)))$ as in $\mathcal{L}(\sigma_{\mathcal{C}}(F(t_1,\ldots,t_n)))$. In the first case we label $\sigma_{\mathcal{C}}(F)$ before instantiating (so we have bound variables in the place of the arguments), whereas in the second case we label $\sigma_{\mathcal{C}}(F)$ after instantiating (so we already have the arguments t_i). As we will see, in both cases the labels we get are the same. The root of one of the arguments t_i can be a left descendant of the @, and its label will depend on such argument. However, if variables do not touch, the head of any argument t_i of F is a constant, and the head of $\mathcal{C}(t_i)$ is either a 0-ary constant a or an @ with label 0. Therefore, the labels of the ancestors of the argument inside $\sigma_{\mathcal{C}}(F)$ will be the same if we replace the argument by a bound-variable, and the label of the corresponding @ inside $\sigma_{\mathcal{L}}(F)$ will be the same.

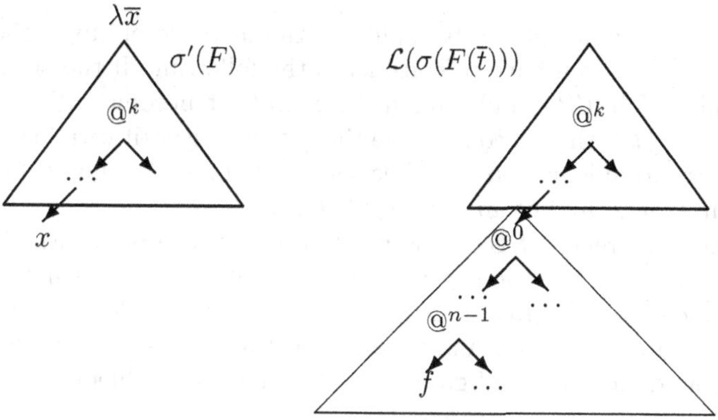

Similarly, we can prove $\sigma_{\mathcal{L}}(\mathcal{L}(\mathcal{C}(u))) = \mathcal{L}(\sigma_{\mathcal{C}}(\mathcal{C}(u)))$. As $\sigma_{\mathcal{C}}(\mathcal{C}(t)) = \sigma_{\mathcal{C}}(\mathcal{C}(u))$, we can conclude that $\sigma_{\mathcal{L}}$ is a unifier of $\mathcal{L}(\mathcal{C}(t)) \stackrel{?}{=} \mathcal{L}(\mathcal{C}(u))$.

Given a unifier of $\mathcal{L}(\mathcal{C}(t)) \stackrel{?}{=} \mathcal{L}(\mathcal{C}(u))$, we can find a unifier of $\mathcal{C}(t) \stackrel{?}{=} \mathcal{C}(u)$ by removing labels. Using this idea, it is easy to prove that, if $\sigma_{\mathcal{C}}$ is most general for $\mathcal{C}(t) \stackrel{?}{=} \mathcal{C}(u)$, then $\sigma_{\mathcal{L}}$ is also most general for $\mathcal{L}(\mathcal{C}(t)) \stackrel{?}{=} \mathcal{L}(\mathcal{C}(u))$. Otherwise, there would be a unifier more general than $\sigma_{\mathcal{L}}$, and removing labels we could obtain a unifier more general than $\sigma_{\mathcal{C}}$. ∎

The following is a technical lemma that we need in the proof of Lemma 5.

Lemma 4. *If the variables of $t \stackrel{?}{=} u$ do not touch, and $\sigma_{\mathcal{C}}$ is a most general unifier of $\mathcal{C}(t) \stackrel{?}{=} \mathcal{C}(u)$, then the arguments t_i of any variable F never occur as left child of an @ in $\sigma_{\mathcal{C}}(\mathcal{C}(t))$.*

Proof: As $\mathcal{C}(t)$ and $\mathcal{C}(u)$ are trivially well-typed, by Lemma 2, $\mathcal{L}(\mathcal{C}(t))$ and $\mathcal{L}(\mathcal{C}(u))$ will not contain @'s with negative labels. Let $\sigma_{\mathcal{L}}$ be the most general unifier of $\mathcal{L}(\mathcal{C}(t)) \stackrel{?}{=} \mathcal{L}(\mathcal{C}(u))$ given by Lemma 3. Now, by Property 1, as $\sigma_{\mathcal{L}}$ is a most general unifier, for any variable F, $\sigma_{\mathcal{L}}(F)$ will not contain @'s with negative labels, either. We can conclude then that the head of any argument t_i of F can not be a left child of an @. As far as the heads of $\sigma_{\mathcal{L}}(t_i)$ have zero label or are 0-ary constants, this situation would introduce a negative label in some @ inside $\sigma_{\mathcal{L}}(F)$. ∎

Lemma 5. *If the variables of $t \stackrel{?}{=} u$ do not touch, and $\sigma_{\mathcal{C}}$ is a most general unifier of $\mathcal{C}(t) \stackrel{?}{=} \mathcal{C}(u)$, then the substitution $\sigma_{\widehat{\mathcal{L}}}$ defined by*

$$\sigma_{\widehat{\mathcal{L}}}(F) = \widehat{\mathcal{L}}(\sigma_{\mathcal{C}}(F))$$

is a most general unifier of $\widehat{\mathcal{L}}(\mathcal{C}(t)) \stackrel{?}{=} \widehat{\mathcal{L}}(\mathcal{C}(u))$, and satisfies

$$\sigma_{\widehat{\mathcal{L}}}(\widehat{\mathcal{L}}(\mathcal{C}(t))) = \widehat{\mathcal{L}}(\sigma_{\mathcal{C}}(\mathcal{C}(t)))$$

Proof: We already know that both terms have the same form and the same labels, thus we only have to prove that they have the same hats. Again, there are two cases:

- If the occurrence of the @ is outside the instance of any variable, then the only situation we have to consider is the following. If the @ has as father a variable F in $\mathcal{C}(t)$, and after instantiation, it becomes a left child of an @ inside $\sigma_{\widehat{\mathcal{L}}}(F)$, then it could loose the hat. However, if variables do not touch, this situation is not possible because, by Lemma 4, arguments t_i of F never occur as left child of an @ in $\sigma_{\widehat{\mathcal{L}}}(F(t_1,\dots,t_n))$.
- If the occurrence of the @ is inside the instance of a variable F, then we have to prove that the fact that @ has a hat or not, does not depend on the arguments of F. This is obvious because this fact does not depend on the descendants of the @. As in Lemma 3, this allows us to replace arguments by bound variables and get a unifier $\sigma_{\widehat{\mathcal{L}}}$ for our problem.

Using the argument of Lemma 3, we conclude that $\sigma_{\widehat{\mathcal{L}}}$ is a most general unifier of $\widehat{\mathcal{L}}(\mathcal{C}(t)) \stackrel{?}{=} \widehat{\mathcal{L}}(\mathcal{C}(u))$. ■

Lemma 6. *If the variables of $t \stackrel{?}{=} u$ do not touch, and $\sigma_{\mathcal{C}}$ is a most general unifier of $\mathcal{C}(t) \stackrel{?}{=} \mathcal{C}(u)$, then there exists a most general unifier σ of $t \stackrel{?}{=} u$ that satisfies*

$$\mathcal{C}(\sigma(t)) = \sigma_{\mathcal{C}}(\mathcal{C}(t))$$

Proof: Let $\sigma_{\widehat{\mathcal{L}}}$ be the most general unifier of $\widehat{\mathcal{L}}(\mathcal{C}(t)) \stackrel{?}{=} \widehat{\mathcal{L}}(\mathcal{C}(u))$ given by Lemma 5. As $\mathcal{C}(t)$ and $\mathcal{C}(u)$ are well-typed, by Lemma 2, they do not contain negative labels nor hats over non-zero labeled @'s. Then, by Property 1, $\sigma_{\widehat{\mathcal{L}}}$ does not introduce such kind of labels or hats. Therefore, as $\sigma_{\widehat{\mathcal{L}}}(F)$ is defined as the labeling of $\sigma_{\mathcal{C}}(F)$, using again Lemma 2, $\sigma_{\mathcal{C}}(F)$ will be well-typed, and we can define:

$$\sigma(F) = \mathcal{C}^{-1}(\sigma_{\mathcal{C}}(F))$$

■

Theorem 1. *Decidability of second-order [context] unification can be NP-reduced to decidability of second-order [context] unification with just one binary function symbol, and constants*

Proof: By Lemmas 1 and 6, we know that, when variables do not touch, satisfiability of second-order and context unification problems is preserved by currying. Now, we will prove that we can NP-reduce solvability of second-order and context unification to solvability of the corresponding problems without touching variables.

For second-order the reduction is as follows. For every n-ary variable F, we conjecture one of the following possibilities:

- Project $F \mapsto \lambda x_1 . \dots \lambda x_n . x_i$, for some $i \in \{1,\dots,n\}$.
- Instantiate $F \mapsto \lambda x_1 . \dots \lambda x_n . f(F_1(x_1,\dots,x_n),\dots,F_m(x_1,\dots,x_n))$, for some constant $f \in \Sigma_m$ occurring in the original unification problem, and being $F_1,\dots,F_m$ fresh free variables.

Obviously, this reduction can be performed in polynomial non-deterministic time. As far as the new problem is an instance of the original one, if the new problem is solvable, so it is the original one. If the original problem is solvable, and σ is a most general unifier, then, for every variable F, let $\sigma(F) = \lambda x_1 \ldots \lambda x_n . t$ be written in normal form. Taking t as a tree, descend from the root to the left-most leave, discarding free variables, until you get a constant f (this must be a constant occurring in the problem, by Property 1), a 0-ary variable, or a bound variable x_i. It is easy to prove that the instantiation $F \mapsto \lambda x_1 \ldots \lambda x_n . x_i$, if we find a bound variable x_i, $F \mapsto \lambda x_1 \ldots \lambda x_n . a$ for some fixed constant a, if we find a 0-ary variable, or $F \mapsto \lambda x_1 \ldots \lambda x_n . f(F_1(x_1, \ldots, x_n), \ldots, F_m(x_1, \ldots, x_n))$, if we find a constant f, results in a solvable problem. In fact, the solution of the new problem is σ composed with a substitution that projects as $G \mapsto \lambda x_1 \cdots \lambda x_n . x_1$ the free variables that we have discarded during the traversal, and as $X \mapsto a$ the 0-ary variables that we have found.

For context unification the reduction is as follows. For every n-ary variable F, we conjecture one of the following possibilities:

- Project $F \mapsto \lambda x . x$, if it is unary.
- Instantiate

$$F \mapsto \lambda x_1 \ldots \lambda x_n . f(F_1(x_{\tau(1)}, \ldots, x_{\tau(r_1)}), \ldots, F_m(x_{\tau(r_{m-1}+1)}, \ldots, x_{\tau(n)}))$$

  for some constant $f \in \Sigma_m$ occurring in the original unification problem, some permutation τ, and being $F_1, \ldots, F_m$ fresh free variables.

As for the second-order case, it can be proved that this nondeterministic reduction preserves satisfiability. However, in this case we have to assume that the original signature (the problem) contains, at least, a binary function symbol and a 0-ary constant.

■

6 Conclusions and Further Work

Currying terms is an standard technique in functional programming and has been used in practical applications of automated deduction. It is also used in higher-order unification via explicit substitutions or explicit unification. However, in these cases not only applications, but also lambda abstractions are made explicit, and unification is made modulo the explicit substitution rules. Here, we propose a partial currying transformation for second-order unification, where the "order" of the unification problem is not reduced, like in explicit unification, but the signature is simplified. The transformation is not trivial, and we prove that, to preserve solvability of the problems, we need to ensure that "variables do not touch". The reduction also works for context unification. This allows us to concentrate on a simpler signature containing constant symbols and just one binary function symbol: the explicit application symbol @.

Decidability of higher-order matching is still an open question. Proving that higher-order matching can be curried, i.e., that we can simplify the signature, could contribute to prove its decidability or undecidability. The extension of our technique to third-order and higher orders is proposed as a further work.

The first difficulty we find trying to apply our transformation to third or higher order matching problems is that we must deal with instances of variables that are not connected. For instance, the following matching problem:

$$\begin{array}{r} f(\ F(\lambda x\,.g(x),a),\ \ F(\lambda x\,.g'(x),a')\ \ ) \stackrel{?}{=} \\ \stackrel{?}{=} f(\ \ f(g(h(a)),a),\ \ \ f(g'(h(a')),a')\ \ ) \end{array}$$

is solved by the substitution:

$$F \mapsto \lambda x\, \lambda y\,.f(x,(h(y)),y)$$

where the instance of F is split into two pieces f and h. In such situations we have to guaranty that these pieces do not touch, to avoid that these "cuts" (in the sense of Plandowski [Pla99]) could cut a left chain of @'s.

Acknowledgments. We acknowledge Roberto Nieuwenhuis for suggesting us the use of currying in second-order unification problems. We also thank all the anonymous referees for their helpful comments.

References

[ACCL98] M. Abadi, L. Cardelli, P.-L. Curien, and J.J. Lévy. Explicit substitutions. *Journal of Functional Programming*, 1(4):375–416, 1998.

[BM00] Nikolaj Bjorner and César Muñoz. Absoulte explicit unification. In *Proceedings of the 11th Int. Conf. on Rewriting Techniques and Applications (RTA'00)*, volume 1833 of *LNCS*, pages 31–46, Norwich, UK, 2000.

[Com98] Hubert Comon. Completion of rewrite systems with membership constraints. *Journal of Symbolic Computation*, 25(4):397–453, 1998.

[DHK00] G. Dowek, T. Hardin, and C. Kirchner. Higher-order unification via explicit substitutions. *Information and Computation*, 157:183–235, 2000.

[Far91] W. M. Farmer. Simple second-order languages for wich unification is undecidable. *Theoretical Computer Science*, 87:173–214, 1991.

[GNN01] Harald Ganzinger, Robert Nieuwenhuis, and Pilar Nivela. Context trees. In *Proceedings of the First Int. Conf. on Automated Reasoning (IJCAR 2001)*, volume 2083 of *LNCS*, pages 242–256, Siena, Italy, 2001.

[Gol81] W. D. Goldfarb. The undecidability of the second-order unification problem. *Theoretical Computer Science*, 13:225–230, 1981.

[Lev96] Jordi Levy. Linear second-order unification. In *Proceedings of the 7th Int. Conf. on Rewriting Techniques and Applications (RTA'96)*, volume 1103 of *LNCS*, pages 332–346, New Brunsbick, New Jersey, 1996.

[LV00] Jordi Levy and Margus Veanes. On the undecidability of second-order unification. *Information and Computation*, 159:125–150, 2000.

[LV01] Jordi Levy and Mateu Villaret. Context unification and traversal equations. In *Proceedings of the 12th Int. Conf. on Rewriting Techniques and Applications (RTA'01)*, volume 2051 of *LNCS*, pages 167–184, Utrecht, The Netherlands, 2001.

[Mak77] G. S. Makanin. The problem of solvability of equations in a free semigroup. *Math. USSR Sbornik*, 32(2):129–198, 1977.

[Pla99] Wojciech Plandowski. Satisfiability of word equations with constants is in pspace. In *Proceedings of the 40th Annual Symposioum on Foundations of Computer Science, FOCS'99*, pages 495–500, New York, NY, USA, 1999.

[SS96] Manfred Schmidt-Schauß. An algorithm for distributive unification. In *Proceedings of the 7th Int. Conf. on Rewriting Techniques and Applications (RTA'96)*, volume 1103 of *LNCS*, pages 287–301, New Jersey, USA, 1996.

[SS98] Manfred Schmidt-Schauß. A decision algorithm for distributive unification. *Theoretical Computer Science*, 208:111–148, 1998.

[SSS98] Manfred Schmidt-Schauß and Klaus U. Schulz. On the exponent of periodicity of minimal solutions of context equations. In *Proceedings of the 9th Int. Conf. on Rewriting Techniques and Applications (RTA'98)*, volume 1379 of *LNCS*, pages 61–75, Tsukuba, Japan, 1998.

[SSS99] Manfred Schmidt-Schauß and Klaus U. Schulz. Solvability of context equations with two context variables is decidable. In *Proceedings of the 16th Int. Conf. on Automated Deduction (CADE-16)*, LNAI, pages 67–81, 1999.

A Decidable Variant of Higher Order Matching

Dan Dougherty[1] and ToMasz Wierzbicki[2,*]

[1] Wesleyan University, USA
[2] University of Wrocław, Poland

Abstract. A lambda term is *k-duplicating* if every occurrence of a lambda abstractor binds at most k variable occurrences. We prove that the problem of higher order matching where solutions are required to be k-duplicating (but with no constraints on the problem instance itself) is decidable. We also show that the problem of higher order matching in the affine lambda calculus (where both the problem instance and the solutions are constrained to be 1-duplicating) is in NP, generalizing de Groote's result for the linear lambda calculus [4].

1 Introduction

A lambda term is called *linear* if every bound variable occurs in it exactly once. In [4] de Groote defined *linear higher order matching* as the problem of deciding if there are linear solutions of a matching problem $M =^? N$, where both terms M and N are also linear, and showed that this problem is NP-complete. This relatively low complexity suggests that this is a quite restricted version of matching. Although de Groote's definition is well-justified by applications he is concerned with, nevertheless it seems interesting to consider another version, which also may be called linear, but which consists in deciding if there are linear solutions of an *arbitrary* rather than linear matching problem (incidentally, Levy [5] defined this way an analogous notion of the linear second order unification, by insisting on linearity of solutions, but allowing to use arbitrary terms in problems). This new linearity condition is much less restrictive than that of de Groote, since the new problem inherits the full complexity of β-reduction, and the best known lower bound for general matching also holds for it: the problem is non-elementary (see 1.1).

In the present paper we show decidability of this new version of matching. In fact we show that a rather more general problem is decidable. Say that a lambda term is k-duplicating if every bound variable occurs in it at most k times (Definition 2). We prove that, for any fixed k, a complete set of k-duplicating solutions of an arbitrary matching problem is always finite, and the problem of finding such a set is decidable.

Finally we strengthen slightly de Groote's result in the following way. Say that a term is *affine*, if it is 1-duplicating (an affine function may, as linear one, use its argument once, but it may also forget it). Using tools developed in

* Supported by KBN grant 8T 11C 04319

S. Tison (Ed.): RTA 2002, LNCS 2378, pp. 340–351, 2002.

the paper we are able to replace "linear" with "affine" in de Groote's theorem and prove that the problem of deciding if an affine matching problem has affine solutions is in NP (thus is NP-complete).

All of our results apply to matching under β-equality as well as $\beta\eta$-equality. The k-duplicating matching is especially appealing in the $\lambda\beta$ calculus, since it turns out to be quite general decidable approximation of β-matching, which has recently been shown to be undecidable [6]. On the other hand the decidability of k-duplicating matching in $\beta\eta$-calculus sheds no light on the status of the full problem in this setting, which is still open.

There have been several other attempts to define some restricted variants of matching, for which decidability could be proved. The most successful include: bounding the order of solutions by 3 [3], 4 [9,10,1], or (in special case) by 5 [12], and restricting the signature to the first order (atomic) constants [8] (requiring at the same time that both sides of an instance are of base type). From the other hand a similar approach to ours has been applied to the problem of higher order unification [11].

1.1 Lower Bounds

To put our results in perspective it may be useful to recall that all of the variants of matching thus far studied in the full lambda calculus are non-elementary, since a straightforward reduction to the problem of checking β-convertibility [14] can be applied to them as well. We shall briefly recall this argument here. It is well known that the problem "given a closed pure term M of type $o^2 \to o$, is it true that $M \xrightarrow{*}_{\beta\eta} \lambda xy.x$?" is non-elementary [13,7]. Since it is required that both terms M and N in the matching problem $M =^? N$ should be in normal form, we cannot just write the equation $M =^? \lambda xy.x$. However we may easily reduce this problem to matching in the following way: replace any redex $(\lambda x_i.P_i)Q_i$ in M with a term $f_i(\lambda x_i.P_i)Q_i$ obtaining this way a term M', which does not contain redexes, where f_i are "fresh" variables of appropriate types. Write equations $f_i =^? \lambda xy.xy$ together with $M' =^? \lambda xy.x$. If one insists on having a single equation instead of a set $\{M_j =^? N_j\}_{j=1}^n$, it suffices to transform it to $\lambda z.zM_1 \dots M_n =^? \lambda z.zN_1 \dots N_n$. The only solution of the set $\{f_i =^? \lambda xy.xy\}_i$ is a substitution $\theta = [f_i/\lambda xy.xy]_i$. Since $M'\theta \xrightarrow{*}_\beta M$, the whole problem has a (unique) solution θ if and only if $M \xrightarrow{*}_\beta \lambda xy.x$, which turns out to be non-elementary. The solution θ is linear. Of course M is *not* linear. In case of a nonempty signature it is sometimes required that both sides of a matching equation are of base type o. Suppose that the signature contains two constants 0 and 1 of type o. Then a reduction from the problem of checking if $M \xrightarrow{*}_{\beta\eta} \lambda xy.x$ to matching proceeds as follows: write a term $M10$, "hide" all β-redexes using fresh variables f_i obtaining a term M', then write the equation $M' =^? 1$ together with equations $f_i(\lambda x.x)0 =^? 0$ and $f_i(\lambda x.x)1 =^? 1$. Again the only solution of the last pair of equations is the substitution $[f_i/\lambda xy.xy]$. Recall, that a matching problem is *atomic* if all right hand sides of equations are single constants of type o. Atomic matching is decidable [8]. Thus pure as well as atomic matching with linear solutions is non-elementary. Hence also pure and atomic k-duplicating, for any $k \geq 1$, as well as general matching is non-elementary.

2 Preliminaries

Let $\mathbb{T}$ be the set of *simple types* built over one base type o and given by the following grammar: $\mathbb{T} ::= o \mid \mathbb{T} \to \mathbb{T}$. We assume that $\to$ associates to the right, and omit parentheses whenever possible. Small Greek letters σ, τ, ρ, etc., denote arbitrary types. The notions of *size* and *order* of types are defined as usual:

$$|o| = 1$$
$$|\sigma \to \tau| = 1 + |\sigma| + |\tau|$$

$$\text{order}(o) = 1$$
$$\text{order}(\sigma \to \tau) = \max(\text{order}(\sigma) + 1, \text{order}(\tau))$$

Let $\langle \mathcal{X}, \text{type}(\cdot)\rangle$ be a set of *variables* together with a function type : $\mathcal{X} \to \mathbb{T}$, which assigns a unique type to each variable $x \in \mathcal{X}$. We assume that there are infinitely many variables of each type. Similarly let $\langle \Sigma, \text{type}(\cdot)\rangle$ be a collection of *constants*. The set Σ is sometimes called *a signature*. This set may be empty, finite or infinite (we say, that the lambda calculus is *pure*, if $\Sigma = \emptyset$, and *atomic*, if $\text{type}(c) = o$ for all $c \in \Sigma$). We use small Latin letters x, y, z, etc., to denote arbitrary variables, and a, c, f, etc., to denote arbitrary constants.

Let $\mathcal{X}_\sigma = \{x \in \mathcal{X} \mid \text{type}(x) = \sigma\}$ and $\Sigma_\sigma = \{c \in \Sigma \mid \text{type}(c) = \sigma\}$ be sets of respectively variables and constants of type σ. For all types $\sigma \in \mathbb{T}$ we simultaneously define sets $\Lambda^{\to}_{\sigma}$ of *lambda terms of type* σ to be the smallest sets satisfying the following conditions:

1. $\mathcal{X}_\sigma \subseteq \Lambda^{\to}_{\sigma}$,
2. $\Sigma_\sigma \subseteq \Lambda^{\to}_{\sigma}$,
3. if $x \in \mathcal{X}_\sigma$ and $M \in \Lambda^{\to}_{\tau}$, then $\lambda x.M \in \Lambda^{\to}_{\sigma \to \tau}$,
4. if $M \in \Lambda^{\to}_{\sigma \to \tau}$ and $N \in \Lambda^{\to}_{\sigma}$, then $MN \in \Lambda^{\to}_{\tau}$.

We use capital letters M, N, P, Q, etc., to denote arbitrary terms. If $M \in \Lambda^{\to}_{\sigma}$, we sometimes write $\text{type}(M) = \sigma$ or $M : \sigma$. The *size* of a term is defined inductively:

$$|x| = 1$$
$$|c| = 1$$
$$|MN| = 1 + |M| + |N|$$
$$|\lambda x.M| = 1 + |M|$$

An *address* is a sequence $p \in \{0,1\}^*$. Empty sequence is denoted by ϵ. A *subterm* $M{\restriction}_p$ *of a term* M *at address* p is defined as follows:

$$M{\restriction}_\epsilon = M$$
$$(\lambda x.M){\restriction}_{0p} = M{\restriction}_p$$
$$(M_1 M_2){\restriction}_{0p} = M_1{\restriction}_p$$
$$(M_1 M_2){\restriction}_{1p} = M_2{\restriction}_p$$

We say that an address p is *valid* in a term M, if $M{\restriction}_p$ is well defined. Note that $(M{\restriction}_p){\restriction}_q = M{\restriction}_{pq}$, where pq is the concatenation of sequences p and q. We say that a term N is a *subterm* of M, and write $N \subseteq M$, if there exists an address p, such that $M\restriction_p = N$. The result of the *substitution of a term N at address p in a term M*, providing that p is valid in M and $\text{type}(M\restriction_p) = \text{type}(N)$, is a term $M[p\restriction N]$ obtained by replacing a subterm at address p in M with N. Similarly we define a *simultaneous* substitution $M[p_i\restriction N_i]_{i=1}^k$, providing there are no two addresses p_i and p_j, such that p_i is a prefix of p_j. The set $\text{Occ}(x, M)$ of *free occurrences* of a variable x in a term M is the set of all addresses in M at which x occurs as a free variable. The set $\text{FV}(M)$ of *free variables* of a term M is the set of those variables, which have at least one free occurrence in M. A term M is *closed*, if $\text{FV}(M) = \emptyset$. The *substitution $M[x/N]$ of a term N for variable x in a term M*, providing that $\text{type}(x) = \text{type}(N)$, is the result of substitution of a term N at all free occurrences of variable x in M. Similarly we define the simultaneous substitution $M[x_i/N_i]_{i=1}^n$.

As usual we consider β- and η-reduction rules:

$$(\lambda x.M)N \rhd_\beta M[x/N] \qquad \lambda y.My \rhd_\eta M$$

for any variables x, y and terms M, N, such that $y \notin \text{FV}(M)$. A binary relation R on $\Lambda_\sigma^{\rightarrow}$ is *monotonic*, if for every every terms M, N, P, Q, and variable x, such that MRN and PRQ, there is $(MP)R(NQ)$ and $(\lambda x.P)R(\lambda x.Q)$. The one step β-reduction $\rightarrow_\beta$ is the monotonic closure of $\rhd_\beta$, the one step η-reduction $\rightarrow_\eta$ is the monotonic closure of $\rhd_\eta$, the (many step) β-reduction $\xrightarrow{*}_\beta$ is the monotonic, reflexive and transitive closure of $\rhd_\beta$, and the (many step) $\beta\eta$-reduction $\xrightarrow{*}_{\beta\eta}$ is the monotonic, reflexive and transitive closure of $\rhd_\beta \cup \rhd_\eta$. The two relations $\xrightarrow{*}_\beta$ and $\xrightarrow{*}_{\beta\eta}$ lead to two slightly different lambda calculi.

A term M is a *β-redex* if it is of the form $(\lambda x.M_1)M_2$; a term is in *β-normal form* if it contains no β-redex. A term M is in $\bar{\eta}$-normal form (called also η-long form), if for every address p in M, such that $M\restriction_p$ is not an abstraction, and $\text{type}(M\restriction_p) \neq o$, there is $p = q0$, and $M\restriction_q$ is an application. $\beta\bar{\eta}$-normal forms are unique modulo $\beta\eta$-conversion and are closed under substitutions and β-reduction.

A β-reduction sequence is a sequence of terms $(M_n)_{i=0}^n$, such that $M_i \rightarrow_\beta M_{i+1}$ for $i = 0, \ldots, n-1$. We say that this reduction sequence *starts* at M_0, *leads* to M_n, and *has length n*.

We say that a term M is an *instance* of N, and write $M \geq N$ (or say, that N is a *generalization* of M, and write $N \leq M$), if there exists a substitution θ, such that $M = N\theta$. The relation $\leq$ is a partial preorder on terms (i.e., is reflexive and transitive). Even though $\leq$ is not an order (is not weakly asymmetric), the notions of the smallest and greatest elements, lower and upper bounds, and suprema and infima of sets of terms with respect to $\leq$ are well defined. Unlike for order relations, these elements are usually not unique (but they are unique modulo renaming of free variables).

Lemma 1 (basic properties of the instance relation).

1. *Every nonempty set $\mathcal{P} \subseteq \Lambda_\sigma^{\rightarrow}$ of terms has an infimum, $\inf \mathcal{P}$. In particular, the set of the smallest elements in $\Lambda_\sigma^{\rightarrow}$ coincides with the set of variables $\mathcal{X}_\sigma$.*

2. *If a nonempty set $\mathcal{P}$ possesses an upper bound, then it has a supremum,* $\sup\mathcal{P}$.
3. *If* $M_1 \le M_2$ *and* p *is a valid address in* M_1*, then* p *is a valid address in* M_2*, and* $M_1\lceil_p \le M_2\lceil_p$.
4. *If there exists a substitution* θ*, such that* $M_1\theta = M_2$*, and* $N_1\theta = N_2$*, and* p *is a valid address in* M_1*, then* p *is a valid address in* M_2*, and* $M_1[p\lceil N_1] \le M_2[p\lceil N_2]$.
5. *If there exists a substitution* θ*, such that* $M_1\theta = M_2$*,* $N_1\theta = N_2$*, and* $x \notin \mathrm{Dom}(\theta)$*, then* $M_1N_1 \le M_2N_2$*, and* $\lambda x.M_1 \le \lambda x.M_2$.
6. *If* $M_1 \le M_2$ *(with additional proviso that there exists substitution* θ*, such that* $M_1\theta = M_2$ *and every* $x \in \mathrm{Dom}(\theta)$ *occurs at most once in* M_1*), and* p *is a valid address in* M_2 *but not in* M_1*, then* $M_1 \le M_2[p\lceil N]$ *for any term* N.
7. *If* $\mathrm{FV}(M) = \emptyset$ *and* $M \le N$*, then* $M = N$.
8. *If* $M \le N$*, and* $M\lceil_p$ *is a* β*-redex, then* $N\lceil_p$ *is a* β*-redex.*
9. *If* $M \le N$*, then* $|M| \le |N|$.
10. *If* $M = \sup\{N_i\}_{i=1}^k$*, where for each* i *there exists a substitution* θ_i *such that* $N_i\theta_i = M$ *and every* $x \in \mathrm{Dom}(\theta_i)$ *occurs at most once in* N_i*, then* $|M| \le 1 - k + \sum_{i=1}^k |N_i|$.

We define a similar instance relation for substitutions: we say that a substitution θ_1 is *at least as general as* θ_2, and write $\theta_1 \le \theta_2$, if there exists a substitution ρ, such that $\theta_2 = \theta_1\rho$ (i.e., if $x\theta_1 \le x\theta_2$ for every variable $x \in \mathcal{X}$).

2.1 Higher-Order Matching

Higher-order $\beta\eta$- (resp. β-) matching is the following problem: "Given two terms M and N of the same type, in $\beta\overline{\eta}$- (resp. β-) normal form, where N is closed, usually written in the form of equation $M =^? N$. Is there any substitution θ for free variables in M, where all terms $x\theta$ for $x \in \mathrm{FV}(M)$ are in $\beta\overline{\eta}$- (resp. β-) normal form, such that $M\theta \xrightarrow{*}_\beta N$?" We say, that θ is a *solution* of the instance $M =^? N$. There is a subtle difference between those two variants of matching [6]. In particular it is known that β-matching is undecidable, while decidability of $\beta\eta$-matching is still open. The results of the present paper apply to both variants. A set $\mathcal{S}$ of substitutions is a *complete set of solutions* of a matching problem $M =^? N$, if for every $\theta \in \mathcal{S}$ and every $\theta' \ge \theta$, the substitution θ' is a solution of $M =^? N$, and for every solution θ' of $M =^? N$ there exists a substitution $\theta \in \mathcal{S}$, such that $\theta \le \theta'$.

It is not the case that if a matching problem $M =^? N$ has a solution then it has a closed solution, since, depending on the signature Σ, the types of some of the free variables of M may not be inhabited. This motivates the following definition and lemma.

Definition 1. *A term is* almost-closed *if its free variables (if any) are all of base type. A substitution* θ *is almost-closed if each* $\theta(x)$ *is an almost-closed term.*

Lemma 2. *If a matching problem has a solution then it has an almost-closed solution.*

Proof. If θ is a solution to $M =^? N$ and $\theta(x)$ has a variable $v : \tau$ free then we may choose some base-type variable z and replace v by the term $\lambda y_1 \dots y_n.z : \tau$; since v does not occur in N this is obviously still a solution.

3 Decidability of k-Duplicating Matching

Definition 2. *A lambda term is* k-duplicating *if for every subterm* $N \subseteq M$*, such that* $N = \lambda x.P$*, there are at most* k *free occurrences of* x *in* P*, i.e.,* $|\operatorname{Occ}(x, P)| \leq k$*.*

In this section we show the decidability of k-duplicating matching, that is, matching with no restriction on the problem instance but with the constraint that solutions must be k-duplicating. It will suffice to show that we can, given a matching problem, effectively generate a finite set of terms which includes all the potential k-duplicating solutions to the problem.

The essential insight is that for each k there are in fact only finitely many *pure* k-duplicating normal forms at each type (independently of the matching problem). When the given matching problem contains constants there is a slight complication since we must allow for the fact that these constants may occur in solutions. But Lemma 3 below shows that we can bound the number of occurrences of constants in potential solutions by inspecting the problem.

Definition 3. *Let* N *be a normal-form term over the signature* Σ *and let* $c \in \Sigma$*. Then* $\#(c, N)$ *is the number of occurrences of* c *in* N*.*

Lemma 3. *Let* $M =^? N$ *be a matching problem and let* c *be some constant. For both* β*- and* $\beta\eta$*-matching: if the matching problem has a solution it has an almost-closed solution in which each solution term* X *satisfies* $\#(c, X) \leq \#(c, N)$*.*

Proof. For simplicity of notation suppose there is only one free variable x in M. Let X be a solution, so that $M[x/X] = N$. By Lemma 2 we may assume that X is almost closed.

Let us define the notion of a c-occurrence in X being *relevant* to the solution, as follows. For each occurrence of c inside of X choose some fresh constant of the same type as c: let us call these new ones $c^1, \dots, c^m$. Build X' by replacing the given occurrences of c by the c^i in X, and let N' be the normal form of $M[x/X']$. Obviously N' will differ from N just in that some of the c occurrences in N will now be certain c^i. The original c occurrences in X whose corresponding c^i appear in N' are declared to be the relevant ones.

Now: obviously no more than $\#(c, N)$ such c^i will occur in N', so there are at most this many relevant occurrences in X. If in X we replace the non-relevant c-occurrences by an arbitrary term of the right type then we still have a solution. If we replace each non-relevant occurrence by an almost closed term, say $\lambda \boldsymbol{y}.z$ where z is some base-type variable, the lemma follows.

Definition 4. *Let* $\sigma \prec \tau$ *mean that* $\operatorname{order}(\sigma)$ *is less than* $\operatorname{order}(\tau)$ *Let* $\prec^*$ *be the multiset extension of* $\prec$*.*

Definition 5. *Given*

$$M = \lambda x_1..\lambda x_n.B,$$

let $\|M\|$ *be the multiset of types formed by including type* τ *with multiplicity* m *if* τ *is* not *a base type and in* B *there are* m *occurrences of free variables or constants of type* τ.

Then write $M \prec^* N$ *if* $|M| \prec^* |N|$.

Note that in computing $\|M\|$ we consider, in B, constants and the x_i as well as variables which are not among the x_i. There should be no ambiguity in using $\prec^*$ to refer to both the relation on type-multisets and the relation on terms. The relations $\prec^*$ are well-founded [2].

Lemma 4. *Let* $M \equiv \lambda x_1 \ldots \lambda x_n.hM_1 \cdots M_p$ *(where the atom* h *is either a constant, one of the* x_i*, or a free variable). Then* $M_j \prec^* M$ *for all* j.

Proof. Let the type M be $\sigma_1 \to \cdots \to \sigma_n \to \sigma$; let the type of M_i be μ_i, for $1 \leq i \leq p$; so that the type of h is $\mu = \mu_1 \to \cdots \to \mu_m \to \sigma$.

Of course if h has base type then $p = 0$ and the lemma is vacuously true. To verify the claim in the non-trivial case: let μ_j be $\delta_1 \to \cdots \to \delta_d \to o$.

In comparing $\|M_j\|$ with $\|M\|$ we have lost one occurrence of μ, the type of h, and possibly added occurrences of the types $\delta_1, \ldots, \delta_d$, due to the fact that M_j may have bound variables of these types (free in the matrix of M_j but not free in M). But each δ_k is of lower order than μ. This establishes the lemma.

The preceding is a general fact about all terms, k-duplicating or not. But the hypothesis of k-duplication and bounds on the occurrences of constants permit us to bound $\|M\|$.

Proposition 1. *Let* w *be a function assigning to each constant* c *in* Σ *a non-negative integer* $w(c)$. *For each* σ *and* k *we can effectively write down finitely many terms* M *comprising all of the almost-closed* β*-normal forms of type* σ *which are* k*-duplicating and satisfy* $\#(c, M) \leq w(c)$.

Proof. Let $M : \sigma_1 \to \cdots \to \sigma_n \to o$ satisfy the hypotheses. Recall that base-type variables do not figure in computing $\|M\|$. Since M is almost-closed we may observe that $\|M\|$ is bounded by the multiset consisting of

- k copies of each σ_i, and
- for each constant c of non-base type τ, $\#(c, M)$ copies of τ.

Together with the fact that $\prec^*$ is well-founded this means that we may effectively bound the depth of all the M satisfying the hypotheses. This establishes the proposition.

So we can solve the general k-duplicating matching problem by an exhaustive search:

Theorem 1. *The problems of* k*-duplicating* β*-matching and* $\beta\eta$*-matching are each decidable.*

Proof. Let problem $M =^? N$ be given. For each free variable $x : \sigma$ of M we may, by Lemma 1, compute the set of almost-closed k-duplicating β-normal forms X of type σ satisfying $\#(c, X) \leq \#(c, N)$ for each c. If we are considering $\beta\eta$ matching we filter out those terms which are not in $\bar{\eta}$-normal form. By Lemma 3 this is a complete search procedure for candidates for a solution.

4 Complexity of Matching in the Affine Lambda Calculus

Definition 6. *A lambda term is* linear *(resp.* affine*, a* λI-term*), if for every subterm* $N \subseteq M$*, such that* $N = \lambda x.P$*, there is exactly one (resp. at most one, at least one) free occurrence of* x *in* P*, i.e.,* $|\operatorname{Occ}(x, P)| = 1$ *(resp.* $|\operatorname{Occ}(x, P)| \leq 1$*,* $|\operatorname{Occ}(x, P)| \geq 1$*).*

Note, that the notion of affinity coincides with 1-linearity from our previous investigations, and that a term is linear exactly when it is an affine λI-term. As opposed to λI-terms, arbitrary terms are sometimes called λK-terms. The sets of linear, affine and λI-terms are closed under β-reduction (as opposed to k-duplicating terms, for $k \geq 2$, which are not). Thus it is possible to investigate theories of lambda conversion for such terms, and to consider the ordinary problem of higher order matching in these calculi, instead of modifying the formulation of the problem itself. Such variants of matching are less general than the variant considered above, since we put restrictions not only on the form of solutions, but also on the instances of the problem. One might expect, that the complexity of such problems should be lower than that of the k-duplicating matching. Indeed, de Groote [4] showed, that matching in the linear lambda calculus is NP-complete. In this section we generalize his proof to the case of affine lambda calculus.

Consider the following *potential function*:

$$\Phi(M) = \sum_{\substack{N \subseteq M \\ N = (\lambda x.N_1)N_2}} |\operatorname{type}(\lambda x.N_1)|.$$

Lemma 5 (de Groote). *If* M *is affine, and* $M \to_\beta N$*, then* $\Phi(M) > \Phi(N)$*.*

Proof. Suppose that M is reduced to N at address p, that is $M\restriction_p = (\lambda x.P)Q$ and $N = M[p \restriction P[x/Q]]$. In a reduction step $(\lambda x.P)Q \rhd_\beta P[x/Q]$ all redexes occurring in P and Q are left intact, and since x occurs in P at most once, there may be created at most one new redex inside P as a result of substitution. This new redex will be created if x occurs at active position (as an operator in an application) in a subterm of the form xP' and Q is a lambda abstraction. Moreover the reduct $P[x/Q]$ may also occur in active position and if it turns out to be a lambda abstraction, it will be a part of another new redex. Let $\operatorname{type}(\lambda x.P) = \sigma \to \tau$. Closing the redex $(\lambda x.P)Q$ decreases the potential by $|\operatorname{type}(\lambda x.P)|$, and adds to it at most $|\operatorname{type}(x)| = |\sigma|$ plus $|\operatorname{type}(P[x/Q])| = |\tau|$. Thus $\Phi(N) \leq \Phi(M) - |\operatorname{type}(\lambda x.P)| + |\operatorname{type}(x)| + |\operatorname{type}(P[x/Q]) = \Phi(M) - |\sigma \to \tau| + |\sigma| + |\tau| = \Phi(M) - 1$.

Definition 7. *We say that a* β*-reduction sequence* $(M_i)_{i=0}^m$ *is* variable-only-forgetting*, if for every* $i = 0, \ldots, m-1$ *and* $p \in \{0,1\}^*$*, if* $M_i\restriction_p = (\lambda x.P)Q \rhd_\beta M_{i+1}\restriction_p$*, and* $x \notin \operatorname{FV}(P)$*, then* Q *is a variable.*

Since if $x \notin \operatorname{FV}(P)$, then $|(\lambda x.P)y| = 3 + |P| = 3 + |P[x/y]|$, and if $x \in \operatorname{FV}(P)$ and x occurs $n > 0$ times in P, then $|P[x/Q]| = |P| + n(|Q| - 1) \geq |P| + |Q| - 1$, so $|(\lambda x.P)Q| = |P| + |Q| + 2 \leq |P[x/Q]| + 3$, we have:

Lemma 6. *If $(M_i)_{i=0}^m$ is variable-only-forgetting, then $|M_m| - |M_0| \leq 3m$.*

The next two lemmas state that every β-reduction sequence containing only affine terms can be replaced with an equivalent (in some sense) variable-only-forgetting sequence. The key idea is to replace any subterm that is forgotten during reduction with a fresh free variable (which is allowed to be forgotten).

Lemma 7. *For any β-reduction sequence $(M_i)_{i=0}^m$ containing only affine terms and any term $N' \leq M_m$ (with additional proviso that there exists substitution θ, such that $N'\theta = M_m$ and every variable from* $\mathrm{Dom}(\theta)$ *occurs at most once in N'), there exists a variable-only-forgetting reduction sequence $(N_j)_{j=0}^n$ leading to N', where $N_0 \leq M_0$, and $n \leq m$.*

Proof. In the whole proof when stating that $M \leq N$ we will tacitly assume, that there exists a substitution θ, such that $M\theta = N$ and every variable from $\mathrm{Dom}(\theta)$ occurs in M at most once. The proof is by induction on the length m of the reduction sequence $(M_i)_{i=0}^m$. For $m = 0$ the conclusion is obvious. Suppose that such a sequence $(N_j)_{j=1}^n$ exists for an arbitrary sequence $(M_i)_{i=1}^m$ of length m (for convenience we shifted indexes i and j by one), and consider a sequence $(M_i)_{i=0}^m$ of length $m+1$. Since $M_0 \to_\beta M_1$, suppose that M_0 is reduced to M_1 at address p, that is $M_0\restriction_p = (\lambda x.P)Q \rhd_\beta P[x/Q] = M_1\restriction_p$. By induction assumption $N_1 \leq M_1$. If p is not a valid address in N_1, then $N_1 \leq M_0$, by Lemma 1.6. Thus $(N_j)_{j=1}^n$ is the desired reduction sequence. Otherwise we need to find a new term N_0 as shown in the following diagram:

$$\begin{array}{ccccc} M_0 & \to_\beta & M_1 & \xrightarrow{*}_\beta & M_m \\ \text{IV} & & \text{IV} & & \text{IV} \\ N_0 & \to_\beta & N_1 & \xrightarrow{*}_\beta & N_n \end{array}$$

in such a way, that the reduction step $N_0 \to_\beta N_1$ is variable-only-forgetting. Fix a fresh free variable y (which does not occur anywhere else in considered terms). Since $N_1 \leq M_1$, then by Lemma 1.3 we have

$$N_1\restriction_p \leq M_1\restriction_p = P[x/Q]. \tag{1}$$

The term M_1 is affine, so $|\,\mathrm{Occ}(x,P)| \leq 1$. If $|\,\mathrm{Occ}(x,P)| = 0$, i.e., $x \notin \mathrm{FV}(P)$, then $P[x/Q] = P$, thus $N_1\restriction_p = P$. Hence, by Lemma 1.5 we have $P' = (\lambda x.(N_1\restriction_p))y \leq (\lambda x.P)Q$, where y is fresh. Let $N_0 = N_1[p\restriction P']$. By Lemma 1.4 we get $N_0 \leq M_1[p\restriction(\lambda x.P)Q] = M_0$. Since $N_1\restriction_p$ does not contain free occurrences of x, then $P' \rhd_\beta N_1\restriction_p$, thus $N_0 \to_\beta N_1$. Moreover $(N_j)_{j=0}^n$ is a desired variable-only-forgetting sequence leading to M_m and not longer than $(M_i)_{i=0}^m$. Now suppose that $|\,\mathrm{Occ}(x,P)| = 1$, and let $\mathrm{Occ}(x,P) = \{q\}$. If q is not a valid address in $N_1\restriction_p$, then, by Lemma 1.6, we have $N_1\restriction_p \leq (M_1\restriction_p)[q\restriction x] = P$. From (1) and Lemma 1.5 we have $P' = (\lambda x.(N_1\restriction_p))y \leq (\lambda x.P)Q$, where y is fresh. So let, as before, $N_0 = N_1[p\restriction P']$. By Lemma 1.4 we get $N_0 \leq M_1[p\restriction(\lambda x.P)Q] = M_0$. Since $N_1\restriction_p$ does not contain free occurrences of x, then $P' \rhd_\beta N_1\restriction_p$, thus $N_0 \to_\beta N_1$. Moreover $(N_j)_{j=0}^n$ is a desired variable-only-forgetting sequence leading to M_m and not longer than $(M_i)_{i=0}^m$. The last case to be considered is when q is a valid address in $N_1\restriction_p$. From (1) and Lemma 1.3 we have $N_1\restriction_{pq} \leq (P[x/Q])\restriction_q = Q$.

By Lemma 1.4 we get $(N_1\restriction_p)[q\restriction x] \leq (P[x/Q])[q\restriction x] = P$, so by Lemma 1.5 we have $P' = (\lambda x.((N_1\restriction_p)[q\restriction x]))(N_1\restriction_{pq}) \leq (\lambda x.P)Q$. Let $N_0 = N_1[p\restriction P']$. Since $N_1 \leq M_1$, then by Lemma 1.4 we have $N_0 \leq M_1[p\restriction(\lambda x.P)Q] = M_0$. Moreover $P' \rhd_\beta (N_1\restriction_p)[q\restriction N_1\restriction_{pq}] = N_1\restriction_p$, thus $N_0 \to_\beta N_1$. The reduction sequence $(N_j)_{j=0}^n$ is not longer than $(M_i)_{i=0}^m$, and, since x occurs in $(N_1\restriction_p)[q\restriction x]$, it is also variable-only-forgetting.

Lemma 8. *If $M \overset{*}{\to}_\beta M'$ and $\mathrm{FV}(M') = \emptyset$, then $N \overset{*}{\to}_\beta M'$ for every $N \geq M$.*

Proof. It is sufficient to show that if $(M_i)_{i=0}^m$ is a reduction sequence such that $\mathrm{FV}(M_m) = \emptyset$, then for any N_0 there exists a reduction sequence $(N_i)_{i=0}^m$, such that $N_m = M_m$. Proof is by induction on m. For $m = 0$ we have $N_0 \geq M_0$, so by Lemma 1.7 we get $N_0 = M_0$, and $(N_i)_{i=0}^0$ is the desired sequence. Now suppose that for any sequence $(M_i)_{i=1}^m$ and any $N_1 \geq M_1$ such a sequence $(N_i)_{i=1}^m$ exists (for convenience, as in previous lemma, we shifted the index i by one), and consider a sequence $(M_i)_{i=0}^m$ and a term $N_0 \geq M_0$. We need to find a term N_1, as shown in the following diagram:

$$\begin{array}{ccccc} M_0 & \to_\beta & M_1 & \overset{*}{\to}_\beta & M_m \\ \text{\rotatebox{90}{$\leq$}} & & \text{\rotatebox{90}{$\leq$}} & \nearrow^{*}_{\beta} & \\ N_0 & \to_\beta & N_1 & & \end{array}$$

Since $M_0 \to_\beta M_1$, there exists an address p, such that $M_0\restriction_p \rhd_\beta M_1\restriction_p$. Since $N_0 \geq M_0$, then by Lemma 1.8 the subterm $N_0\restriction_p$ is a β-redex. Suppose $N_0\restriction_p = (\lambda x.P)Q$. Let N_1 be the term obtained by contracting a redex at address p, i.e., $N_1 = N_0[p\restriction P[x/Q]]$. By Lemma 1.3 we have $(\lambda x.P)Q \geq M_0\restriction_p$. Thus $P[x/Q] \geq M_1\restriction_p$, so $N_1 \geq M_1$. By induction hypothesis there exists a reduction sequence $(N_i)_{i=1}^m$ leading to M_m. Then $(N_i)_{i=0}^m$ is the desired reduction sequence leading to M_m and starting from N_0.

Using Lemmas 7 and 8 we are able to prove the following:

Proposition 2. *For any matching problem $M =^? N$ in the affine lambda calculus there always exists a finite complete set of solutions $\mathcal{S}$, such that if*

$$[x/N_x]_{x\in\mathrm{FV}(M)} \in \mathcal{S},$$

then

$$|N_x| \leq |N| + 3 \sum_{x\in\mathrm{FV}(M)} |\operatorname{type}(x)| \cdot |\operatorname{Occ}(x, M)|, \tag{2}$$

for any $x \in \mathrm{FV}(M)$. (In particular $\mathcal{S}$ may be empty, if $M =^? N$ has no solution.)

Proof. Let θ be an arbitrary solution of $M =^? N$. It may be too big to satisfy (2), since it may contain huge terms that are forgotten during reduction of $M\theta$ to N. Suppose $(M_i)_{i=0}^m$ is a reduction sequence starting from $M\theta$ and leading to N. By Lemma 7 there exists a variable-only-forgetting sequence $(N_j)_{j=0}^n$, such that $N_0 \leq M_0$, $N_n = M_m = N$, and $n \leq m$. Now N_0 contains only necessary fragments of $M\theta$, so let

$$N_x = \sup(\{N_0\restriction_p \mid p \in \operatorname{Occ}(x, M) \text{ and } p \text{ valid in } N_0\} \cup \{x\}).$$

Since $N_0 \leq M\theta$, then $N_0\lceil_p \leq (M\theta)\lceil_p = x\theta$ by Lemma 1.3, for any $p \in \mathrm{Occ}(x, M)$. Thus the above set possesses an upper bound $x\theta$, and by Lemma 1.2 there exists its supremum, so N_x is well-defined. Let $\theta' = [x/N_x]_{x\in \mathrm{FV}(M)}$. We have immediately $\theta' \leq \theta$. Moreover for any substitution $\theta'' \geq \theta'$, since $M\theta'' \geq M\theta' \geq N_0$ and N is closed, then by Lemma 8 there exists a reduction sequence starting from $M\theta''$ and leading to N. Thus θ'' is a solution of $M =^? N$. From the other hand, by Lemma 6 we have $|N_0| - |N| \leq 3n$, while by Lemma 5 we have $n \leq \Phi(N_0) - \Phi(N)$. Note that $\Phi(N) = 0$, since N is in normal form. Comparing the last two inequalities we get

$$|N_0| \leq |N| + 3n \leq |N| + 3\Phi(N_0).$$

By Lemmas 1.9 and 1.10 we have $|N_x| \leq |N_0|$. From the other hand it is easy to see, that since $N_0 \leq M\theta$, then $\Phi(N_0) \leq \Phi(M\theta)$. Moreover, since M and N_x are in normal form, then all redexes in $M\theta$ are those of the form $N_x P'$, created as a result of substitution of terms N_x for variables from $\mathrm{FV}(M)$. Since $|\mathrm{type}(N_x)| = |\mathrm{type}(x)|$, we get (2).

We have shown that for any solution θ of $M =^? N$ there exists a substitution $\theta' \leq \theta$ satisfying (2), such that any $\theta'' \geq \theta'$ is a solution of $M =^? N$. Let $\mathcal{S}$ be the set of such substitutions θ' for all solutions θ of $M =^? N$. The set $\mathcal{S}$ is finite, since the size of θ' is bounded.

Theorem 2. *The problem of checking if an instance of a matching problem in the affine lambda calculus has a solution is NP-complete.*

Proof. The inequality (2) gives an upper bound on the size of solution, which depends only on terms M and N, and is a polynomial function of the length of any reasonable encoding of the problem $M =^? N$ as a word over a finite alphabet. Thus it suffices to guess terms N_x and check if $\theta = [x/N_x]_{x\in \mathrm{FV}(M)}$ is a solution. Since a normal form of a linear term may be found in a polynomial number of steps, the problem of checking if an instance $M =^? N$ has a solution is in NP.

From the other hand the simplest proof of NP-hardness of the second order matching [1] works also in the affine case. Also de Groote's proof of NP-hardness of matching in the linear lambda calculus can be applied here.

Again, as in the proof of Theorem 1, if the $\beta\eta$-matching is considered, then, when guessing a solution we omit terms, which are not in $\bar{\eta}$-normal form, so the theorem is valid for β- as well as $\beta\eta$-lambda calculus.

References

1. Hubert Comon, Yan Jurski, Higher-Order Matching and Tree Automata, *Proc. 11th Int'l Workshop Computer Science Logic*, CSL'97, Mogens Nielsen, Wolfgang Thomas, *eds.*, Aarhus, Denmark, August 1997, LNCS **1414**, Springer-Verlag, 1998, 157–176.
2. Nachum Dershowitz, Zohar Manna, Proving termination with multiset orderings, *Comm. Assoc. for Computing Machinery*, **6** (22), 465–476, 1979.

3. Giles Dowek, Third order matching is decidable, *Proc. 7th IEEE Symp. Logic in Computer Science*, LICS'92, IEEE Press, 1992, 2–10, also in *Annals of Pure and Applied Logic*, **69**, 1994, 135–155.
4. Philippe de Groote, Linear Higher-Order Matching Is NP-Complete, *Proc. 11th Int'l Conf. Rewriting Techniques and Applications*, RTA 2000, Leo Bachmair, *ed.*, Norwich, UK, July 10–12, 2000, LNCS **1833**, Springer-Verlag, 2000, 127–140.
5. Jordi Levy, Linear Second Order Unification, *Proc. 7th Int'l Conf. Rewriting Techniques and Applications*, RTA'96, H. Ganzinger, *ed.*, New Brunswick, NJ, 1996, LNCS **1103**, Springer-Verlag, 1996, 332–346.
6. Ralph Loader, *Higher Order β Matching is Undecidable*, October 2001, manuscript.
7. Harry G. Mairson, A Simple Proof of a Theorem of Statman, *Theoretical Computer Science*, **103**, 1992, 213–226.
8. Vincent Padovani, Decidability of All Minimal Models, *Proc. 3rd Int'l Workshop Types for Proofs and Programs*, TYPES'95, Stefano Berardi, Mario Coppo, *eds.*, Torino, Italy, 1995, LNCS **1158**, Springer-Verlag, 1996, 201–215.
9. Vincent Padovani, On equivalence classes of interpolation equations, *Proc.Int'l Conf. Typed Lambda Calculi and Applications*, TLCA'95, M. Dezani-Ciancaglini, G. Plotkin, *eds.*, LNCS **902**, Springer-Verlag, 1995, 335–349.
10. Vincent Padovani, Decidability of fourth-order matching, *Mathematical Structures in Computer Science*, **3** (10), 2000, 361–372.
11. Manfred Schmidt-Schauß, Klaus U. Schulz, *Decidability of bounded higher order unification*, technical report Frank-report-15, Institut für Informatik, J. W. Goethe-Universität, Frankfurt am Main, 2001.
12. Aleksy Schubert, Linear interpolation for the higher order matching problem, *Proc. 7th Int'l Joint Conf. Theory and Practice of Software Development*, TAPSOFT'97, M. Bidoit, M. Dauchet, *eds.*, LNCS **1214**, Springer-Verlag, 1997.
13. Richard Statman, The Typed λ-Calculus is Not Elementary Recursive, *Theoretical Computer Science*, **15**, 1981, 73–81.
14. Sergei Vorobyov, The"Hardest" Natural Decidable Theory, *Proc. 12th Annual IEEE Symp. Logic in Computer Science*, LICS'97, IEEE Press, 1997, 294–305.

Combining Decision Procedures for Positive Theories Sharing Constructors

Franz Baader[1] and Cesare Tinelli[2]

[1] TU Dresden, Germany, baader@inf.tu-dresden.de
[2] University of Iowa, USA, tinelli@cs.uiowa.edu

Abstract. This paper addresses the following combination problem: given two equational theories E_1 and E_2 whose positive theories are decidable, how can one obtain a decision procedure for the positive theory of $E_1 \cup E_2$? For theories over disjoint signatures, this problem was solved by Baader and Schulz in 1995. This paper is a first step towards extending this result to the case of theories sharing constructors. Since there is a close connection between positive theories and unification problems, this also extends to the non-disjoint case the work on combining decision procedures for unification modulo equational theories.

1 Introduction

Built-in decision procedures for certain types of theories (like equational theories) can greatly speed up the performance of theorem provers. In many applications, however, the theories actually encountered are combinations of theories for which dedicated decision procedure are available. Thus, one must find ways to combine the decision procedures for the single theories into one for their combination. In the context of *equational theories over disjoint signatures*, this combination problem has been thoroughly investigated in the following three instances:[1] the word problem, the validity problem for universally quantified formulae, and the unification problem. For the word problem, i.e., the problem whether a single (universally quantified) equation $s \equiv t$ follows from the equational theory, the first solution to the combination problem was given by Pigozzi [9] in 1974. The problem of combining decision procedures for universally quantified formulae, i.e., arbitrary Boolean combinations of equations that are universally quantified, was solved by Nelson and Oppen [8] in 1979. Work on combining unification algorithms started also in the seventies with Stickel's investigation [12] of unification of terms containing several associative-commutative and free symbols. The first general result on how to combine *decision procedures* for unification was published by Baader and Schulz [1] in 1992. It turned out that decision procedures for unification (with constants) are not sufficient to allow for a combination result. Instead, one needs decision procedures for unification *with linear constant restrictions* in the theories to be combined. In 1995, Baader and Schulz

[1] Some of the work mentioned below can also handle more general theories. To simplify the presentation, we restrict our attention in this paper to the equational case.

S. Tison (Ed.): RTA 2002, LNCS 2378, pp. 352–366, 2002.

[2] described a version of their combination procedure that applies to *positive theories*, i.e., positive Boolean combinations of equations with an arbitrary quantifier prefix. They also showed [3] that the decidability of the positive theory is equivalent to the decidability of unification with linear constant restrictions.

Since then, the main open problem in the area was how to extend these results to the combination of *theories having symbols in common.* In general, the existence of shared symbols may lead to undecidability results for the union theory (see, e.g., [6,5] for some examples). This means that a controlled form of sharing of symbols is necessary. For the word problem and for universally quantified formulae, a suitable notion of *shared constructors* has proved useful. In [5], Pigozzi's combination result for the word problem was extended to theories all of whose shared symbols are constructors. A similar extension of the Nelson-Oppen combination procedure can be found in [13].

In a similar vein, we show in this paper that the combination results in [2] for positive theories (and thus for unification) can be extended to theories sharing constructors. We do that by extending the combination procedure in [2] with an extra step that deals with shared symbols and proving that the extended procedure is sound and complete. Since this extra step is not finitary, the new procedure in general yields only a semi-decision procedure for the combined theory. Under some additional assumptions on the equational theory of the shared symbols, the procedure can, however, be turned into a decision procedure. Although the combination procedure described here differs from the one in [2] by just one extra step, proving its correctness is considerably more challenging, due to the non-disjointness of the theories. A major contribution of this work is a novel algebraic construction of the free algebra of the combined theory. As in the non-disjoint case [2], this construction is vital for the correctness proof of the procedure, and we believe that it will prove helpful also in future research on non-disjoint combination.

The paper is organized as follows. Section 2 contains some formal preliminaries. Section 3 defines our notion of constructors and presents some of their properties, which will be used later to prove the correctness of the combination procedure. Section 4 describes our extension of the Baader-Schulz procedure to component theories sharing constructors. It then introduces a straightforward condition on the component theories under which the semi-decision procedure obtained this way can in fact be used to decide the positive consequences of their union. Finally, it proves that the general procedure is sound and complete. We conclude the paper with a comparison to related work and suggestions for further research. Space constraints prevent us from providing all the proofs of the results in the paper. The missing proofs can be found in [4].

2 Preliminaries

In this paper we will use standard notions from universal algebra such as formula, sentence, algebra, subalgebra, generators, reduct, entailment, model, homomorphism and so on. Notable differences are reported in the following.

We consider only first-order theories (with equality) over a functional signature. A *signature* Σ is a set of *function symbols*, each with an associated *arity*, an integer $n \geq 0$. A *constant* symbol is a function symbol of zero arity. We use the letters Σ, Ω, Δ to denote signatures. Throughout the paper, we fix a countably-infinite set V of *variables*, disjoint with any signature Σ. For any $X \subseteq V$, $T(\Sigma, X)$ denotes the set of Σ*-terms* over X, i.e., first-order terms with variables in X and function symbols in Σ. Formulae in the signature Σ are defined as usual. We use $\equiv$ to denote the equality symbol. We also use the standard notion of substitution, with the usual postfix notation. We call a substitution a *renaming* iff it is a bijection of V onto itself. We say that a subset T of $T(\Sigma, V)$ is *closed under renaming* iff $t\sigma \in T$ for all terms $t \in T$ and renamings σ.

If A is a set, we denote by A^* the set of all finite tuples made of elements of A. If $\boldsymbol{a}$ and $\boldsymbol{b}$ are two tuples, we denote by $\boldsymbol{a}, \boldsymbol{b}$ the tuple obtained as the concatenation of $\boldsymbol{a}$ and $\boldsymbol{b}$. If φ is a term or a formula, we denote by $\mathcal{V}ar(\varphi)$ the set of φ's free variables. We will often write $\varphi(\boldsymbol{v})$ to indicate, as usual, that $\boldsymbol{v}$ is a tuple of variables with no repetitions and all elements of $\mathcal{V}ar(\varphi)$ occur in $\boldsymbol{v}$. A formula is *positive* iff it is in prenex normal form and its matrix is obtained from atomic formulae using only conjunctions and disjunctions. A formula is *existential* iff it has the form $\exists \boldsymbol{u}.\ \varphi(\boldsymbol{u}, \boldsymbol{v})$ where $\varphi(\boldsymbol{u}, \boldsymbol{v})$ is a quantifier-free formula.

If $\mathcal{A}$ is an algebra of signature Ω, we denote by A the universe of $\mathcal{A}$ and by $\mathcal{A}^{\Sigma}$ the reduct of $\mathcal{A}$ to a given subsignature Σ of Ω. If $\varphi(\boldsymbol{v})$ is an Ω-formula and α is a valuation of $\boldsymbol{v}$ into A, we write $(\mathcal{A}, \alpha) \models \varphi(\boldsymbol{v})$ iff $\varphi(\boldsymbol{v})$ is satisfied by the interpretation $(\mathcal{A}, \alpha)$. Equivalently, where $\boldsymbol{a} = \alpha(\boldsymbol{v})$, we may also write $\mathcal{A} \models \varphi(\boldsymbol{a})$. If $t(\boldsymbol{v})$ is an Ω-term, we denote by $[\![t]\!]^{\mathcal{A}}_{\alpha}$ the interpretation of t in $\mathcal{A}$ under the valuation α of $\boldsymbol{v}$. Similarly, if T is a set of terms, we denote by $[\![T]\!]^{\mathcal{A}}_{\alpha}$ the set $\{[\![t]\!]^{\mathcal{A}}_{\alpha} \mid t \in T\}$.

A *theory* of signature Ω, or an Ω-theory, is any set of Ω-sentences, i.e., closed Ω-formulae. An algebra $\mathcal{A}$ is *a model of a theory* $\mathcal{T}$, or *models* $\mathcal{T}$, iff each sentence in $\mathcal{T}$ is satisfied by the interpretation $(\mathcal{A}, \alpha)$ where α is the empty valuation. Let $\mathcal{T}$ be an Ω-theory. We denote by $Mod(\mathcal{T})$ the class of all Ω-algebras that model $\mathcal{T}$. The theory $\mathcal{T}$ is *satisfiable* if it has a model, and *trivial* if it has only *trivial models*, i.e., models of cardinality 1. For all sentences φ (of any signature), we say as usual that $\mathcal{T}$ *entails* φ, or that φ *is valid in* $\mathcal{T}$, and write $\mathcal{T} \models \varphi$, iff $\mathcal{T} \cup \{\neg\varphi\}$ is unsatisfiable. We call *(existential) positive theory of* $\mathcal{T}$ the set of all (existential) positive sentences *in the signature of* $\mathcal{T}$ that are entailed by $\mathcal{T}$.

An *equational theory* is a set of (universally quantified) equations. If E is an equational theory of signature Ω and Σ is an arbitrary signature, we denote by E^{Σ} the set of all (universally quantified) Σ-equations entailed by E. When $\Sigma \subseteq \Omega$ we call E^{Σ} the Σ*-restriction of* E. For all Ω-terms $s(\boldsymbol{v}), t(\boldsymbol{v})$, we write $s =_E t$ and say that s and t are *equivalent in* E iff $E \models \forall \boldsymbol{v}.\ s \equiv t$.

We will later appeal to the two basic model theory results below about subalgebras (see [7] among others).

Lemma 1. *Let $\mathcal{B}$ be a Σ-algebra and $\mathcal{A}$ a subalgebra of $\mathcal{B}$. For all quantifier-free formulae $\varphi(v_1, \ldots, v_n)$ and individuals $a_1, \ldots, a_n \in A$, $\mathcal{A} \models \varphi(a_1, \ldots, a_n)$ iff $\mathcal{B} \models \varphi(a_1, \ldots, a_n)$.*

Lemma 2. *For all equational theories E, $Mod(E)$ is closed under subalgebras.*

Similarly to [2], our procedure's correctness proof will be based on free algebras. Instead of the usual definition of free algebras, we will rely on the following characterization [7].

Proposition 3. *Let E be a Σ-theory and $\mathcal{A}$ a Σ-algebra. Then, $\mathcal{A}$ is free in E over some set X iff the following holds:*

1. *$\mathcal{A}$ is a model of E generated by X;*
2. *for all $s, t \in T(\Sigma, V)$ and injections α of $\mathcal{V}ar(s \equiv t)$ into X, if $(\mathcal{A}, \alpha) \models s \equiv t$ then $s =_E t$.*

When $\mathcal{A}$ is free in E over X we will also say that $\mathcal{A}$ *is a free model of E (with basis X)*. We will implicitly rely on the well-known fact that every non-trivial equational theory E admits a free model with a countably infinite basis, namely the quotient term algebra $T(\Sigma, V)/\!=_E$. We will also use the following two results from [2] about free models and positive formulae.

Lemma 4. *Let $\mathcal{B}$ be an Ω-algebra free (in some theory E) over a countably infinite set X. For all positive Ω-formulae $\varphi(\boldsymbol{v}_1, \boldsymbol{v}_2, \ldots, \boldsymbol{v}_{2m-1}, \boldsymbol{v}_{2m})$ the following are equivalent:*

1. *$\mathcal{B} \models \forall \boldsymbol{v}_1 \exists \boldsymbol{v}_2 \cdots \forall \boldsymbol{v}_{2m-1} \exists \boldsymbol{v}_{2m}.\ \varphi(\boldsymbol{v}_1, \boldsymbol{v}_2, \ldots, \boldsymbol{v}_{2m-1}, \boldsymbol{v}_{2m})$;*
2. *there exist tuples $\boldsymbol{x}_1, \ldots, \boldsymbol{x}_m \in X^*$ and $\boldsymbol{b}_1, \ldots, \boldsymbol{b}_m \in B^*$ and finite subsets $Z_1, \ldots, Z_m$ of X such that*
   a) *$\mathcal{B} \models \varphi(\boldsymbol{x}_1, \boldsymbol{b}_1, \ldots, \boldsymbol{x}_m, \boldsymbol{b}_m)$,*
   b) *all components of $\boldsymbol{x}_1, \ldots, \boldsymbol{x}_n$ are distinct,*
   c) *for all $n \in \{1, \ldots, m\}$, all components of $\boldsymbol{b}_n$ are generated by Z_n in $\mathcal{B}$,*
   d) *for all $n \in \{1, \ldots, m-1\}$, no components of $\boldsymbol{x}_{n+1}$ are in $Z_1 \cup \cdots \cup Z_n$.*

Lemma 5. *For every equational theory E having a countable signature and a free model $\mathcal{A}$ with a countably infinite basis, the positive theory of E coincides with the set of positive sentences true in $\mathcal{A}$.*

In this paper, we will deal with *combined* equational theories, that is, theories of the form $E_1 \cup E_2$, where E_1 and E_2 are two *component* equational theories of (possibly non-disjoint) signatures Σ_1 and Σ_2, respectively. Where $\Sigma := \Sigma_1 \cap \Sigma_2$, we call *shared symbols* the elements of Σ and *shared terms* the elements of $T(\Sigma, V)$. Notice that, when Σ_1 and Σ_2 are disjoint, the only shared terms are the variables.

Most combination procedures, including the one described in this paper, work with $(\Sigma_1 \cup \Sigma_2)$-formulae by first "purifying" them into a set of Σ_1-formulae and a set of Σ_2-formulae. There is a standard *purification procedure* that, when Σ_1 and Σ_2 are disjoint, can convert any set S of equations of signature $\Sigma_1 \cup \Sigma_2$ into a set S' of *pure* equations (that is, each of signature Σ_1 or Σ_2) such that S' is satisfiable in a $(\Sigma_1 \cup \Sigma_2)$-algebra $\mathcal{A}$ iff S is satisfiable in $\mathcal{A}$. As we show in [5], a similar procedure also exists for the case in which Σ_1 and Σ_2 are not disjoint.

3 Theories with Constructors

The main requirement for our generalization of the combination procedure described in [2] to apply is that the symbols shared by the two theories are *constructors* as defined in [5,13]. For the rest of the section, let E be an non-trivial equational theory of signature Ω. Also, let Σ be a subsignature of Ω.

Definition 6 (Constructors). *The signature Σ is a* set of constructors for E *iff for every free model $\mathcal{A}$ of E with a countably infinite basis X, $\mathcal{A}^\Sigma$ is a free model of E^Σ with a basis Y including X.*

It is usually non-trivial to show that a signature Σ is a set of constructors for a given theory E by using just the definition above. Instead, using a syntactic characterization of constructors given in terms of certain subsets of $T(\Omega, V)$ is usually more helpful. Before we can give this characterization, we need a little more notation.

Given a subset G of $T(\Omega, V)$, we denote by $T(\Sigma, G)$ the set of Σ-terms over the "variables" G. More precisely, every member of $T(\Sigma, G)$ is obtained from a term $s \in T(\Sigma, V)$ by replacing the variables of s with terms from G. To express this construction, we will denote any such term by $s(\boldsymbol{r})$ where $\boldsymbol{r}$ is a tuple collecting the terms of G that replace the variables of s. Note that $G \subseteq T(\Sigma, G)$ and that $T(\Sigma, V) \subseteq T(\Sigma, G)$ whenever $V \subseteq G$.

Definition 7 (Σ-base). *A subset G of $T(\Omega, V)$ is a* Σ-base of E *iff*

1. *$V \subseteq G$;*
2. *for all $t \in T(\Omega, V)$, there is an $s(\boldsymbol{r}) \in T(\Sigma, G)$ such that $t =_E s(\boldsymbol{r})$;*
3. *for all $s_1(\boldsymbol{r}_1), s_2(\boldsymbol{r}_2) \in T(\Sigma, G)$, $s_1(\boldsymbol{r}_1) =_E s_2(\boldsymbol{r}_2)$ iff $s_1(\boldsymbol{v}_1) =_E s_2(\boldsymbol{v}_2)$, where $\boldsymbol{v}_1$ and $\boldsymbol{v}_2$ are tuples of fresh variables abstracting the terms of $\boldsymbol{r}_1, \boldsymbol{r}_2$ so that two terms in $\boldsymbol{r}_1, \boldsymbol{r}_2$ are abstracted by the same variable iff they are equivalent in E.*

We say that E admits a Σ-base *if some $G \subseteq T(\Omega, V)$ is a Σ-base of E.*

Theorem 8 (Characterization of constructors). *The signature Σ is a set of constructors for E iff E admits a Σ-base.*

A proof of this theorem and of the following corollary can be found in [5].

Corollary 9. *Where $\mathcal{A}$ is a free model of E with a countably-infinite basis X, let α be an arbitrary bijection of V onto X. If G is a Σ-base of E, then $\mathcal{A}^\Sigma$ is free in E^Σ over the superset $[\![G]\!]^{\mathcal{A}}_\alpha$ of X.*

In the following, we will assume that the theories we consider admit Σ-bases *closed under renaming*. This assumption is necessary for technical reasons. It is used in the long version of this paper in the proof of a lemma (Lemma 4.18 in [4]; omitted here) needed to prove the soundness of the combination procedure described later. Although we do not know whether this assumption can be made with no loss of generality, it is not clear how to avoid it and it seems to be satisfied by all "sensible" examples of theories admitting constructors. Also note that the same technical assumption was needed in our work on combining decision procedures for the word problem [5].

It is shown in [5] that, under the right conditions, constructors and the property of having Σ-bases closed under renaming are modular with respect to the union of theories.

Proposition 10. *For $i = 1,2$ let E_i be a non-trivial equational Σ_i-theory. If $\Sigma := \Sigma_1 \cap \Sigma_2$ is a set of constructors for E_1 and for E_2 and ${E_1}^{\Sigma} = {E_2}^{\Sigma}$, then Σ is a set of constructors for $E_1 \cup E_2$. If both E_1 and E_2 admit a Σ-base closed under renaming, then $E_1 \cup E_2$ also admits a Σ-base closed under renaming.*

A useful consequence of Proposition 10 for us will be the following.

Proposition 11. *Let E be an Ω-theory and let E' be the empty Δ-theory for some signature Δ disjoint with Ω. If $\Sigma \subseteq \Omega$ is a set of constructors for E, then it is a set of constructors for $E \cup E'$. Furthermore, if E admits a Σ-base closed under renaming, then so does $E \cup E'$.*

4 Combining Decision Procedures

In this section, we generalize the Baader-Schulz procedure [2] for combining decision procedures for the validity of positive formulae in equational theories from theories over disjoint signatures to theories sharing constructors. More precisely, we will consider two theories E_1 and E_2 that satisfy the following assumptions for $i = 1, 2$, which we fix for the rest of the section:

- E_i is a non-trivial equational theory of some countable signature Σ_i;
- $\Sigma := \Sigma_1 \cap \Sigma_2$ is a set of constructors for E_i, and E_i admits a Σ-base closed under renaming;
- ${E_1}^{\Sigma} = {E_2}^{\Sigma}$.

Let $E := E_1 \cup E_2$. Under the assumptions above, $E^{\Sigma} = {E_1}^{\Sigma} = {E_2}^{\Sigma}$ (see [5]). In the following then, we will use E^{Σ} to refer indifferently to ${E_1}^{\Sigma}$ or ${E_2}^{\Sigma}$.

The combination procedure will use two kinds of substitutions that we call, after [13], *identifications* and *Σ-instantiations*. Given a set of variables U, an *identification of U* is a substitution defined by partitioning U, selecting a representative for each block in the partition, and mapping each element of U to the representative in its block. A *Σ-instantiation of U* is a substitution that maps some elements of U to non-variable Σ-terms and the other elements to themselves. For convenience, we will assume that the variables occurring in the terms introduced by a Σ-instantiation are *always fresh*.

4.1 The Combination Procedure

The procedure takes as input a positive existential $(\Sigma_1 \cup \Sigma_2)$-formula $\exists \boldsymbol{w}.\ \varphi(\boldsymbol{w})$ and outputs, non-deterministically, a pair of sentences: a positive Σ_1-sentence and a positive Σ_2-sentence. It consists of the following steps.

1. **Convert into DNF.** Convert the input's matrix φ into the disjunctive normal form $\psi_1 \vee \cdots \vee \psi_n$ and choose a disjunct ψ_j.
2. **Convert into Separate Form.** Let S be the set obtained by purifying, as mentioned in Section 2, the set of all the equations in ψ_j. For $i = 1, 2$, let $\varphi_i(\boldsymbol{v}, \boldsymbol{u}_i)$ be the conjunction of all Σ_i-equations in S,[2] with $\boldsymbol{v}$ listing the variables in $\mathcal{V}ar(\varphi_1) \cap \mathcal{V}ar(\varphi_2)$ and $\boldsymbol{u}_i$ listing the remaining variables of φ_i.
3. **Instantiate Shared Variables.** Choose a Σ-instantiation ρ of $\mathcal{V}ar(\boldsymbol{v}) = \mathcal{V}ar(\varphi_1) \cap \mathcal{V}ar(\varphi_2)$.
4. **Identify Shared Variables.** Choose an identification ξ of $\mathcal{V}ar(\varphi_1\rho) \cap \mathcal{V}ar(\varphi_2\rho) = \mathcal{V}ar(\boldsymbol{v}\rho)$. For $i = 1, 2$, let $\varphi'_i := \varphi_i\rho\xi$.
5. **Partition Shared Variables.** Group the elements of $V_s := \mathcal{V}ar(\boldsymbol{v}\rho\xi) = \mathcal{V}ar(\varphi'_1) \cap \mathcal{V}ar(\varphi'_2)$ into the tuples $\boldsymbol{v}_1, \ldots, \boldsymbol{v}_{2m}$, with $2 \leq 2m \leq |V_s| + 1$, so that each element of V_s occurs exactly once in the tuple $\boldsymbol{v}_1, \ldots, \boldsymbol{v}_{2m}$.[3]
6. **Generate Output Pair.** Output the pair of sentences

$$(\exists \boldsymbol{v}_1 \forall \boldsymbol{v}_2 \cdots \exists \boldsymbol{v}_{2m-1} \forall \boldsymbol{v}_{2m} \exists \boldsymbol{u}_1.\, \varphi'_1,\ \forall \boldsymbol{v}_1 \exists \boldsymbol{v}_2 \cdots \forall \boldsymbol{v}_{2m-1} \exists \boldsymbol{v}_{2m} \exists \boldsymbol{u}_2.\, \varphi'_2).$$

Ignoring inessential differences and our restriction to functional signatures, this combination procedure differs from Baader and Schulz's only for the presence of Step 3. Note however that, for component theories with disjoint signatures (the case considered in [2]), Step 3 is vacuous because Σ is empty. In that case then the procedure above reduces to that in [2]. Correspondingly, our requirements on the two component theories also reduce to that in [2], which simply asks that E_1 and E_2 be non-trivial. In fact, when Σ is empty it is always a set of constructors for E_i $(i = 1, 2)$, with $T(\Sigma_i, V)$ being a Σ-base closed under renaming. Moreover, ${E_1}^{\Sigma} = {E_2}^{\Sigma}$ because they both coincide with the theory $\{v \equiv v \mid v \in V\}$.

As will be shown in Section 4.3, our combination procedure is sound and complete in the following sense.

Theorem 12 (Soundness and Completeness). *For all possible input sentences $\exists \boldsymbol{w}.\, \varphi(\boldsymbol{w})$ of the combination procedure, $E_1 \cup E_2 \models \exists \boldsymbol{w}.\, \varphi(\boldsymbol{w})$ iff there is a possible output (γ_1, γ_2) such that $E_1 \models \gamma_1$ and $E_2 \models \gamma_2$.*

Unlike the procedure in [2], the combination procedure above does not necessarily yield a decision procedure. The reason is that the non-determinism in Step 3 of the procedure is not finitary since in general there are infinitely-many possible Σ-instantiations to choose from. One viable, albeit strong, restriction for obtaining a decision procedure is described in the next subsection.

4.2 Decidability Results

In order to turn the combination procedure from above into a decision procedure, we require that the equivalence relation defined by the theory $E^{\Sigma} = {E_1}^{\Sigma} = {E_2}^{\Sigma}$ be *bounded* in a sense described below.

2 Where Σ-equations are considered arbitrarily as either Σ_1- or Σ_2-equations.

3 Note that some of the subtuples $\boldsymbol{v}_i$ may be empty.

Definition 13. *Let E be an equational Ω-theory. We say that equivalence in E is* finitary modulo renaming *iff there is a finite subset R of $T(\Omega, V)$ such that for all $s \in T(\Omega, V)$ there is a term $t \in R$ and a renaming σ such that $s =_E t\sigma$. We call R a set of* E-representatives.

When Ω in the above definition is empty, equivalence in E is trivially finitary—with any singleton set of variables being a set of E-representatives. A non-trivial example is provided at the end of this section.

If E^Σ is finitary modulo renaming, then it is easy to see that it suffices to consider only finitely many instantiations in Step 3 of the procedure, which leads to the following decidability result.

Proposition 14. *Assume that Σ, E_1, E_2 satisfy the assumptions stated at the beginning of Section 4, and that equivalence in E^Σ is finitary modulo renaming. If the positive theories of E_1 and of E_2 are both decidable, then the positive existential theory of $E_1 \cup E_2$ is also decidable.*

Using a Skolemization argument together with Proposition 11, the result above can be extended from positive existential input sentences to arbitrary positive input sentences. The main idea is to Skolemize the universal quantifiers of the input sentence and then expand the signature of one the theories, E_2 say, to the newly introduced Skolem symbols. Proposition 11 and the combination result in [2] for the disjoint case imply that the pair E_1, E_2', where E_2' is the conservative extension of E_2 to the expanded signature, satisfies the assumptions of Proposition 14.

Theorem 15. *Assume that E_1, E_2 satisfy the assumptions of Proposition 14. If the positive theories of E_1 and of E_2 are both decidable, then the positive theory of $E := E_1 \cup E_2$ is also decidable.*

The following example describes one theory satisfying all the requirements on the component theories imposed by Theorem 15.

Example 16. Consider the signature $\Omega := \{0, s, +\}$ and, for some $n > 1$, the equational theory E_n axiomatized by the identities

$$\begin{array}{lll} x + (y + z) \equiv (x + y) + z, & x + y \equiv y + x, & \\ x + \mathsf{s}(y) \equiv \mathsf{s}(x + y), & x + 0 \equiv x, & \mathsf{s}^n(x) \equiv x. \end{array}$$

where as usual $\mathsf{s}^n(x)$ stands for the n-fold application of s to x. We show in [4] that, for E_n and the subsignature $\Sigma := \{0, \mathsf{s}\}$ of Ω, all the assumptions of Theorem 15 on the component theories are satisfied.

4.3 Soundness and Completeness of the Procedure

The soundness and completeness proof for the disjoint case in [2] relies on an explicit construction of the free model of $E = E_1 \cup E_2$ as an *amalgamated product* of the free models of the component theories. A direct adaptation of the free amalgamation construction of [2] to the non-disjoint case has so far proven elusive. An important technical contribution of the present work is to provide

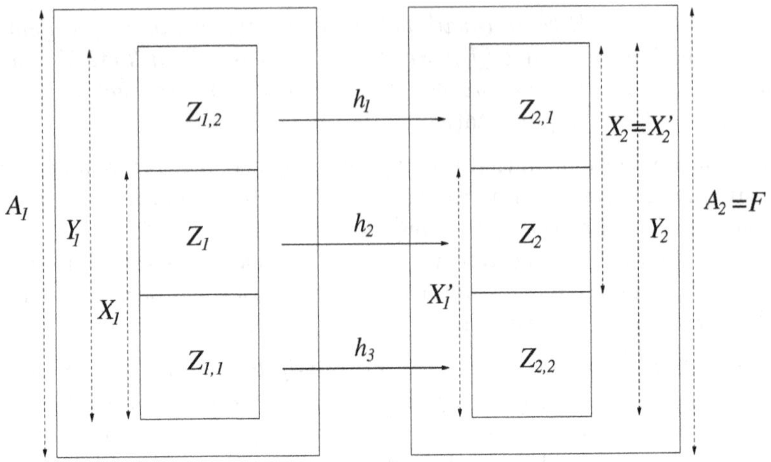

Fig. 1. The Fusion $\mathcal{F}$ of $\mathcal{A}_1$ and $\mathcal{A}_2$.

an alternative way to obtain an appropriate amalgamated free model in the case of theories sharing constructors. We obtain this model for the union theory E indirectly, by first building a simpler sort of amalgamated model as a *fusion* (defined below) of the free models of the two component theories. Contrary to Baader and Schulz's free amalgamated product, the fusion model we construct is not free in E. However, it has a subalgebra that is so. That subalgebra will serve as the free amalgamated model of E.

Definition 17 (Fusion [5,13]). *A $(\Omega_1 \cup \Omega_2)$-algebra $\mathcal{F}$ is a* fusion *of a Ω_1-algebra $\mathcal{A}_1$ and a Ω_2-algebra $\mathcal{A}_2$ iff $\mathcal{F}^{\Omega_1}$ is Ω_1-isomorphic to $\mathcal{A}_1$ and $\mathcal{F}^{\Omega_2}$ is Ω_2-isomorphic to $\mathcal{A}_2$.*

It is shown in [13] that two algebras $\mathcal{A}_1$ and $\mathcal{A}_2$ have fusions exactly when they are isomorphic over their shared signature, and that every fusion of a model of a theory $\mathcal{T}_1$ with a model of a theory $\mathcal{T}_2$ is a model of the theory $\mathcal{T}_1 \cup \mathcal{T}_2$.

In the following, we will construct a model of $E = E_1 \cup E_2$ as a fusion of free models of the theories E_1 and E_2 fixed earlier, whose shared signature Σ was a set of constructors for both. We start by fixing, for $i = 1, 2$,

- a free model $\mathcal{A}_i$ of E_i with a countably infinite basis X_i,
- a bijective valuation α_i of V onto X_i,
- a Σ-base G_i of E_i closed under renaming, and
- the set $Y_i := [\![G_i]\!]^{\mathcal{A}_i}_{\alpha_i}$.

We know from Corollary 9 that $X_i \subseteq Y_i$ and ${\mathcal{A}_i}^{\Sigma}$ is free in $E^{\Sigma} = {E_1}^{\Sigma} = {E_2}^{\Sigma}$ over Y_i. Observe that $\mathcal{A}_i$ is countably infinite, given our assumption that X_i is countably infinite and Σ_i is countable. As a consequence, Y_i is countably infinite as well.

Now let $Z_{i,2} := Y_i \setminus X_i$ for $i = 1, 2$, and let $\{Z_{1,1}, Z_1\}$ be a partition of X_1 such that Z_1 is countably infinite and $|Z_{1,1}| = |Z_{2,2}|$.[4] Similarly, let $\{Z_{2,1}, Z_2\}$ be a partition of X_2 such that $|Z_{2,1}| = |Z_{1,2}|$ and Z_2 is countably infinite (see Figure 1). Then consider 3 arbitrary bijections

$$h_1 \colon Z_{1,2} \longrightarrow Z_{2,1}, \;\; h_2 \colon Z_1 \longrightarrow Z_2, \;\; h_3 \colon Z_{1,1} \longrightarrow Z_{2,2},$$

as shown in Figure 1. Observing that $\{Z_{i,1}, Z_i, Z_{i,2}\}$ is a partition of Y_i for $i = 1, 2$, it is immediate that $h_1 \cup h_2 \cup h_3$ is a well-defined bijection of Y_1 onto Y_2. Since $\mathcal{A}_i{}^{\Sigma}$ is free in E^{Σ} over Y_i for $i = 1, 2$, we have that $h_1 \cup h_2 \cup h_3$ extends uniquely to a (Σ)-isomorphism h of $\mathcal{A}_1{}^{\Sigma}$ onto $\mathcal{A}_2{}^{\Sigma}$. The isomorphism h induces a fusion of $\mathcal{A}_1$ and $\mathcal{A}_2$ whose main properties are listed in the following lemma, taken from [5].

Lemma 18. *There is a fusion $\mathcal{F}$ of $\mathcal{A}_1$ and $\mathcal{A}_2$ having the same universe as $\mathcal{A}_2$ and such that*

1. *h is a (Σ_1)-isomorphism of $\mathcal{A}_1$ onto $\mathcal{F}^{\Sigma_1}$;*
2. *the identity map of A_2 is a (Σ_2)-isomorphism of $\mathcal{A}_2$ onto $\mathcal{F}^{\Sigma_2}$;*
3. *$\mathcal{F}^{\Sigma_i}$ is free in E_i over $X_i' := Z_{2,j} \cup Z_2$ for $i, j = 1, 2$, $i \neq j$;*
4. *$\mathcal{F}^{\Sigma}$ is free in E^{Σ} over $Y_2 = Z_{2,1} \cup Z_2 \cup Z_{2,2}$;*
5. *$Y_2 = [\![G_2]\!]^{\mathcal{F}^{\Sigma_2}}_{\alpha_2} = [\![G_1]\!]^{\mathcal{F}^{\Sigma_1}}_{h \circ \alpha_1}$.*

We will now consider the theory $E = E_1 \cup E_2$ again, together with the algebras $\mathcal{F}$, $\mathcal{F}_1$, $\mathcal{F}_2$ and $\mathcal{A}$ where:

- $\mathcal{F}$ is the fusion of $\mathcal{A}_1$ and $\mathcal{A}_2$ from Lemma 18;
- $\mathcal{F}_i := \mathcal{F}^{\Sigma_i}$ for $i = 1, 2$;[5]
- $\mathcal{A}$ is the subalgebra of $\mathcal{F}$ generated by Z_2.

Both $\mathcal{F}$ and $\mathcal{A}$ are models of E. In fact, $\mathcal{F}$ is a model of $E = E_1 \cup E_2$ for being a fusion of a model of E_1 and a model of E_2, whereas $\mathcal{A}$ is a model of E by Lemma 2. We prove in [4] that $\mathcal{A}$ is in fact a free model of E. To do that we use the following sets of terms, which will come in handy later as well.

Definition 19 ($G_1^{\infty}, G_2^{\infty}, G^{\infty}$). *Let $G^{\infty} := G_1^{\infty} \cup G_2^{\infty}$ where for $i = 1, 2$, $G_i^{\infty} := \bigcup_{n=0}^{\infty} G_i^n$ and $\{G_i^n \mid n \geq 0\}$ is the family of sets defined as follows:*

$$\begin{aligned} G_i^0 &:= V, \\ G_i^{n+1} &:= G_i^n \cup \{r(r_1, \ldots, r_m) \mid r(v_1, \ldots, v_m) \in G_i \setminus V,\ r \neq_E v \text{ for all } v \in V, \\ &\qquad r_j \in G_k^n \text{ with } k \neq i, \text{ for all } j = 1, \ldots, m, \\ &\qquad r_j \neq_E r_{j'} \text{ for all distinct } j, j' = 1, \ldots, m\,\}. \end{aligned}$$

As proved in [5], the sets $G_1^{\infty}, G_2^{\infty}, G^{\infty}$ satisfy the following two properties.

Lemma 20. *Let $i \in \{1, 2\}$. For any bijection α of V onto Z_2 the following holds:*

[4] This is possible because $Z_{2,2}$ is countable (possibly finite).
[5] These algebras are defined just for notational convenience.

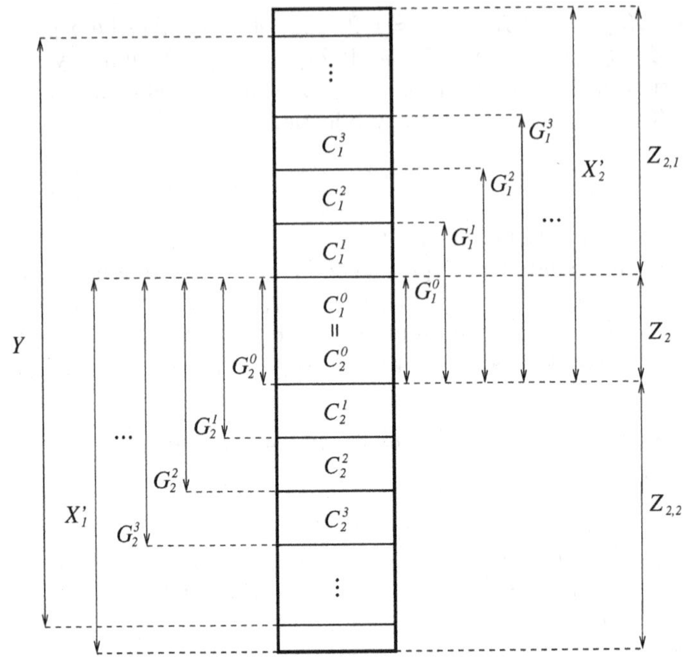

Fig. 2. The families $\{[\![G_i^n]\!] \mid n \geq 0\}$ and $\{C_i^n \mid n \geq 0\}$.

1. $[\![G_i^\infty \setminus V]\!]_\alpha^{\mathcal{F}} \subseteq Z_{2,i}$;
2. *for all* $t_1, t_2 \in G_i^\infty \setminus V$, *if* $[\![t_1]\!]_\alpha^{\mathcal{F}} = [\![t_2]\!]_\alpha^{\mathcal{F}}$ *then* $t_1 =_E t_2$.

Proposition 21. *The set* G^∞ *is* Σ*-base of* $E = E_1 \cup E_2$.

Note that this proposition entails by Theorem 8 that Σ is a set of constructors for E. Using these two properties (and Proposition 3) we can show the following.

Proposition 22. $\mathcal{A}$ *is free in* E *over* Z_2.

Corollary 23. *For every bijection* α *of* V *onto* Z_2, $\mathcal{A}^\Sigma$ *is free in* E^Σ *over* $Y := [\![G^\infty]\!]_\alpha^{\mathcal{A}}$, *and* $Y \subseteq Y_2$.

For the rest of the section, let us fix a bijection α of V onto Z_2 and the corresponding set $Y := [\![G^\infty]\!]_\alpha^{\mathcal{A}}$.

To prove the completeness of the combination procedure we will need two families $\{C_1^n \mid n \geq 0\}$ and $\{C_2^n \mid n \geq 0\}$ of sets partitioning the set Y above. To build these families we use the denotations in $\mathcal{A}$ of the sets G_1^n and G_2^n introduced in Definition 19. More precisely, for $i = 1, 2$, we consider the family $\{[\![G_i^n]\!]_\alpha^{\mathcal{A}} \mid n \geq 0\}$ of subsets of Y. Since $\mathcal{A}$ is the subalgebra of $\mathcal{F}$ generated by Z_2 and α is a valuation of V into Z_2, it is easy to see that $[\![G_i^n]\!]_\alpha^{\mathcal{A}} = [\![G_i^n]\!]_\alpha^{\mathcal{F}}$ for all $n \geq 0$. Therefore, we will write just $[\![G_i^n]\!]$ in place of either $[\![G_i^n]\!]_\alpha^{\mathcal{A}}$ or $[\![G_i^n]\!]_\alpha^{\mathcal{F}}$.

Observe that $[\![G_1^0]\!] = [\![G_2^0]\!] = Z_2$ and $[\![G_i^n]\!] \subseteq [\![G_i^{n+1}]\!]$ for all $n \geq 0$ and $i = 1, 2$. Given that $[\![G_i^n]\!] \setminus Z_2 \subseteq [\![G_i^n \setminus V]\!]_\alpha^{\mathcal{A}}$, we can conclude by Lemma 20 that $[\![G_i^n]\!] \setminus Z_2 \subseteq Z_{2,i}$.[6] By Corollary 23 we have that

$$\bigcup_{n\geq 0} ([\![G_1^n]\!] \cup [\![G_2^n]\!]) = [\![\bigcup_{n\geq 0} (G_1^n \cup G_2^n)]\!] = [\![G_1^\infty \cup G_2^\infty]\!] = [\![G^\infty]\!] = Y.$$

Now consider the family of sets $\{C_i^n \mid n \geq 0\}$, depicted in Figure 2 along with $\{[\![G_i^n]\!] \mid n \geq 0\}$ and defined as follows:

$$C_i^0 := [\![G_i^0]\!] \quad \text{and} \quad C_i^{n+1} := [\![G_i^{n+1}]\!] \setminus [\![G_i^n]\!] \quad \text{for all } n \geq 0.$$

First note that $\bigcup_{n\geq 0} (C_1^n \cup C_2^n) = \bigcup_{n\geq 0} ([\![G_1^n]\!] \cup [\![G_2^n]\!]) = Y$. Then note that, for all $n \geq 0$ and $i = 1, 2$, the elements of C_i^n are individuals of the algebras $\mathcal{F}_1$ and $\mathcal{F}_2$ (which have the same universe). By Lemma 20, $C_1^n \subseteq [\![G_1^n]\!] \subseteq Z_{2,1} \cup Z_2 = X_2'$; in other words, every element of C_1^n is a generator of $\mathcal{F}_2$. Similarly, $C_2^n \subseteq [\![G_2^n]\!] \subseteq Z_{2,2} \cup Z_2 = X_1'$, that is, every element of C_2^n is a generator of $\mathcal{F}_1$. In addition, we have the following.

Lemma 24. *For all distinct $m, n \geq 0$ and distinct $i, j \in \{1, 2\}$,*

1. *$C_i^m \cap C_i^n = \emptyset$ and*
2. *C_i^{n+1} is Σ_i-generated by $[\![G_j^n]\!]$ in $\mathcal{F}_i$.*

Now, Theorem 12 is an easy consequence of the following proposition.

Proposition 25. *For $i = 1, 2$, let $\varphi_i(\boldsymbol{v}, \boldsymbol{u}_i)$ be a conjunction of Σ_i-equations where $\boldsymbol{v}$ lists the elements of $\mathcal{V}ar(\varphi_1) \cap \mathcal{V}ar(\varphi_2)$ and $\boldsymbol{u}_i$ lists the elements of $\mathcal{V}ar(\varphi_i)$ not in $\boldsymbol{v}$. The following are equivalent:*

1. *There is a Σ-instantiation ρ of $\boldsymbol{v}$, an identification ξ of $\mathcal{V}ar(\boldsymbol{v}\rho)$ and a grouping $\boldsymbol{v}_1, \ldots, \boldsymbol{v}_{2m}$ of $\mathcal{V}ar(\boldsymbol{v}\rho\xi)$ with each element of $\mathcal{V}ar(\boldsymbol{v}\rho\xi)$ occurring exactly once in $\boldsymbol{v}_1, \ldots, \boldsymbol{v}_{2m}$ such that*

$$\mathcal{A}_1 \models \exists \boldsymbol{v}_1 \, \forall \boldsymbol{v}_2 \cdots \exists \boldsymbol{v}_{2m-1} \, \forall \boldsymbol{v}_{2m} \, \exists \boldsymbol{u}_1. \, (\varphi_1 \rho\xi) \quad \textit{and}$$
$$\mathcal{A}_2 \models \forall \boldsymbol{v}_1 \, \exists \boldsymbol{v}_2 \cdots \forall \boldsymbol{v}_{2m-1} \, \exists \boldsymbol{v}_{2m} \, \exists \boldsymbol{u}_2. \, (\varphi_2 \rho\xi).$$

2. *$\mathcal{A} \models \exists \boldsymbol{v} \exists \boldsymbol{u}_1 \exists \boldsymbol{u}_2. \, (\varphi_1 \wedge \varphi_2)$.*

Proof. The proof of $(1 \Rightarrow 2)$ is similar to the corresponding proof in [2], although it requires some additional technical lemmas (see [4] for details). We concentrate here on the proof of $(2 \Rightarrow 1)$.

Assume that $\mathcal{A} \models \exists \boldsymbol{v}, \boldsymbol{u}_1, \boldsymbol{u}_2. \, (\varphi_1(\boldsymbol{v}, \boldsymbol{u}_1) \wedge \varphi_2(\boldsymbol{v}, \boldsymbol{u}_2))$. Let α be the bijection of V onto Z_2 and Y the subset of Y_2 that we fixed after Corollary 23. Since the reduct $\mathcal{A}^\Sigma$ of $\mathcal{A}$ is Σ-generated by Y by the same corollary, there is a Σ-instantiation ρ of $\boldsymbol{v}$, an identification ξ of $\mathcal{V}ar(\boldsymbol{v}\rho)$, and an injective valuation β

[6] This entails that $[\![G_1^m]\!] \setminus Z_2$ is disjoint with $[\![G_2^n]\!] \setminus Z_2$ for all $m, n > 0$.

of $\boldsymbol{v}'$ into Y such that, for $\varphi_i' := \varphi_i \rho \xi$ $(i = 1, 2)$ and $\boldsymbol{v}'$ listing the variables of $\boldsymbol{v}\rho\xi$, we have

$$(\mathcal{A}, \beta) \models \exists \boldsymbol{u}_1, \boldsymbol{u}_2.\ (\varphi_1'(\boldsymbol{v}', \boldsymbol{u}_1) \wedge \varphi_2'(\boldsymbol{v}', \boldsymbol{u}_2)).$$

From this, recalling that $\mathcal{A}$ is $(\Sigma_1 \cup \Sigma_2)$-generated by Z_2 by construction and Y is included in Y_2, we can conclude that there is a tuple $\boldsymbol{a}$ of *pairwise distinct* elements of Y_2, all $(\Sigma_1 \cup \Sigma_2)$-generated by Z_2, such that

$$\mathcal{A} \models \exists \boldsymbol{u}_1, \boldsymbol{u}_2.\ \varphi_1'(\boldsymbol{a}, \boldsymbol{u}_1) \wedge \varphi_2'(\boldsymbol{a}, \boldsymbol{u}_2).$$

Since $\mathcal{A}$ is a subalgebra of $\mathcal{F}$ and $\varphi_1' \wedge \varphi_2'$ is quantifier-free, it follows by Lemma 1 that $\mathcal{F} \models \exists \boldsymbol{u}_1, \boldsymbol{u}_2.\ \varphi_1'(\boldsymbol{a}, \boldsymbol{u}_1) \wedge \varphi_2'(\boldsymbol{a}, \boldsymbol{u}_2)$ as well. Given that each φ_i' is a Σ_i-formula and $\boldsymbol{u}_1$ and $\boldsymbol{u}_2$ are disjoint, we have then that

$$\mathcal{F}_1 \models \exists \boldsymbol{u}_1.\ \varphi_1'(\boldsymbol{a}, \boldsymbol{u}_1) \text{ and } \mathcal{F}_2 \models \exists \boldsymbol{u}_2.\ \varphi_2'(\boldsymbol{a}, \boldsymbol{u}_2). \tag{1}$$

We construct a partition of the elements of $\boldsymbol{a}$ that will induce a grouping of $\boldsymbol{v}'$ having the properties listed in Point 1 of the proposition. For that, we will use the families $\{C_1^n \mid n \geq 0\}$ and $\{C_2^n \mid n \geq 0\}$ defined before Lemma 24.

First, let $\boldsymbol{a}_1$ be a tuple collecting the components of $\boldsymbol{a}$ that are in $C_1^0 \cup C_1^1$. Then, for all $n > 1$, let $\boldsymbol{a}_n$ be a tuple collecting the components of $\boldsymbol{a}$ that are in C_1^n. Finally, for all $n > 0$, let $\boldsymbol{b}_n$ be a tuple collecting the components of $\boldsymbol{a}$ that are in C_2^n.[7]

Since $\boldsymbol{a}$ is a (finite) tuple of Y^* and $Y = \bigcup_{n \geq 0} (C_1^n \cup C_2^n)$ as observed earlier, there is a smallest $m > 0$ such that every component of $\boldsymbol{a}$ is in $\bigcup_{n=0}^{m} (C_1^n \cup C_2^n)$. Let $n \in \{0, \ldots, m-1\}$. By Lemma 24(2), $\boldsymbol{b}_{n+1}$ is Σ_2-generated by $[\![G_1^n]\!]$ in $\mathcal{F}_2$. Let Z_{n+1} be any finite subset of $[\![G_1^n]\!]$ that generates $\boldsymbol{b}_{n+1}$. Now recall that $\mathcal{F}_2$ is free over the countably-infinite set X_2'. We prove that $\boldsymbol{a}_1, \ldots, \boldsymbol{a}_m, \boldsymbol{b}_1, \ldots, \boldsymbol{b}_m$, and $Z_1, \ldots, Z_m$ satisfy Lemma 4(2).

To start with, we have that $\boldsymbol{a}_n \in (X_2')^*$ for all $n \in \{1, \ldots, m\}$ because $C_1^n \subseteq [\![G_1^n]\!] \subseteq X_2'$ by construction of C_1^n. From Lemma 24(1) it follows that the tuples $\boldsymbol{a}_n$ and $\boldsymbol{a}_{n'}$ are pairwise disjoint for all distinct $n, n' \in \{1, \ldots, m\}$, which means that all components of $\boldsymbol{a}_1, \ldots, \boldsymbol{a}_m$ are distinct. Now let $n \in \{1, \ldots, m-1\}$. Observe that the set $Z_1 \cup \cdots \cup Z_n$ is included in $[\![G_1^{n-1}]\!] = C_1^0 \cup \cdots \cup C_1^{n-1}$ whereas every component of $\boldsymbol{a}_{n+1}$ belongs to C_1^{n+1}. It follows that no components of $\boldsymbol{a}_{n+1}$ are in $Z_1 \cup \cdots \cup Z_n$. Finally, where f is the bijection that maps, in order, the components of $\boldsymbol{a}$ to those of $\boldsymbol{v}'$, let $\boldsymbol{v}_1, \boldsymbol{v}_2, \ldots, \boldsymbol{v}_{2m-1}, \boldsymbol{v}_{2m}$ be the rearrangement of $\boldsymbol{v}'$ corresponding to $\boldsymbol{a}_1, \boldsymbol{b}_1, \ldots, \boldsymbol{a}_m, \boldsymbol{b}_m$ according to f. From (1) above we know that $\mathcal{F}_2 \models \exists \boldsymbol{u}_2.\ \varphi_2'(\boldsymbol{a}_1, \boldsymbol{b}_1, \ldots, \boldsymbol{b}_m, \boldsymbol{a}_m, \boldsymbol{u}_2)$. By Lemma 4 we can then conclude that $\mathcal{F}_2 \models \forall \boldsymbol{v}_1 \exists \boldsymbol{v}_2 \cdots \forall \boldsymbol{v}_{2m-1} \exists \boldsymbol{v}_{2m} \exists \boldsymbol{u}_2.\ \varphi_2'$.

Almost symmetrically, we can prove $\mathcal{F}_1 \models \exists \boldsymbol{v}_1 \forall \boldsymbol{v}_2 \cdots \exists \boldsymbol{v}_{2m-1} \forall \boldsymbol{v}_{2m} \exists \boldsymbol{u}_1.\ \varphi_1'$. The claim then follows from the fact that $\mathcal{F}_i$ is Σ_i-isomorphic to $\mathcal{A}_i$ for $i = 1, 2$ by Lemma 18. □

[7] Each tuple above is meant to have no repeated components, and may be empty.

5 Related Research

From a technical point of view, this work strongly depends on previous research on combining decision procedures for unification in the disjoint case and on research on combining decision procedures for the word problem in the non-disjoint case. The combination procedure as well as the proof of correctness are modeled on the corresponding procedure and proof in [2]. The only extension to the procedure is Step 3, which takes care of the shared symbols. In the proof, one of the main obstacles to overcome was to find an amalgamation construction that worked in the non-disjoint case. Several of the hard technical results used in the proof depend on results from our previous work on combining decision procedures for the word problem [5]. The definition of the sets G_i, which are vital for proving that the constructed algebra $\mathcal{A}$ is indeed free, is also borrowed from there. It should be noted, however, that this definition can also be seen as a generalization to the non-disjoint case of a syntactic amalgamation construction originally due to Schmidt-Schauß [11]. As already mentioned in the introduction, the notion of constructors used here is taken from [5,13].

The only other work on combination methods for unification in the non-disjoint case is due to Domenjoud, Ringeissen and Klay [6]. The main differences with our work are that (i) their notion of constructors is considerably more restrictive than ours; and (ii) they combine algorithms computing complete sets of unifiers, and so their method cannot be used to combine decision procedures. On the other hand, Domenjoud, Ringeissen and Klay do not impose the strong restriction that the component theories be finitary modulo renaming, which we need for our decidability result. However, it was recently discovered [10] that termination of the combination algorithm in [6] is actually not guaranteed with the conditions given in that paper.

6 Conclusion

We have extended the Baader-Schulz combination procedure [2] for positive theories to the case of component theories over non-disjoint signatures. The main contribution of this paper is the formulation of appropriate restrictions under which this procedure is sound and complete, and the proof of soundness and completeness itself. This proof depends on a novel construction of the free model of the combined theory, which is *not* just a straightforward extension of the free amalgamation construction used in [2] in the disjoint case. Regarding the generality of our restriction to theories sharing constructors, we believe that the notion of constructors is as general as one can get, a conviction that is supported by the work on combining decision procedures for the word problem and for universal theories [5,13].

Unfortunately, our combination procedure yields only a semi-decision procedure since it incorporates an infinitary step. The restriction to equational theories that are finitary modulo renaming overcomes this problem, but it is probably too strong to be useful in applications. Thus, the main thrust of further research will be to remove or at least relax this restriction. We believe that the overall framework introduced in this paper and the proof of soundness and completeness of the semi-decision procedure (or at least the tools used in this proof) will help us obtain more interesting decidability results in the near future. One direction to follow could be to try to impose additional algorithmic requirements on the theories to be combined or on the constructor theory, and exploit those requirements to transform the infinitary step into a series of finitary ones. For this, the work in [6], which assumes algorithms computing complete sets of unifiers for the component theories, could be a starting point. Since the combination algorithm presented there has turned out to be non-terminating [10], that work needs to be reconsidered anyway.

Another direction for extending the results presented here is to withdraw the restriction to functional signatures. As a matter of fact, the combination results in [2] apply not just to equational theories, but to arbitrary atomic theories, i.e., theories over signatures also containing relation symbols and axiomatized by a set of (universally quantified) atomic formulae. Since the algebraic apparatus employed in the present paper (in particular, free algebras) is also available in this more general case (in the form of free structures), it should be easy to generalize our results to atomic theories.

References

1. F. Baader and K.U. Schulz. Unification in the union of disjoint equational theories: Combining decision procedures. In *Proc. CADE-11*, Springer LNCS 607, 1992.
2. F. Baader and K.U. Schulz. Combination of constraint solving techniques: An algebraic point of view. In *Proc. RTA'95*, Springer LNCS 914, 1995.
3. F. Baader and K.U. Schulz. Unification in the union of disjoint equational theories: Combining decision procedures. *J. Symbolic Computation* 21, 1996.
4. F. Baader and C. Tinelli. Combining decision procedures for positive theories sharing constructors. Technical report no. 02-02, Dept. of Computer Science, University of Iowa, 2002 (available at `http://cs.uiowa.edu/~tinelli/papers.html`).
5. F. Baader and C. Tinelli. Deciding the word problem in the union of equational theories. *Information and Computation*, 2002. To appear (preprint available at `http://cs.uiowa.edu/~tinelli/papers.html`).
6. E. Domenjoud, F. Klay, and Ch. Ringeissen. Combination techniques for non-disjoint equational theories. In *Proc. CADE-12*, Springer LNCS 814, 1994.
7. W. Hodges. *Model Theory*, Cambridge University Press, 1993.
8. G. Nelson and D.C. Oppen. Simplification by cooperating decision procedures. *ACM Trans. on Programming Languages and Systems* 1, 1979.
9. D. Pigozzi. The join of equational theories. *Colloquium Mathematicum* 30, 1974.
10. Ch. Ringeissen. Personal communication, 2001.
11. M. Schmidt-Schauß. Unification in a combination of arbitrary disjoint equational theories. *J. Symbolic Computation* 8, 1989.
12. M. E. Stickel. A unification algorithm for associative commutative functions. *J. ACM* 28, 1981.
13. C. Tinelli and Ch. Ringeissen. Unions of non-disjoint theories and combinations of satisfiability procedures. *Theoretical Computer Science*, 2002. To appear (preprint available at `http://cs.uiowa.edu/~tinelli/papers.html`).

JITty: A Rewriter with Strategy Annotations

Jaco van de Pol

Centrum voor Wiskunde en Informatica
P.O.-box 90.079, 1090 GB Amsterdam, The Netherlands

1 Introduction

We demonstrate JITty, a simple rewrite implementation with strategy annotations, along the lines of the Just-In-Time rewrite strategy, explained and justified in [4]. Our tool has the following distinguishing features:

- It provides the flexibility of user defined strategy annotations, which specify the order of normalizing arguments and applying rewrite rules.
- Strategy annotations are checked for correctness, and it is guaranteed that all produced results are normal forms w.r.t. the underlying TRS.
- The tool is "light-weight" with compact but fast code.
- A TRS is interpreted, rather than compiled, so the tool has a short start-up time and is portable to many platforms.

We shortly review strategy annotations in Section 2. JITty is available via `http://www.cwi.nl/~vdpol/jitty/` together with a small demonstrator (Section 3) that can be used to experiment with strategy annotations. The rewrite engine has also been integrated in the μCRL tool set [1]. Although performing a rewrite step takes more time in JITty than in the standard compiling rewriter of the μCRL toolset, the former is often preferred, owing to avoidance of compilation time, and better normalization properties of the just-in-time strategy. In Section 4 we emphasize certain requirements on the rewriter imposed by the μCRL toolset. This leads to an unconventional application programmer's interface, which is described in Section 5.

2 User Defined and Predefined Strategy Annotations

A strategy annotation for a function symbol f is a list of integers and rule labels, where the integers refer to the arguments of f ($1 \leq i \leq arity(f)$) and the rule labels to rewrite rules for f, i.e., rules whose left hand side have top symbol f. For instance, the annotation $f : [1, \alpha, \beta, 2]$ means that a term with top symbol f should be normalized by first normalizing its first argument, then trying rule α and β; if both fail the second argument is normalized.

To normalize correctly, a strategy annotation must be *full* and *in-time*. It is full if all arguments and rules for f are mentioned. It is *in-time* if arguments are mentioned before rules that need them. A rule needs an argument if either the argument starts with a function symbol, or the argument is a non-linear variable.

S. Tison (Ed.): RTA 2002, LNCS 2378, pp. 367–370, 2002.

```
signature
 T(0)      or(2)      loop(0)
 F(0)      and(2)
rules
 a1([x], and(x,T), x)     o1([x], or(T,x), T)     l([], loop, loop)
 a2([x], and(x,F), F)     o2([x], or(F,x), x)
default justintime
strategies
 and([2,a1,a2,1])
end
  rewrite( and(loop,F) )
  rewrite( or(T,loop) )
  rewrite( or(and(loop,F),or(T,loop)) )
stop
```

Fig. 1. Boolean example of a demonstrator file.

It has been proved in [4] that if a normal form is computed under a strategy annotation satisfying the above restrictions, then the result is a normal form of the original TRS *without strategies*. Therefore `JITty` checks these criteria. The following strategies are predefined: *leftmost innermost*, which first normalizes all arguments, and subsequently tries all rewrite rules; and *just-in-time*, which also normalizes its arguments from left to right, but tries to apply rewrite rules as soon as their needed arguments have been evaluated.

`JITty`'s strategy annotations are similar to OBJ's annotations (e.g. [3]). The annotations of `JITty` are more refined, because rules can be mentioned individually, but less sophisticated, because laziness annotations are not supported. See [4] for other rule based systems with user-controlled strategies (ELAN, Maude, OBJ-family, Stratego).

Although, strictly speaking, evaluation can be done when strategy annotations are *not* in-time or full, an interesting optimization can be applied if they are. In particular, for in-time annotations we have that all subterms of a normal form are in normal form. Hence, in $\alpha : f(g(x)) \to h(x)$, with $f : [1, \alpha]$, it is guaranteed that the argument of h will be normal, so it will not be traversed. Therefore, `JITty` currently only supports "correct" annotations.

3 Simple Demonstrator

We provide a simple demonstrator, which reads a file containing a signature, a number of rules, a default strategy, a number of user defined strategy annotations, and a number of commands. It has a fixed structure, as shown in Figure 1. The signature consists of a number of function symbols with their arity. A rule consists of a label, a list of variables, a left hand side and a right hand side. The default strategy should be either *innermost* or *justintime*. The default strategy can be overwritten for each function symbol, by an annotation, being a mixed list of integers and rule labels.

The commands are of the form `rewrite(term)`, to start rewriting. The three examples in Figure 1 are carefully chosen to terminate. Other examples may loop for ever. After replacing *justintime* by *innermost*, the first term will terminate, but the last two will not. The website shows more examples, such as the following rule for division, which terminates for closed x and y by virtue of the just-in-time strategy: `div(x,y)` $\rightarrow$ `if(lt(x,y),0,S(div(minus(x,y),y)))`.

4 Embedding in the μCRL Toolset

A μCRL specification consists of an equational data theory and a process part. The μCRL toolset [1] contains a.o. an automated theorem prover for the equational theory, and a simulator for the process part which serves as the front end of visualization and model checking tools. Both tools depend heavily on term rewriting in order to decide the equational theory. Therefore we need that the strategy annotations always yield normal forms of the TRS.

The simulator has to normalize the guards of transition rules (i.e., terms with state variables). The same guard is rewritten in many states. For efficiency reasons, `JITty` maintains a current environment, which is a normalized substitution. Given global environment σ, rewriting a term t now means to get the normal form of t^σ, assuming that the substitution σ is normalized. This can be exploited in the implementation: t has to be traversed only once, and x^σ (for variables x in t) is not traversed at all, because it is supposed to be in normal form. The user can modify the current environment by assigning a term to a variable, provided this term is in normal form. To resolve name conflicts, the user can enter and leave blocks. Entering a block doesn't change the global environment, but leaving a block restores the previous environment.

The theorem prover is based on binary decision diagrams (BDD) with equations instead of proposition symbols. BDDs are nothing but highly shared if-then-else trees. An important optimization, crucial when rewriting BDDs, is that the rewriter can be put in "hash mode". In this case, each computed result is stored in a look-up table. So each sub-computation is performed only once.

5 Application Programmer's Interface

`JITty` is implemented in the programming language `C`, and relies on the `ATerm` library [2]. This library is supposed to have an efficient term implementation. It guarantees that terms are always stored in maximally shared form. Moreover, the μCRL toolset uses `ATerms` as well, so at term level there is no translation between μCRL and `JITty`. Therefore, `ATerms` also show up in the Application Programmer's Interface of `JITty`, shown in Figure 2. A complete C program using the basic functionality is shown in Figure 3.

`JIT_init` is used to initialize (or reset) the rewriter. At initialization, the following information is needed: lists of function symbols, rewrite rules and strategy annotations, an indication of the default strategy (currently one of the constants `INNERMOST` or `JUSTINTIME`) and an indication whether hash tables should be used

```
#include "aterm2.h"
#define INNERMOST 1
#define JUSTINTIME 2
void JIT_init(ATermList funs, ATermList rules, ATermList strategy,
        int default_strat, char withhash);
ATerm JIT_normalize(ATerm t);
void JIT_flush(void);
void JIT_assign(Symbol v, ATerm t);
void JIT_enter(void);
void JIT_leave(void);
void JIT_clear(void);
int  JIT_level(void);
```

Fig. 2. JITty – Application Programmer's Interface

```
#include "jitty.h"
int main(int argc, char* argv[]) {
 ATinitialize(argc,argv);                         /* initialize ATerms    */
 JIT_init(ATparse("[f(1),g(2),a(0),b(0)]"),      /* signature            */
        ATparse("[frule([x],f(x),g(x,b)),"      /* rule for f           */
                " grule([x],g(x,x),f(a))]"),    /* rule for g           */
        ATparse("[f([frule,1])]"),              /* strategy annotation */
        INNERMOST,                               /* default strategy     */
        0);                                      /* without hashing      */
 ATprintf("%t\n",JIT_normalize(ATparse("f(b)"))); }
```

Fig. 3. Example program of using JITty

(0 = no, 1 = yes). `JIT_normalize(t)` returns the normal form of t in the current environment. `JIT_flush()` is used to clear the hash table, in case it becomes too memory consuming. The other functions are used to manipulate the global environment, as explained in Section 4. `JIT_enter()` and `JIT_leave()` can be used to enter or leave a new block. `JIT_clear()` undoes all bindings in the current block. `JIT_assign(v,t)` assigns term t to variable v (represented as Symbol from the `ATerm` library). Finally, `JIT_level()` returns the current level.

References

1. S. Blom, W. Fokkink, J. Groote, I. van Langevelde, B. Lisser, and J. van de Pol. μCRL: A toolset for analysing algebraic specifications. In *Proceedings of CAV 2001*, LNCS 2102, pages 250–254, 2001. See also `http://www.cwi.nl/~mcrl/`.
2. M. van den Brand, H. de Jong, P. Klint, and P.A. Olivier. Efficient Annotated Terms. *Software – Practice & Experience*, 30:259–291, 2000.
3. J. Goguen, T. Winkler, J. Meseguer, K. Futatsugi, and J.-P. Jouannaud. Introducing OBJ. In J. Goguen and G. Malcolm, editors, *Software Engineering with OBJ: algebraic specification in action*. Kluwer, 2000.
4. J. van de Pol. Just-in-time: On strategy annotations. In B. Gramlich and S. Lucas, editors, *Electronic Notes in TCS*, volume 57, 2001. (Proc. of WRS 2001, Utrecht).

Autowrite: A Tool for Checking Properties of Term Rewriting Systems

Irène Durand

LaBRI, Université de Bordeaux I
33405 Talence, France
idurand@labri.fr

1 Introduction

Huet and Lévy [6] showed that for the class of orthogonal term rewriting systems (TRSs) every term not in normal form contains a needed redex (i.e., a redex contracted in every normalizing rewrite sequence) and that repeated contraction of needed redexes results in a normal form if it exists. However, neededness is in general undecidable. In order to obtain a decidable approximation to neededness Huet and Lévy introduced the subclass of *strongly sequential* TRSs and showed that strong sequentiality is a decidable property of orthogonal TRSs.

Several authors [2,3,7,9,10,11,12] proposed decidable extensions of the class of strongly sequential TRSs. Since Comon's [1,2] and Jacquemard's [7] work, all the corresponding decidability proofs have been expressed using tree automata techniques: typically a property is satisfied if and only if some associated automaton recognizes the empty language. A uniform framework for the study of sequentiality is described in [3] where classes of TRSs are parameterized by approximation mappings. Nowdays the best known approximation for which sequentiality is decidable is the growing approximation studied in [10]. We assume that the reader is familiar with the basics of term rewriting [8]. Familiarity with the theory of sequentiality ([1,3,4,5,6,7,10,11]) is helpfuf.

Autowrite is an experimental tool written in Common Lisp for checking properties of TRSs. It was initially designed to check sequentiality properties of TRSs. For this purpose, it implements the tree automata constructions used in [7,3,4,10] and many useful operations on terms, TRSs and tree automata (unfortunaletly not all yet integrated into the graphical interface). A graphical interface (still under construction) is written using FreeCLIM, the free implementation of the CLIM specification. Using this interface, one can check sequentiality of the different approximations of a given system. Many other functionalities are available directly from the Common Lisp interactive environment and could easily be integrated into the graphical interface. The Autowrite tool was used to check sequentiality for most of the examples presented in [5].

Our term rewriting systems consist of *left-linear* rewrite rules $l \to r$ that satisfy $\mathsf{root}(l) \notin \mathcal{V}$ and $\mathcal{V}\mathsf{ar}(r) \subseteq \mathcal{V}\mathsf{ar}(l)$. The following example is adapted from [9].

S. Tison (Ed.): RTA 2002, LNCS 2378, pp. 371–375, 2002.

$$\mathcal{R}_1 = \begin{cases} \mathsf{f}(\mathsf{a}, \mathsf{g}(x_0, \mathsf{a})) & \rightarrow \mathsf{a} \\ \mathsf{f}(\mathsf{b}, \mathsf{g}(\mathsf{a}), x_0) & \rightarrow \mathsf{a} \\ \mathsf{f}(x_0, \mathsf{a}) & \rightarrow x_0 \\ \mathsf{g}(\mathsf{b}, \mathsf{b}) & \rightarrow \mathsf{a} \end{cases}$$

If the condition $\mathcal{V}\mathsf{ar}(r) \subseteq \mathcal{V}\mathsf{ar}(l)$ is not imposed, we speak of *extended* TRSs (eTRSs). Such TRSs arise naturally when we approximate TRSs, as explained below. Let $\mathcal{R}$ be an eTRS over a signature $\mathcal{F}$. The set of ground normal forms of $\mathcal{R}$ is denoted by $\mathsf{NF}(\mathcal{R})$. Let $\mathcal{R}_\bullet$ be the eTRS $\mathcal{R} \cup \{\bullet \rightarrow \bullet\}$ over the extended signature $\mathcal{F}_\bullet = \mathcal{F} \cup \{\bullet\}$. We say that redex Δ in $C[\Delta] \in \mathcal{T}(\mathcal{F})$ is *$\mathcal{R}$-needed* if there is no term $t \in \mathsf{NF}(\mathcal{R})$ such that $C[\bullet] \rightarrow^*_{\mathcal{R}} t$. Finally, we say that $\mathcal{R}$ is *sequential* if every reducible term in $\mathcal{T}(\mathcal{F})$ contains a $\mathcal{R}$-needed redex.

Let $\mathcal{R}$ and $\mathcal{S}$ be eTRSs over the same signature. We say that $\mathcal{S}$ approximates $\mathcal{R}$ if $\rightarrow^*_{\mathcal{R}} \subseteq \rightarrow^*_{\mathcal{S}}$ and $\mathsf{NF}(\mathcal{R}) = \mathsf{NF}(\mathcal{S})$. An approximation mapping is a mapping α from TRSs to eTRSs with the property that $\alpha(\mathcal{R})$ approximates $\mathcal{R}$, for every TRS $\mathcal{R}$. Given a TRS $\mathcal{R}$, we say that $\mathcal{R}$ is *α-sequential* if $\alpha(\mathcal{R})$ is sequential. Next we define the approximation mappings s, nv, and gr. Let $\mathcal{R}$ be a TRS. The *strong* approximation $\mathrm{s}(\mathcal{R})$ is obtained from $\mathcal{R}$ by replacing the right-hand side of every rewrite rule by a variable that does not occur in the corresponding left-hand side. The *nv* approximation $\mathrm{nv}(\mathcal{R})$ is obtained from $\mathcal{R}$ by replacing the variables in the right-hand sides of the rewrite rules by pairwise distinct variables that do not occur in the corresponding left-hand sides. The *growing* approximation $\mathrm{gr}(\mathcal{R})$ is obtained from $\mathcal{R}$ by renaming the variables in the right-hand sides that occur at a depth greater than 1 in the corresponding left-hand side. Given a TRS $\mathcal{R}$ and $\alpha \in \{\mathrm{s}, \mathrm{nv}, \mathrm{gr}\}$, it is decidable whether $\alpha(\mathcal{R})$ is sequential. However the decision procedures are very complex (exponential for s, doubly exponential for nv and gr [4]) so it is impossible to show by hand that a particular system is sequential, even for very small systems. However showing that a system is not sequential can be done more easily by exibiting a term with no needed redex. This later property of a term can be proved in polynomial time. For $\mathcal{R}_1$, we obtain the following approximated TRSs:

$$\mathrm{s}(\mathcal{R}_1) = \begin{cases} \mathsf{f}(\mathsf{a}, \mathsf{g}(x_0, \mathsf{a})) & \rightarrow y_0 \\ \mathsf{f}(\mathsf{b}, \mathsf{g}(\mathsf{a}, x_0)) & \rightarrow y_0 \\ \mathsf{f}(x_0, \mathsf{a}) & \rightarrow y_0 \\ \mathsf{g}(\mathsf{b}, \mathsf{b}) & \rightarrow y_0 \end{cases} \qquad \mathrm{nv}(\mathcal{R}_1) = \begin{cases} \mathsf{f}(\mathsf{a}, \mathsf{g}(x_0, \mathsf{a})) & \rightarrow \mathsf{a} \\ \mathsf{f}(\mathsf{b}, \mathsf{g}(\mathsf{a}, x_0)) & \rightarrow \mathsf{a} \\ \mathsf{f}(x_0, \mathsf{a}) & \rightarrow y_0 \\ \mathsf{g}(\mathsf{b}, \mathsf{b}) & \rightarrow \mathsf{a} \end{cases}$$

$$\mathrm{gr}(\mathcal{R}_1) = \mathcal{R}_1$$

The next example ($\mathcal{R}_2$) is adapted from [11].

$$\mathcal{R}_2 = \begin{cases} \mathsf{f}(\mathsf{g}(\mathsf{a}, x_0), \mathsf{a}) & \rightarrow \mathsf{a} \\ \mathsf{f}(\mathsf{g}(x_0, \mathsf{a}), \mathsf{k}(\mathsf{a})) & \rightarrow \mathsf{a} \\ \mathsf{f}(\mathsf{k}(\mathsf{c}), x_0) & \rightarrow \mathsf{a} \\ \mathsf{g}(\mathsf{k}(\mathsf{a}), \mathsf{k}(\mathsf{a})) & \rightarrow \mathsf{h}(\mathsf{k}(\mathsf{a})) \\ \mathsf{k}(\mathsf{k}(\mathsf{a})) & \rightarrow \mathsf{a} \\ \mathsf{h}(x_0) & \rightarrow \mathsf{k}(x_0) \end{cases} \qquad \mathrm{nv}(\mathcal{R}_2) = \begin{cases} \mathsf{f}(\mathsf{g}(\mathsf{a}, x_0), \mathsf{a}) & \rightarrow \mathsf{a} \\ \mathsf{f}(\mathsf{g}(x_0, \mathsf{a}), \mathsf{k}(\mathsf{a})) & \rightarrow \mathsf{a} \\ \mathsf{f}(\mathsf{k}(\mathsf{c}), x_0) & \rightarrow \mathsf{a} \\ \mathsf{g}(\mathsf{k}(\mathsf{a}), \mathsf{k}(\mathsf{a})) & \rightarrow \mathsf{h}(\mathsf{k}(\mathsf{a})) \\ \mathsf{k}(\mathsf{k}(\mathsf{a})) & \rightarrow \mathsf{a} \\ \mathsf{h}(x_0) & \rightarrow \mathsf{k}(y_0) \end{cases}$$

$$\mathrm{gr}(\mathcal{R}_2) = \mathcal{R}_2$$

The last example ($\mathcal{R}_3$) is adapted from Berry's example (the first three rules).

$$\mathcal{R}_3 = \begin{cases} \mathsf{f}(x_0,\mathsf{a},\mathsf{b}) & \rightarrow \mathsf{h}(x_0) \\ \mathsf{f}(\mathsf{b},x_0,\mathsf{a}) & \rightarrow \mathsf{h}(x_0) \\ \mathsf{f}(\mathsf{a},\mathsf{b},x_0) & \rightarrow \mathsf{h}(x_0) \\ \mathsf{h}(\mathsf{k}(x_0)) & \rightarrow \mathsf{g}(x_0,x_0) \\ \mathsf{g}(\mathsf{a},\mathsf{a}) & \rightarrow \mathsf{g}(\mathsf{a},\mathsf{a}) \\ \mathsf{g}(\mathsf{a},\mathsf{b}) & \rightarrow \mathsf{a} \\ \mathsf{g}(\mathsf{b},\mathsf{a}) & \rightarrow \mathsf{b} \end{cases} \qquad \mathrm{gr}(\mathcal{R}_3) = \begin{cases} \mathsf{f}(x_0,\mathsf{a},\mathsf{b}) & \rightarrow \mathsf{h}(x_0) \\ \mathsf{f}(\mathsf{b},x_0,\mathsf{a}) & \rightarrow \mathsf{h}(x_0) \\ \mathsf{f}(\mathsf{a},\mathsf{b},x_0) & \rightarrow \mathsf{h}(x_0) \\ \mathsf{h}(\mathsf{k}(x_0)) & \rightarrow \mathsf{g}(y_0,y_0) \\ \mathsf{g}(\mathsf{a},\mathsf{a}) & \rightarrow \mathsf{g}(\mathsf{a},\mathsf{a}) \\ \mathsf{g}(\mathsf{a},\mathsf{b}) & \rightarrow \mathsf{a} \\ \mathsf{g}(\mathsf{b},\mathsf{a}) & \rightarrow \mathsf{b} \end{cases}$$

Autowrite computes $\alpha(\mathcal{R})$ for any approximation α in $\{\mathrm{s}, \mathrm{nv}, \mathrm{gr}\}$.

2 Automata

Here are the different kinds of automata that Autowrite can build:

1. Given a set of ground terms $\mathcal{L}$: an automaton $\mathcal{A}_{\mathcal{L}}$ such that $L(\mathcal{A}_{\mathcal{L}}) = \mathcal{L}$.
2. Given a set of linear terms $\mathcal{L}$: an automaton $\mathcal{A}_{\mathcal{L}}$ such that $L(\mathcal{A}_{\mathcal{L}}) = \{\sigma(t) \mid t \in \mathcal{L}$ and σ is a ground substitution$\}$.
3. Given the set of left-hand sides of a left-linear TRS $\mathcal{R}$: an automaton $\mathcal{A}_{\mathsf{NF}(\mathcal{R})}$ such that $L(\mathcal{A}_{\mathsf{NF}(\mathcal{R})}) = \mathsf{NF}(\mathcal{R})$.
4. Given any regular tree language $\mathcal{L}$ recognized by an automaton $\mathcal{A}_{\mathcal{L}}$ and a left-linear growing e-TRS $\mathcal{R}$: a deterministic automaton $\mathcal{C}_{\mathcal{R},\mathcal{L}}$ (as described in [10]) such that $L(\mathcal{C}_{\mathcal{R},\mathcal{L}}) = (\xleftarrow{*})(\mathcal{L})$.
5. Given a left-linear growing e-TRS $\mathcal{R}$: an automaton $\mathcal{D}_{\mathcal{R}}$ such that $L(\mathcal{D}_{\mathcal{R}}) = \emptyset$ is equivalent to $\mathcal{R}$ is sequential [4].

Autowrite may check the following properties about tree automata:

1. Given an automaton $\mathcal{A}$: does $L(\mathcal{A}) = \emptyset$? If not, it may exhibit ground terms belonging to the language.
2. Given a ground term t and an automaton $\mathcal{A}$: does $t \in L(\mathcal{A})$?

To decide whether a TRS $\mathcal{R}$ is α-sequential, Autowrite computes $\alpha(\mathcal{R})$ then builds the automaton $\mathcal{D}_{\alpha(\mathcal{R})}$ and finally checks whether $L(\mathcal{D}_{\alpha(\mathcal{R})}) = \emptyset$. If so it concludes that $\mathcal{R}$ is α-sequential otherwise it exhibits a ground term of $L(\mathcal{D}_{\alpha(\mathcal{R})})$ which is a term with no $\alpha(\mathcal{R})$-needed redex. Note that to build $\mathcal{D}_{\alpha(\mathcal{R})}$, Autowrite must previously compute $\mathcal{C}_{\alpha(\mathcal{R})} = \mathcal{C}_{\alpha(\mathcal{R}),\mathsf{NF}(\mathcal{R})}$. For $\mathcal{C}_{\alpha(\mathcal{R})}$, both Jacquemard's algorithm for growing-linear systems and Toyama and Nagaya's for linear-growing systems have been implemented. For $\mathcal{D}_{\alpha(\mathcal{R})}$, we have implemented the algorithm presented in [4]. In fact these three algorithms have been adapted in order to compute directly automata with only accessible states. This complicates the code but reduces considerably the size of the constructed automata.

3 Experimental Results

The graphical version of Autowrite can be found at
`http://dept-info.labri.u-bordeaux.fr/~idurand/autowrite`

Autowrite has been used to check α-sequentiality for many linear TRSs. It can also check many other properties of TRSs and perform computations on tree automata; although these are not yet accessible from the graphical interface, they can be called using Lisp. However, with regard to sequentiality, one cannot hope to use Autowrite for big TRSs because the size of the constructed automaton $\mathcal{D}_{\mathcal{R}}$ is in $\mathcal{O}(2^{2^{\|\mathcal{R}\|}})$ as shown in [4].

With Autowrite, we were able to check sequentiality for many examples found in the literature. In the following table we present a few sequentiality results for some TRSs $\mathcal{R}$ with different approximations. We present the number of states (st) and rules (rl) of the automata $\mathcal{C}_{\alpha(\mathcal{R})}$ and $\mathcal{D}_{\alpha(\mathcal{R})}$ built to decide α-sequentiality. If the TRS is not α-sequential we present the first term with no $\alpha(\mathcal{R})$-needed redex computed by Autowrite.

	α	Time in sec	$\mathcal{C}_{\alpha(\mathcal{R})}$ st	$\mathcal{C}_{\alpha(\mathcal{R})}$ rl	$\mathcal{D}_{\alpha(\mathcal{R})}$ st	$\mathcal{D}_{\alpha(\mathcal{R})}$ rl	Sequentiality results
$\mathcal{R}_1$	s	13	20	412	107	2181	$\mathsf{f(g(b,b),g(g(b,b),g(b,b)))}$
	nv	101	25	742	243	7463	$\mathsf{f(f(a,a),g(f(a,a),f(a,a)))}$
	gr	98	15	182	114	16477	Sequential
$\mathcal{R}_2$	s	19	24	573	109	2460	$\mathsf{f(g(h(a),h(a)),h(a))}$
	nv	69	33	1329	400	4391	$\mathsf{f(g(h(a),h(a)),h(h(a)))}$
	gr	145	19	292	117	17469	Sequential
$\mathcal{R}_3$	s	3	12	435	15	179	$\mathsf{f(g(b,a),g(b,a),g(b,a))}$
	nv	16	14	857	27	716	$\mathsf{f(f(b,b,a),f(b,b,a),f(b,b,a))}$
	gr	263	11	293	30	8187	Sequential

References

1. H. Comon. Sequentiality, second-order monadic logic and tree automata. In *Proc. 10 th LICS*, pages 508–517, 1995.
2. H. Comon. Sequentiality, monadic second-order logic and tree automata. *Information and Computation*, 1999. Full version of [1].
3. I. Durand and A. Middeldorp. Decidable call by need computations in term rewriting (extended abstract). In Proc. *14th CADE*, volume 1249 of *LNAI*, pages 4–18, 1997.
4. Irène Durand and Aart Middeldorp. On the complexity of deciding call-by-need. Technical Report 1196–98, LaBRI, 1998.
5. Irène Durand and Aart Middeldorp. On the modularity of deciding call-by-need. In *Foundations of Software Science and Computation Structures*, volume 2030 of *Lecture Notes in Computer Science*, pages 199–213, Genova, 2001. Springer-Verlag.
6. G. Huet and J.-J.Lévy. Computations in orthogonal rewriting systems, i and ii. In *Computational Logic, Essays in Honor of Alan Robinson*, pages 396–443. The MIT Press, 1991. Original version: Report 359, Inria, 1979.
7. F. Jacquemard. Decidable approximations of term rewriting systems. In *Proc. 7th RTA*, volume 1103 of *LNCS*, pages 362–376, 1996.
8. J.W. Klop. Term rewriting systems. *In Handbook of Logic in Computer Science, Vol. 2*, pages 1–116. Oxford University Press, 1992.
9. J.W. Klop and A. Middeldorp. Sequentiality in orthogonal term rewriting systems. *Journal of Symbolic Computation*, 12:161–195, 1991.

10. Takashi Nagaya and Yoshihito Toyama. Decidability for left-linear growing term rewriting systems. In *Proc. 10th RTA*, volume 1631 of *LNCS*, 1999.
11. M. Oyamaguchi. NV-sequentiality: A decidable condition for call-by-need computations in term rewriting systems. *SIAM Journal on Computation*, 22:114–135, 1993.
12. Y. Toyama. Strong sequentiality of left-linear overlapping term rewriting systems. In *Proc. 7th LICS*, pages 274–284, 1992.

TTSLI: An Implementation of Tree-Tuple Synchronized Languages

Benoit Lecland and Pierre Réty

LIFO - Université d'Orléans B.P.6759,45067 Orléans cedex 2, France
{lecland,rety}@lifo.univ-orleans.fr
http://www.univ-orleans.fr/SCIENCES/LIFO/Members/{lecland,rety}

1 Introduction

Tree-Tuple Synchronized Languages have first been introduced by means of Tree-Tuple Synchronized Grammars (TTSG) [3], and have been reformulated recently by means of (so-called) Constraint Systems (CS), which allowed to prove more properties [2,1]. A number of applications to rewriting and to concurrency have been presented (see [5] for a survey).

TTSLI is an implementation of constraint systems, together with the main operations. It is written in Java, and is available on Lecland's web page.

2 Constraint Systems for Tuple Synchronized Languages

We just give examples to recall what a CS is. Formal definitions are given in [2].

Example 1. In the signature $\Sigma = \{f^{\backslash 2}, b^{\backslash 0}\}$, let $L_{id} = \{(t,t) \mid t \in T_\Sigma\}$ be the set of pairs of identical terms. L_{id} can be defined by the following grammar, given in the form of a constraint system :

$$X_{id} \supseteq (b,b)$$
$$X_{id} \supseteq (f(1_1,2_1),\ f(1_2,2_2))\ (X_{id}, X_{id})$$

where $1_1, 2_1, 1_2, 2_2$ abbreviate pairs (for readability). For example 2_1 means $(2,1)$, which denotes the first component of the second argument (the second X_{id}). Note that since 1_1 and 1_2 come from the same X_{id}, they represent two identical terms, in other words they are linked (synchronized), whereas for example 1_1 and 2_1 are independent.

Example 2. Now if we consider the slightly different constraint system :

$$X_{sym} \supseteq (b,b)$$
$$X_{sym} \supseteq (f(1_1,2_1),\ f(2_2,1_2))\ (X_{sym}, X_{sym})$$

We get the set $L_{sym} = \{(t, t_{sym}) \mid t_{sym}$ is the symmetric tree of $t\}$.

S. Tison (Ed.): RTA 2002, LNCS 2378, pp. 376–379, 2002.

Example 3. In the signature $\Sigma = \{s^{\backslash 1}, b^{\backslash 0}\}$ let $L_{dble} = \{(s^n(b), s^{2n}(b))\}$. It can be defined by the constraint system :

$$X_{dble} \supseteq (b, b)$$
$$X_{dble} \supseteq (s(1_1),\ s(s(1_2)))\ X_{dble}$$

Constraint systems are closed under union, projection, and (assuming a restriction) join. Membership and emptiness are decidable. They are not closed under intersection, except for a subclass [4]. The implementation also includes an extension of membership: language matching.

3 Implementation

3.1 Language Matching

Example 4. Consider again the CS C_{sym} given in Example 2, which generates $L_{sym} = \{(t, t_{sym})\}$. We want to find a CS that generates the triple language U (U for Unknown) s.t.:

$$(t_1, t_2, t_3) \in U \iff (f(t_1, b),\ f(t_2, t_3)) \in L_{sym}$$

This language matching problem is denoted by $P = (f(1_1, b),\ f(1_2, 1_3))_{U, X_{sym}}$, which means that if we replace 1_1, 1_2, 1_3 by the components of U, we should get the language L_{sym} (whose axiom is X_{sym}).

Obviously, $U = \{(t, b, t_{sym}) \mid t \in T_\Sigma\}$ and can be generated by the CS $\{U \supseteq (1_1, b, 1_2) X_{sym}\} \cup C_{sym}$.

Note that the language matching problem is an extension of the membership problem: for example, if (t, t') is a pair of ground terms[1], $(t, t') \in L_{sym} \iff (t, t')_{U, X_{sym}} = true$ where U is a 0-tuple language.

The proof of the algorithm is written and give a new result about CS's : the closure under language matching.

Algorithm. Starting from the initial problem P, the algorithm works in several steps. A step generates one constraint and a number of language matching subproblems to be solved. When no more subproblems are left, the algorithm stops. The size of each subproblem is not greater than the size of the initial problem. Therefore there are at most finitely many subproblems (up to a variable renaming). Thus, if the algorithm pays attention not to solve the same subproblem twice, it will terminate.

3.2 Projection

Let X be the axiom of a constraint system C that recognizes a language of n-tuples, and let $\{x_1, \ldots, x_p\} \subseteq \{1, \ldots, n\}$. To obtain the projection of C on components $x_1, \ldots, x_p$, one can add to C the constraint $X_{proj} \supseteq (1_{x_1}, \ldots, 1_{x_p})(X)$, and consider that the axiom is X_{proj}.

[1] Recall that CS's express only ground terms.

Example 5. Let C be the following CS:

$$X \supseteq (s(1_1),\ p(1_2),\ f(1_3))\ X_1$$
$$X_1 \supseteq (s(1_1),\ p(1_2),\ f(1_3))\ X$$
$$X \supseteq (a,\ b,\ c)$$

C generates the language $\mathcal{L} = \{(s^{2n}(a),\ p^{2n}(a),\ f^{2n}(a)) \mid n \in \mathbb{N}\}$.
The projection on components 1 and 3, i.e. the language $\mathcal{L}' = \{(s^{2n}(a),\ f^{2n}(a)) \mid n \in \mathbb{N}\}$, is generated by the CS $C' = \{X_{proj} \supseteq (1_1,\ 1_3)X\} \cup C$. However, to generate $\mathcal{L}'$, a better CS obviously is:

$$X \supseteq (s(1_1),\ f(1_2))\ X_1$$
$$X_1 \supseteq (s(1_1),\ f(1_2))\ X$$
$$X \supseteq (a,\ c)$$

The following algorithm returns a CS that generates only useful intermediate components.

Algorithm. Let X be the axiom of a constraint system C that recognizes a language L of n-tuples, and let $E = \{x_1, \ldots, x_p\} \subseteq \{1, \ldots, n\}$. Let $\Pi_{X,E}(C)$ denote a CS that generates the projection of L on components $x_1, \ldots, x_p$. Let $\pi_{X,E}(C)$ be the operation that for each constraint $X \supseteq (t_1, \ldots, t_n)(X_1, \ldots, X_k) \in C$ generates $X \supseteq (t_{x_1}, \ldots, t_{x_p})(X_1, \ldots, X_k) \in C$, hence that removes useless components (but not useless non-terminals).

$\Pi_{X,E}(C)$ is generated by a recursive function:

- $C' = \emptyset$
- $D = \pi_{X,E}(C)$
- for each constraint $X \supseteq (t'_1, \ldots, t'_p)(X_1, \ldots, X_k) \in D$, for each $i \in \{1, \ldots, k\}$, let $C' = C' \cup \Pi_{X_i,E_i}(C)$, where $E_i = \{x \mid i_x \text{ occurs in } (t'_1, \ldots, t'_p)\}$
- return $C' \cup D$

At the end, we clean the resultant CS by removing useless non-terminals (if any).

3.3 Join

Join is natural join, like in data-bases. The general algorithm is given in [1].

Given a language S of l-tuples, and T of n-tuples, and for $i \in \{1, \ldots, l\}$, $j \in \{1, \ldots, n\}$ the *i, j-join* of S and T is a $l + n - 1$-tuple language, denoted $S \bowtie_{i,j} T$, and defined by:

$$\{(s_1, \ldots, s_l, t_1, \ldots, t_{j-1}, t_{j+1}, \ldots, t_n) \mid (s_1, \ldots, s_l) \in S \wedge (t_1, \ldots, t_n) \in T \wedge s_i = t_j\}$$

$S \bowtie T$ stands for $S \bowtie_{l,1} T$.

Example 6. Let the language $L_1 = \{(s^n(a), s^{2n}(a)) \mid n \in \mathbb{N}\}$ and the language $L_2 = \{(s^{3n}(a), s^n(a)) \mid n \in \mathbb{N}\}$
The join of L_1 and L_2 respectively on components 2 and 1 is
$L_1 \bowtie_{2,1} L_2 = \{(s^{3n}(a), s^{6n}(a), s^{2n}(a)) \mid n \in \mathbb{N}\}$

3.4 Other Operations

Checking emptiness for a CS is very similar to checking emptiness for a regular tree language defined by a regular grammar. See [2]. Union, as well as a membership test faster than running the language matching test on a ground tuple, are also implemented.

4 Experimentations

We have run an application to rewriting as described in [1] : given regular tree languages L_1, L_2, and a tree binary relation R defined by a CS,

$$R(L_1) \cap L_2 = \emptyset \Longleftrightarrow \Pi_2(L_1 \bowtie R) \bowtie L_2 = \emptyset$$

Let R be the relation defined by the only rewrite rule $f(x, y, z, u) \rightarrow f(u, z, x, y)$. Let $L_1 = (s^*(a), s^*(b), s^*(b), s^*(b))$. To show that $R_3(L_1) \cap L_1 = \emptyset$ where R_3 is the rewriting in three steps, we check that

$$\Pi_2(L_1 \bowtie R_3) \bowtie L_1 = \Pi_2(\Pi_2(\Pi_2(L_1 \bowtie R) \bowtie R) \bowtie R) \bowtie L_1 = \emptyset$$

We have successfully run this example on a Sun Ultra 5 computer in 50".

5 Conclusion

Thanks to this implementation, we also hope to run the other existing known applications of synchronized languages, as well as the further ones.

References

1. V. Gouranton, P. Réty, and H. Seidl. Synchronized Tree Languages Revisited and New Applications. Research Report 2000-16, LIFO, 2000. http://www.univ-orleans.fr/SCIENCES/LIFO/Members/rety/publications.html.
2. V. Gouranton, P. Réty, and H. Seidl. Synchronized Tree Languages Revisited and New Applications. In *Proceedings of FoSSaCs*, volume 2030 of *LNCS*. Springer-Verlag, 2001.
3. S. Limet and P. Réty. E-Unification by Means of Tree Tuple Synchronized Grammars. In *Proceedings of 6th Colloquium on Trees in Algebra and Programming*, volume 1214 of *LNCS*, pages 429–440. Springer-Verlag, 1997. Full version in DMTCS (http://dmtcs.loria.fr/), volume 1, pages 69-98, 1997.
4. S. Limet, P. Réty, and H. Seidl. Weakly Regular Relations and Applications. In *Proceedings of 12th Conference on Rewriting Techniques and Applications, Utrecht (The Netherlands)*, volume 2051 of *LNCS*. Springer-Verlag, 2001.
5. P. Réty. Langages synchronisés d'arbres et applications. *Habilitation Thesis (in French). LIFO, Université d'Orléans*, June 2001.

in²: A Graphical Interpreter for Interaction Nets

Sylvain Lippi

Institut de Mathématiques de Luminy, Marseille, France
lippi@iml.univ-mrs.fr

in² can be considered as an attractive and didactic tool to approach the interaction net paradigm. But it is also an implementation in C of the core of a real programming language featuring a user-friendly graphical syntax and an efficient garbage collector free execution.

1 Introduction

Interaction nets introduced by Yves Lafont [3] are a generalization of Proof nets of Jean-Yves Girard [1]. It was originally presented as a model of computation but we rather focus on the programming language viewpoint. Indeed we present a concrete implementation of an interpreter for the interaction nets: in². So we first shortly recall the main principles of the interaction nets and give an example of interaction system. Then we give a small description of in² and show how it can be used to execute a program written with interaction nets.

2 The Interaction Net Paradigm

We present the interaction net paradigm in a practical way by introducing a simple example: the interaction combinators. For a more detailed presentation see [3].

2.1 Graphical Syntax

- The basic ingredient is a *symbol* with its *arity*. Our example system has three symbols: δ and γ of arity 2 and ε of arity 0.
- Occurrences of symbols are called *cells*. Cells have one *principal port* and their number of *auxiliary ports* is given by the arity of their corresponding symbol. The γ-, δ- and ε-cells are pictured like this:

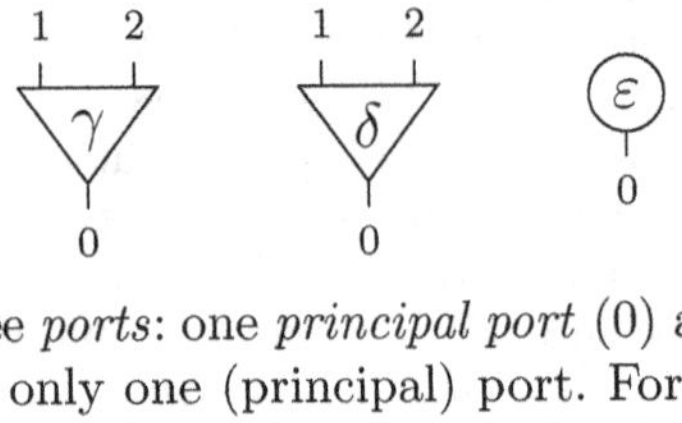

γ- and δ-cells have three *ports*: one *principal port* (0) and two *auxiliary ports* (1 and 2). ε-cells have only one (principal) port. For esthetic reasons, cells with no auxiliary port are pictured with a circle; the important point is to distinguish the principal port of a cell.

S. Tison (Ed.): RTA 2002, LNCS 2378, pp. 380–385, 2002.

Convention. Auxiliary ports are not interchangeable. For instance, one can number them from 1 to n, keeping 0 for the principal port. In practice, the ports will always be implicitly numbered in clockwise order.

- A *net* is a graph built with cells and *free ports.* Ports (principal ports, auxiliary ports or free ports) are connected pairwise by *wires.* An example of a net built with the combinators can be found in figure 1

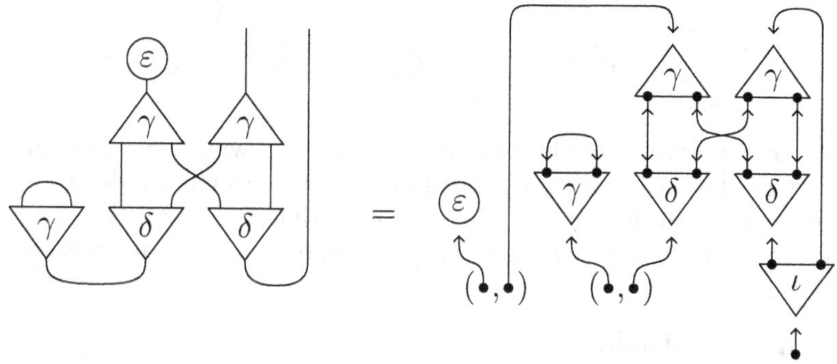

Fig. 1. *A net and its concrete representation.*

- A *rule* is a pair of nets (*left member* $\longrightarrow$ *right member*) with the same number of free ports. The important restriction is that the left member must be built with two cells connected by their principal ports; such a net is called a *cut.* [1] There is, at most, one rule for each pair of symbols. Figure 2 gives the rules for the combinators.

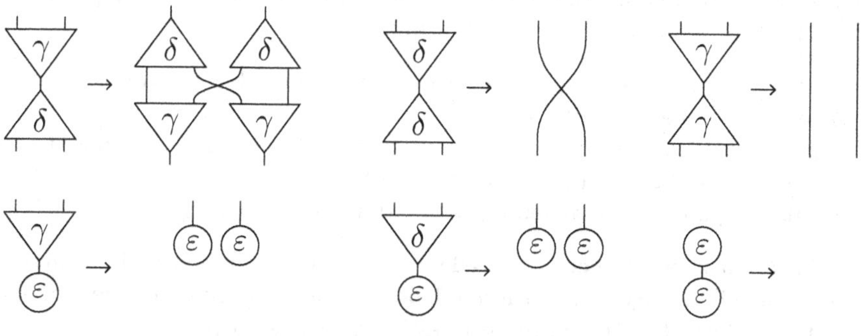

Fig. 2. *Interaction rules for the combinators*

[1] if the two cells of the left member share the same symbol, there is also a symmetry condition on the right member (see [4]).

2.2 Execution

Once a set of symbols and rules has been fixed, we can apply one of those rules to a net obtaining another net and so on until we have reached an irreducible one. The *reduction* relation is denoted by $\rightarrow$ and its reflexive and transitive closure by $\xrightarrow{*}$. Here is an example of reduction where a net is duplicated by a δ-cell:

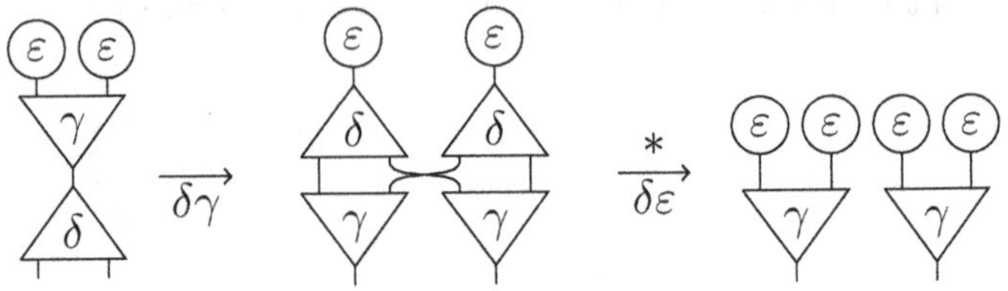

There may be several cuts in a net but the order in which they are eliminated does not matter. All reduction strategies lead to the same irreducible net and have the same length (see [4]). The reason is that two instances of a cut are necessarily disjoint, so we can apply the corresponding rules independently.

3 Implementation

The interpreter can be freely downloaded at `http://iml.univ-mrs.fr/~lippi`. It is written in *C* in order to obtain an efficient execution and to benefit from the multiple program libraries: *flex* and *bison* for parsing, the *Gnu Library* for memory allocation and the *Gimp Tool Kit* [7] for graphical layout. Nevertheless, the implemented algorithms are widely independant of the programming language.

Let us have a look at the "concrete" representation of a net. Surprisingly, it is closer from a tree structure than from a graph. Basically, the main difference between the graphical representation and the concrete one is that links are oriented in the second case. All the connections start from an auxiliary port. The most important data types are the following:

- A `cell` of arity n is composed of a pointer in the table of symbols and an array of n `addresses`.
- A `port address` (or simply `address`) represented by $\updownarrow$ is composed of a pointer to a `cell` and a port number $n \geq 0$.
- A `cut` is a pair of `addresses` represented as follows: $(\updownarrow, \updownarrow)$. [2]

Let us remark that the `cell` and `address` data types are mutually recursive. Moreover, a cell of arity n *i.e* with $n+1$ ports contains only n `addresses`; this is due to the fact that there are no links starting from a principal port. Cuts are stored separately is a stack. Instead of defining another data type for free ports, we simply introduce an *interface cell* ι that "collect" all the free ports of

[2] In fact, we could simply use a pair of pointers to a cell since the cells of a cut are always connected through their principal ports and this is how cuts are really implemented.

a net. Figure 1 presents a net and its concrete representation. The same ideas are used for the concrete representation of a rule; we consider the net obtained by connecting the corresponding free ports of the left and the right member.

4 User Manual

An interaction net program (for short interaction program) is completely described by the following data:

- *symbol*: the symbols with their respective arities;
- *rule*: the interaction rules (this is the core of the program)
- *net*: the net from which the reduction starts.

Similarly to PROLOG programs, the difficult part to write is the set of rules and *not* the starting net to be reduced (or the set of goals to be eliminated). An interaction program is written with a textual syntax. Let us begin with the syntax of a net; it is defined by the following grammar:

```
NET   ::= empty | CUT  NET
CUT   ::= TERM -- TERM
TERM  ::= variable | symbol  LIST
LIST  ::= empty | ( LISTE1 )
LIST1 ::= TERM | TERM , LIST1
```

The simplest idea is to start from scratch and introduce cells and connections between the ports (auxiliary ports, principal ports and free ports) using variables to name the ports:

- Let us take the symbol δ (of arity 2) for example. A δ-cell with the principal port named x and the auxiliary ports respectively named y and z is denoted by `x -- δ(y,z)` or `δ(y,z) -- x`.
- A connection between two ports named x and y is denoted by `x -- y` or `y -- x`.

Definition 1. *Variables that occur only once correspond to free ports and are called* free variables. *The other variables appear twice; they correspond to connections and are called* bound variables.

This syntax is quite simple but it also requires a large number of variable names: one for each port. So there is a syntactic shortcut to lighten the notation; sentences of the form `x -- T` where `x` is a bound variable and `T` a term can be removed by replacing the other occurrence of `x` by `T`. By repeating this process we obtain a condensed syntax (Lafont's syntax) where bound variables are needed only for connections between two auxiliary ports. With a bit of training, one can write a net directly with the condensed syntax: putting principal ports down and auxiliary ports up, we obtain trees with connections between leaves. Nevertheless, the user can write a net with the "naïve" syntax.

The syntax of a rule is about the same: the left member is described by two cells separated by >< and the right member is written like a net. The corresponding free ports of the two members must share the same names. Rules are separated by commas. The $\delta\gamma$ rule can be written:

```
δ(x,y) >< γ(r,s)
        r -- δ(a,b)
        s -- δ(c,d)      or      δ(γ(a,c),γ(b,d)) >< γ(δ(a,b),δ(c,d))
        x -- γ(a,c)
        y -- γ(b,d)
```

Figure 3 is a screen shot of in^2 launched on the combinators system. One window shows the "source code" of the program written with the (textual) condensed syntax; the user can specify a display color for each symbol. Two other windows are displayed by the interpreter: the *net* window shows the net which is currently reduced and the *next rule* window shows the rule that is going to be applied if the user presses the *reduce* button.

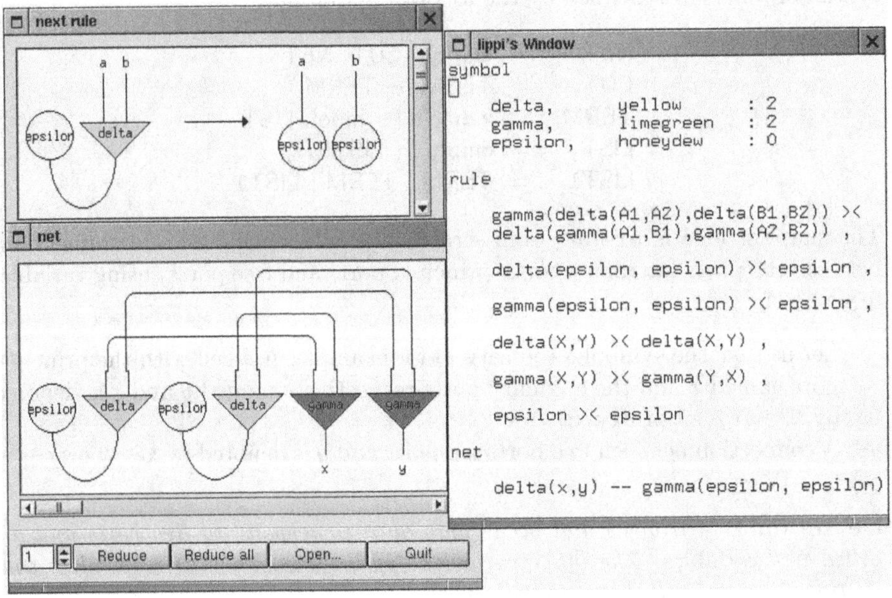

Fig. 3. *A screenshot of* in^2

References

1. J.-Y. Girard. Linear logic. *Theoretical Computer Science 50*, 50(1):1–102, 1987.
2. Y. Lafont. Interaction nets. *In proceedings of the 17th Annnual ACM Symposium on Principles of Programming Languages, Orlando (Fla., USA)*, pages 95–108, 1990.

3. Y. Lafont. From proof-nets to interaction nets. In *Advances in Linear Logic*, London Mathematical Society Lecture Note Series 222. Cambridge University Press, 1995.
4. Y. Lafont. Interaction combinators. *Information and Computation*, 137(1):69–101, 1997.
5. Sylvain Lippi. Interaction nets as a programming language. Technical report, Institut de Mathématiques de Luminy, 2001.
6. Sylvain Lippi. Encoding left reduction in the lambda-calculus with interaction nets. *To appear in Mathematical Structures in Computer Science*, 2002.
7. Peter Mattis and the GTK+ team. *The GIMP Toolkit*. `http://www.gtk.org`, 1998.

Author Index

Lecture Notes in Computer Science

For information about Vols. 1–2302
please contact your bookseller or Springer-Verlag

Vol. 2303: M. Nielsen, U. Engberg (Eds.), Foundations of Software Science and Computation Structures. Proceedings, 2002. XIII, 435 pages. 2002.

Vol. 2304: R.N. Horspool (Ed.), Compiler Construction. Proceedings, 2002. XI, 343 pages. 2002.

Vol. 2305: D. Le Métayer (Ed.), Programming Languages and Systems. Proceedings, 2002. XII, 331 pages. 2002.

Vol. 2306: R.-D. Kutsche, H. Weber (Eds.), Fundamental Approaches to Software Engineering. Proceedings, 2002. XIII, 341 pages. 2002.

Vol. 2307: C. Zhang, S. Zhang, Association Rule Mining. XII, 238 pages. 2002. (Subseries LNAI).

Vol. 2308: I.P. Vlahavas, C.D. Spyropoulos (Eds.), Methods and Applications of Artificial Intelligence. Proceedings, 2002. XIV, 514 pages. 2002. (Subseries LNAI).

Vol. 2309: A. Armando (Ed.), Frontiers of Combining Systems. Proceedings, 2002. VIII, 255 pages. 2002. (Subseries LNAI).

Vol. 2310: P. Collet, C. Fonlupt, J.-K. Hao, E. Lutton, M. Schoenauer (Eds.), Artificial Evolution. Proceedings, 2001. XI, 375 pages. 2002.

Vol. 2311: D. Bustard, W. Liu, R. Sterritt (Eds.), Soft-Ware 2002: Computing in an Imperfect World. Proceedings, 2002. XI, 359 pages. 2002.

Vol. 2312: T. Arts, M. Mohnen (Eds.), Implementation of Functional Languages. Proceedings, 2001. VII, 187 pages. 2002.

Vol. 2313: C.A. Coello Coello, A. de Albornoz, L.E. Sucar, O.Cairó Battistutti (Eds.), MICAI 2002: Advances in Artificial Intelligence. Proceedings, 2002. XIII, 548 pages. 2002. (Subseries LNAI).

Vol. 2314: S.-K. Chang, Z. Chen, S.-Y. Lee (Eds.), Recent Advances in Visual Information Systems. Proceedings, 2002. XI, 323 pages. 2002.

Vol. 2315: F. Arhab, C. Talcott (Eds.), Coordination Models and Languages. Proceedings, 2002. XI, 406 pages. 2002.

Vol. 2316: J. Domingo-Ferrer (Ed.), Inference Control in Statistical Databases. VIII, 231 pages. 2002.

Vol. 2317: M. Hegarty, B. Meyer, N. Hari Narayanan (Eds.), Diagrammatic Representation and Inference. Proceedings, 2002. XIV, 362 pages. 2002. (Subseries LNAI).

Vol. 2318: D. Bošnački, S. Leue (Eds.), Model Checking Software. Proceedings, 2002. X, 259 pages. 2002.

Vol. 2319: C. Gacek (Ed.), Software Reuse: Methods, Techniques, and Tools. Proceedings, 2002. XI, 353 pages. 2002.

Vol.2320: T. Sander (Ed.), Security and Privacy in Digital Rights Management. Proceedings, 2001. X, 245 pages. 2002.

Vol. 2321: P.L. Lanzi, W. Stolzmann, S.W. Wilson (Eds.), Advances in Learning Classifier Systems. Proceedings, 2002. VIII, 231 pages. 2002. (Subseries LNAI).

Vol. 2322: V. Mařík, O. Štěpánková, H. Krautwurmová, M. Luck (Eds.), Multi-Agent Systems and Applications II. Proceedings, 2001. XII, 377 pages. 2002. (Subseries LNAI).

Vol. 2323: À. Frohner (Ed.), Object-Oriented Technology. Proceedings, 2001. IX, 225 pages. 2002.

Vol. 2324: T. Field, P.G. Harrison, J. Bradley, U. Harder (Eds.), Computer Performance Evaluation. Proceedings, 2002. XI, 349 pages. 2002.

Vol 2326: D. Grigoras, A. Nicolau, B. Toursel, B. Folliot (Eds.), Advanced Environments, Tools, and Applications for Cluster Computing. Proceedings, 2001. XIII, 321 pages. 2002.

Vol. 2327: H.P. Zima, K. Joe, M. Sato, Y. Seo, M. Shimasaki (Eds.), High Performance Computing. Proceedings, 2002. XV, 564 pages. 2002.

Vol. 2328: R. Wyrzykowski, J. Dongarra, M. Paprzycki, J. Waśniewski (Eds.), Parallel Processing and Applied Mathematics. Proceedings, 2001. XIX, 915 pages. 2002.

Vol. 2329: P.M.A. Sloot, C.J.K. Tan, J.J. Dongarra, A.G. Hoekstra (Eds.), Computational Science – ICCS 2002. Proceedings, Part I. XLI, 1095 pages. 2002.

Vol. 2330: P.M.A. Sloot, C.J.K. Tan, J.J. Dongarra, A.G. Hoekstra (Eds.), Computational Science – ICCS 2002. Proceedings, Part II. XLI, 1115 pages. 2002.

Vol. 2331: P.M.A. Sloot, C.J.K. Tan, J.J. Dongarra, A.G. Hoekstra (Eds.), Computational Science – ICCS 2002. Proceedings, Part III. XLI, 1227 pages. 2002.

Vol. 2332: L. Knudsen (Ed.), Advances in Cryptology – EUROCRYPT 2002. Proceedings, 2002. XII, 547 pages. 2002.

Vol. 2333: J.-J.Ch. Meyer, M. Tambe (Eds.), Intelligent Agents VIII. Revised Papers, 2001. XI, 461 pages. 2001. (Subseries LNAI).

Vol. 2334: G. Carle, M. Zitterbart (Eds.), Protocols for High Speed Networks. Proceedings, 2002. X, 267 pages. 2002.

Vol. 2335: M. Butler, L. Petre, K. Sere (Eds.), Integrated Formal Methods. Proceedings, 2002. X, 401 pages. 2002.

Vol. 2336: M.-S. Chen, P.S. Yu, B. Liu (Eds.), Advances in Knowledge Discovery and Data Mining. Proceedings, 2002. XIII, 568 pages. 2002. (Subseries LNAI).

Vol. 2337: W.J. Cook, A.S. Schulz (Eds.), Integer Programming and Combinatorial Optimization. Proceedings, 2002. XI, 487 pages. 2002.

Vol. 2338: R. Cohen, B. Spencer (Eds.), Advances in Artificial Intelligence. Proceedings, 2002. XII, 373 pages. 2002. (Subseries LNAI).

Vol. 2340: N. Jonoska, N.C. Seeman (Eds.), DNA Computing. Proceedings, 2001. XI, 392 pages. 2002.

Vol. 2342: I. Horrocks, J. Hendler (Eds.), The Semantic Web – ISCW 2002. Proceedings, 2002. XVI, 476 pages. 2002.

Vol. 2345: E. Gregori, M. Conti, A.T. Campbell, G. Omidyar, M. Zukerman (Eds.), NETWORKING 2002. Proceedings, 2002. XXVI, 1256 pages. 2002.

Vol. 2346: H. Unger, T. Böhme, A. Mikler (Eds.), Innovative Internet Computing Systems. Proceedings, 2002. VIII, 251 pages. 2002.

Vol. 2347: P. De Bra, P. Brusilovsky, R. Conejo (Eds.), Adaptive Hypermedia and Adaptive Web-Based Systems. Proceedings, 2002. XV, 615 pages. 2002.

Vol. 2348: A. Banks Pidduck, J. Mylopoulos, C.C. Woo, M. Tamer Ozsu (Eds.), Advanced Information Systems Engineering. Proceedings, 2002. XIV, 799 pages. 2002.

Vol. 2349: J. Kontio, R. Conradi (Eds.), Software Quality – ECSQ 2002. Proceedings, 2002. XIV, 363 pages. 2002.

Vol. 2350: A. Heyden, G. Sparr, M. Nielsen, P. Johansen (Eds.), Computer Vision – ECCV 2002. Proceedings, Part I. XXVIII, 817 pages. 2002.

Vol. 2351: A. Heyden, G. Sparr, M. Nielsen, P. Johansen (Eds.), Computer Vision – ECCV 2002. Proceedings, Part II. XXVIII, 903 pages. 2002.

Vol. 2352: A. Heyden, G. Sparr, M. Nielsen, P. Johansen (Eds.), Computer Vision – ECCV 2002. Proceedings, Part III. XXVIII, 919 pages. 2002.

Vol. 2353: A. Heyden, G. Sparr, M. Nielsen, P. Johansen (Eds.), Computer Vision – ECCV 2002. Proceedings, Part IV. XXVIII, 841 pages. 2002.

Vol. 2355: M. Matsui (Ed.), Fast Software Encryption. Proceedings, 2001. VIII, 169 pages. 2001.

Vol. 2358: T. Hendtlass, M. Ali (Eds.), Developments in Applied Artificial Intelligence. Proceedings, 2002 XIII, 833 pages. 2002. (Subseries LNAI).

Vol. 2359: M. Tistarelli, J. Bigun, A.K. Jain (Eds.), Biometric Authentication. Proceedings, 2002. X, 197 pages. 2002.

Vol. 2360: J. Esparza, C. Lakos (Eds.), Application and Theory of Petri Nets 2002. Proceedings, 2002. X, 445 pages. 2002.

Vol. 2361: J. Blieberger, A. Strohmeier (Eds.), Reliable Software Technologies – Ada-Europe 2002. Proceedings, 2002 XIII, 367 pages. 2002.

Vol. 2363: S.A. Cerri, G. Gouardères, F. Paraguaçu (Eds.), Intelligent Tutoring Systems. Proceedings, 2002. XXVIII, 1016 pages. 2002.

Vol. 2364: F. Roli, J. Kittler (Eds.), Multiple Classifier Systems. Proceedings, 2002. XI, 337 pages. 2002.

Vol. 2366: M.-S. Hacid, Z.W. Raś, D.A. Zighed, Y. Kodratoff (Eds.), Foundations of Intelligent Systems. Proceedings, 2002. XII, 614 pages. 2002. (Subseries LNAI).

Vol. 2367: J. Fagerholm, J. Haataja, J. Järvinen, M. Lyly. P. Råback, V. Savolainen (Eds.), Applied Parallel Computing. Proceedings, 2002. XIV, 612 pages. 2002.

Vol. 2368: M. Penttonen, E. Meineche Schmidt (Eds.), Algorithm Theory – SWAT 2002. Proceedings, 2002. XIV, 450 pages. 2002.

Vol. 2369: C. Fieker, D.R. Kohel (Eds.), Algebraic Number Theory. Proceedings, 2002. IX, 517 pages. 2002.

Vol. 2370: J. Bishop (Ed.), Component Deployment. Proceedings, 2002. XII, 269 pages. 2002.

Vol. 2372: A. Pettorossi (Ed.), Logic Based Program Synthesis and Transformation. Proceedings, 2001. VIII, 267 pages. 2002.

Vol. 2373: A. Apostolico, M. Takeda (Eds.), Combinatorial Pattern Matching. Proceedings, 2002. VIII, 289 pages. 2002.

Vol. 2374: B. Magnusson (Ed.), ECOOP 2002 – Object-Oriented Programming. XI, 637 pages. 2002.

Vol. 2375: J. Kivinen, R.H. Sloan (Eds.), Computational Learning Theory. Proceedings, 2002. XI, 397 pages. 2002. (Subseries LNAI).

Vol. 2378: S. Tison (Ed.), Rewriting Techniques and Applications. Proceedings, 2002. XI, 387 pages. 2002.

Vol. 2380: P. Widmayer, F. Triguero, R. Morales, M. Hennessy, S. Eidenbenz, R. Conejo (Eds.), Automata, Languages and Programming. Proceedings, 2002. XXI, 1069 pages. 2002.

Vol. 2382: A. Halevy, A. Gal (Eds.), Next Generation Information Technologies and Systems. Proceedings, 2002. VIII, 169 pages. 2002.

Vol. 2383: M.S. Lew, N. Sebe, J.P. Eakins (Eds.), Image and Video Retrieval. Proceedings, 2002. XII, 388 pages. 2002.

Vol. 2384: L. Batten, J. Seberry (Eds.), Information Security and Privacy. Proceedings, 2002. XII, 514 pages. 2002.

Vol. 2385: J. Calmet, B. Benhamou, O. Caprotti, L. Henocque, V. Sorge (Eds.), Artificial Intelligence, Automated Reasoning, and Symbolic Computation. Proceedings, 2002. XI, 343 pages. 2002. (Subseries LNAI).

Vol. 2386: E.A. Boiten, B. Möller (Eds.), Mathematics of Program Construction. Proceedings, 2002. X, 263 pages. 2002.

Vol. 2389: E. Ranchhod, N.J. Mamede (Eds.), Advances in Natural Language Processing. Proceedings, 2002. XII, 275 pages. 2002. (Subseries LNAI).

Vol. 2391: L.-H. Eriksson, P.A. Lindsay (Eds.), FME 2002: Formal Methods – Getting IT Right. Proceedings, 2002. XI, 625 pages. 2002.

Vol. 2392: A. Voronkov (Ed.), Automated Deduction – CADE-18. Proceedings, 2002. XII, 534 pages. 2002. (Subseries LNAI).

Vol. 2393: U. Priss, D. Corbett, G. Angelova (Eds.), Conceptual Structures: Integration and Interfaces. Proceedings, 2002. XI, 397 pages. 2002. (Subseries LNAI).

Vol. 2398: K. Miesenberger, J. Klaus, W. Zagler (Eds.), Computers Helping People with Special Needs. Proceedings, 2002. XXII, 794 pages. 2002.

Vol. 2399: H. Hermanns, R. Segala (Eds.), Process Algebra and Probabilistic Methods. Proceedings, 2002. X, 215 pages. 2002.

Vol. 2405: B. Eaglestone, S. North, A. Poulovassilis (Eds.), Advances in Databases. Proceedings, 2002. XII, 199 pages. 2002.